I dedicate this book to my wife, Mara.

If you want to find out if your spouse loves you, write a book. I am a lucky person: my wife loves me, and I have this book and our still-successful marriage to prove it. If she didn't love me, I still might have this book, but the marriage would have been gone long ago.

Contents

Foreword

A science department that has the good fortune to have, as a member of its staff, an experienced and inventive equipment designer and glassblower is thrice blessed. First, because the students and faculty, instead of having to rely on standard commercial designs for equipment, can themselves be inventive, and have their designs custom-made. Second, because there can be rapid availability of a wide range of apparatus. And third, because the members of the department can draw upon that staff member's experience and ingenuity to avoid costly and possibly hazardous mistakes in the design and application of equipment.

The Department of Chemistry and Biochemistry at California State University, Los Angeles, is fortunate indeed to have Gary Coyne on its staff, and he helps the department in all the ways I have indicated above—and more. If your department is not so fortunate, or even if it is, but wants to learn more about the applications of materials commonly used in constructing scientific equipment, buy this book. It is an unusual exposition of the properties of a wide range of materials, including glass, that make an important contribution to the fabrication of scientific equipment. In it you will find the fundamentals of equipment design, detailed discussions of measurement basics, and the techniques of manipulating materials. Additionally, it has a full presentation of the principles and parts of practical laboratory vacuum systems.

This book tells you not only how things are done, but why they are done. I recommend it to any creative scientist, and I thank Gary Coyne for having had the idea and the perseverance to write it.

Harold Goldwhite
Professor of Chemistry
California State University, Los Angeles

Preface

I am a scientific glassblower. Although that may not turn many heads, my profession indirectly forms the basis of how and why this book began. As a scientific glassblower at a university, I have two primary functions: the first is to make research apparatus, and the second is to repair broken research apparatus. In addition to my formal glassblowing duties, I am often a middle person in the academic hierarchy, since students often find staff less intimidating than professors. As such, students are likely to come to me with their "dumb questions" on how to use a piece of laboratory equipment.

When students don't bother (or know) to ask "dumb questions" before proceeding with their laboratory work, they inevitably come to me with pieces of apparatus for repair. After repairing the damage caused by the students' ignorance, I talk to the students (and, occasionally, the professors) to see what went wrong (although I usually know from the nature of the damage), and guide them toward safer laboratory procedures. From these experiences, I've gained knowledge of the problems that inexperienced people have in laboratories.

This book actually started when I got into a discussion with a professor who had the mistaken belief that the number designation for an O-ring joint referred to the outside diameter of the connecting tube. I took a caliper and showed him that the number actually referred to the inside, not outside, diameter of the hole at the O-ring fixture. This incident inspired me to write a simple monograph on the identification of standard taper joints, ball-and-socket joints, and O-ring joints.

I wrote several more monographs, on various subjects, until another faculty member suggested I assemble them for publication in a journal. After some consideration, I wrongly assumed I had enough information for a book. At the time, I really didn't know enough to fill a book, but because I was ignorant of that fact, I proceeded, expecting to finish the book in short order. Some five years later (after much research and learning) I have a book, a bit more wisdom, and hopefully more knowledge.

The purpose of this book is to provide some basics on the materials, equipment, and techniques required in a laboratory. In addition to the information on how various procedures are done, I've also added historical and other background information to better explain how and why these procedures, equipment, and theory evolved. Some readers may not find all the answers they need, whereas other readers may wonder why *obvious* information is included. Unfortunately, no manual of this type can be all-inclusive, and what may be obvious for one may be new for another. I apologize for omissions of information which you hoped to find, and I encourage those of you who found little new to please share your knowledge.

I am indebted to many of the Chemistry department faculty at California State University, Los Angeles, not only for their willingness to answer many of my "dumb questions," but for their support throughout this whole process. I would be in error if I did not single out one special faculty member, Dr. Cathy Cobb, who proofread and served as a sounding board on various aspects of this book, with whom I co-wrote a paper, and who has been a special friend. I am also indebted to many strangers, all experts in their fields, who answered their phones and my "dumb questions."

Please note:

Proper operation of all equipment should be taught to all potential users. Such knowledge should *never* be assumed. The most dangerous person in a laboratory, to both equipment and other personnel, is the person who through pride, ego, or ignorance, claims knowledge that he or she does not have. It is up to the professor, group leader, or research director to monitor the quality of technical support and provide additional training as required.

A simple laboratory procedure to provide information to those who need it can be provided by photocopying all equipment manuals, no matter how seemingly trivial. These photocopies should be placed in binders and stored in a specific location in the lab where equipment is used. The originals should be placed in the research director's office and not removed unless new copies are made as needed.

I would be honored if a copy of this book were placed next to those binders, available to all.

Chapter 1

Materials in the Lab

1.1 Glass

1.1.1 Introduction

When one talks of chemistry (in general), or working in a laboratory (in specific), the average person usually envisions an array of glassware—test tubes, beakers, coiled tubes, and the like. This list obviously is not the sum and substance of laboratory equipment (or chemistry for that matter), but items made of glass will always play a predominant role in most laboratories.

Glass is used in the laboratory because it has three important properties. First, it is transparent and thus allows the user to observe a reaction taking place within a container. Second, it is very stable and is nonreactive to many materials used in the laboratory. Third, it is (relatively) easily malleable, allowing new designs and shapes of apparatus to be created and produced, and broken items to be repaired.

The nature of glass, its structure and chemistry as well as the shapes and designs made from it, has come from years of development. The standard shapes of the beaker, Erlenmeyer flask, and round bottom flask were each developed to meet specific needs and functions. Variations in wall thicknesses, angles, and height-to-width ratios are all critical for specific functions of each container.

Before discussing any container however, it may be best to talk of the containment material: what it is, what makes it unique, and what are its strengths and weaknesses. We can then relate these properties to specific types of containers and know why they are made as they are. By understanding the shape, design, and function of laboratory glassware, one can achieve greater efficiency, and proficiency, in the laboratory.

1.1.2 Structural Properties of Glass

The standard definition of glass is a 'noncrystalline solid.' In 1985, the ASTM (American Society for Testing and Materials) referred to glass as "... an inorganic product of fusion that has cooled to a rigid state without crystallizing."[1] However, since 1985, glasses of organic and metallic materials have been produced. Therefore, a more accurate definition of a glass would be any material that is cooled fast enough to prevent the development of crystals. Any further reference to the term "glass" in this book refers specifically to the inorganic glasses that we normally associate as glass. As we shall see, this specific type of glass has properties that facilitate the creation of the glass state.

The key structural concept of glass is that it is not a crystal. This concept holds true for high-temperature quartz glass and for ornate lead crystal glass. A crystal, by definition, is a collection of atoms in a repeating sequence and form that develops or has symmetry in its structure. By comparison, the structure of glass at the atomic level is a three-dimensional web of inter-connecting oxides of materials, frozen in place by rapid cooling. Thus, glass made of inorganic materials may be defined as a super-cooled liquid composed of a mixture of oxides in a solution.

To better understand what glass is, and why different glasses have different properties, one should first compare crystals and glass at the molecular level. This comparison begins with the molecular structure of a quartz crystal. The quartz crystal is composed of (essentially) pure SiO_2, the same molecular composition of quartz glass, and the chief component material of most glass. The quartz crystal is a silicon tetrahedra composed of one atom of silicon surrounded by four oxygen atoms, in a tight three-dimensional network of high energy bonds. The extra two oxygen atoms, *bridging* oxygens, interconnect the silicon atoms into a three-dimensional network. The left side of Fig. 1-1 illustrates a two-dimensional representation of one plane of such a network.

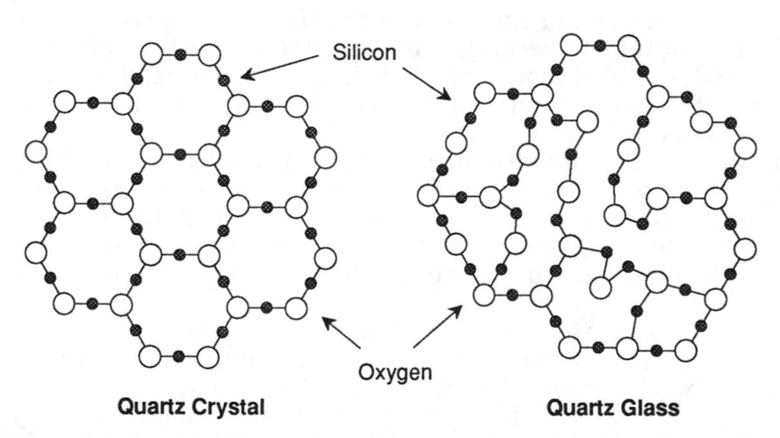

Fig. 1-1 The molecular structure of a quartz crystal and quartz glass.

If you take a quartz crystal, heat it until it melts, and allow it to rapidly cool, the even network of rings is broken into nonuniform irregular chains (see right side of Fig. 1-1). These glass chains have no discrete molecular form as they interconnect throughout the entire glass item. The quartz crystal is now called either "fused quartz", "fused silica", or "quartz glass."

The bonds between the Si and O atoms are still high energy bonds, and tremendous amounts of thermal energy are still required to disturb and break those bonds.

When another oxide molecule replaces a silicon oxide tetrahedron, the connecting oxygen atom is considered a *non-bridging* oxygen. The inclusion of these materials breaks up the continuity of the structural network and provides the glass with significant changes in properties. Thus, the uniformity of the bonds within this glass soup does not exist as it once did with pure SiO_2. Therefore, much less energy is required to break the bonds, and less heat is required to melt the new glass. In addition, all properties of the glass change as the composition of the glass changes, including optical, thermal, and expansion properties. By changing the materials of a glass, various properties can be enhanced or suppressed by a manufacturer to satisfy the needs of the user.

When a crystal breaks, it tends to fracture on a cleavage plane based on its molecular structure. That is why a large table salt crystal will break into fragments that maintain the original geometric shape and surface angles of the original crystal. When glass is broken, only amorphous, irregular shapes remain, because glass has no structural geometric consistency. This lack of structural consistency occurs in all glasses, and it is impossible to distinguish one type of glass from another based on a fracture pattern. On the other hand, because glass fracture patterns are consistent throughout all glass types, these markings can help provide clues to the origin and cause of glass fracture.

Another difference between a glass and a crystal is that if you take a cubic crystal of table salt and heat it until it melts, the crystal cube will start to slump into a puddle at a specific temperature (801°C). The melting and/or freezing temperature of table salt, as with most materials, is usually considered the point at which the solid and liquid form of the material can exist together. Glass, on the other hand, has no single fixed melting point. It maintains its physical shape after it begins to soften. External forces such as gravity will cause the glass to sag under its own weight once temperatures are above the softening point. Gradually as the temperature continues to rise, the surface will begin to lose form. Then, internal forces, such as surface tension, cause sharp corners on the glass to round as the glass 'beads up' on itself. Eventually, higher temperatures will cause the glass to collect into a thick, liquid puddle. The term 'liquid' here is rather nebulous, because the viscosity of glass can be as thick as honey. The hotter a glass 'liquid' gets, the less viscous it becomes. It is, in fact, the high viscosity of this puddle that helps prevent glass from crystalizing: As glass cools past the crystallization temperature, its high viscosity inhibits atomic mobility, preventing the atoms from aligning themselves into a crystalline form.

We've already mentioned that glass does not have a specific melting temperature. Rather, its viscosity gradually changes as the temperature varies. The viscosity decreases until the glass is identified as being in a melted state. Thus, scientists define important transitional changes based on specific viscosity ranges of the glass in question. There are four significant viscosities of glass (in comparison to the viscosity [Poises] numbers presented below, the viscosity of glass at room temperature is 10^{22+} Poises):

1) **Strain Point** (≈$10^{14.5}$ Poises) Anything above this temperature may cause strain in glass. Internal stresses may be relieved if the glass is baked in an oven for several hours at this temperature.

2) **Annealing Point** (≈10^{13} Poises) If an entire item were uniformly baked at this temperature, the item would be relieved of strain in about fifteen minutes. The annealing range is considered to be the range of temperatures between the *strain point* and the *annealing point.*

3) **Softening Point** (≈$10^{7.6}$ Poises) Glass will sag under its own weight at this temperature. The surface of the glass is tacky enough to stick (but not fuse*) to other glass. The specific viscosity for the softening point depends on both the density and surface tension of the glass.

4) **Working Point** (≈10^{4} Poises) Glass is a very thick liquid at this point (like honey) and can be worked by most conventional glassblowing techniques. At the *upper end* of the working range, glass can be readily fused together or worked (if the glass is heated too high, it will boil and develop characteristics that are different from the original glass). At the *lower end* of the working range, glass begins to hold its formed shape.

The fact that glass is a solid at room temperatures should not be underestimated, and the statement "solids cannot flow" cannot be overemphasized. There is a romantic notion that windows in old churches in Europe (or old colonial homes in the U.S.) have sagged over time. The common belief that the glass is thicker on the bottom than on the top because of such sagging is incorrect. Studies by F.M. Ernsberger,[2] an authority on visco-elastic behavior of glass, showed that glass will not permanently sag under its own weight at room temperature.

Ernsberger took several ¼ in glass rods and bent them 1.7 cm off-center over a 20 cm span. The amount of bending stress was calculated to be about 150,000 lbs/in^2,** significantly greater than any stress received from simple sagging. After 26 years, one of the rods was released from the strain: within 48 hours it returned to its original shape.[3]

R.C. Plumb offers an excellent theory[4] as to why old windows are sometimes thicker on the bottom than on top. He reports that the old technique of manufacturing windows involved collecting a large amount of melted glass at the end of a metal blowpipe, blowing a vase, and attaching the vase bottom to a solid metal rod called a ponty. By rapidly spinning the hot (still soft) vase, the glassblower would use centrifugal force to make the open end flair out, thus transforming the vase into a flat circular pane some five feet in

* Metals can be welded or stuck together so that they cannot be separated. Glass, on the other hand, must be fused by heating two separate pieces of glass to the point at which they flow together and become one piece. If glass pieces are simply 'stuck' together, it is relatively simple to break them apart.

** This amount of stress without fracture was accomplished by covering just newly-made glass with lacquer, which prevented surface flaws and surface hydration. Normally glass cannot receive this amount of stress without breaking.

diameter. From this pane (or "table," as it was called), the glassblower would cut square sections. The sections would have varying thicknesses depending on how far from the center of the 'table' they were cut.

Plumb does not offer a strong reason as to why he believes the thicker sections were placed on the bottom. He states, "It would certainly make good sense to install the glass with the thick edge down!" I am unaware of anyone acknowledging if any windows have ever been found to be thicker on the top than on the bottom. It is conceivable that if any such windows were ever noticed, they were disregarded because they did not fit into the expected pattern of being thicker on the bottom.

Because of the belief that glass may sag under its own weight, there has been concern about the storage of glass tubing and rods at an angle. The only danger to glass being stored at an angle is fear of damage to the ends of the glass. Otherwise, there is no problem with storing glass vertically, at an angle, or on its side.

1.1.3 Devitrification

Devitrification is the re-crystallization of glass. Glass that is devitrified appears frosty and is no longer transparent, is structurally weaker, and is more vulnerable to chemical attack. If a glass is held within its *crystallization temperature** for a sufficiently long time, the atoms will have time to align themselves. Once the temperature is allowed to drop, the glass becomes increasingly more viscous, until it cannot further vitrify Eventually, it will maintain its own shape.

There are several ways to force glass to devitrify (whether devitrification is desired or not). One technique is to heat the glass until it begins to soften, then mechanically work, or flex, the glass while it cools. Eventually a whitish frost will appear on the surface in the region of compression (see Fig. 1-2).

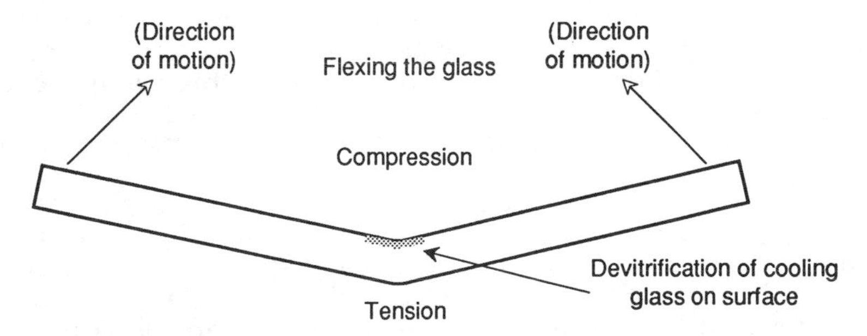

Fig. 1-2 Creating devitrification in glass.

The risk of devitrification rises the longer a glass is kept in a softened or melted state, and is also linked to how dirty the glass is. Devitrification typically begins as a surface phenomenon, using either dirt or some other surface defect as a nucleation point.[5] The devitrification process may be assisted by variations in the exterior composition (which is typically differ-

* The specific crystallization temperature is not commonly identified, but typically is between the annealing and softening points.

ent from the interior) of a glass object.[6] These variations may be the result of flame-working the glass or chemical attack. Devitrification has never been observed originating from the inside of a glass mass.

Devitrification can often be removed by re-heating a glass up to its melting temperature and avoiding any mechanical action while it re-cools to a rigid state. However, if a glass is over-worked or dirty when originally flame-heated, removal of the devitrification may be impossible.

Mechanical stress is not a requirement for devitrification. The phenomenon is also common in quartz glass furnace tubes maintained at high temperatures for extended periods of time. Devitrification of silica occurs at increasing rates from 1000°C to 1710°C, which is the crystallization temperature and melting point range of β-cristobalite.* Insufficient surface cleaning and very slow cool-down times typically facilitate devitrification on these tubes. Early-stage devitrification on a quartz glass furnace tube may be removed by a hydrofluoric acid dip. This cleaning procedure can remove only surface cristobalite. Devitrification deeper than surface level cannot be removed in this or any other fashion.

The best way to limit or prevent devitrification on quartz glass is to ensure that it is maintained scrupulously clean: no fingerprints, oils, dirt, or chemicals of any kind should get on the surface. Also, temperatures above 1200°C should be limited as much as possible.

Note that devitrification is a nucleation process. That is, devitrification does not confine itself within the area where it begins. Rather, once started, it spreads like mold on a piece of fruit. Thus, a drop of tap water, a fingerprint, or any other localized contaminant can initiate devitrification, which can then spread over the entire surface of the container.

1.1.4 Different Types of Glass Used in the Lab

Different properties of glass can be exploited by combining different oxides in a glass mix. There are thousands of different commercially-made glasses (although many of these glasses overlap in type and characteristics). The glass used in a laboratory can be divided into three major categories: soft glasses, hard glasses, and high temperature and UV-transmission glasses. Table 1-2 in the next section lists selected properties of these various glasses.

In this section, I will occasionally list the contents of glass types. The glass industry typically lists the component chemicals of glass in their oxide states as percentages of total weight. Chemically speaking, this practice does not make sense because when you add CaO and Na_2O to SiO_2, you are actually adding Ca + 2Na + Si + 4O. Some chemists do not consider this industrial approach proper because it does not look at the reaction from a proper chemical perspective. Regardless, I shall defer to the industry standard simply because it is easier to explain variations in glass properties according

* β-cristobalite is transparent. We do not normally consider devitrified glass as a transparent material. However, once fused silica has cooled below 250°C, β-cristobalite is transformed into α-cristobalite. This substance is the white opaque material we usually associate with devitrified silica. When fused silica is reheated into the devitrification range, the α-cristobalite turns back into β-cristobalite. However, because α-cristobalite has many fissures and cracks, the opacity remains when it is reheated back into β-cristobalite.

to changes in concentration percentages. In addition, this convention is used because these materials must be added to a glass melt in their oxide states or they will not 'glass.'

Please note: Glasses of different compositions should never be mixed together for recycling. Therefore, you should never take a box of used and/or broken glassware from the lab to the local recycling yard. For one thing, you risk contamination by toxic materials left on inadequately washed broken glassware (which could violate environmental laws). Equally important, if glass of the wrong type should be mixed with commercially recyclable types of glass, an entire glass melt would be wasted. From an environmental standpoint, the best thing you can do in the lab is to avoid waste and carelessness. Damaged glassware should be discarded in a proper receptacle. Final disposal of glassware depends on its history and contaminants. For example, non-toxic materials can be sent directly to a landfill, but bio-organic residue may need to be sterilized before disposal. Glassware contaminated with heavy metals may require disposal at an appropriate hazardous waste disposal site.

The Soft Glasses. The soft glasses receive their name identification by the fact that they are physically softer than other glasses; that is, they abrade more easily. In addition, these glasses tend to maintain their soft working properties over a greater temperature range. Thus, they remain 'soft,' or workable, longer than hard glasses.

Commercially soft glass is used in items such as window panes, bottles, jars, and drinking glasses. Soft glasses can take colors readily (as brown and green bottles show and as do the many hues and colors that are created by artisans such as the Italian glass masters).*

The most common soft glass is soda-lime glass. The soda is a carbonate of sodium or sodium oxide (Na_2O), and the lime is calcium oxide (CaO) or magnesium oxide (MgO). Soda-lime glass is used in soda-pop bottles, drinking glasses, and windows. The materials for this glass are relatively inexpensive, and because of its lower energy demands (lower melting temperatures) and long working times, it is the most inexpensive to manufacture. This glass is also easily recycled.

Not all soft glasses are of equal quality: some are significantly inferior to others. The quality of a soft glass may be based on the proportions of its constituent materials. Typically, lime is present in concentrations of 8% to 12% (by weight) and soda in concentrations of 12% to 17% (by weight).[7] If lime concentration levels are too high, devitrification can occur during the manufacturing process. Conversely, if the lime is held too low (or the alkali concentration is too high), the glass is subject to easy weathering and attack by water. Thus, drinking glasses that spot easily and are difficult to clean *may* be the result of a low lime or high alkalinity concentration glass rather than bad soap or hard water.

An extreme example of a glass that can be easily weathered is *sodium silicate*, also known as *soluble silicate* or, formerly, *waterglass*. The sodium content in sodium silicate is so high that the glass can be dissolved in water and shipped as a liquid glass. Sodium silicate is used as an adhesive, cleaner, and protective coating material.

* Often, the colors you see on glass animals (made out of borosilicate glass) are acrylic paint.

Sodium silicate is stored and shipped as a liquid, but when exposed to air, it will turn into a hard, glass-like material. In reality, it is a glass in a totally anhydrous form; the sodium hydroxide content is 34% by weight the remaining 66% is silicon dioxide. These proportions are theoretical, however, as an anhydrous sodium silicate is not really possible as it would always be absorbing water from the air.

When sodium silicate is shipped, its contents are typically:

≈27% SiO_2
≈14 % NaOH
≈59 % H_2O.

These proportions are all approximate because the water percentage can vary significantly. Once the water has evaporated from the sodium silicate, there is still a high level of water left behind. The dried sodium silicate will have the following approximate contents (by weight), although the actual percentages will again vary depending on the atmospheric water:

≈49% SiO_2
≈26% NaOH
≈25 % H_2O.

Note that although dry soda content would normally be listed as Na_2O, in sodium silicate all the soda is hydrated and therefore, is identified in the form NaOH.

Many disposable glass laboratory items are made of soft glass. If a laboratory glass item does not specifically bear one of the words Pyrex®, Kimax®, or Duran® in printed or raised letters, it is likely to be a soda-lime type of glass.

Beginning chemistry students usually receive their first glassblowing lesson using soft glass tubes over a Bunsen burner to make bends or eye droppers, because only soft glass is malleable at Bunsen burner temperatures. Currently, only Kimble Glassware makes soda-lime tubing (Kimble #R-6). Thus, if you have a glass tube soft enough to bend over a Bunsen burner, it is a soft glass.

The other common soft glass is lead glass. Commercially, we see this glass in lead crystal glass, where 22-25% (by weight) of the component material is PbO_2. Although misnamed as a crystal, lead crystal glass nevertheless has beautiful optical properties that are often associated with crystals. The high refractive index of lead glass (one of the highest refractive indexes of all glasses) is the foundation of its effects with light. An additional property of lead crystal glass is its resonance: decorative bells of lead glass have a beautiful chime, while those of borosilicate glass often have a dull, dead 'clank.'

The most important application in the laboratory for lead glass is as a shield to protect workers from the harmful rays produced by x-ray machines. For example, the glass window behind which an x-ray technician stands is made of lead glass. These windows can be made of up to 75% PbO_2. This greater quantity of lead oxide in this glass (6.0 $^{gm}/_{cm^3}$) makes the glass three times heavier than standard laboratory borosilicate glass (2.1 $^{gm}/_{cm^3}$).

Lead glass is very vulnerable to impact abrasion (it is almost twice as soft as soda-lime) and can be extremely sensitive to temperature changes. On the other hand, lead glass is among the most electrically resistive glasses available, and because of that property is one of the chief glasses used in electrical components. Neon signs and TV tubes are made almost exclusively of lead glass.

Another term for lead (or lead-potash-silica) glass is *flint glass*. This name can be confusing, because the term has two other meanings in the glass industry: It is used by the container industry to connote colorless glass. The term flint glass is also used by the optical glass industry to connote glass that has a high refractive index and dispersion* of optical rays. (Soda-lime [or potash-lime-silica] glass has a low refractive index and dispersion of optical rays. It is called *crown glass*.)

In addition to soda-lime and lead glass, there are several 'speciality' soft glasses. One of the most notable is Exax® made by the Kimble Corporation. This glass repels a static charge, making it particularly useful for holding or weighing powders.

The biggest drawback with all soft glasses is that any apparatus made with these glass types are essentially non-repairable. Because of their relatively high thermal coefficients of expansion, they are likely to shatter when the flame of a gas–oxygen torch touches them. Additionally, replacement soft glass components such as stopcocks and joints are generally not easily available. Thus, unless an item has particularly unique qualities or value, the cost of component repair is far greater than wholesale replacement. Although soft glass items may be less expensive initially, their long-term costs can be much greater than the more expensive (but less expensively repaired) borosilicate ware.

The Hard Glasses. Borosilicate glass** is the most common glass found in the laboratory is one example of a hard glass. It is considered a hard glass for two reasons. First, its ability to resist impact abrasion is over three times the level of soft glass. Second, it sets at a higher temperature and thereby, gets '*harder*' faster. (This second quality was a physical characteristic that scientific glassblowers found particularly challenging in the early days of borosilicate glass.) Finally, because hard glasses have much lower thermal coefficients of expansion than do soft glasses, they can withstand much greater thermal shocks than soft glasses. The hard glasses are more chemically resistant to alkaline solutions and many other chemicals.

Commercially, borosilicate glasses are found in many consumer products. In the kitchen, oven windows, baking containers, and cooking pot lids all take advantage of the thermal strength of borosilicate glass. In addition, measuring cups are also made out of borosilicate glass, not only for their thermal abilities (like pouring boiling water into a cold measuring cup), but also for their ability to withstand abrasion and impact. Measuring cups are typically nested (smaller cups placed within the bigger cups) and banged around in cupboards and drawers. Borosilicate glass is also used for automobile headlights, and floodlights used in indoor/outdoor lighting.

* The dispersion of glass is related to its index of refraction and is based on an analysis of the passage through the glass of a yellow helium line (587.6 nm), blue and red hydrogen lines (486.1 and 656.3 nm, respectively), and green mercury line (546.1 nm).

** Any glass containing 5% or more (by weight) of B_2O_3 (boron oxide) is considered a borosilicate glass.

There is more than one brand and type of borosilicate glass. Pyrex® (by Corning), Kimax® (by Kimble), and Duran® (by Schott) are all brand names of particular borosilicate glasses of similar composition made by different companies. The term "Pyrex" is to borosilicate glass as "Xerox" is to photocopying equipment: it is an almost generic term. There is little difference between the products of these three major companies as far as the chemical makeup of the glass is concerned; in theory, they are all interchangeable. What is important to the user, however, is the quality of the glass itself and the quality and uniformity of its manufacturing.

Borosilicate glass is chemically more resistant than soft glasses and will therefore resist the weathering effects of standing water better than soft glasses. Although they are resistant to most chemicals, hard glasses are still susceptible to damage from hydrofluoric acid, hot phosphoric acid, and strong alkali solutions. **You should never store any of these solutions (even a weak alkali solution) in a glass container**.

Although the susceptibility of a glass container to chemical attack is a common consideration, there is another issue which is seldom raised: What effect does the consumed glass from a container have on the solution inside the container? A. Smith studied the effects of a borosilicate glass container (an Erlenmeyer flask) on sodium and potassium hydroxide solutions of varying molarities at both room temperature and boiling point.[8] The percentage of molarity change was greatest (approximately a -17% change)* when a low molarity solution of 0.001148 M aqueous NaOH was boiled. Higher molarity solutions (0.1116 M aqueous NaOH) still provided an impressive change when boiled (approximately -3%). Changes were even evident (approximately -0.02%) when the solution (0.1029 M aqueous NaOH) was maintained at room temperature.

Other common borosilicate glasses are used on pipettes. These borosilicate glasses are more chemically inert than common laboratory borosilicate glass. Their different composition makes them unable to be fused directly to common laboratory borosilicate glass; therefore, they require use of what is called *grading glass* to make a graded seal.

Grading glass, yet another type of borosilicate glass (although not all grading glass is borosilicate glass), is used to seal glasses of different thermal coefficients of expansion. For example, you cannot fuse a pipette directly onto a round bottom flask because the thermal coefficient of expansion for the pipette is about 51 x 10^{-7} Δcm/cm/°C and the thermal coefficient of expansion for the round bottom flask is about 32 x 10^{-7} Δcm/cm/°C. This range of thermal coefficients of expansion is too great, and the stress created at the seal makes the seal nonviable. However, if a third glass, with a thermal coefficient of expansion of about 40 x 10^{-7} Δcm/cm/°C, is introduced, the stress between the round bottom flask and the third (grading) glass is within tolerable limits, as is the stress between the third (grading) glass and the pipette. The strain is split into smaller "steps," which will permit the final product.

Graded seals can be used to join not only different borosilicate glasses, but different types of glasses (and even metals) of different thermal coeffi-

* The approximate percent change is used because four tests were presented for each value. I have presented average figures for these tests.

cients of expansion. Obviously, it is most efficient and desirable to select glasses and metals with close thermal coefficients of expansion, because it reduces the amount of stress with which the glassblower must deal.

Metals can be sealed directly onto glasses even though they may have radically different thermal coefficients of expansion. For example, copper or stainless steel can be sealed directly to standard laboratory borosilicate glass. However, to do this sealing, the metal must be machined so thin that it is mechanically weak. Therefore, it is often preferable to use Kovar®, an alloy composed of cobalt, iron, and nickel with a relatively low thermal coefficient of expansion. Although it requires two grading glasses to seal it to common laboratory borosilicate glass, Kovar does not have to be machined thin and therefore maintains greater strength than do machined metals. The advantages far outweigh the extra effort involved in making the graded seal. Kovar is easily soldered or welded to other metals for vacuum-tight seals.

One special category of hard glasses is the *aluminosilicate glasses.* They are made by the addition of alumina (aluminum oxide) to the glass mix. Aluminosilicate glasses maintain higher viscosities at higher temperatures, thus, they maintain their shapes at temperatures that would cause borosilicate to sag. Aluminosilicate glasses are principle components in many ceramics and fiberglass, and (along with the lead glasses) they are used in many electronic applications.

Although most glass containers cannot contain helium gas because the small atoms leak past the glass' molecular network, one type of aluminosilicate glass, Corning #1720, is considered helium leak-proof and is used to contain helium for long periods in a laboratory glass system.

The High-Temperature and UV-transmission Glasses. The last type of glass found in the laboratory is the quartz glasses. From the many names that are used to describe this type of material, there may be confusion as to what to call it. It is often just called *quartz*, but this name can be deceiving because the term *quartz* could equally refer to the mineral quartz* (which is a crystal) or amorphous silica (which is a glass).

Historically, *fused quartz* referred to transparent products produced from quartz crystal rock, and *fused silica* referred to opaque products produced from sand. With the advent of new manufacturing techniques, transparent products can now be produced from sand, so the old distinction is no longer applicable. Currently, the term *fused quartz* is used whenever the raw product is either quartz rock or sand. The term *fused silica* is used whenever the raw product is synthetically derived (from $SiCl_4$). Generically, the term *quartz glass* or, better yet, *vitreous silica* can be used to cover the whole range of materials.

The average consumer is only likely to come across vitreous silica when purchasing special high-intensity, high-brightness light bulbs for specialized lamps such as stadium lights, flash lamps (for cameras), or stroboscopes. Because of its purity, vitreous silica is essential in the manufacture of many consumer products that would otherwise be impossible to manufacture. One such product is the ubiquitous silicon chip that controls everything from computers and calculators to toys and cars. To maintain the purity of chips

* Silica, or SiO_2, occurs in nature as quartz, crystobalite, or tidymite.

during construction, the silicon wafers are baked in large tubes of ultra-high-purity fused silica.*

Whereas other glasses obtain their unique properties by the addition of various oxides, the lack of other materials provides quartz glasses with their unique properties. Most distinctive of all quartz glass characteristics are their thermal and UV-transmission abilities. Unlike other glasses that deform and/or melt at temperatures in excess of 1200°C, quartz glass maintains a rigid shape. In addition, quartz glass has an extremely low thermal coefficient of expansion (approximately 5.0×10^{-7} Δcm/cm/°C), meaning that it can withstand thermal shock that would likely shatter all other glasses. Further, although all transparent materials limit various frequencies of light, quartz glass has the potential to transmit the broadest spectrum of light frequencies. However, not all types of quartz glass transmit light equally well: basically, the purer the material, the better the UV-transmission.

A little known fact of quartz glasses is that not all quartz glasses are alike. Although quartz glass is often called pure SiO_2, it can contain impurities such as alkali metals, hydroxyls, and oxides. These impurities come from the raw materials and/or manufacturing process. Although these impurities typically are less than 1% and can extend down to the ppb (parts per billion) range, they affect the characteristics of the quartz glass.

The manufacturing processes that limit and/or eliminate these impurities are costly. Thus you select quartz glass first on a basis of intended use, then on the basis of cost. For example, fused silica with a high hydroxyl (OH^-) content will have a significant transmission drop at 2.73 µm in the infra-red range. However, this infra-red range is being absorbed, less energy is required to fabricate items from this type of quartz glass, resulting in lower manufacturing costs.

Currently, either extremely pure quartz crystals (sand) or silicon tetrachloride ($SiCl_4$) are the raw materials from which quartz glass is made commercially. Sand must be separated, sometimes by hand, to exclude any particles with obvious impurities. Then, through one of four heating techniques, raw SiO_2 is melted and formed directly into tubes, rods, or crucibles, or it is formed into large solid ingots of quartz for later manufacturing.

There are four types of manufacturing processes for quartz glass:

1) **Type I**: Natural quartz is electrically heated in a vacuum or inert atmosphere (at low pressure). This glass is low in hydroxyl content but high in metal impurities.

2) **Type II**: Natural quartz is heated in a flame. This glass has about the same metal impurity levels as Type I but a much higher hydroxyl content.

3) **Type III**: Synthetic quartz is heated in a flame (for example, a oxy-hydrogen flame). This glass is extremely high in hydroxyl content but very low in metal impurities (except Cl, which can be as high as 50 ppm).

* Fused quartz is likely to have sufficiently high impurity levels to jeopardize the purity (and thereby the quality) of the final product

4) **Type IV**: Synthetic quartz is electrically heated. This glass has an extremely low hydroxyl content, but with the absence of hydroxyl, chlorine is increased.

Table 1-1 shows a categorization of commercially available quartz glasses.

Table 1-1[9, 10, 11]

Categorization of Commercially Available Vitreous Silicas					
Type	**Raw Material**	**Manufacturing Method**	**Impurity Levels (ppm)**	**OH^{-1} Content (ppm)**	**Examples**
Natural					
I	Natural Quartz	Electrical Fusion	Al: 20-100 Na: 1-5	<3	GE 214,510® Infrasil® IR-Vitreosil®
II	Quartz Powder	Flame Fusion (Verneuille-Process)	≈ Type I	100-400	Herasil® Homosil® Optosil® O.G. & O.H. Vitreosil®
Synthetic					
III	$SiCl_4$	Flame Hydrolysis	Metals: <1	1000	Suprasil® Spectrosil® CGW 7940® Dynasil® Synsil®
IV	$SiCl_4$	Plasma Oxidation	Cl: 200 Metals: <1	<1	Suprasil W® Spectrosil WF® Corning 7943

There is one unique type of high-temperature glass, manufactured by Corning, called Vycor®. Rather than starting out with (essentially) pure silicon as the core material, Vycor® starts out with a special type of borosilicate glass. Then, in a special process, Corning chemically leaches out almost everything that is not silicon dioxide, leaving 96% pure SiO_2. Vycor will deform and melt about 100°C lower than fused silica, and it is a poor transmitter of UV light. In the early years of Vycor production, Vycor was significantly less expensive than pure quartz glass. However, as the manufacturing techniques of pure quartz glass have become more efficient, Vycor is now the more expensive material.

It would seem that there would not be much of a market for Vycor because fused silica has (seemingly) better properties and is less expensive. However, 'better' has always been a relative term, and this case is a classic one of such relativity. It turns out that some of the undesirable properties of Vycor actually become assets. The lower temperature required to soften and/or melt Vycor means that less energy is required to form and shape the material. Also Vycor maintains its liquid state over a wider temperature range than fused silica,* thus making it easier to fuse it to other Vycor or fused silica.

Vycor has a thermal coefficient of expansion of 7.5×10^{-7} $\Delta cm/cm/°C$ and can be fused directly to fused silica. This property has provided an

* In a relatively narrow temperature range, fused silica becomes soft, melts, and then volatilizes. Because it maintains a relatively high viscosity once melted, it does not fuse easily.

excellent technique for fusing a borosilicate glass to a quartz glass. Normally, to make such a seal, a combination of three intermediate glasses fused between the outer two is required. The union of each intermediate glass is under strain, albeit within a tolerable range. However, during Vycor manufacturing it is possible to remove the non-SiO_2 materials using a controlled, tapered process. This creates a section of glass which can provide an infinitely graded seal between fused silica and borosilicate glass with no significant strain.

In addition, Vycor can be shaped and/or formed while in its borosilicate state *before* it is transformed into Vycor. Thus, molded, pressed, tapered, and other shapes that would otherwise be very difficult and/or expensive in fused silica can be done (relatively) easily with Vycor. Because Vycor already carries about 4% impurities, it is safe to 'dope' Vycor to obtain characteristics such as color or UV opacity. Any similar doping of fused silica would alter the characteristics that pure silica strives to achieve.

Finally, Vycor devitrifies far less than fused silica. Therefore, if you do not require ultra-pure baking environments (similar to those demanded in the silicon industry), furnace tubes made from Vycor may be cheaper in the long run than those made from less expensive fused silica.

1.1.5 Separating Glass by Type

Because laboratory glassware may be manufactured from a variety of different types of glass, such varied glass can become mixed up—potentially leading to confusion or later problems. It is important to either maintain different types of glassware separately, or be able to tell them apart. Although the former approach is preferred, the ability to identify and separate glass is important not only to save time, but also for safety and even the integrity of your experimentation.

Typically, only a few commercial soft glass items may work their way into a research lab. Such items as student ware graduated cylinders or burettes are readily identifiable, and since these items shouldn't be heated, are unlikely to cause damage. Specialized or custom-made glassware may not be as easily identifiable and therefore requires analysis.

Soft glasses, hard glasses, and high temperature glasses all look the same—like clear glass—and therefore may be hard to separate. The process to separate glass may be either destructive or nondestructive. Destructive techniques involve doing some permanent physical change to the glass, after it which will never be the same. Nondestructive techniques, obviously, are preferred if you wish to preserve all of your laboratory glassware.

Logical Deduction and Observation (Nondestructive). By logically deducing what you are likely to have in your laboratory, it is often simple to separate different types of glass and glass apparatus. For example, fused silica glassware is expensive. Unless there is a specific demand or need for fused silica in your lab, it is extremely unlikely that you have it. Thus, there is generally no need to look for things that are not likely to be there.

Glassware identified with ceramic decals or raised letters saying "Pyrex", "Kimex", or "Duran" are exactly what they say they are.* Pipettes

* Prolonged cleaning by base baths or HF can remove ceramic markings.

Table 1-2

Characteristics of Specific Glass Types[a]

Glass Type	Thermal Coefficient of Expansion (0-300°C) x 10^{-7} Δcm/cm/°C	Strain Point (≈°C)	Annealing Point (≈°C)	Softening Point (≈°C)	Working Point (≈°C)	Density (g/cm³)	Refractive Index
High lead[b]	104	400	435	600	860	6.22	1.97
Potash soda lead[c]	89-93	395	435	625-630	975-985	2.86-3.05	1.539-1.560
Soda-lime[d]	93	486	525	700	985	2.53	1.52
Common lab borosilicate[e]	32	510	560	821	1252	2.23	1.474
Soda barium borosilicate[f]	50	533	576	795	1189	2.36	1.491
Aluminosilicate[g]	42-45	665-735	710-785	915-1015	1200	2.52-2.77	1.530-1.536
Vycor®[h]	7.5	820	910	1500	—	2.18	1.458
Fused silica[i]	5.5	990	1050	1580	—	2.20	1.459

a Except where otherwise noted, these data are compiled from Properties of Corning's Glass and Glass Ceramic Families by Corning Glass Works, Corning, New York 14831, ©1979. Ranges indicate a small family of glass types with similar characteristics.

b This glass is for radiation shielding. The data are from Engineering with Glass by Corning Glass Works, Corning, New York 14831, ©1963.

c This type of tubing is commonly used on neon signs.

d This information is based on Kimble Glass type R-6, compiled from Kimble Glass Technical Data by Owens-Illinois Inc., Toledo, OH 43666, ©1960.

e Most laboratory glassware is made from this glass.

f Pipettes and other pharmaceutical items are made from this glass.

g Aluminosilicates are used for ignition tubes and containers for helium.

h Vycor® is 96% silica glass.

i Fused silica is essentially pure silica with few impurities.

from Kimble may be identified as "Kimex-51®," and are not the same as Kimble's common laboratory borosilicate glass (KG 33). Pipettes from Corning use a lab glass called "Corex®," which is also different from Corning's common laboratory glass (7740).

Glassware with bends and/or has been fused to other pieces of glass will be either common laboratory borosilicate glass or fused silica. Any soft glass made into laboratory apparatus is for special application, or more likely very old (and ought to be in a museum).

Sighting Down the End of a Glass Tube or Rod (Nondestructive). Even though most glasses will look the same when observed from the side, looking at glass end-on will exaggerate the different colors inherent within the glass. Common laboratory borosilicate glass will show a soft pale blue or green shade of color. Soft glass (soda-lime) will typically exhibit a fuller deep blue. It is recommended that you obtain a 15- to 20-cm rod of each type of glass to keep on hand for use as comparison samples. Absolutely clear (water white) glass is high-temperature fused silica *or* Vycor. By shining a deep UV [254nm] lamp on the side of the glass and sighting down the end of the tube, Vycor will display a yellow/green, or colorless, fluorescence while fused quartz will be strong blue[12] [**do not stare directly at the UV light**]).

Matching the Index of Refraction (**Nondestructive**). To match the index of refraction for two or more glasses, obtain a liquid with the same index of refraction as one type of glass. When an unknown glass is put into the liquid, the glass will disappear if it has the identical index of refraction. Two standard solutions for identifying Pyrex or Kimex glass (their refractive index is 1.474) are:

1) 16 parts methyl alcohol
   84 parts benzene

and,

2) 59 parts carbon tetrachloride
   41 parts benzene.

The major problem with both of these solutions is their toxicity. The first solution should only be used in a fume hood, and the second is so toxic that it should not be used at all. An alternative solution, albeit a bit messier, is common kitchen corn oil. It does not match the index of refraction as closely as do the two solutions mentioned above, but it does the job and is safe to use. The corn oil can be removed with soap and water.

Checking the Sodium Content of the Glass (**Semi-destructive**). Take the bottom unglazed side of a white ceramic tile and evaporate onto it a small quantity of a phenolphthalein-ethanol solution. To test a piece of glassware, drip a little water onto the plate and draw a line across it with the glassware. The action will lightly scratch the surface, so choose a non-significant spot. By scratching the surface, the alkalis of a soda-lime glass will be exposed to the phenolphthalein solution and a pink line will form across the plate. Borosilicate glass will exhibit no significant color change. The plate can be reused repeatedly.[13]

Using Linear Expansion (**Destructive**). This techniques requires that the end of known glass be fused to the end of an unknown glass (with a gas–oxygen torch). While still hot and soft, the glass is taken out of the flame and the ends are squeezed together with a pair of tweezers. This union is then placed back in the flame and reheated until it is soft again. It is then removed from the flame and, while still soft, the end is taken in the tweezers, and pulled out to a thread about 10-15 cm long. When the glass hardens, fire-cut off the end piece held in the tweezers. If the glass curls off to one side, the glasses are of different composition (see Fig. 1-3). Incidentally, if one of the two glasses resists melting and emits a very bright white light while being heated, it is very likely a high temperature fused silica, and no further identification techniques are required.

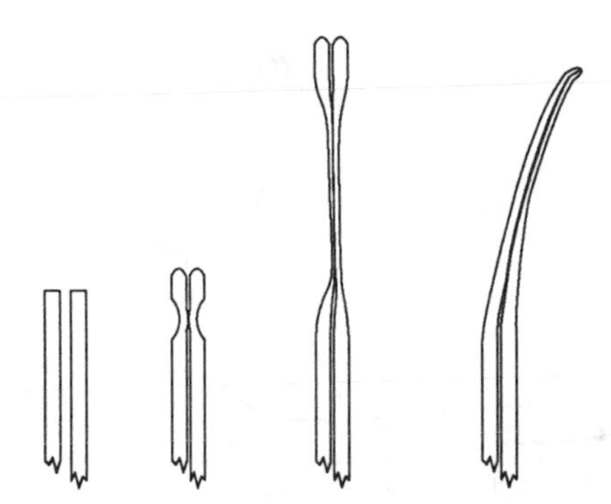

Fig. 1-3 Using the thermal coefficient of expansion to separate glass types.

1.1.6 Physical Properties of Glass and Mechanisms of Glass Fracture

Glass is one of the strongest materials on earth. This statement is a seemingly bizarre one to make concerning a material generally known for its fragility. Surprisingly, freshly-drawn, water-free glass can exhibit strengths of over 1,500,000 psi (pounds per square inch). However, once its surface has been hydrated and the uniformity of that surface has been broken (i.e., by a flaw), its potential strength can decrease dramatically.*

Glass is perfectly elastic until failure. That is, no matter how little or how much you flex glass, it will always return to its original shape when flexed at room temperatures. If a glass is flexed in the region of a flaw, it will flex until it breaks (fails). On the other hand, metals (even spring metals) can *remember* a new shape or position if flexed beyond a certain point. This quality is true for tangential as well as axial rotations. However, unlike metals, glass gives no indication that it is about to break, nor does glass provide any indication of where it is receiving stress** (for example, glass does not begin to fold prior to fracture).

Once flexed beyond a given point, glass breaks. The amount of flexure required to achieve that given point is dependent the nature of the flaws in

* Newly formed glass has high surface energy. Adsorption of water into the surface of the glass facilitates the lowering of this energy. Water contamination on, and in, a glass surface weakens the tensile strength of that item. The amount of water which potentially may be adsorbed depends the type of glass and relative atmospheric humidity. For example, soda-lime glass may adsorb water in as deep as several hundred molecules at 75% relative humidity. Common laboratory borosilicate glass may hydrate to between 50 and 100 molecules of depth, and fused silica may exhibit 10 to 50 molecules of depth of adsorbed water (at 75% relative humidity).[13] The only mechanisms to dry glass are high heat (approximately 400°C) or a combination of high heat and high vacuum.

Because much of chemistry is wet chemistry or is done in the ambient atmosphere (which has an inherent amount of humidity), it is neither practical nor relevant to try to duplicate the theoretical strength limits of dry glass. For some lamp manufacturing and most high vacuum work, these theoretical limits are relevant, but beyond the scope of this book. Although I may refer to wet or dry glass, I'm not expecting anyone to try to 'dry' their glassware.

** An exception to this statement would be viewing glass through polarized light, which provides excellent visualization of strains within glass. However, this requires special equipment and the glassware must be in a particular orientation with respect to the polariscope.

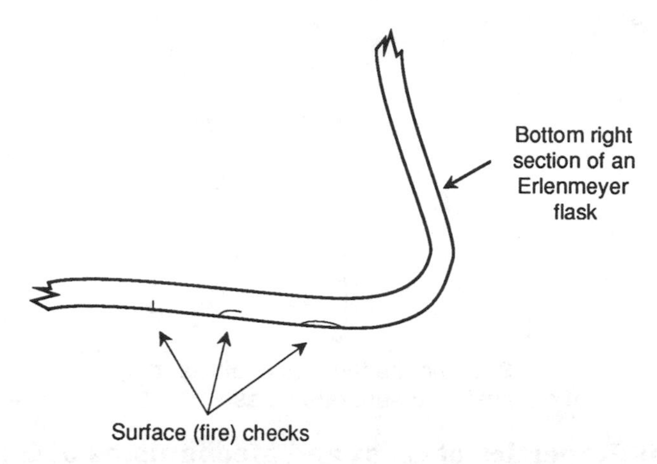

Fig. 1-4 Surface (fire) checks caused by placing hot glass on a cold surface.

the region being flexed. Glass can only be broken if two conditions are present: *flaw* and *stress* (more specifically, the stress of tension, not compression). Because glass is perfectly elastic until the point of failure, the location of the failure will occur at the most susceptible flaw. This property is a sort of *weak link* principle: glass under tension will break at the weakest link, which, by definition, is its most vulnerable flaw.

Tests demonstrate[14] that a one-quarter inch thick piece of glass that has received normal handling can withstand pressures of 6,000 psi. Sandblasting the surface to provide large numbers of flaws drops the potential strength to some 2,000 psi. However, if a fresh surface is acid-polished and then coated with lacquer (to prevent further abrasion by handling and to limit water content), strengths of up to 250,000 psi have been achieved.

Despite its tremendous strength, glass can be fairly easy to abrade. On the Mohs' scale of hardness,* glass is between #5 (apatite) and #7 (quartz). So, glass can scratch materials with lower numbers (for example, copper, aluminum, and talc). Likewise, glass can be scratched by materials that have higher numbers (for example, sand, hard steel, and diamond).

We know that diamond can scratch glass, but so can hard metals such as the hardened steel of a file or a tungsten-carbide glass knife. Glass slid or rolled across a dirty benchtop can be scratched by dirt particles. Also, glass can scratch glass. Because glass is often stored with other glass, it is constantly being scratched and abraded. A sharp surface is not necessary for glass to rub and cause flaws on the surface of other glass.

Another type of flaw that can develop on glass is a *surface check*. These flaws are micro- or macroscopic cracks that lie just on the surface of the glass. They can be caused by laying a very hot glass item on a cold surface. For example, taking a glass item from a hot plate and laying it on a benchtop can cause surface checks. In addition, surface checks can also be caused by brushing a hot gas–oxygen torch flame across a glass surface. The thicker

* A ten-scale division of geologic minerals used by geologists to help type rocks, with talc as #1, and diamond as #10.

the glass, the more likely surface checks may develop. Surface checks can aim straight into the glass, curve into the surface, or even curve back out to form a scallop-like crack (see Fig. 1-4).

As can be seen in Fig. 1-4, glass in normal use is soon covered with flaws. Most of these flaws are microscopic but any one of them could result in glass failure when subjected to sufficient stress.

The depth of a flaw is not as critical as is the ratio of the depth over the radius at the root of the flaw. An equation indicating the relative degree of stress based on this ratio is given in Eq. (1.1), and provides the Stress Concentration Factor, "K." The greater the value of K, the less stress is required to break the glass. When K is equal to 3, the depression can be considered not a flaw and is as strong as the surrounding glass.

$$K = 1 + 2\frac{A}{B} \qquad \text{Eq. (1.1)}$$

where A = depth of flaw,
B = radius of the flaw at its root.

Note that K, A, and B are dimensionless. The only criteria is that A and B be in the same measurement units. Some sample calculations using Eq. (1.1) can be seen in Table 1-3

Table 1-3

Samples of Stress Concentration Factors

1 2 3 4

1	$K = 1+2(2/2.00)$	K = 3
2	$K = 1+2(2/0.250)$	K = 17
3	$K = 1+2(2/0.032)$	K = 126
4	$K = 1+2(2/0.001)$	K = 4001

Glass can be broken at a specified location by creating a flaw where the break is desired and then applying stress (tension) to that flaw. However, an improperly made flaw can easily result in a flaw of undesirable quality. For example, inexperienced *flat glass* cutter will take a wheeled glass knife, aggressively bear down on the glass, and push the cutter back and forth several times. While a single scratch could have broken cleanly and easily, the repeated scratches create a heavy, round bottom groove with many side fissures. The resulting crack is likely to drift off to the side and not follow the heavily gouged crevasse. Alternatively, an inexperienced glass *rod and tube* cutter will take a triangular file and (again) bear down on the glass and saw the file back and forth for a "real good scratch." Unfortunately, the nicely rounded fissure will also not break easily. The extra force required to break the heavily gouged flaw is likely to cause the glass to break into many fragments, creating a serious risk of injury.

The key to successful glass breaking is a thin, deep scratch. However, you do not achieve effective depth by excessive force or repeated scratching.

To cut a soda-lime flat glass pane, you must maintain a firm pressure on a rotating wheel cutter and make a single continuous scratch toward you. Using a yardstick or meter stick as a straight line guide is recommended. If you bear down extra hard on any part of the swipe, you are likely to create side fissures that may cause the crack to sweep off to one side. The same technique applies to borosilicate flat glass, although more pressure will have to be exerted to achieve the same depth of flaw. Regardless, it is important to maintain even pressure throughout and to make only a single stroke.

Although not as effective as a tungsten-carbide glass knife blade, a triangular file can be used on glass tubing. The trick is to make a single, fast swipe of the file toward you and not to saw back and forth. In addition, the sharper the file's edge, the narrower the scratch. One way to ensure a sharp edge on the file is to use a grinding wheel to remove one face of the file (see Fig. 1-5). During the grinding process, constantly lower the file into a water bucket to cool the metal. If the file get too hot, it will turn a bluish color, indicating the file has lost its temper and will dull faster.

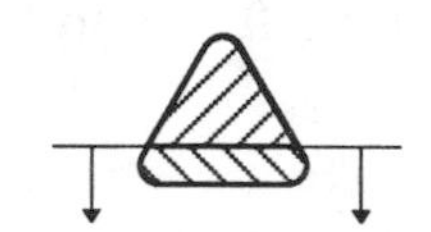

Grind a face from one side
of the file to obtain two
sharp edges

Fig. 1-5 Altering a triangular file for more efficient glass cutting.

1.1.7 Stress in Glass

There are two different mechanisms that produce stress (or strain) in glass: physical and thermal. Either can cause a distortion of a glass surface that will create compression and tension at one or more points. If sufficient tension develops in the region of a flaw, the glass will fracture.

Physical stress is easy to understand because we can feel, observe, or relate to the events that develop physical stresses. Pulling hard on a plastic tube attached to a hose connection, bending glass, and watching an item fall to the floor, are all easily understood (or observed) physical stresses.

When using glass, we want to reduce the amount of physical stress placed on the glass. Anything we do that limits any overt bending and twisting on glass is important. The key phrases here are REDUCE THE LEVER ARM and/or ELIMINATE THE RESISTANCE. The smaller the lever arm, the less torque and therefore the less stress you place on the glass. Equally important, when sliding or rotating glass against other objects, use a lubricant to limit torque.

Examples of these limitations would be:

1) When placing a glass tube into a stopper, be sure the glass tube and the hole are lubricated with glycerin, soapy water, or plain water to reduce friction. Never hold the glass tube more than two to four (tube) diameters from the rubber stopper to reduce the lever arm. Always wear leather gloves.
2) When removing a plastic or rubber hose from a glass hose connection, do not try to pull the hose off in one piece. Cut the plastic or rubber hose off with a razor blade to avoid torque.

3) When rotating a stopcock that is free-standing (i.e., one end of the stopcock is attached to glassware and the other is neither attached to other glassware nor supported by a clamp), eliminate the lever arm by holding the stopcock with one hand and rotating the plug with the other hand. If the plug requires grease, be sure to clean and change the stopcock grease at regular intervals. Old grease is harder to rotate and creates more torque when rotated. Be sure to use stopcock grease and not joint grease: greases made for stopcocks have more lateral 'slip' than do joint greases and therefore, they develop less torque when rotated.

4) In assembling and disassembling distillation apparatus (typically composed of sections of varying lengths, curved pieces, and pieces of different shapes connected by joints with no lateral movement), tremendous torque can be created when the piece is secured by support clamps. Use caution and common sense when attaching and tightening support clamps so as not to torque pieces toward or away from each other. An additional problem involves joints, or stopcocks, which have become frozen because either distillation solvents have stripped them of grease, or improper cleaning has left dirt on joint or stopcock members. Tremendous torque can be created when trying to disassemble stuck joints and stopcocks on a distillation assembly. Prevention of stuck joints and stopcocks is achieved by selecting greases that will not be affected by the solvents used and/or by using Teflon sleeves (these sleeves are excellent because they cannot be affected by most solvents). Always be sure that joint members and stopcocks are clean and dust-free. Wipe both members with acetone on a Kimwipe® before applying grease or Teflon sleeves and assembling the joint or stopcock.

In summary, *limit physical abrasion* to glassware, *reduce the lever arm*, and *provide adequate lubrication* to moving items. Your foresight will reduce glass breakage.

Fracturing glass at the site of a flaw is based entirely on the amount of tensile stress (or deformation) at a specific location, rather than on the amount of stress the glass was put under to achieve that level of tensile stress. More stress must be placed on thick glass than on thin glass to achieve a given amount of flexing (and resultant tensile stress on the surface). In other words, thick glass is physically stronger than thin glass because it is harder to flex. Unfortunately, the risk of broken glass in the lab cannot be resolved by making all glass thick because of thermal stresses.

Thermal stresses are more difficult to imagine than physical stresses, but we can observe what causes them anytime we look at a liquid thermometer. When materials get hot, they expand. When glassware becomes uniformly hot, it expands uniformly. When glassware becomes non-uniformly hot (above its strain point), there is uneven expansion of the glass. The regions between the uneven expansion will develop strain and the strain will remain after the glass has cooled. This strain can be great enough to cause fracture if there is a flaw in the region of the strain. We've all seen the effects of thermal stress when we've poured a hot liquid (i.e., freshly brewed tea) on ice cubes. The warmer exterior of the cube expands faster than the cooler interior, and trapped air bubbles provide the flaw that causes the ice to fracture.

Different materials expand at different rates when heated. Different types of the same materials can have different rates of expansion as well. Thus, different types of glass have radically different expansion properties. The measure of the rate of expansion of materials is the thermal coefficient of expansion stated in Δcm/cm/°C. Because the thermal coefficient of expansion of any material varies as the temperature varies, the thermal coefficient of expansion that is attributed to any particular glass is based on an average of the expansion figures compiled from a 0-300°C range. Table 1-2 includes the thermal coefficients of expansion for a variety of glasses.

The larger the thermal coefficient of expansion number, the less radical a temperature change the glass can withstand and vise versa. However, a thin piece of glass with a high thermal coefficient of expansion may be able to withstand a more radical temperature change than a very thick piece of glass with the same, or lower, thermal coefficient of expansion.

We see the effects of a high thermal coefficient of expansion when we pour boiling water in a cold glass: the glass breaks. We see the effects of a low thermal coefficient of expansion when we take quartz glass at melting temperatures and plunge it into ice water with no problems. Telescope mirrors are made out of materials such as quartz glass because they do not distort with changes of temperature to the degree that glasses of higher thermal coefficients of expansion will.

1.1.8 Managing Thermal and Physical Stress in the Laboratory

Glassware manufacturers must determine whether an item is more likely to be used in a *physically* stressing environment or a *thermally* stressing environment. For use in a more physically stressing situation glass must be made thicker to achieve physical strength, and for use in a more thermally stressing situation, glass must be made thinner to achieve thermal strength. It is not possible to make glass optimally strong for both physically and thermally stressful environments: a compromise must always be struck.

We see very thick glass used for kitchenware such as measuring cups and bakeware. Kitchenware receives minimal heat stress (baking is a slow heating process and therefore is not a thermally stressing activity) and is more likely to receive physical stress as it is banged around in cupboards and drawers or nested within other kitchenware. A glass coffee pot, however, is expected to withstand rapid heat changes (and is not generally stacked or banged against other objects), and is therefore made out of thinner glass. Kitchenware and coffee pots are made from exactly the same type of glass (common laboratory borosilicate glass), but are designed to withstand radically different types of stress.

We also see thick and thin glass in the laboratory. Because their concave bottoms could not otherwise withstand the force of a vacuum, filter flasks are made of thick glass. However, do not place a filter flask on a heating

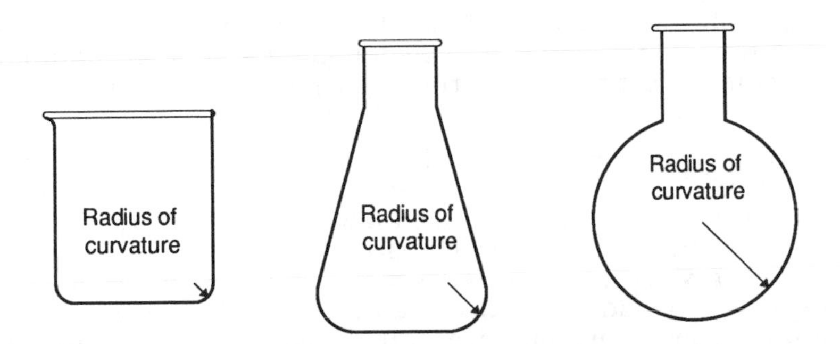

Fig. 1-6 The radius of curvature on the bottom of a container is smallest on the beaker, greater on the Erlenmeyer, and largest on the round bottom flask.

plate: it cannot tolerate the stress. The standard Erlenmeyer, by comparison, is thin-walled, designed to withstand thermal stress. However, a standard Erlenmeyer flask cannot tolerate the physical stresses of a vacuum: the flask's concave bottom will flex (stress) and is likely to implode in regions of flaws.

The shape of glassware can be a clue as to how and/or where it can be used. The more rounded its corners, the better it can diffuse thermal stress. This idea is similar in concept to the sharpness of flaws and can be compared to Eq. (1.1), the Stress Concentration Factor. Although that equation is intended to be used for surface flaws on glass, the principle is the same.

The three most frequently used laboratory containers are the beaker, the Erlenmeyer, and the round bottom flasks. Their shapes, radii of curvature at their bases (see Fig. 1-6), and functions all differ. If you measure the radii of curvature at the bases of 250-ml beakers, Erlenmeyer flasks, and round bottom flasks, you will find that they are approximately 6 mm, 12 mm, and 42 mm, respectively. Because the bottom of a beaker is essentially a right angle, as the bottom expands, when subjected to heat, the walls receive stress at right angles. There is little, if any, means to diffuse the stress. Because the Erlenmeyer has a larger curvature on its base corners, it can diffuse thermal stress, but only to a limited degree. Because the round bottom flask is all curvature, it is best suited for diffusing thermal stress. If you are using a Bunsen burner flame as a heat source, the safest heating vessel to use is clearly a round bottom flask. If you must use an Erlenmeyer flask or beaker, you can diffuse the intensity of the flame by placing a wire square between it and the container (a hot plate, on the other hand, is not a direct heat source and therefore is a safer heat source for an Erlenmeyer flask or a beaker).

Before heating glass apparatus, be sure to:

1) Examine the surface of the glassware for obvious flaws that may cause fracture when heated.
2) Heat only borosilicate or fused silicate ware. Never heat a soft glass container.
3) Try to avoid using flat bottom containers such as beakers or Erlenmeyer flasks.

4) Preferably, avoid using a direct flame. An open flame has inherent dangers (i.e., can ignite reactant gases and other materials that come in contact with the flame) that electrical and steam heating do not have.

The use of boiling chips is, and should be, standard practice when heating fluids. However, careless placement of boiling chips in a container can severely reduce the life span of the container. Never dump boiling chips into an empty glass container. Boiling chips have sharp edges that can scratch glass. By placing your liquid solution into the container before introducing boiling chips, you provide a lubricant and reduce the freedom of movement of the boiling chips. This practice can decrease the potential wear (i.e., flaws) that boiling chips could otherwise inflict on the surface of the glass.

Liquid will not provide adequate lubrication or protection against the scraping of a glass rod across the bottom of a glass container. Glass rods are often used for mixing, creating bubbles (boiling), or scraping material from the sides of a container. All three objectives can be met with a *rubber policeman* or a plastic stirring rod, both of which cause no damage to the inside surface of the container.

Heating is usually only half the story. After you have heated the material in a container, there is the problem of what to do with the container while it cools. A radical drop in temperature can be as damaging as a radical rise in temperature. Never heat or rest a round bottom flask on a non-cushioned metal support ring. The metal may scratch the glass (providing flaws) or provide pressure (causing strain) and/or radical temperature changes (providing stress). For the round bottom flask, a cork ring provides a safe stand and avoids radical temperature change. Similarly, never place a hot beaker or Erlenmeyer flask on a cold lab bench surface. Always place such a container on a wire gauze square or other 'trivet' type of resting place.

1.1.9 Tempered Glass

It is possible to use thermal stress to physically strengthen glass by processes known as *tempering glass* and *heat-strengthening glass*. Both processes depend on rapid cooling to produce strain within the glass. The difference between the two depends on the initial temperature before cooling.

Both tempering and heat-strengthening use the two specific properties of glass to a) expand when heated and b) be a poor conductor of heat. Tempered glass is heated near its softening point and then rapidly chilled by blowing cold air on the surface (see Fig. 1-7). The technique for heat-strengthening glass is similar to tempering glass, but the glass is not preheated as high and the cooling rates are slower.

Regardless of the strengthening technique, the rapid chilling contracts the surface glass. However, the slow heat conduction of glass prevents the inside glass from cooling at the same rate. In time, the inner glass cools and contracts, which causes tension on the inside of the glass and compression on the outside. Because it is not very likely for there to be flaws on the inside of the glass, this situation demonstrates 'safe' tension.

To break tempered or heat-strengthened glass requires the placement of stress on the surface that is greater than the thousands of pounds per square inch of compression created by the strengthening process. That is why these glasses are so physically strong. The surface compression of heat-strengthened glass is between 3,000 and 10,000 lbs/in^2. The surface compression of tempered glass is greater than 10,000 lbs/in^2.

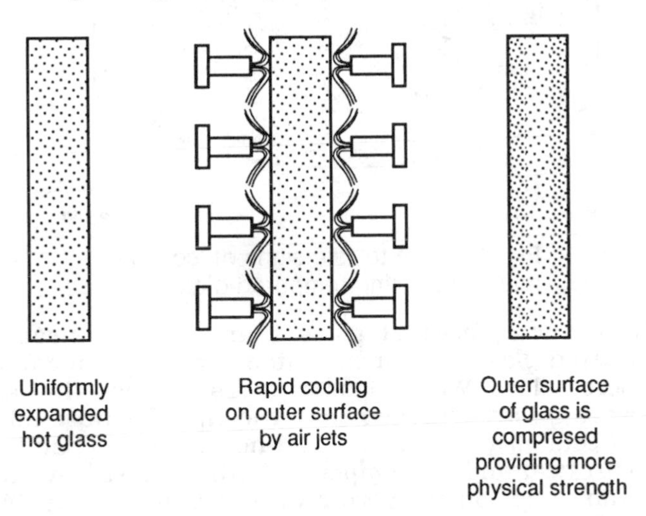
Uniformly expanded hot glass
Rapid cooling on outer surface by air jets
Outer surface of glass is compresed providing more physical strength

Fig. 1-7 Production of tempered glass.

Tempering glass is only viable for physically strengthening moderate to heavy walled glass. It cannot be used to strengthen thin glass because the internal regions cool too fast preventing the compression and tension forces to develop. In addition, tempered glass cannot be used in high thermal environments because the constant heating and cooling cycles would eventually untemper the glass.

Tempering and heat-strengthening glass decrease its flexibility as well as increase its ability to resist impact abrasion. Heat-tempered or heat-strengthened glass cannot be cut once it is tempered. However, both types of glass-strengthening processes can be reversed by simply annealing the glass. Once the strengthening has been reversed, the glass can be cut in any normal fashion.

If you have a pair of polarized sunglasses, you are likely to have seen the effects of tempered glass on rear and side windows of cars. The telltale white or gray spot array pattern is an indication of where the cold air jets were located (see Fig. 1-8).

There are other techniques used to temper glass. One technique is called *ion exchange*. In this process glass is 'boiled,' or soaked, in liquid potassium, during which time the surface sodium ions exchange places with the potassium ions. The potassium atom is larger than the sodium atom, so during ion exchange, the potassium must 'squeeze' itself into positions that formerly held the sodium. This exchange produces the same compression effect on the glass surface as rapid cooling. Although it is a more expensive process, ion exchange tempering produces a more uniform compressed surface than the air jet process. In addition, the ion exchange process can

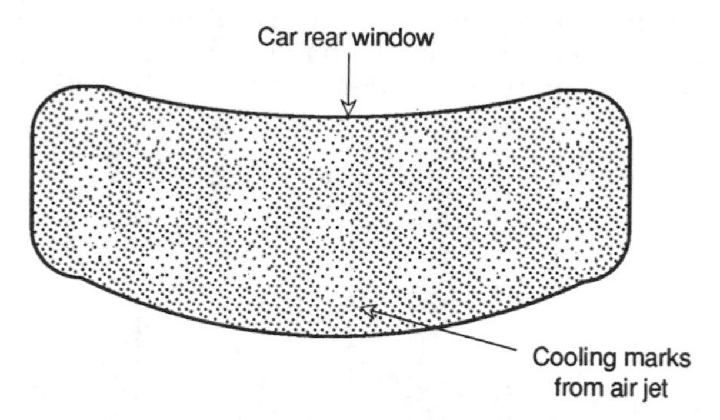

Fig. 1-8 The telltale signs of 'cooling' marks indicating tempered glass.

temper much thinner pieces of glass than using a heat approach. Ion exchanged tempered glass cannot be untempered by annealing the glass as with heat-tempered glass. When glass catalogs mention that all (or part) of a piece of glassware is strengthened (like the tip of a burette), it is typically done by blowing cold air on hot glass, and not by ion exchange.

Although both heat and ion tempering both successfully strengthen glass, heat-strengthening and ion tempering do not satisfy either ANSI or federal specifications for safety glazing.

A third technique to produce tempered ware uses two glasses (or ceramics) formed together, each with different thermal coefficient of expansion. This is used to make Correll® dishes by Corning. For this tableware, a pyroceramic* material with one type of thermal coefficient of expansion is covered with another pyroceramic with a greater thermal coefficient of expansion and baked until the outer layer melts uniformly. Materials with greater thermal coefficients of expansion will expand more when heated as well as contract more when cooled. The greater contraction (once cooled) of the outside material causes compression.

1.1.10 Glass and Internal Pressure

Because tubing walls** on a vacuum system are in compression, vacuum systems do not require heavily walled glass. On the other hand, pressure on the inside of a tube creates tension on the walls of the tube. If the tension is great enough, the tube will explode from internal pressure. The strength of tubing to withstand a given amount of pressure is called its *bursting strength*. To increase the bursting strength and withstand internal pressure requires heavy wall or small diameter tubing. This requirement is based on the ratio of the diameter of the tubing to the wall thickness. For example, 2-mm tubing has an inside diameter of 1.0 mm. For 100-mm tubing to have the same percentage wall thickness, it would need an internal diameter of 50 mm. Standard 100-mm wall tubing has an inside diameter of 95.2 mm.

* A pyroceramic material is a special type of glass that has been devitrified by controlled nucleation of crystals. It has certain properties that are like glass and other properties that are like ceramic.

** Standard round tubing, not square, flat, or distorted walls.

Wall thickness is not the only factor that determines the potential bursting strength of tubing. Glass strength is also based on surface quality and preexistent strains. Either of these properties can significantly jeopardize the theoretical bursting strength of tubing.

The quality of the surface is based on the amount and degree of surface flaws. Obviously abusive handling of glassware can cause flaws, but so can normal glass handling, such as laying a piece of glass on a table or against other glass. Any visible surface markings can significantly decrease the potential strength of a glass, but not all flaws are readily visible.

A preexistent strain is any stress within the tubing prior to receiving internal pressure. These flaws can come from poor-quality annealing or torquing the glass during assembly resulting in physical strain. The extent to which flaws and inherent strain will affect a glass' bursting strength cannot be predetermined. Therefore, any calculations determining bursting strength must have built-in safety tolerances to account for the unknown.

The glass industry assumes that new tubing, with no observable flaws, can withstand stresses of up to 4,800 psi. It has been calculated that the maximum amount of stress applied to lightly-used glassware should not be greater than 1,920 psi. However, typical laboratory glass receives great amount of surface abuse so an even greater safety factor should be considered. Unstrained laboratory glassware should not receive stress greater than 960 psi. If preexisting stress in the glassware is considered as a safety factor, the tolerable stress should again be lowered to 750 psi.

Despite such a large safety factor, any pressurization of glassware should be done behind a safety shield. In addition, glassware should be wrapped in a threaded tape such as cloth surgical tape or fibered packaging tape. Both of these tapes lose their adhesive ability and strength as they age, so periodically check the quality of the wrap. The tape serves an additional function by protecting the surface from abrasion which could weaken it. Masking tape is not satisfactory for either purpose and should not be used.

To calculate the tangential or *hoop** strength, Shand[15] offers Eq. (1.2). When stress created by internal pressure is greater than the strength of the tubing (based on the location of a potentially susceptible flaw) the tubing will fail (break).

$$\sigma_a = \frac{p\ d}{D\ d} \qquad \text{Eq. (1.2)}$$

where σ_a = average stress (psi)
p = pressure
d = inside diameter
D = outside diameter.

Unfortunately, this simple equation is only good when a tubing's wall thickness is 5% or less of the total outside diameter, thus omitting most of the tubing found in a laboratory. Shand offers the somewhat more complicated following equation (Eq. [1.3]) for calculating the hoop strength for all sizes of laboratory tubing:

* The area around the circumference of the tube is called the hoop, similar in concept to the hoop surrounding wooden wine barrel.

$$\sigma_i = \frac{\left(1 + \frac{d}{D}\right)^2}{\left(1 \frac{d}{D}\right)^2} p \qquad \text{Eq. (1.3)}$$

The average stress (σ_a) is the amount of stress that the system is receiving with any given tubing dimensions and under the pressure loads provided. The internal stress (σ_i) is the amount of stress placed on the internal wall by any pressure on the inside of the glass.

If Eq. (1.4) is solved for *pressure*, and we substitute for σ_i the allowable bursting pressure of 750 psi, we can calculate the acceptable pressure limits for commercially-available tubing.

$$p = \sigma_i \Big/ \frac{\left(1+ \frac{d}{D}\right)}{\left(1- \frac{d}{D}\right)} \qquad \text{Eq. (1.4)}$$

As an example of safe pressure limits of a standard commercial tubing, consider one-half inch tubing and let σ = 750 psi. Standard wall tubing* can withstand a pressure of 173.1 psi, one-half inch medium wall tubing can withstand pressures of 268.3 psi, and one-half inch heavy wall tubing can withstand pressures of 420.4 psi.

Shand[16] also provides Eq. (1.5) for flat plates.

$$\sigma_m = k_1 \frac{d^2}{t^2} p \qquad \text{Eq. (1.5)}$$

where
σ_m = maximum stress
k_1 = Stress concentration factor
(when edges are free, k_1 = 0.3025, and
when edges are clamped, k_1 = 0.1875)
d = diameter
t = thickness of plate.

This equation can also be used on rectangular plates by substituting the narrow dimension of a rectangle for "d."

WARNING: Observing of a flaw on the surface, or within, a piece of glass that is to be used in any pressure situation warrants discarding that piece and replacing it with a new, flawless, piece.

1.1.11 Limiting Broken Glass in the Lab

If you work with glass in a laboratory, you will break glass. Accidents happen. However, by limiting the causes, you can dramatically decrease the

* Medium and heavy wall tubing were originally measured in English measurements and now use the metric equivalent. Standard wall tubing has always been measured in metric measurements. Because there is no one-half inch standard wall tubing, twelve mm tubing was used for this example.

incidents of breakage. First, limit abrasion of glass surfaces. This precaution can be taken by limiting the amount that glass may roll, slid, and/or, bounce into other glass. Similarly, try to prevent, limit, or control the amount of contact with other hard surfaces such as metals and ceramics (including boiling chips).

The second level of defense against glass breakage is to limit stress. For instance, physical stress can be limited by shortening the lever arm and providing proper lubrication. Thermal stress can be limited by selecting appropriately thin walled glass and/or selecting glass with an appropriate thermal coefficient of expansion. In addition, limit the amount that glassware is subjected to radical temperature changes.

Despite the fact that common laboratory borosilicate glass is capable of withstanding high-heat conditions, laboratory containers should never be left empty on a heat source. Although this oversight is not made intentionally, it can occur if, for example, a boiling solution is left unattended. The liquid in the container performs an important *heat-sink* function without which the heated glass may achieve temperatures sufficient to cause thermal strains. There are three options for dealing with glass left empty on a heat source:

1) Play it safe and throw the item away. Although this option may seem wasteful, it is the safest thing you can do if you have no other means to verify safety or eliminate possible strain (see point #3).
2) Examine the glass with polarized light (such as a polariscope)* to look for strain. If no strain is evident, the item should be safe to use. If strain is evident and you have no means to relieve the strain, throw the item away.
3) If strain in the glassware is evident, or if you cannot verify whether strain is present, oven anneal the entire item by bringing it to a temperature of 565°C and holding it at that temperature for 15 minutes. Then slowly (> four hours) let the temperature drop to room temperature. Alternatively, bring the entire item to a temperature of about 600°C with no holding time and let the temperature slowly (> four hours) drop to room temperature. The second technique is more likely to cause thin diameter tubing to sag.

Of the three options, the second and third are limited by available equipment. Without a polariscope and/or glass annealing oven, you cannot do them. If you do not have this equipment, play it safe and use the first option.

* A polariscope is an optical device which shines a light through two polarizing filters. The filters are crossed, which prevents all light from passing except the rays that are aimed straight at the viewer. When a glass object is placed between the two filters, the light passes through the glass. If the glass object has no strain, the light passes through with no deflection. If the glass object has strain, the specific strain regions deflect and twist the light so it passes through the second filter in a different orientation than light which has not passed through strain. Regions of strain in glassware are therefore easily observed as changes in light intensity or color (if the polariscope is adapted for color).[17]

1.1.12 Storing Glass

When not in use, glassware should be cleaned and put away. Glassware that is scattered on benchtops and out in the open clutters working areas, is easily broken, will not stay clean, and if dirty, may be confused for clean glassware. In other words, glassware that is not cleaned and put away is a toxic and/or physical danger that is likely to undermine and potentially negate any viable research. There is a phenomenal amount of wasted glassware and research solely due to glassware that was not cleaned and put away. The techniques of cleaning glassware are discussed in Chap. 4.

Once cleaned, glassware should be dried before it is placed in storage. If possible, a section in your lab should be reserved for cleaning and drying glassware so that you 1) do not contaminate glassware with cleaning materials, and 2) do not have glassware that is drying be in your way.

The best place to store laboratory glassware and equipment is in covered or enclosed storage areas. Open shelving and other areas where dust can settle on (and in) apparatus should be avoided. Glass-enclosed storage cabinets are excellent as they provide opportunity for visual inspection of available items, and reduce unnecessary door opening. Place strips of tape across large glass doors to prevent accidents by people who may not see the glass.

The most common place for glassware to be stored is in a drawer. Drawers provide many of the recommended requirements for glassware and equipment storage except visibility. In fact, a drawer's biggest liability is that it can be opened and closed quickly by someone trying to locate a particular item. The rattling of glassware rolling into other glassware or apparatus portends glass repair.

There are several ways to protect glassware stored in drawers:

1) Label all drawers. Self-sticking labels are sufficient for most labs. However, if the contents of the drawers are constantly being transferred, metal or plastic card holders may be more practical.
2) Encourage users of the lab not to jerk drawers open or slam them shut. I do not know of any way to prevent these actions, but continued abuse may be thwarted by demanding financial remuneration of broken glassware (if the culprit is caught).
3) Limit free movement of glass items. **a)** Small items should be kept in small boxes with cut off tops (to facilitate observation of the contents). **b)** Line drawers with plastic mesh or plastic bubble packing material to limit both movement as well as cushion impact against the walls of the drawer. **c)** Be sure to prevent items from twisting or tilting within the drawer. Such movement may cause part of an apparatus to stick up and either prevent the drawer from opening or be broken off (if the drawer opener is strong and persistent).

Nesting glassware is a good space saver, but be sure there is adequate room for the glassware to nest. Evaporation dishes are not a problem, but the pour spouts of small beakers tend to wedge within larger beakers.

Obviously the best solution to prevent this situation from happening is to insure adequate room for smaller beakers. A simple technique may be used to separate beakers. Squeeze the larger beaker at right angles to the jammed unit (see Fig. 1-9). It does not take a great amount of pressure, but you must remove the smaller beaker *as* you squeeze. For safety's sake, wear leather or Kevlar® gloves when squeezing glassware. Although the amount of flexure needed is very small, if there is a flaw in the right place, a piece could break in your hand.

Limit dust within glass items such as pipettes by plugging the hole with a Kimwipe® or wrapping the ends in a Kimwipe held by a rubber band or tape. Do not tape directly on the glassware as it may be difficult to remove after the tape has aged.

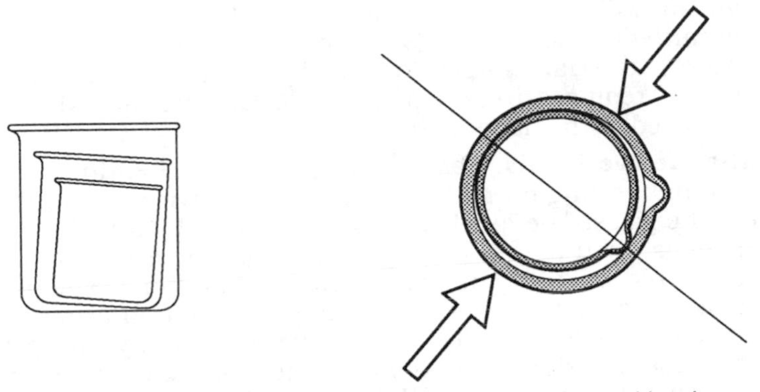

Fig. 1-9 Nesting beakers and removing stuck nested beakers.

1.1.13 Marking Glass

The permanent marks on glassware are made of a special type of glass-ceramic that fuses onto glass surfaces above 500°C. They are applied either by a silk-screening process, or by decal. Ceramic decals are permanent and will resist attacks by most chemicals. They are susceptible to chemical attack and can be removed by alkalis and hydrofluoric acid.

Old glassware used to provide a frosted circular spot for marking the glass. This frosted zone was easier to write on than the ceramic spot, but was a flawed surface. These flawed surfaces were more easily fractured by stress and included one more step to the manufacturing process, and therefore were phased out.

The ceramic white dot on most glassware can be written on by pencils for identification. Although it is not easy to write on glass, there are five techniques that allow one to identify and mark on glass:

1) **Alcohol-based pens**. These pens, usually fiber tipped, can write on most surfaces and are water-resistant. They can be removed with any hydrocarbon solvent and will burn off in a drying oven. Although they are not likely to smear from the glass onto fingers, they may. These pens can quickly dry out if not recapped after use. They are available in a variety of colors.

2) **Waxed pencil**. Like a crayon, waxed pencils can easily write on glass and are subject to the same removal techniques as alcohol pens. They are likely to smear onto fingers and other equipment. They are available in a variety of colors.

3) **Soft pencil**. The company Schwan - STABILO GERMANY makes a line of pencils that are identified as '*All* - STABILO.' These pencils are so soft they can easily write on glassware. The markings from these pencils can withstand high heat (600 to 700°C) but can be wiped off by the rub of a finger.

4) **Titanium writing**. Titanium dipped in water (or writing on wet glass) will leave a permanent marking on glass. Although admittedly the markings are not easy to see (only ≈ 0.25mm wide*), they are impervious to hydrocarbon solvents and can resist temperatures up to 1500°C. The markings can be removed with nitric acid.

5) **Self-adhesive labels**. This method is sort of cheating because you are not writing on glass. Rather you are writing on labels stuck to the glass. The labels can be written on by standard pens and pencils, then placed on any type of glassware. Because they are available in different colors, the labels can assist identifying glassware by type, then write more specific information on the labels. It is possible to remove these labels by lifting them off, but sticky residue may remain. The residue may be removed by acetone. If you write on the label with an ink pen, the ink may smear or bleed if chemicals leak onto the label.

1.1.14 Consumer's Guide to Purchasing Laboratory Glassware

Laboratory glassware can be expensive. As far as the user is concerned, the chemistry of the glass by the three major companies is equally good. However, considerable ranges of quality can be exhibited in glass quality and items made out of a particular glass. If a glass or the products made from that glass are not acceptable, do not accept the item(s).

Glass companies do not try to make poor-quality glassware, nor are they intent on selling their mistakes. However, mistakes happen and sometimes the mistakes get by quality control. Regardless, the final quality control person is the customer. If you receive poor quality glassware, by working with your supplier and/or the glassware manufacturer, the problem can be resolved.

Faults in glass can include the following:

1) **Seeds**** are like specs of sand in a glass that did not properly melt. They are particularly susceptible to thermal strain that would not affect regular glass.

2) **Bubbles** are trapped air within glass.

3) **Airlines** are bubbles that got stretched during the manufacturing process.

* The width is related to the amount of contact surface the titanium has with the glass.
** This flaw can become a focal point for fracture.

4) **Blisters** are bubbles that are very close to the surface and are likely to break open. If they partially break open, they may hold liquids or particulate material that can contaminate current or future work.

5) **Cords*** are glass inclusions of a bad melt. They may appear like sections or spots of glass that did not properly melt.

6) **Chill mark** is a wrinkled surface caused by poor forming in a construction technique called 'mold pressing.' The bases of graduated cylinders and the bodies of funnels are typically made with this process.

There are more companies that make glassware than make glass. Each is dependent upon the glass manufacturing companies to provide glass with a minimum of the problems mentioned above. Ideally, such glass flaws are sorted out before, or during the manufacturing process. In addition, there are flaws that can be created during, the manufacturing or shipping of glass items. Ideally these problems should be caught before reaching the consumer. However, because a number of manufacturing processes are automated, errors in production may not be caught until complaints start to come in from the field (i.e., you).

No manufacturer can be considered totally dependable. Some manufacturers make some products better than other manufacturers, and some items better than other items. Therefore, in addition to the flaws that can be found in glass as just listed, do not accept the following flaws in manufactured glass items:

1) **Chipped, scratched, or cracked glass out of the box**. There are ample opportunities for these flaws to happen in any lab, so you do not need an early assist from the manufacturer, the warehouse, or the shipper. Because any of these three may be equally culpable in causing physical damage, it is best to not assume who might be to blame. Let the supplier worry about it.

2) **New stopcocks or joints that leak**. Do not 'repair' them by adding more stopcock grease than should normally be required. Such practice can easily exacerbate the problem and cause other problems as well. However, before you complain, be sure that pieces are assembled correctly and matching parts are assembled with their correct pairs.

3) **Seals with folds and/or large quantities of internal bubbles**. When glass is first attached to other glass, it will show folds, or ripples, that need to be worked smooth. Sometimes in production, speed overtakes quality and these flaws are not adequately worked out. These folds can be weak spots requiring little stress for fracture. One approach a glassblower may use to speed the 'working out' of these folds is to use a hot flame. If the glass is inadequately cleaned (another victim of production speed) the intense heat will cause the glass in

* This flaw can become a focal point for fracture.

the region of the seal to develop many bubbles. These bubbles contribute to (and are an indication of) weak glass. Although many seals may have bubbles, it is the overabundance of bubbles that should elicit concern.

4) **'Burnt' glass**. The 'burning' of glass is caused by severe overheating of glass during glassblowing operations. This process removes some of the component materials. Because this new glass has a different chemical composition than the glass it was made from, it cannot flow or mix properly with the surrounding glass. It can be identified by a spot of glass that does not look as though it has properly fused with the surrounding glass and may be confused with a fold.

1.2 Flexible Tubing

1.2.1 Introduction

There is an enormous variety of flexible tubing available for use in the laboratory because no single tubing type, or size, is right for all purposes. Within most laboratories, flexible tubing is used for such purposes as connecting vacuum systems to mechanical pumps and manometers, transferring nitrogen or argon gas around the lab, and connecting water lines to condensers, coolers, and constant temperature baths. In addition, you may connect a source for natural gas (or propane) and oxygen to a torch and even temporarily connect a mechanical pump to drain used pump oil.

Comparing the various brands and types of flexible tubing is not unlike comparing a variety of stereos at competing stores. Because of the many different brands and models of tubing available, and because not every manufacturer uses the same analysis parameters, a cross comparison of features (or prices) from one type of flexible tubing to another is very difficult.

Purchasing flexible tubing has a few other complications. First, few laboratory supply-house catalogs identify the manufacturers of the tubings listed. Furthermore, just as stereo stores do not carry all brands of stereos, no single laboratory supply company carries all manufacturers' tubing. Therefore, if a particular type of tubing has features that you require, you may need to ask your laboratory supply-house if it carries (or can obtain) a particular manufac-

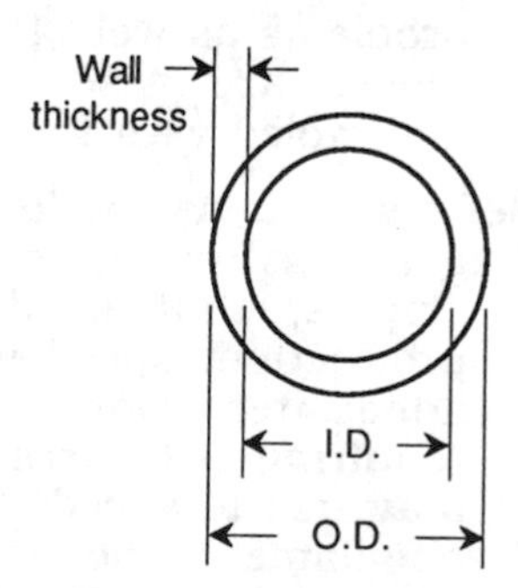

Fig. 1-10 The measurements of flexible tubing.

turer's product. This section will detail the differences and similarities between the types of tubing that may be used in the chemical laboratory.

Flexible tubing is identified by its *inside diameter* and *wall thickness* (see Fig. 1-10). This identification is unlike glass tubing, which is identified by its *outside diameter* and *wall thickness*. When ordering flexible tubing, specify the inside diameter *and* the wall thickness.

In ordering tubing, size is only one of several variables to consider. Two others are the tubing's physical characteristics (see Table 1-4) and its chemical resistance properties (see Table 1-5). Table 1-6 compares the advantages and disadvantages of the various types of flexible tubing.

IMPORTANT NOTICE

Unless otherwise stated, all data in this section (test conditions) are based on tubing that has 1/4" I.D. and 1/16" wall thickness.

1.2.2 Physical Properties of Flexible Tubing

The following physical characteristics may be important in your selection of flexible tubing (see Table 1-4).

Color and/or Transparency. If it is important to see a solution flowing through tubing, then transparency or translucency of the tubing is important. Color cannot be used as an indicator of physical properties or chemical resistance. Color can be used by manufacturers to distinguish between various tubing types however. For example, the manufacturer of Nalgene® tubing uses blue imprinting on its clear tubing to designate its 8000 line of tubing, red imprinting to designate its 8000-vacuum line, and black imprinting to designate its 8007 line of tubing. Color can also aid in laboratory setup. For example, labs that use color to identify operations may use orange tubing for natural gas, green tubing for oxygen, blue tubing for water going into a condenser, and red tubing for water leaving a condenser.

Durometer Range. The durometer range is a measure of a tubing's physical hardness, which is indirectly related to the tubing's flexibility and resilience. The harder a tubing's composition, the less flexible and resilient it is likely to be.

The hardness of tubing material is calibrated with an instrument called a durometer. The durometer is a device that measures the amount of reflected bounce by a special hammer off the material being tested. Most tubing is tested using what is called the *Shore A* technique. When very firm tubing is tested, the *Shore D* technique is used. Shore D measurements can be interpolated to provide approximate Shore A measurements for comparison. For simplicity's sake, I have chosen to interpolate all Shore D measurements to Shore A. Except where advised to the contrary, please assume that all durometer readings preceded with a "≈" symbol are Shore D measurements that have been interpolated to Shore A.

Flame Resistance. Some tubing materials are naturally resistant to flames, while others are flammable to various degrees. This quality must be considered if the tubing may be exposed to a high, or direct, heat source. Unfortunately, although we know that a given tubing may be flammable, flame-resistant, or nonflammable, there is no information currently available by which one can quantitatively compare the flammability of tubing.

Flexibility. Is the tubing flexible or stiff? Is the tubing prone to kinking? Transport of gases or fluids may be impaired by non-flexible tubing.

Gas Impermeability. The permeability of tubing depends on the gas being used: tubing that is fine for nitrogen may be totally unacceptable for helium. Unfortunately, comprehensive data on the permeability of various gases through tubing is not readily available. In most cases, it is necessary to contact the manufacturer for information.

Resilience (*Memory*). Over time, some tubings mold themselves to new shapes, whereas others will always return to their original shape. For example, rubber has the best resiliency of any tubing, and thus can maintain a constant grip on a hose connection. On the other hand, Tygon® tubing, which is not very resilient, will 'learn' the new shape of a hose connection and will eventually lose its grip on the nipple. Internal pressure (i.e., water pressure) on a non-resilient hose can cause it to slip off its connection. Low-resilience tubing should be attached with screw clips (see Fig. 1-11), for a more secure attachment.

Temperature Range. As temperature increases, a tubing's ability to withstand internal pressure decreases. Conversely, if the temperature drops below the recommended temperature range, a normally flexible tubing may crack when flexed.

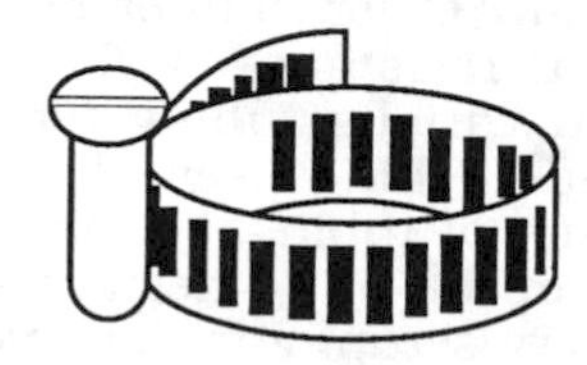

Fig 1-11 A typical screw clamp.

Vacuum. Under vacuum conditions, thin wall tubing will collapse. In general, it is safe to use tubing for vacuum work if the tubing satisfies Eq. (1.6):

$$\text{I.D.} \leq 2 \text{ X wall thickness} \qquad \text{Eq. (1.6)}$$

In many labs it is standard to use red heavy-wall rubber tubing to connect mechanical pumps to vacuum systems. However, there is no technical reason for using red (colored) tubing. Any tubing that meets the qualifications for vacuum work (or your specific vacuum work) should suffice, regardless of color.

Pressure. There are many variables that interact to affect the maximum potential pressure tolerance of any given tubing. These include:

Inside Diameter: The smaller the I.D., the greater the potential pressure tolerance.
Wall Thickness: The greater the wall thickness, the greater the tolerable pressure.
Temperature: The higher the temperature, the less the tolerable pressure.*

* This statement does not necessarily mean that the lower the temperature, the greater the tolerable pressure because at very low temperatures, tubing becomes brittle and prone to failure.

Time: If you are working over the recommended pressure, it is only a matter of time before the tube will burst.

Material Transmitted: Most types of tubing can handle most materials for at least a short time. However, if tubing is undergoing chemical attack, the tubing will eventually fail.

Braiding: Internal or external braiding provides extra strength for pressure systems.

External Surface. If you wear gloves and/or work in a glove box, you may want a tubing with friction rather than a smooth and/or slippery surface. To improve its handling capabilities, natural rubber tubing is sometimes *cloth wrapped* during the curing process. Once the curing process is completed and the cloth is removed, the cloth impression remains on the tubing, providing a sure grip.

1.2.3 Chemical Resistance Properties of Flexible Tubing

A tubing's resistance to chemical attack depends on the nature, quantities, and length of time it is exposed to particular liquids or gases. Some tubing manufacturers have tested their tubings against a variety of chemicals and gases. Some manufacturers have even made these studies at various temperatures. The Nalgene Corporation has listed the effects of many chemicals on a variety of its polymers, and this list is included in Appendix B. If you have a question about the resistance of a particular type of tubing to a given material, contact the manufacturer. However, remember that there are many possible combinations and permutations of chemicals. Unless you are sure that a given chemical is not likely to affect your tubing, it is best to test the tubing with the chemical in an environment that duplicates your conditions. Then you can properly determine if your system, set-up, and/or chemicals will affect your tubing (or vice versa).

If in doubt, test it!

In general, the following materials are those you should consider to be potential reactants with flexible tubing:

Acids (weak)
Acids (strong)
Bases (weak)
Bases (strong)
Oils
Organic Solvents
Oxygen
Salt Solutions

Table 1-4‡

Physical Characteristics of Flexible Tubing

Tubing Type	Manufacturer	Durometer	Flame Resistance	Flexibility	Maximum Pressure (psig) [a]	Resilience	Temperature Range (°C)
Fluran® F-5500-A (Black Fluoroelastomeric Compound)	1	65	no	yes	?	good	-40° - 204°
Nalgene® 8000 (Clear Plastic Tubing)	2	55	yes	yes	38	poor	-35° - 105°
Nalgene® 8005 (Clear Braided Plastic Tubing)	2	65	yes	yes	150-220	poor	-35° - 82°
Nalgene® 8007 (Clear Plastic Tubing with Higher Durometer)	2	75	yes	less	50	poor	-35° - 102°
Nalgene® 8010 (Low Density Clear Polyethylene Tubing)	2	95	no	yes	120	none	-100° - 80°
Nalgene® 8020 (Clear Polypropylene Tubing)	2	99	no	not	300	none	-46° - 149°
Nalgene® 8030 (Pure Translucent Polyurethane Tubing)	2	85	no	less	54	little	-70° - 93°
Norprene® (Black Thermoplastic Elastomer Tubing)	1	60-73	yes	less	10	good	-60° - 135°
Natural Rubber	—	30-90	no	yes		excellent	- 80°
Tygon® R-3400 (Black Plastic Tubing)	1	64	?	yes	38	some	-21° - 165°
Tygon® R-3603 (Clear Plastic Tubing)	1	55	yes	yes	25	poor	-58° - 165°
Tygon® F-4040-A (Transparent Yellow Plastic Tubing)	1	57	no	yes	?	?	-37° - 74°

a The manufacturers list is in Sec. 8.4
b The smaller the I.D., the greater the potential pressure.
‡ Material for this table came from:
Gates® Specialized Products, circular 63009-C, © 1976, Printed in U.S.A.
Nalgene Labware 1988, from Nalge Company, #188, © 1988, Printed in U.S.A.
Norton Performance Plastics, circular #10M0035 1184, © 1983, Printed in U.S.A.
Norton Performance Plastics, circular #5M-321023-485R, © 1978, Printed in U.S.A.

Table 1-5

Chemical Resistance Characteristics of Flexible Tubing

Tubing Type	Acids (weak)	Acids (strong)	Bases (weak)	Bases (strong)	Oils	Organic Solvents	Oxygen	Salt Solutions
Fluran® F-5500-A (Black Fluoroelastomeric Compound)	E	E	P-F	U	G-E	G-E	E	E
Nalgene® 8000 (Clear Plastic Tubing)	G	P	F	P	P	U	G	E
Nalgene® 8005 (Clear Braided Plastic Tubing)	E	P	E	P	P	U	G	E
Nalgene® 8007 (Clear Plastic Tubing)	G	P	F	P	P	U	G	E
Nalgene® 8010 (Low Density Polyethylene Tubing)	E	E	E	E	G	F	G	E
Nalgene® 8020 (Polypropylene Tubing)	E	E	E	E	G	F	G	E
Nalgene® 8030 (Pure Polyurethane Tubing)	E	F	E	G	G	P	G	E
Natural Rubber (Pure amber colored (or dyed any color) Latex Tubing)	F-G	F	G	G	?	P	G	E
Norprene® (Black Thermoplastic Elastomer Tubing)	E	G-E	E	G-E	E	U-P	E	E
Tygon® R-3400 (Black Plastic Tubing)	G-E	P-F	E	E	P	U	E	E
Tygon® R 3603 (Clear Plastic Tubing)	G-E	F-G	E	E	U-P	U	E	E
Tygon® F-4040-A (Transparent Yellow Plastic Tubing)	F-G	P	P	U-P	E	G-E	E	E

Code for Resistance Characteristics

E = Excellent
G = Good
F = Fair
P = Poor
U = Unsatisfactory

Table 1-6

Comparison of Flexible Tubing Characteristics		
Tubing Material:	**Especially Good Because:**	**Watch Out For:**
Fluran® F-5500-A	Resistant to a broad range of corrosive materials: oils, fuels, lubricant, most mineral acids and some aliphatic and aromatic hydrocarbons. Has excellent weather resistance.	Not good with low molecular weight esters, ethers, amines, hot anhydrous HF, or chlorosulfonic acids.
Nalgene® (8000)	General laboratory use, good flexibility, nonflammable, clear.	Not good with organic solvents and most oils, OK with weak acids, but best to avoid strong acids and alkalis, contains plasticisers that can leach out during operations such as distillation.
Nalgene® (8005)	This is braided tubing and can withstand high pressure applications.	(Same as Nalgene® 8000).
Natural Rubber	Outstanding resilience and electrical resistivity. It is resistant to tearing and remains flexible in low temperature situations. Can easily be colored, and the exterior surface can be impressed with a cloth wrap to insure a good grip in all conditions.	Not good for high pressure. Poor resistance to flame, all hydrocarbons, and ozone. Standard thin wall is prone to kinking. If the tubing was sulfur vulcanized (check with manufacturer) it should not be used with any catalytic experiments.
Neoprene	Good resistance to weather, oxidation, ozone, oils, and flame.	Does not stand up well to aromatic hydrocarbons or phosphates. Best not to let remain in water.
Nitrile (Butadiene Acrylonitrile Copolymer)	Excellent resistance to water, alcohol, and aliphatic hydrocarbons.	Poor electrical resistivity and flame resistivity. Should not be used with halogenated hydrocarbons.
Norprene®	Ozone resistant, good aging resistance, good resistance to oils, comes in a variety of stiffness, heat-sealable, formable.	Not recommended for use with any solvents.
Polyethylene Nalgene 8010	Economical, good general chemical resistance, contains no plasticisers.	Not autoclavable, very stiff, translucent, flammable.
Polypropylene Nalgene 8020	Unaffected by most solvents at ambient temperatures.	Not flexible. It is flammable.
Polyurethane (Polyester/Polyether Urethane) Hygenic Corp. (HC480AR,) Nalgene 8030	Contains no plasticisers. Can be used both for vacuum or pressure systems. Has higher chemical resistance to fuels, oils, and some solvents than does PVC tubing.	Not autoclavable, stiffer than PVC, flammable, not recommended with strong acids or alkalis.
PVC (Polyvinyl Chloride) HC6511	Clear, very flexible, comes in varying degrees of durometer hardness. Excellent resistance to water and oxidation.	Contains plasticisers (if leached out, will cause tubing to harden).
Tygon® R-3400	General laboratory use, good flexibility, nonflammable, clear.	Not recommended with any organic solvents and most oils, OK with weak acids, but best to avoid strong acids and alkalis, contains plasticisers that can leach out.
Tygon® R-4040	Resistant to gasoline, lubricants, coolants, heating fuels, and industrial solvents.	Should not be used with strong acids, food, beverages, or drugs. Strong alkalis can harden the tubing, Very low maximum temperature (74°C).
SBR (Styrene-Butadiene Copolymer)	Similar in many respects to natural rubber, except cheaper.	The resilience and % elongation is not as good as natural rubber.

1.3 Corks, Rubber Stoppers, and Enclosures

1.3.1 Corks

Cork, a natural product, has been used in the laboratory for years in many ways. Typically, it is used as seals for glassware, rings for supporting round bottom flasks, and as sheets to protect surfaces from impact shock. Despite the incredible variety of plastics and other elastomers available, cork is still the material of choice for many operations in the laboratory.

Corks (stoppers made from cork) are still widely used to cap many materials within glassware. They are essential when storing organic solvents or other materials that could react with rubber stoppers. Each cork typically fits several different size tubes. An examination of various cork sizes is displayed in Table 1-7.

Corks are graded into five levels of quality: X, XX, XXX, XXXX, and Select. The grades are an assessment of the number and degree of irregular cavities and cracks on the walls of the cork. Grade X is the lowest quality and Select is the highest. As would be expected, the better the quality, the more expensive the cork. Regardless of the stated quality, always examine the integrity of the cork before assuming that it will contain your solution. Cork quality can vary significantly, even in better grades.

1.3.2 Rubber Stoppers

Rubber stoppers are also efficient, simple, temporary system closures. Properly drilled, stoppers can support thermometers or funnels. Because new stoppers have no surface cavities or cracks, there is no risk of leaks as with corks. Like corks, stoppers are used to keep things in (or out) of container. Table 1-7 includes a list of rubber stopper sizes.

Stoppers, as opposed to corks, can react with a number of organic chemicals used in the laboratory. Likewise, a concern when using stoppers is that the closure not adversely affect the contained sample. For this reason, cork stoppers are often preferred when containing organic solvents. If you do not have any corks, you may use a rubber stopper that has been enclosed with an aluminum foil cover. This technique has a few limitations: the covered rubber stopper will not grip into the seal (there is no friction to hold it in) and the crevices of the foil provide many potential small leaks.

Although rubber stoppers are normally intended for use with water-based solutions, do not use the aluminum foil technique with an acid or strong base solution as it will destroy the foil.

If you look in the average laboratory supply catalog, you may see listings for amber (or natural rubber), white, black, and/or green stoppers. Ignoring the stoppers with pre-made single or double holes, there is not much difference between the various colored stoppers.

Amber or Natural Rubber. These colored rubber stoppers are simply, pure rubber. They are the most flexible of all the stoppers. There once was concern that sulfur in the stopper would affect catalytic reactions. This concern is now unlikely because most stoppers should be peroxide-cured. If there is any question, check with the manufacturer.

White. These colored stoppers are essentially natural rubber stoppers dyed white.

Black. These colored stoppers are natural rubber stoppers with a mixture of chemical agents and dyes to give black color. Black stoppers are somewhat more durable than natural rubber stoppers.

Green. These colored stoppers are made of neoprene (a synthetic rubber) and resist the deteriorating effects of oils better than natural rubber. Neoprene also has a wider temperature range than natural rubber. These advantages come at a price, as neoprene stoppers are somewhat more expensive than the other stoppers.

1.3.3 Pre-holed Stoppers

Stoppers can be ordered with none, one, or two holes. As can be seen in Table 1-8, holes are sized according to the size of the stopper. One brand of white stoppers (Twistit®, made by the J.P. Stevens Co.) has three "pre-holes." You simply twist/tear off small nipples on the bottom of the stopper to make the holes as needed.

Obviously, a hole of 3 to 5 mm cannot take a very large tube. In general, it is safe to insert a tube which is about one to two mm in diameter larger than the original hole size. As can be seen in Table 1-8, you can only insert 6 to 8 mm diameter tubes or rods into the holes of size 2 or greater stoppers (about the diameter of the average glass thermometer).

If you find that the holes in your "pre-holed" stopper are the wrong size or in the wrong location, you can take a solid stopper and custom-drill holes which meet your specific requirements. There are several techniques for drilling holes in stoppers, but regardless of the technique used, always start with a solid stopper. Because rubber distorts with pressure, it is more difficult to alter a holed stopper than to drill a new hole. In addition, because elastomers distort and compress, drill hole(s) the same size (rather than slightly smaller) as the tubing you wish to insert. This practice will allow easier insertion of the tube into the hole, yet will still provide a satisfactory leak-proof seal. When the stopper is placed into its final location, compression from the container neck will further improve the seal.

Freezing and Drilling. Because of the flexibility and friction of rubber, it is essentially impossible to use a standard drill bit on a rubber stopper. However, if you freeze the rubber stopper in liquid nitrogen, you can then drill a hole with a standard drill bit and drill press (or a machinist's lathe). Never hold the frozen rubber stopper with your hands while drilling. The stopper is extremely cold, and you will injure your fingers just by holding the rubber stopper. Furthermore, the rubber stopper may start to defrost during the drilling (the drilling friction will cause heat), or the rubber could grab the drill bit and spin out of your hand. It is equally recommended to use a drill press or machinist's lathe and not a hand held drill. You cannot drill straight with a hand drill and the potential for slipping is far too great. Be sure the drill bit is new or newly sharpened. Dull bits cause more friction, causing the stopper to heat up faster. Drilling speed should be at the same slow speed used for plastics. Let the bit cut its way into the stopper by itself, do not force the bit into the stopper while drilling.

Table 1-7

Comparisons of Cork and Rubber Stopper Sizes							
Cork #'s and Sizes				Rubber Stopper #'s and Sizes			
#	Top	Bottom	Length	#	Top	Bottom	Length
0000	5	3	12				
000	6	4	12				
00	8	6	13				
0	10	7	13				
1	11	8	16	000	13	8	21
2	13	9	17				
				00	15	10	25
3	14	11	18				
4	16	12	20				
5	17	13	22	0	17	13	25
6	19	14	24	1	19	14	25
7	21	16	25	2	20	16	25
8	22	17	26				
9	24	18	28	3	24	18	25
10	25	20	31	4	26	20	25
11	27	21	31				
12	28	23	31	5	27	23	25
				5 1/2	28	24	25
13	30	25	31				
14	32	25	31				
				6	32	26	25
15	33	27	31	6 1/2	34	27	25
16	35	28	38				
17	36	29	38				
18	38	30	38	7	37	30	25
				7 1/2	39	31	25
19	40	32	38				
20	41	33	38	8	41	33	25
				8 1/2	43	36	25
				9	45	37	25
22	44	38	38	9 1/2	46	38	25
24	47	40	38				
				10	50	42	25
26	50	44	38				
				10 1/2	53	45	25
28	54	47	38				
				11	56	48	25
30	56	50	38	11 1/2	63	50	25
				12	64	54	25
				13	68	58	25
				13 1/2	75	62	35
				14	90	75	39
				15	100	81	38

Table 1-8

Pre-drilled Stopper Dimensions					
Stopper Size	top diam (mm)	bottom diam (mm)	length (mm)	1 Hole (mm)	2 Hole (mm)
000	13	8	21	-	-
00	15	10	25	3.0	3.0
0	17	13	25	3.0	3.0
1	19	14	25	4.0	4.0
2	20	16	25	5.0	5.0
3	24	18	25	5.0	5.0
4	26	20	25	5.0	5.0
5	27	23	25	5.0	5.0
$5^1/_2$	28	24	25	5.0	5.0
6	32	26	25	5.0	5.0
$6^1/_2$	34	27	25	5.0	5.0
7	37	30	25	5.0	5.0
$7^1/_2$	39	31	25	5.0	5.0
8	41	33	25	5.0	5.0
$8^1/_2$	43	36	25	5.0	5.0
9	45	37	25	5.0	5.0
$9^1/_2$	46	38	25	5.0	5.0
10	50	42	25	5.0	5.0
$10^1/_2$	53	45	25	5.0	5.0
11	56	48	25	5.0	5.0
$11^1/_2$	63	50	25	5.0	5.0
12	64	54	25	5.0	5.0
13	68	58	25	5.0	5.0
$13^1/_2$	75	62	35	5.0	5.0
14	90	75	39	5.0	5.0
15	103	81	38	5.0	5.0

When drilling rubber stoppers with a drill press, do not use a rag to hold the frozen rubber stopper. If the rubber grabs the drill bit, the rag could catch rings, watches, or other equipment. Leather gloves provide good protection and dexterity, and should be used for all similar operations.

Using a Hand Cork Borer. A six piece graduated set of hand cork borers (see Fig. 1-12) is relatively inexpensive, under $20.00 in brass and under $35 in steel. If the bit is properly sharpened, a hand cork borer can make a fairly clean hole. If you are cutting into a rubber stopper, it is best to use a lubricant such as glycerin to keep the rubber from grabbing the borer. When

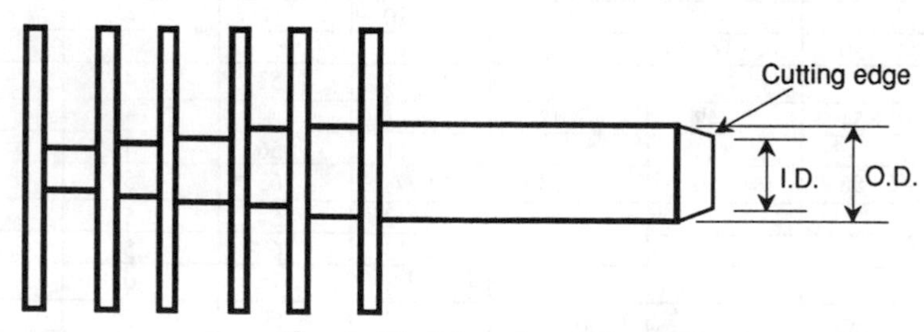

Fig. 1-12 A hand cork borer.

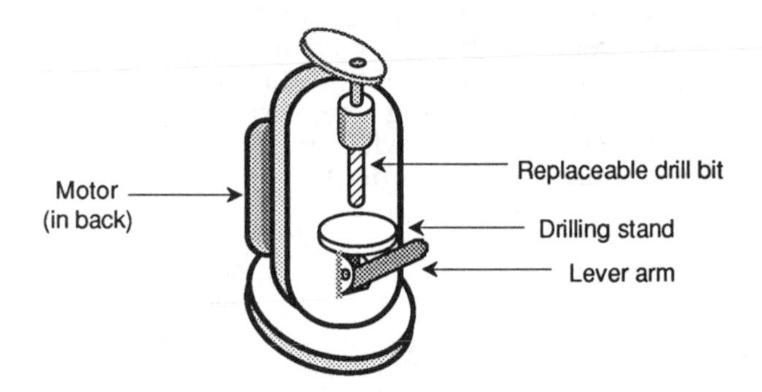

Fig. 1-13 Motorized cork borer.

selecting the proper bit to drill, use the cutting edge, (the inside diameter), not the outside diameter as a guide for selecting the proper boring bit.

Using a Motorized Cork Borer. See Fig. 1-13. Motorized cork borers can be relatively expensive (about $700) as compared to hand cork borers. However, if there is constant demand for custom-sized rubber stopper holes in your laboratory, it can be worth the expense. When selecting the proper bit with which to drill, use the cutting edge (inside diameter), not the outside diameter, as your guide for size.

Motorized cork boring requires a lubricant. There are two choices: glycerin, which is also used with a hand cork borer, and beeswax (or paraffin). To use beeswax, or paraffin, you must to momentarily press the cutting edge of the bit against the stopper so that the friction heats up the bit. Then, press a block of wax against the edge of the turning bit, allowing the melted wax to coat the drill. Now you can begin drilling the hole using an even, light to moderate pressure. After the beeswax cools and hardens, it can be easily picked off. Although it may seem like a bit more work to use paraffin than glycerin, I prefer it because there is less to clean up afterward, the clean-up is easier, and the wax is less messy than glycerin. Glycerin can be wiped off with a rag, although some soap and water may be necessary for a more complete cleaning.

Never force a bit through a stopper, but rather let a bit cut into a stopper on its own. It may be necessary to add more beeswax onto a bit while you drill. This addition can be made by extracting the bit from the stopper, pressing the block of beeswax against the bit, then continuing the drilling process. If you force the bit through the stopper, the resulting hole will lose its continuity and generally will decrease in diameter, as shown in Fig. 1-14.

Fig. 1-14 An example of forced core drilling of a rubber stopper.

When using a motorized borer machine on cork, use a cork bit that has an edge with saw blade teeth rather than a knife-edged rubber stopper bit (see Fig. 1-15). Cork tends to grab and pinch a knife edge, whereas the tooth-edged bit, which removes the cork as it bores, cuts neatly. When selecting cork boring bits, match the outside diameter of the bit to the piece you wish to insert. During the drilling operation, occasionally lift the blade up to help remove cork shards. Otherwise these shards will fill up the channel being cut and will make drilling difficult.

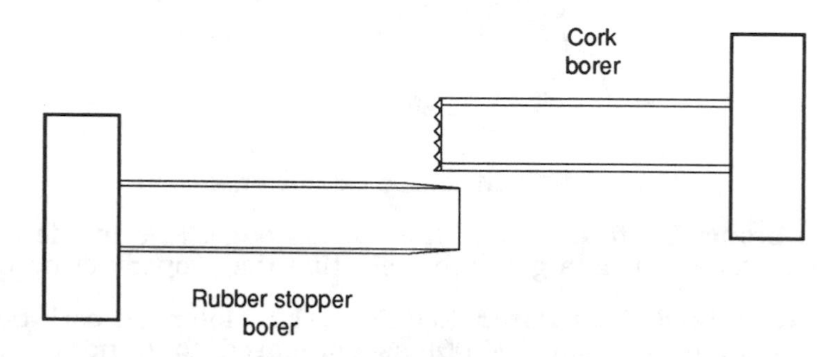

Fig. 1-15 Two different types of core bits.

1.3.4 Inserting Glass Tubing into Stoppers

The most common reason to bore holes in rubber stoppers is to insert glass tubing. Safe and proper techniques are simple and easy. Before inserting the glass, be sure that both ends of the glass tube are fire polished. Fire-polishing removes sharp ends or chipped edges. To fire-polish the end of a tube, place it in the flame of a gas-oxy torch (see the chapter on gas-oxy torches) and rotate the tube until the edge just starts to melt. Do not overdo it. If you let the tube remain in the flame too long, the end could close up. Because glass is a poor conductor of heat, the piece must be rotated while in the flame. Otherwise only one side will be softened and surface tension will

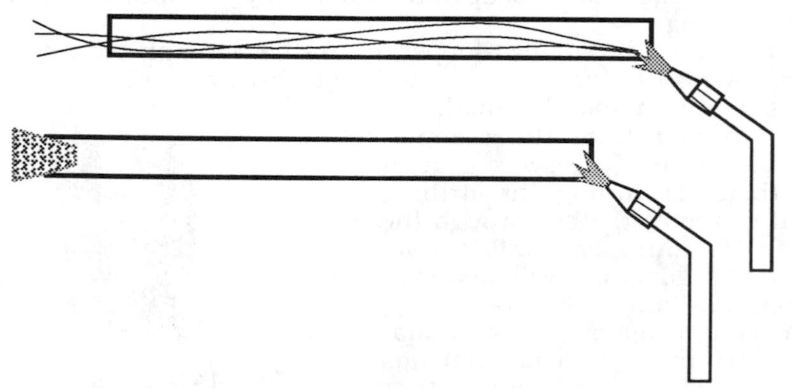

Fig. 1-16 Preventing hot gases from 'tunneling' up a tube by placing a cork in the end.

cause it to sag in and distort. It may be possible to fire-polish a small tube over a Bunsen burner, but it will take a long time for borosilicate glass to get sufficiently hot to melt. Tunneling (the transport of hot gases from the flame traveling down the tube, causing the tube to become excessively hot) can be prevented by placing a cork in the opposite end of the tube (see Fig. 1-16).

After fire-polishing a glass tube, let it cool before trying to place it in a stopper. Do not try to speed up the cooling process by placing the tube under running water, because the rapid temperature change is likely to cause a crack on the newly-prepared end.

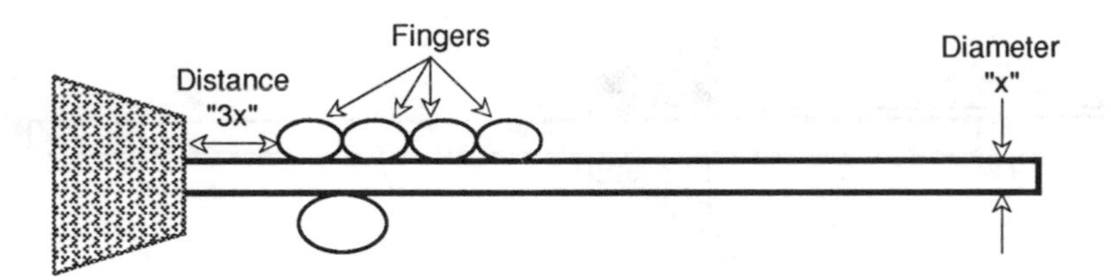

Fig. 1-17 Insertion of a tube into a rubber stopper.

Use some lubrication on the tube and rubber stopper. This lubrication can be glycerin, soapy water, water, or even saliva (but do not lick the rubber stopper or glass rod, as either could be contaminated). Hold the glass rod close to the rubber stopper and keep the glass rod as short as possible. The preferable distance should be no greater than about three diameters of the tubing away from the rubber stopper (see Fig. 1-17). The longer the distance, the greater the torque that can be created. As the tensile forces increase, the chances increase that you will break the glass tubing. Use a rotating motion to guide (not force) the tubing into the stopper. For safety's sake, wear leather gloves. Leather gloves are preferred because they provide a good amount of tactile control, a reasonable amount of friction, and excellent protection. A paper towel is not sufficient to meet any of these three criteria.

1.3.5 Removing Glass from Stoppers and Flexible Tubing

After a glass tube has been left in flexible tubing or a rubber stopper for a period of time it is typically hard to remove. The easiest and safest way to remove flexible tubing from a glass tube or rod is to cut it off with a razor blade and discard the destroyed end of the flexible tubing. This method is not always possible with a rubber stopper because the stopper is too big to cut through. However, whenever cutting tubing or a stopper is possible, it is the recommended and safest procedure.

The reason for cutting off flexible tubing or a rubber stopper is because the force to remove the tubing or stopper may be greater than the tensile strength of the glass. Generally, it is much more difficult and costly to repair and/or replace glass items than it is to replace a stopper or flexible tube. When the hazard of sharp glass is considered, the flexible tubing and rubber stopper become clearly expendable.

There is a trick for removing a stopper from glass tubing or a rod that uses a cork borer[18] (see Fig. 1-18). Select a borer size that is just greater than the size of your glass rod or tube. Then pre-lubricate the outside of the borer with glycerin (preferred), or soap and water, and ease the borer between the stopper and glass. When the borer is inserted all the way into the stopper, the glass tube or rod can be easily pulled out from the borer's hole. This trick can also be used to insert a tube or rod into a stopper by reversing the process.

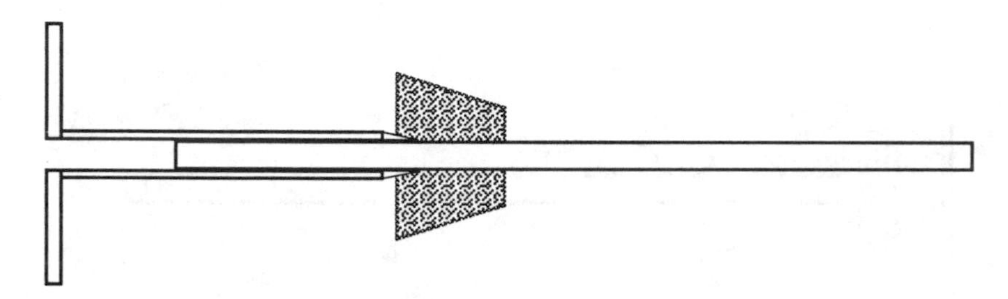

Fig. 1-18 Using a cork borer to insert or remove a stopper from a glass tube or rod.

1.3.6 Film Enclosures

Ground-joint stoppers and snap-on plastic caps are commonplace, but many laboratory containers do not come with built-in stoppers or seals. In the absence of a formal closure, or closures such as corks or rubber stoppers, there is the covering film, PARAFILM "M"®.*

PARAFILM "M"® is a thermoplastic film. Depending on its width, it can come in rolls of 50 to 250 feet in length. A portion of the plastic sheet is cut off the roll, laid over the clean dry top of a container, and stretched over and downward. The paper support film usually tears off during this process. The film forms a closure that can prevent spillage from the container during normal use.

PARAFILM is best used over water solutions, but may be used for short periods over polar-hydrocarbon solutions. Other organic solutions will dissolve the film. It is not designed to prevent spills from tipped-over containers, nor can it contain pressure. If you need to shake a container, do not depend on PARAFILM to maintain its seal. When agitating a container, leave a thumb, finger, or even the palm of your hand (depending on the size of the opening) over the seal to ensure an effective seal.

PARAFILM is excellent for keeping air and dust out of containers, and can be used to maintain containers clean when stored. Although PARAFILM does not wet (liquids run right off it), once used, it should be discarded to avoid contaminating other work. If it has been in contact with toxic materials, it should be thrown into a proper hazardous-waste receptacle.

* PARAFILM "M"® is a product of American National Can, Greenwich, CT. 06836.

1.4 O-rings

1.4.1 O-rings in the Laboratory

O-rings are commonly found on mechanical vacuum pumps, rotary valves, and O-ring joints. O-rings are used to separate environments. If an O-ring is attacked by the chemicals from one or both of these separated environments and fails, it will lose its protective sealing capabilities. Similarly, if an O-ring is left in a chemically destructive environment, it may become dysfunctional without ever having been used.

1.4.2 Chemical Resistance of O-ring Material

When an O-ring needs replacement (such as in a mechanical vacuum pump), the manufacturer can provide, or recommend, an O-ring that will be resistant to the pump's vacuum fluid. On the other hand, when an O-ring is being used in varying conditions, you will be responsible for the selection of the proper O-ring material to maintain the integrity of your system and the health and safety of the operators.

As is typical for most polymers, the material composition of an O-ring may be suitable for one chemical environment and unsuitable for another. There are seven primary materials from which O-rings are made. Table 1-9 catalogs these different materials, listing suitable and unsuitable chemical contacts and properties for each. Also included is a single O-ring price comparison (these 1991 prices are not meant to be absolute and are only offered to provide comparison).

By its design, an O-ring should not require stopcock grease to improve its seal, although the grease may provide peripheral assistance. For example, stopcock grease may be used on rotary valves to facilitate the axial movement of O-rings against the glass barrel. In addition, it may also be used as an extra protective barrier against solvents. For example, if you are using Viton O-rings in a ketone environment (i.e., acetone), you could lay a thin film of silicone grease on the O-ring to protect the surface. The easiest way to apply a thin film of grease is to rub a bit of grease on your fingertips,* then rub it onto the O-ring. This method will limit the contact between the O-ring and the ketone, which in turn will increase the longevity of the O-ring. Do not, however, depend on this technique as a standard, or long-term, O-ring protection procedure.

Incidentally, several companies cover one type of O-ring material (i.e., Buna-N) with a Teflon sheath. These O-rings have the resiliency of less expensive O-rings with the chemical inertness of Teflon.

1.4.3 O-ring Sizes

There are almost 400 *standard* size O-rings. This number does not take into consideration military sizes, special orders, and unique shapes. Most standard size O-rings are for use in reciprocating seals, static seals, and rotary seals, each of which makes contact on the inside or outside diameter of the O-ring.

* Be sure your hands are clean, to minimize contamination as much as possible.

Table 1-9

Comparison of Primary O-ring Material				
Name	**Suitable for ...**	**Unsuitable for ...**	**Comments**	**Price for (1) size 001 O-ring**
BUNA N (Nitrile)	aromatic hydrocarbons, dilute acids and bases, silicones, helium & hydrogen	halogen compounds, halogenated hydrocarbons (carbon tetrachloride, trichlorethylene), ketones (acetone), nitro compounds, or strong acids	Typical color: black. Temperature range: -50 to 120°C. Easily compressed. Density: 1.00.	$0.25
E.P. (Ethylene Propylene)	dilute acids and alkalies, ketones, alcohols, phosphate ester base fluids, and silicone oils	petroleum oils or Di-ester base lubricants	Typical color: purple. Temperature range: -54 to 149°C. Easily compressed. Density: 0.86.	$0.85
FETFE (Fluoroelastomer with TFE additives)	alcohols, aldehydes, chlorinated organics, paraffins, concentrated mineral acids, and mild bases	ketones and ethers	Typical color: black Temperature range: -23 to 240°C. Firm compression. Density: 1.85	$0.85
Kalrez (perfluoroelastomer)	all chemicals	alkali metals and fluorine	Typical color: black. Temperature range: -37 to 260°C. Firm compression. Density: 2.02. Chemically inert properties similar to Teflon, but mechanically similar to Viton. Very expensive.	$21.50
Silicone	alcohols, aldehydes, ammonia, dry heat, chlorinated di-phenyls, and hydrogen peroxide	petroleum oils or fuels, aldehydes, concentrated mineral acids, ketones, esters, and silicone fluids	Typical color: brick red Temperature range: -60 to 260°C. Easily compressed. Density: 1.15-1.32	$0.90
Teflon (PTFE)	all chemicals	alkali metals and fluorine	Typical color: white. Temperature range: -180 to 260°C. Firm compression, but poor resiliency. Density: 2.20.	$0.85
Viton A (hexafluoroprpylene and 1, 1-difluoroethylene)	acids, halogenated aromatic and aliphatic hydrocarbons, alcohols, concentrated bases, nonpolar compounds, oxidizing agents, and metalloid halides	aldehydes, ketones, ammonia, fluorides/acetates, acrylonitrile, hydrozine/analine, and concentrated mineral acids	Typical color: brown. Temperature range: -30 to 200°C. Firm compression. Density: 1.85. Probably the best all round O-ring material.	$ 0.90

O-ring dimensions are based on the ring's *internal diameter* (I.D.) and its *wall thickness** (W) (see Fig. 1-19), which are typically measured in English measurements (in thousandths of an inch). However, O-rings are not

* Occasionally, the outside diameter (O.D.) is also referred to, but such references are redundant.

ordered by outside diameter or wall thickness. Rather, you order by a standardized size code called a *Dash Number*. O-ring sizes are grouped into common thicknesses, and the first number of the Dash Number represents a wall thickness group. All O-rings with the same first number have common wall thicknesses. Table 1-10 shows the Dash Number and sizes of *metric* dimensions of four commonly-used O-ring thicknesses.

Table 1-10

Representative Dash Numbers and Dimensions of O-rings in Metric Sizes

Dash #	I.D. mm	W mm	Dash #	I.D. mm	W mm	Dash #	I.D. mm	W mm	Dash #	I.D. mm	W mm
001	0.74	1.52				201	4.34	3.53			
002	1.07	1.52	102	1.25	2.62	202	5.94	3.53			
003	1.42	1.52	103	2.06	2.62	203	7.52	3.53			
004	1.78	1.78	104	2.85	2.62	204	9.12	3.53			
005	2.57	1.78	105	3.63	2.62	205	10.69	3.53			
006	2.90	1.78	106	4.42	2.62	206	12.29	3.53			
007	3.68	1.78	107	5.23	2.62	207	13.87	3.53			
008	4.47	1.78	108	6.02	2.62	208	15.47	3.53			
009	5.28	1.78	109	7.59	2.62	209	17.04	3.53	309	10.46	5.33
010	6.07	1.78	110	9.19	2.62	210	18.64	3.53	310	12.07	5.33
011	7.65	1.78	111	10.77	2.62	211	20.22	3.53	311	13.64	5.33
012	9.25	1.78	112	12.37	2.62	212	21.82	3.53	312	15.24	5.33
013	10.82	1.78	113	13.90	2.62	213	23.39	3.53	313	16.81	5.33
014	12.42	1.78	114	15.50	2.62	214	24.99	3.53	314	18.42	5.33
015	14.00	1.78	115	17.12	2.62	215	26.57	3.53	315	19.99	5.33
016	15.60	1.78	116	18.72	2.62	216	28.17	3.53	316	21.59	5.33
017	17.17	1.78	117	20.30	2.62	217	29.74	3.53	317	23.16	5.33
018	18.77	1.78	118	21.89	2.62	218	31.34	3.53	318	24.77	5.33
019	20.35	1.78	119	23.47	2.62	219	32.92	3.53	319	26.34	5.33
020	21.95	1.78	120	25.07	2.62	220	34.52	3.53	320	27.94	5.33
021	23.52	1.78	121	26.65	2.62	221	36.09	3.53	321	29.51	5.33
022	25.12	1.78	122	28.25	2.62	222	37.69	3.53	322	31.12	5.33
023	26.70	1.78	123	29.82	2.62	223	40.87	3.53	323	32.69	5.33
024	28.30	1.78	124	31.42	2.62	224	44.04	3.53	324	34.29	5.33
025	29.87	1.78	125	33.00	2.62	225	47.22	3.53	325	37.47	5.33
026	31.47	1.78	126	34.59	2.62	226	50.39	3.53	326	40.64	5.33
027	33.05	1.78	127	36.17	2.62	227	53.57	3.53	327	43.82	5.33
028	34.65	1.78	128	37.77	2.62	228	56.74	3.53	328	46.99	5.33
029	37.82	1.78	129	39.35	2.62	229	59.92	3.53	329	50.17	5.33
030	41.00	1.78	130	40.95	2.62	230	59.92	3.53	330	53.34	5.33
031	42.52	1.78	131	42.52	2.62	231	66.27	3.53	331	56.52	5.33
032	47.35	1.78	132	44.12	2.62	232	69.44	3.53	332	59.69	5.33
033	45.70	1.78	133	45.69	2.62	233	72.62	3.53	333	62.87	5.33
034	53.70	1.78	134	47.29	2.62	234	75.79	3.53	334	66.04	5.33
035	56.87	1.78	135	48.90	2.62	235	78.97	3.53	335	69.22	5.33
036	60.05	1.78	136	50.47	2.62	236	82.14	3.53	336	72.39	5.33
037	63.22	1.78	137	52.07	2.62	237	85.32	3.53	337	75.57	5.33
038	66.40	1.78	138	53.64	2.62	238	88.49	3.53	338	78.74	5.33
039	69.57	1.78	139	55.24	2.62	239	91.67	3.53	339	81.92	5.33
040	72.75	1.78	140	56.82	2.62	240	94.84	3.53	340	85.09	5.33

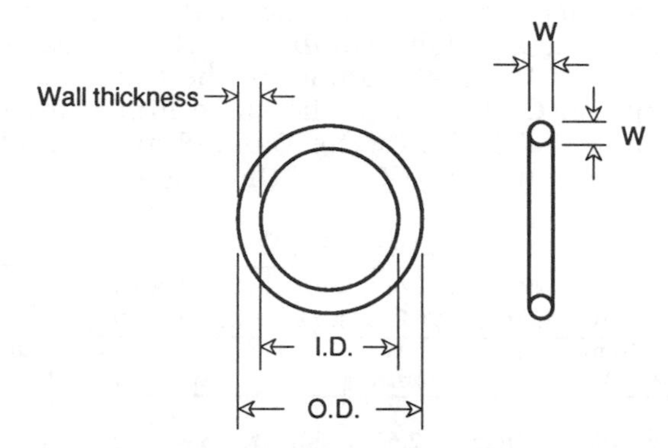

Fig. 1-19 The O-ring dimensions.

References

1 ASTM Designation C 162-85a, "Standard Definitions of Terms Relating to Glass and Glass Products," Annual Book of ASTM Standards, Vol. 15.02.

2 F.M. Ernsberger, *Glass: Science and Technology*; eds., D.R. Uhlmann; N.J. Dreidle, New York: Acad., 1980; Vol. V, Chapter 1.

3 G.W. McLellan and E.B. Shand, *Glass Engineering Handbook,* 3rd Ed. (New York, N.Y.: McGraw-Hill Book Co., Inc., 1984), pp. 2-20.

4 R.C. Plumb, "Antique Windowpanes and the Flow of Supercooled Liquids," *J. of Chemical Education*, 66 (1989), pp. 994-6.

5 D.C. Holloway, *The Physical Properties of Glass,* (London: Wykeham Publications LTD, 1973), p. 205.

6 J. E. Stanworth, *Physical Properties of Glass,* (London: Oxford University Press, 1953), p. 209.

7 G.W. McLellan and E.B. Shand, *Glass Engineering Handbook,* 3rd Ed. (New York: McGraw-Hill Book Co., Inc. 1984), pp. 1-7.

8 A.A. Smith, "Consumption of Base by Glassware," *J. of Chemical Education*, 63 (1986), pp. 85-86.

9 G. Hetherington, K.H. Jack, and M.W. Ramsay, "The High Temperature Electrolysis of Vitreous Silica, Part I. Oxidation, Ultra-violet Induced Fluroescence, and Irradiation Colour," *Physics and Chemistry of Glasses*, 6 (1965), pp. 6-15.

10 R. Brückner, "Properties and Structure of Vitreous Silica. I," *Journal of Non-Crystalline Solids*, 5 (1970), pp. 123-175.

11 *ibid* 177-216.

12 Don Kempf, personal conversation.

13 W.H. Brown, "A Simple Method of Distinguishing Borosilicate and Soda Lime Glass," *J. of Chem. Ed.*, 56 (1979), p. 692.

14 *Kimble Glass Technical Data*, Owens-Illinois Inc., Toledo, Ohio 43666, (1960), p. G-3.

15 G.W. McLellan and E.B. Shand, *Glass Engineering Handbook,* 3rd Ed. (New York: McGraw-Hill Book Co., Inc., 1984, pp. 6-4.

16 E.B. Shand, *Glass Engineering Handbook,* 2nd Ed. (New York: McGraw-Hill Book Co., Inc., 1958), p. 141.

17 *ibid* p. 143.

18 J. Walker, "What Causes the Color in Plastic Objects Stressed Between Polarizing Filters?," *J. of Chem. Ed.*, 246 (1983), pp. 146-52

19 I.C.P. Smith, "Safety Letter; ref: Insertion or Removal of Glass Tubes in Rubber Bungs by Use of Cork Borers," *Journal of the B.S.S.G.*, 12 (1975), p. 62.

Chapter 2

Measurement

2.1 Measurement: The Basics

2.1.1 Uniformity, Reliability, and Accuracy

We do not measure things. Rather, we measure *properties* of things. For example, we cannot measure a box, but we can measure its properties such as its mass, length, and temperature.

To properly compare and analyze the things in our universe, we need to compare and analyze their properties. Because people all around the world are making measurements, we must ensure there is agreement on all the various types of measurements used. Difficulties arise because measurements have both quantitative and qualitative aspects. The fact that the two lines on the international prototype platinum-iridium bar in France are one meter apart is quantitative; how you measure other objects with that bar is qualitative. The ability to match up two lines may seem simple, but depending the desired accuracy, such simple operations are in fact difficult. This difficulty is why *using* measuring equipment is a qualitative art.

No one uses the prototype meter as an actual measuring tool; rather, copies are made from the original prototype, and these copies are used as masters to make further copies. By the time you purchase a meter stick, it is a far distant cousin from the original meter prototype. However, despite the length of the progeny line, you hope that the copy you have is as good as the original. Depending on the expertise of the engineers and machinists involved, it should be very close. To obtain that quality, the engineers and machinists were guided by three factors; *uniformity, reliability,* and *accuracy.* Without these basic tenets, the quality of the meter stick you use would be in doubt. Likewise, the quality of the use of the meter stick is also dependent the same three factors, without which all readings made would be in doubt.

Uniformity requires that all people use the same measurement system (i.e., metric vs. English) **and** that all users intend that a given unit of measurement represents the same amount and is based on the same measurement standard used everywhere else. It is the user's responsibility to select equipment that provides measurements that agree with everyone else.

Reliability requires the ability to consistently read a given measurement device and for a given measurement device to perform equally well, test after test. It is the user's responsibility to know how to achieve repeatable data from the equipment being used.

Accuracy refers to how well a measurement device is calibrated and how many significant figures one can reliably expect. It is the user's responsibility to know how to read his equipment and not interpolate data to be any more accurate (i.e., significant figures) than they really are.

That laboratory research is dependent on reliable quality measurements, and the use of *uniformity*, *reliability*, and *accuracy* to achieve this goal cannot be emphasized enough. Poor or inaccurate measurements can only lead to poor or inaccurate conclusions. A good theory can be lost if an experiment produces bad data.

Commerce is equally dependent on *uniformity*, *reliability*, and *accuracy* of measurement systems. The potential economic liabilities of mis-measurements and misunderstanding are easy to understand. In fact, it was the economic advantages of uniformity that led to the metric system's expansion after the Napoleonic wars. What can upset and/or confuse consumers and businesses is when the same word (which has varying meanings) is applied to different weights. We still confuse the weight value of a *ton* and the volume of a *gallon*: in the U.S., the ton is equal to 2,000 lbs; however, in Britain, it is equal to 2240 lbs. Similarly, in the U.S., the gallon is established as 231 cubic inches; however, in Britain, it is 277.42 cubic inches. If you are aware of these differences, you can make the appropriate mathematical corrections. But realistically, it should not be a problem to be dealt with. Rather the problem of different measurement systems should be avoided in the first place. That is specifically was what the metric system was designed to do.

2.1.2 History of the Metric System

Overcoming the incongruities of inconsistent measurement systems on an international basis was considered for centuries. The basics of the metric system were first proposed by Gabriel Mouton of Lyon, France, in 1670. This vicar (of St. Paul's Church) proposed three major criteria for a universal measurement system: decimalization, rational prefixes, and using parts of the earth as a basis of measurement (length was to be based on the arc of one minute of the earth's longitude). There was also a desire to find a relationship between the foot and gallon (i.e., a cubic foot would equal one gallon). Unfortunately, these measurement units were already in use, and because there was no basis for these measurements to have any easy mathematical agreement, they did not. No simple whole number could be used to correct the discrepancy.

Gabriel Mouton's ideas were discussed, amended, changed, and altered for over 120 years. Eventually, a member of the French assembly, Charles Maurice Talleyrand-Périgord, requested the French Academy of Sciences to formalize a report. The French Academy of Sciences decided to start from scratch and develop all new units. They defined the *meter* as one ten-millionth of the distance from the north pole to the equator. They also decided that the unit for weight would be based on the weight of a cubic meter of water in its most dense state (at 4°C). This plan allowed the interlinking of mass and length measurement units for the first time. In addition, they proposed prefixes for multiples and sub-multiples of length and mass measurements, eliminating the use of different names for smaller and larger units (i.e., inch - foot or pint - quart units).

On the eve of the French Revolution, June 19, 1791, King Louis XVI of France gave his approval of the system. The next day, Louis tried to escape France but was arrested and jailed. A year later from his jail cell, Louis directed two engineers to make the measurements necessary to implement the metric system. Because of the French Revolution, it took six years to complete the required measurements. Finally, in June 1799 the "Commission sur l'unité de poids du Systeme Metrique décimal" met and adopted the metric system. It was based on the gram as the unit of weight and the meter as the unit of length. All other measurements were to be derived from these units. The metric system was adopted "For all people, for all time."

The metric system sought to establish simple numerical relationships between the various units of measurement. To accomplish this goal, the Commission took a cubic decimeter* of water at its most dense state (4°C), designated this volume as one "liter," and designated its mass (weight) as one "kilogram." In so doing, the commission successfully unified mass, length, and volume into a correlated measurement system for the first time. Official prototypes of the meter and kilogram were made and stored in Paris.

Because of Napoleon's conquests, the metric system spread rapidly throughout Europe. However, it was not in common usage in many areas (even in France) until international commerce took advantage of its simplicity and practicality. By the mid 1800's, it was the primary measurement system in most of Europe. In 1875 the International Bureau of Weights and Measures was established near Paris, France. It formed a new international committee, called the General Conference on Weights and Measures (CGPM), whose goal was to handle international matters concerning the metric system. The CGPM meets every six years to compare data and establish new standards. Every member country of this committee** receives a copy of the meter and kilogram prototype with which to standardize their own country's measurement system.

Over the years there has been ongoing fine-tuning of the measurement system because the greater the precision with which our measurement units can be ascertained, the better we can define our universe. A new era in the measurement system came in 1960, at the 11th meeting of the CGPM, when the International System of Units (SI) was established. This system established four Base Units. the *meter*, *kilogram*, *second*, and *ampere*. They are collectively known as the *MKSA system*. Later, three more Base Units were added: *kelvin* (for temperature), *candela* (for luminous intensity), and *mole* (for the amount of substance) (see Table 2-1). In addition, two Supplementary Base Units (which are dimensionless) were added: *radian* (plane angles) and *steradian* (solid angles) (see Table 2-2).

* For a description of how the meter was derived, see Sec. 2.1.3a.

** The National Institute of Standards and Technology (formally the National Bureau of Standards) represents the United States at the CGPM. They store the United States' copies of the original measurement prototypes.

Table 2-1‡

Base SI Units		
Quantity	**Unit**	**Symbol**
length	meter	m
mass	kilogram	kg
time	second	s
electric current	ampere	A
thermodynamic temperature	kelvin	K
amount of substance	mole	mol
luminous intensity	candela	cd

‡ From the ASTM document E380-86, Table 1. With permission.

Table 2-2‡

Supplementary SI Units		
Quantity	**Unit**	**Symbol**
plane angle	radian	rad
solid angle	steradian	sr

‡ From the ASTM document E380-86, Table 2. With permission.

Table 2-3‡

Derived SI Units with Special Names			
Quantity	**Unit**	**Symbol**	**Formula**
frequency (of a periodic phenomenon)	hertz	Hz	$1/s$
force	newton	N	$kg \cdot m/s^2$
pressure, stress	pascal	Pa	N/m^2
energy, work, quantity of heat	joule	J	$N \cdot m$
power, radiant flux	watt	W	J/s
quantity of electricity, electric charge	coulomb	C	$A \cdot s$
electrical potential, potential difference, electromotive force	volt	V	W/A
electric capacitance	farad	F	C/V
electric resistance	ohm	Ω	V/A
electric conductance	siemens	S	A/V
magnetic flux	weber	Wb	$V \cdot s$
magnetic flux density	tesla	T	Wb/m^2
inductance	henry	H	Wb/A
Celsius temperature	degree	°C	K[a]
luminous flux	lumen	lm	$cd \cdot sr$
illuminance	lux	lx	lm/m^2
activity (of a radionuclide)	becquerel	Bq	$1/s$
absorbed dose[b]	gray	Gy	J/kg
dose equivalent	sievert	Sv	J/kg

[a] Celcius temperature (t) is related to thermodynamic temperature (T) by the equation:

$$t = T - T_0$$

where T_0 = to 273.15 K by definition.

[b] Related quantities using the same unit are: specific energy imparted, kerma, and absorbed dose index.

‡ From the ASTM document E380-86, Table 3. With permission.

Table 2-4‡

Some Common Derived Units of SI		
Quantity	**Unit**	**Formula**
absorbed dose rate	gray per second	Gy/s
acceleration	metre per second squared	m/s^2
angular acceleration	radian per second squared	rad/s^2
angular velocity	radian per second	rad/s
area	square metre	m^2
concentration (of amount of substance)	mole per cubic metre	mol/m^2
current density	ampere per square metre	A/m^2
density, mass	kilogram per cubic metre	kg/m^2
electric charge density	coulomb per cubic metre	C/m^3
electric field strength	volt per metre	V/m
electric flux density	coulomb per square metre	C/m^2
energy density	joule per cubic metre	J/m^3
entropy	joule per kelvin	J/K
exposure	coulomb per kilogram	C/kg
heat capacity	joule per kelvin	J/K
heat flux density irradiance	watt per square metre	W/m^2
luminance	candela per square metre	cd/m^2
magnetic field strength	ampere per metre	A/m
molar energy	joule per mole	J/mol
molar entropy	joule per mole kelvin	$J/(mol \cdot K)$
molar heat capacity	joule per mole kelvin	$J/(mol \cdot K)$
moment of force	newton metre	$N \cdot m$
permeability (magnetic)	henry per metre	H/m
permittivity	farad per metre	F/m
power density	watt per square metre	W/m^2
radiance	watt per square metre steradian	$W/(m^2 \cdot sr)$
radiant intensity	watt per steradian	W/sr
specific heat capacity	joule per kilogram kelvin	$J/(kg \cdot K)$
specific energy	joule per kilogram	J/kg
specific entropy	joule per kilogram kelvin	$J/(kg \cdot K)$
specific volume	cubic metre per kilogram	m^3/kg
surface tension	newton per metre	N/m
thermal conductivity	watt per metre kelvin	$W/(m \cdot K)$
velocity	metre per second	m/s
viscosity, dynamic	pascal second	$Pa \cdot s$
viscosity, kinematic	square metre per second	m^2/s
volume	cubic metre	m^3
wave number	1 per metre	$1/m$

‡ From the ASTM document E380-86, Table 4. With permission.

From the nine Base Units, over 58 further units have been derived and are known as *Derived Units*. There are two types of Derived Units: those that have special names (see Table 2-3) and those that have no special names (see Table 2-4). An example of a Derived Unit with a special name is *force*, which has the unit *newton* (the symbol 'N') and is calculated by the formula

"N = $kg.m/s^2$." An example of a Derived Unit that does not have a special name is *volume*, which has the unit of *cubic meter* (no special symbol), and is calculated by the formula "volume = m^3."

Among the advantages of the International System of Units system is that there is one, and only one, unit for any given physical quantity. Power, for instance, will always have the same unit, whether it has electrical or mechanical origins.

In the U. S. measurements made with metric units were not legally accepted in commerce until 1866. In 1875 the U. S. became a signatory to the Metric Convention, and by 1890 it received copies of the International Prototype metre and kilogram. However, rather than converting our measurement system to metric, in 1893 Congress decided that the International Prototype units were acceptable as fundamental standards on which to base our English units. Thus, the metric system was not implemented by the government and was not adopted by the nation at large.

By 1968 the U. S. was the only major country that had not adopted the SI system. Congress ordered an investigation to determine whether the U. S. should follow the International Standard. In its 1971 report Congress recommended that we have a coordinated program of metric conversion such that within ten years we would be a "metric country." With this hope in mind, President Ford signed the Metric Conversion Act of 1975. Unfortunately, as of the writing of this book, in 1991, we see little evidence of "metrification." In the near future there may be difficulties for U.S. manufacturers because the European Common Market has decided that by 1992, it will no longer accept items not made to metric standards.

If you look at the history of measurement, it shows constant attempts to better define and refine our measurement units. Early measurement standards were arbitrary such as the Egyptian cubit (the tip of the finger to the elbow). Since then we have tried to base measurement standards on nonchanging, consistent, and repeatable standards, some of which turned out to be inconsistent and/or impractical. The work of the metrologist will never be complete. The more accurately we can define our measurement standards, the better we can measure the properties of our universe.

2.1.3 The Base Units

This book uses the original Base Units (length, mass, and temperature), plus one of the derived units (volume) because these measurements are the most commonly used measurements in the lab. Time is included in this section only as it provides an interesting comment on the metrologists' desire to split hairs in their endeavor to achieve accuracy.

Length. The original metric standard for the length of a meter was "one ten-millionth part of a quadrant of the earth's meridian." Astronomical measurements (at that time) indicated that one tenth of a quadrant of the earth's meridian was measured between Dunkirk, in France, and a point near Barcelona, Spain. This distance was dutifully measured and divided by one million to obtain the meter. It is fortunate that the meter was later redefined, because if you take what we now call a meter and measure the distance between the Dunkirk and Barcelona points used, you get the length space of 1,075,039 meters. This space is an accuracy of only 7.5 in 100 meters.

Reproducible accuracy was increased substantially by the development and use of the *International Prototype Metre*, the platinum-iridium bar. By physically comparing the lines on this bar with a secondary prototype, an accuracy was achieved to within two parts in 10 million.

In 1960 (at the same meeting during which the International System of Units was accepted) the meter was redefined as being 1,650,763.73 wavelengths (in vacuum) of the orange-red spectral line of krypton-86. This system had several problems, chief of which was that this figure could only be reached by extrapolation because it was not possible to extend an accurate quantity of waves beyond some 20 centimeters. To obtain the required number of wavelengths for a meter, several individual measurements were taken in succession and added together. This measurement procedure, because of its nature, increased the likelihood for error. Despite the limitations of the krypton-86 based meter, it brought the accuracy to within two parts in 100 million.

In 1983 yet another definition of the meter was adopted: it was the distance that light would travel (in a vacuum) during 1/299,792,458TH of a second. This attempt provided a measurement ten times more accurate than that obtained through krypton-86 techniques. This meter, accurate to within four parts in one billion, is still in use today.

Mass. The kilogram is unique in several aspects within the metric system. For one, it is the only Base Unit that includes a prefix in its name (the *gram* is not a Base Unit of the International System of Units). Also, it is the only Base Unit defined by a physical object as opposed to a reproducible physical phenomenon. The kilogram was based on the weight of one cubic decimeter of water at its most dense state (4°C). The original "Standard Mass Kilogram" was made by constructing a brass weight, making careful weighings (using Archimedes' Principle*), and then making an equivalent platinum weight. The platinum weight is called the *Kilogramme des Archives* and is kept at the International Bureau of Weights and Measures near Paris, France.**

About a hundred years later during studies verifying previous data, it was discovered that a small error in the determination of the kilogram had been made. The problem was caused by the consequences of very small variations of temperature causing very small variations in density, which in turn caused inaccurate weighings. The CGPM decided against changing the mass of the kilogram to a corrected amount; rather, it accepted the mass of the kilogram as being that obtained from the original Kilogramme des Archives.

It has been hoped that there will be some naturally occurring phenomenon in nature to which we can ascribe the value of the kilogram, thus

* Simply stated, a submerged body is buoyed up by a force equal to the weight of the water that the submerged body displaces. In other words, if an object weighs 1.5 kilograms in air, and the weight of the water it displaces is equal to 1.0 kilogram, the object will weigh 0.5 kilograms in water.

** A copy of the prototype is held by the National Institute of Standards and Technology (formally the National Bureau of Standards) and by all other countries who have signed agreements following the International System of Units.

allowing the kilogram to be based on a reproducible phenomenon rather than relying on a physical artifact. Although there have been several efforts toward this goal, such as counting molecules, our current technological level cannot achieve any greater accuracy than that obtained by simply comparing weights of unknowns against our current prototypes.

Volume. As previously mentioned, the unit for volume (the liter) was to be one cubic decimeter of pure water at its most dense state (4°C). Later analysis determined that errors were made in the determination of the kilogram and that the mass of one decimeter of water was slightly less than the prototype kilogram. However, the use of the kilogram as it had already been defined was so well established that change became impossible. Thus, in 1872, the "Commission Internationale du Métre" met to redefine the kilogram as the mass of a particular standard weight (the Prototype Kilogram) instead of the weight of a liter of water.

The concept of the liter was cast into doubt. Was it to be based on the weight of a standard volume of dense water as before, or was it to be an alternate name of the cubic decimeter?

The first attempt to resolve the conflict was in 1901 when the "Comité International des Poinds et Mesures" resolved that "The unit of volume for determinations of high precision is the volume occupied by a mass of 1 kilograms of pure water at its maximum density and under normal atmospheric pressure." This volume, so defined, was identified as a "liter."[1] However, this designation still left the discrepancy that this definition actually redefined the liter to be equivalent to 1.000027 dm^3. There was once again a discrepancy as to what a liter was. So, at the 12th conference of the "Comité International des Poinds et Mesures," the unit of volume was redefined for the last time as the *cubic decimeter*. The liter, no longer the official unit of volume was nevertheless so close to the cubic decimeter in size that unofficially, it was deemed acceptable to continue its use; however, for situations that required extremely high quality measurements, only the cubic decimeter was acceptable. The cubic decimeter remains the standard of volume today.

Temperature. Temperature was not one of the original properties that the French academy deemed necessary to include in the metric system. In fact, as late as 1921, members of the 6th General Conference of the International System of Weights and Measures were still objecting to the inclusion of measurements (other than length and mass) seemingly for no other reason other than to keep the Base Units "pure."

The measurement of temperature is a measurement of energy and therefore has different measurement characteristics than other properties. The primary difference in temperature measurement is that it is not cumulative. You can take two different meter sticks and place them end to end to measure something longer than one meter. The comparable action can be made for the measurement of mass. However, you cannot take two thermometers and add the temperature of the two if a material's temperature is greater than the scale of either thermometer.

The first thermometer is believed to have been created by Galileo sometime between 1592 and 1598. It was a glass sphere attached to one end of a long thin glass tube. The other end of the glass tube was lowered vertically into a bowl of colored water (or perhaps wine). Using the warmth of his hands, Galileo heated the glass sphere, causing bubbles to come out of the glass tube. Then, by letting the sphere cool to room temperature, Galileo caused the liquid to be drawn up into the tube. By heating and/or cooling the glass sphere, the liquid would ride up and down the tube.

Although there is no record that he ever calibrated the tube, he used it in temperature study. Galileo's thermometer was impossible to calibrate even if he had decided on fixed points with which to establish specific temperatures because it was exposed to the atmosphere and subject to variations in atmospheric pressure. By 1640, it was realized that the "air thermometer" was subject to variations of barometric pressure and the sealed thermometer was created. However, the need to establish fixed points of reference had still not been addressed.

The need to establish fixed points to provide uniformity between thermometers is no different than the need for uniformity with any measurement system. Many thermometers were built in the following years, but there was either no calibration, or the calibration was not based on any repeatable fixed point. One of the first attempts of establishing calibration points was made in 1693 by Carlo Renaldini of Padua, who set the low temperature point with ice. For the next point, he took eleven ounces of boiling water and mixed it with one ounce of cold water. Next he mixed ten ounces of boiling water with two ounces of ice water. The process continued until twelve divisions were established on the thermometer. Although Renaldini had an interesting approach to establishing fixed points on the thermometer, it was neither practical nor accurate.

The first commonly-used temperature measurement scale was devised in 1724 by the Dutch scientist D. Gabriel Fahrenheit. It took sixteen years for Fahrenheit to devise a process for calibrating his scales. Fahrenheit used three points for calibration. The lowest, 0°, was the lowest temperature he could create by mixing ice, water, and ammonium chloride (a slush bath of sorts). The second temperature was ice, at 32°, and the high temperature point was the human body temperature, 96°. The choice of identifying the low temperature was logical, but the choice of the high temperature seems illogical. Lindsay[2] assumed that 96 was chosen because it was an even multiple of twelve. However, I haven't seen any evidence that the Fahrenheit scale was ever divided into multiples of twelve. Also, the human body is closer to 98°. If the scale was originally based on a duodecimal system and changed to a decimal system, the human body would be 100° Fahrenheit. Regardless, if the human body were fixed at 96°, that fixed point would not change as thermometers were made more accurate. Regardless of the logic Fahrenheit used for his scale, his quality was excellent and his thermometers became very popular.

In 1742 Anders Celsius, a Swedish astronomer, developed the mercury *Centigrade* thermometer. He chose the boiling and freezing points of water as calibration points. Curiously, he chose 0° for the high temperature and 100° for the low temperature.* His choices were reversed in 1850 by

* The Royal Society of London in the early 1700's used a reversed scale that describes 0° as "extreme hot" and 90° as "extreme cold."[3]

Märten Strömer, also a Swedish astronomer. In 1948 the centigrade scale was officially renamed the *Celsius* scale.

During the early 19th century, Lord Kelvin theorized that as temperature drops, so does thermal motion. Thus, 0° should be the point at which there is zero motion. This new 0° temperature would be equal to -273.15 degrees Celsius. Fortunately, Kelvin had the foresight to keep things simple, and made a 1-degree increment Kelvin exactly equal to a 1-degree increment Celsius. Originally, the temperature *Kelvin* was capitalized. Now, it is not capitalized, but should be written as kelvin. The abbreviation of kelvin is capitalized and written as K.*

There are two other temperature scales that still may be seen in old texts or journals, but are not acceptable for any current scientific work. Perhaps the rarer is the *Réaumur* scale (°Ré). It separated the range between freezing and boiling of water into 80 units and was used in parts of Europe. The other temperature scale, the Rankine, may be referred to in old books on thermodynamics. It was named after W. J. M. Rankine, who did early research in that field. The Rankine is to Fahrenheit what Kelvin is to Celsius. In other words, just as one degree K = one degree C, one degree F = one degree R. Thus, 0 K = 0° R = -273.15°C = -459.67°F.

In 1954 the General Conference wanted to redefine the temperature scale using various primary points in addition to the two points of freezing and boiling water. The triple point of water (at 273.16 K) proved easy to obtain and very accurate (one part in a million). In 1960 the triple point of water and five other fixed points were accepted for an *International Practical Temperature Scale*.** This scale was superseded in 1968 by the International Practical Temperature Scale (IPTS 1968), which added eight more fixed points. The current scale is shown in Table 2-34.

Time. How long is a second, and how do you store that measurement? Various attempts at measuring the length of a pendulum swing proved inadequate for the task of accurately measuring the points of a given swing. The people who devised very complex and systematic methods for defining the metric system did nothing for time. During that period of time, $1/_{86,400}$ of a mean solar day was considered to be an adequate definition of one second. However, by the mid-1900's the mean solar day was found to vary by as much as three seconds per year. In 1956 the International System of Weights and Measures defined the second to be "$1/_{31,556,925.9747}$ for the tropical year 1900 January 0 at 12 hours ephemeris time."***[4] In 1960 this standard was accepted by the General Conference with the caveat that work continue toward development of an atomic clock for the accurate measurement of time.

The atom exhibits very regular, hyperfine energy level transitions and it is possible to count these "cycles" of energy. In 1967 the General Conference accepted 9,192,631,770 cycles of cesium 133 as the measure-

* Measurements in kelvin are not preceded by the "°" symbol.

** Six different points were adopted because no one thermometer can read a full range of temperatures, and no one thermometer can read a wide range of temperatures accurately. Thus, different fixed points allowed for thermometers measuring different ranges to be accurately calibrated.

*** Ephemeris time is a uniform measure of time defined by the orbital motions of the planets.

ment of one second, making the atomic clock the true international timekeeper. The *cesium clock* is maintained in Boulder, Colorado, in the offices of the National Institute of Standards and Technology (formally the National Bureau of Standards). Its accuracy is one part in 1,000,000,000,000 (10^{-12}). It will not gain or lose a second in 6,000 years.

2.1.4 The Use of Prefixes in the Metric System

One of the strengths of the metric system is the consistency of its terms. In other measurement systems, the names for measures change as the size of the measurements change. For instance, consider the changes in the English measurement system for length (inch-foot-yard), weight (ounce-pound-ton), and volume (ounce-quart-gallon). In the metric system, all measurement names consist of a root term that, by use of prefixes, yields fractions and multiples of the base measurement unit.

The metric system is a decimal system, based on powers of ten. Table 2-5 is a list of the prefixes for the various powers of ten. Between scientific notation and the prefixes shown below, it is very simple to identify, name, read, and understand thirty-six decades of power of any given Base or Derived Unit.

Table 2-5‡

Powers of the Metric System					
Name	**Symbol**	**Power**	**Name**	**Symbol**	**Power**
exa	E	10^{18}	deci[a]	d	10^{-1}
peta	P	10^{15}	centi[a]	c	10^{-2}
tera	T	10^{12}	milli	m	10^{-3}
giga	G	10^{9}	micro	μ	10^{-6}
mega	M	10^{6}	nano	n	10^{-9}
kilo	k	10^{3}	pico	p	10^{-12}
hecta[a]	h	10^{2}	femto	f	10^{-15}
deca[a]	da	10^{1}	atto	a	10^{-18}

‡ From the ASTM document E380-86, Table 5 . With permission.

a To be avoided where practical.*

2.1.5 Measurement Rules

There are several general rules about making measurements. These rules are standard regardless of the type of equipment being used, material being studied, or the measurement units being used.

1) **The quality of the measurement is only as good as the last clearly read number**. No matter how obvious an estimated number may seem, it is not as reliable as the one your equipment actually provides for you. For example, see Fig. 2-1(a) and (b). If you want greater accuracy in your readings, you need better equipment, which provides more precise measurements.

* Although the SI prefers avoidance of these "power" names, the centi is still commonly used.

One safe way to indicate greater accuracy is to qualify a number. In the case of Fig. 2-1(a), you could write 6.2±0.2. The selection of ±0.2 is somewhat arbitrary, as the length is obviously greater than 6.0, yet smaller than 6.5.

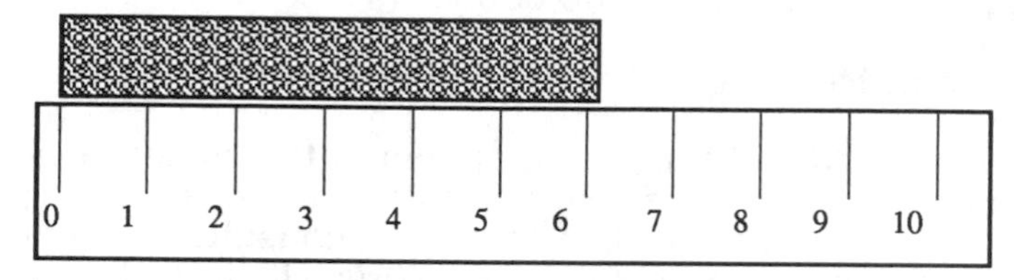

It may be tempting to read the length of the block as 6.2 units, but an accurate measurement would be 6 units.

Fig. 2-1(a) The degree of accuracy obtainable for any given reading is limited by the quality of your equipment.

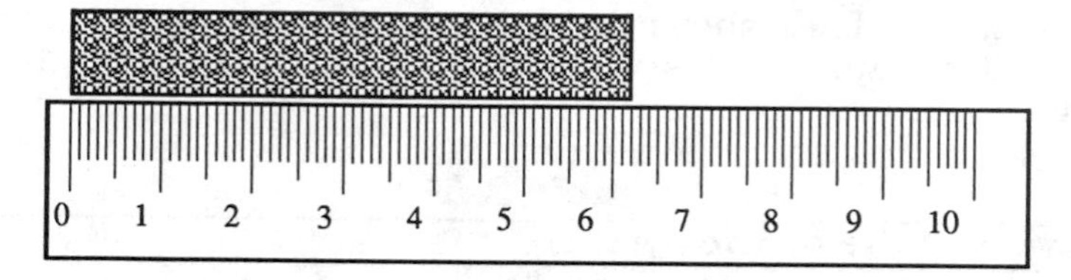

Using this ruler, it is correct and accurate to measure the block at 6.2 units.

Fig. 2-1(b) It is important to read only to the limits of your equipment.

2) **Remember the limits of significant figures**. The numbers 8.3 and 0.00083 both have two significant figures. The number 8.30 has three. Any calculations made can only be as good as the minimum number of significant figures. Before the mid 1970's, three significant figures were often the standard, because most calculations were made on slide rules which had only three significant figures. With the advent of digital readouts and calculators, the use of more significant figures is common.

3) **Be aware of equipment error**. Measuring tools can be improperly calibrated or otherwise inaccurate. This problem can be the result of improper use or storage, or faulty manufacturing. To guard against inaccuracy of tools, maintain periodic records of calibration tests. When the results are off the nominal values, your records will help pinpoint the time frame when the inaccuracies started. All data made prior to the last verification test should be discarded or held in question.

4) **Be sure the equipment is used in the proper environment**. Some equipment is designed to be used in limited environmental conditions. For example, a vacuum thermocouple gauge is cali-

brated for readings made in specific gaseous environments. If the vacuum system contains gases other than those the gauge was designed to read, inaccurate pressure data will result. In addition, variables such as temperature, elevation, drafts, and whether the equipment is level can all affect accurate performance.

5) **Be aware of user error**. Many errors are caused by the operator, through sloppy readings or lack of experience with particular equipment. Be sure you know how to read your equipment and how to make any necessary conversions of your readings to the units that you are using (i.e., microns to torr).

6) **Avoid parallax problems**. A "parallax" problem occurs when the object being observed is not placed directly between the eye and measuring device, but rather at an angle (see Fig. 2-2) between the two. An incorrect sighting is easy to avoid, but requires conscious effort. Parallax problems show up with linear as well as liquid measurements.

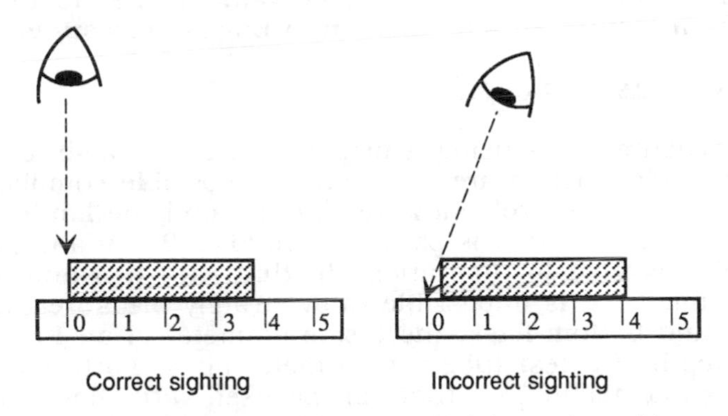

Fig. 2-2 Correct and incorrect sightings of an object.

7) **Make many readings**. This form of cross-checking proves that your work is reliable and reproducible. It is unscientific to make conclusions based on a limited amount of data. On the other hand, there has to be a limit on the number of data collected because too much data wastes time and money. Collect a statistically viable amount of data, but be guided by common sense.

8) **Keep your instruments and equipment clean**. Filth cannot only chemically affect the materials of your experiment, it can also affect the operation of your measuring equipment. Therefore, just as your experimental equipment must be kept clean, it is equally important that your analysis and recording equipment also be kept clean. For example, if the mercury in a McLeod gauge or manometer is dirty, the manometer or McLeod gauge will not provide accurate readings. Similarly, if pH electrodes

are not stored in a proper, standard solution, they will not provide accurate readings. Even fingerprints on objects used for weighing can affect the final weight. You are not likely to produce results worthy of note (let alone reproducible results) if your equipment is so filthy that the mess is obscuring or affecting your readings.

2.2 Length

2.2.1 The Ruler

It is appropriate that the object used to measure length is called a ruler. The dictionary's definition of a ruler is "a person who rules by authority."[5] It therefore makes sense that the final arbiter of an object's length should be known as a ruler.

In the laboratory, the *ruler* is the meter stick. The meter stick may have inch measurements on one side (1 meter = 39$^3/_8$ inches), but a yardstick, or any ruler of 6, 12, or 18 inches has limited value in the laboratory unless it also has metric measurements in addition to English measurements.

2.2.2 How to Measure Length

Because measurements using a meter stick are straightforward, no explanation is provided. There are however, two possible complications when using a meter stick. One problem is reading beyond the limits of the measurement device, and the other is parallax (see Figs. 2-1(a) and (b) and 2-2).

Meter sticks have major limitations in that they can only measure flat objects. For example, it is impossible to accurately measure the outside diameter of a round bottom flask, the inside diameter of an Erlenmeyer flask neck, or the depth of a test tube with a meter stick. Fortunately, there are various devices and techniques that can be used with the meter stick. The three limitations just mentioned can be resolved with the aid of other tools, as shown in Figs. 2-3, 2-4, and 2-5.

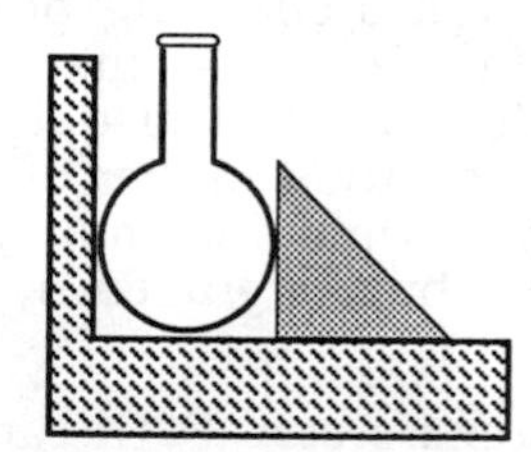

(a)
A right-angle triangle pressed against a round surface that is up against another right-angled surface will provide a linear surface for a meter stick to make a measurement.

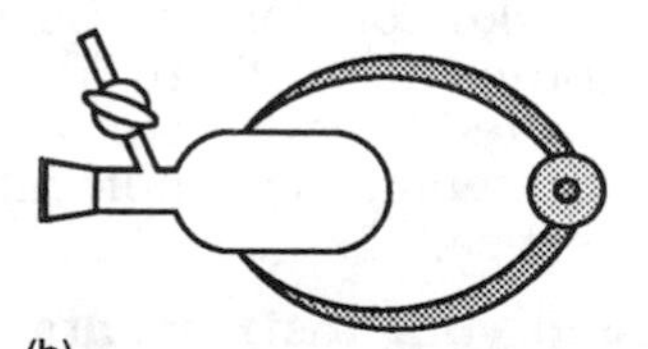

(b)
A pair of dividers can be placed on the outside edge of a round surface. The dividers can then be placed on a meter stick for diameter measurement.

Fig. 2-3 How to measure the outside diameter of rounded objects.

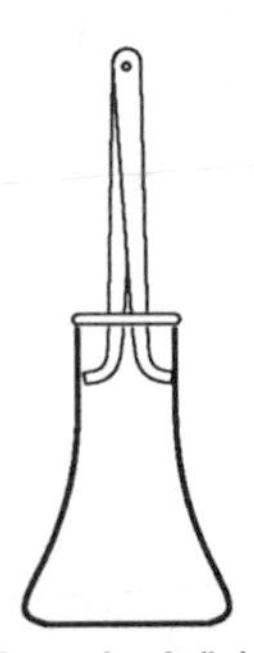

The ends of a pair of dividers can be inserted into an object to obtain an internal measurement. The dividers are then removed and laid on a meter stick to obtain the measurement.

Fig. 2-4 How to measure the inside diameter of an object.

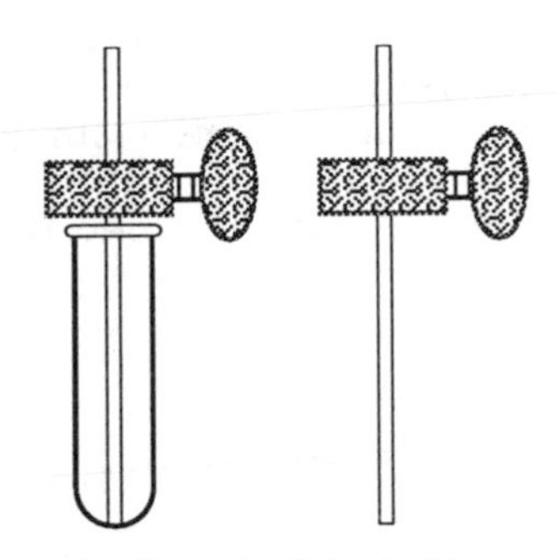

Once depth is established with a depth gauge, measurement can be read directly, or the device is removed and laid onto a meter stick to determine depth measurement.

Fig. 2-5 How to measure the inside length of a narrow object.

Aside from meter sticks, there are other instruments used for linear measurement such as the caliper and micrometer. Their designs and mechanisms for use are very different, but both meet specific needs and have specific capabilities.

2.2.3 The Caliper

The caliper is most commonly used by the machining industry. This industry in the U. S. has consistently resisted the transition toward the metric system and thus most of the tools for machining are made in English measurements. It is possible to obtain calipers in metric measurements, otherwise conversion may be necessary.

A good quality caliper can provide an accurate measurement of up to ± 0.05 mm (± 0.002 in). For more accurate measurements (± 0.0003 mm (± 0.0001 in)), a micrometer is used. The micrometer is discussed in Sec. 2.2.4.

The caliper has one moving part, called the *slide piece*, with three extensions. The extensions of the slide piece are machined so that they can make an inside, outside, or depth measurement (See Fig. 2-6(a)). To achieve greater precision, all calipers include vernier calibration markings on their slide piece. The vernier calibration markings allow measurements one decade greater than the markings on the caliper. The length of the vernier's ten units on the side piece are only nine units of the ruler markings on the body of the caliper (see Fig. 2-6(b)). Fortunately, you do not need to know the geometry or mathematics of how a vernier system works to use one.

A measurement reading on a caliper is made in two parts: the main unit and the tenth unit are made on the body of the caliper. The hundredth part is read on the slide piece. For example, to read the distance 'X' in Fig. 2-7, first read the units to the left of the *index line* on the body's ruler. (The index line is after the 4.4 unit line.) Second, continue the reading to the 100th place by noting where the measurement lines of the slide piece best align with the measurement lines of the body. In this example, they align on the 6th unit of the slide piece. Therefore the 100th place is 0.06, and the total measurement of X is 4.46 units.

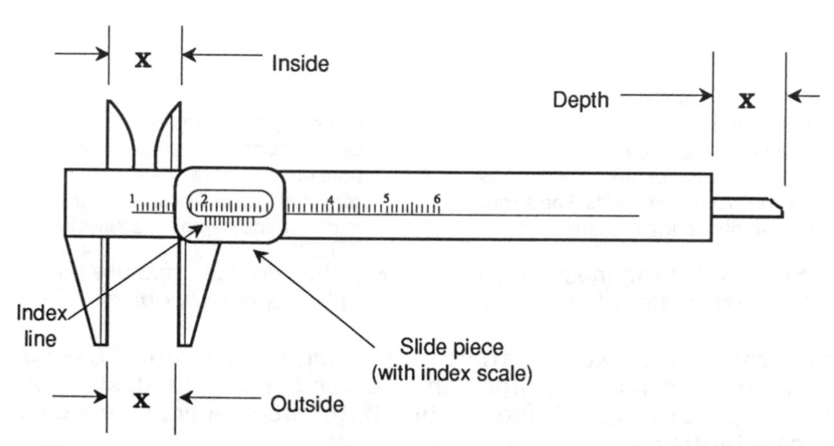

Fig. 2-6(a) The caliper and its three measurement capabilities.

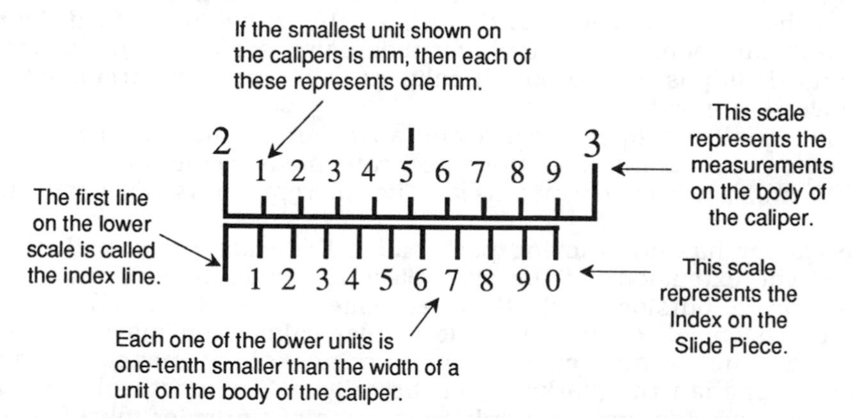

Fig. 2-6(b) This representation of the marks on a caliper body and slide piece shows that ten units on the slide piece are as long as nine units on the body of the caliper.

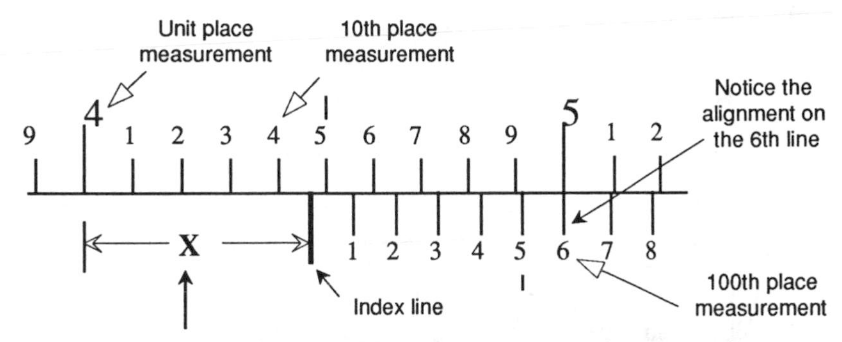
Unit place measurement
10th place measurement
Notice the alignment on the 6th line
9 4 1 2 3 4 5 6 7 8 9 5 1 2
X
1 2 3 4 5 6 7 8
Index line
100th place measurement
The distance "X" is determined where the "index" line touches the upper scale (4.4) PLUS the number of the unit on the slide piece that best alligns with a line on the main scale (the 6th line) which provides the hundredth place measurement (0.06). Therefore "X" is 4.46 caliper units.

Fig. 2-7 How to read a vernier caliper.

A dial caliper (see Fig. 2-8) is easier to read than a vernier caliper, but this ease comes at a price. An acceptable vernier caliper can be purchased for \$30 – \$50, and can be found with metric and English measurements on the same tool. An acceptable dial caliper may cost between \$75 – \$150, and shows only metric *or* English measurements. On the dial caliper, the 10th place is read on the numbers of the dial itself and the 100th place is read on the lines between the numbers.

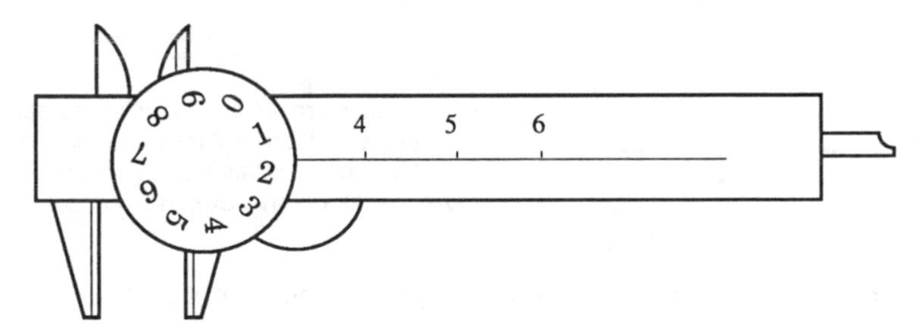
0 1 2 3 4 5 6 7 8 9
4 5 6

Fig. 2-8 The dial caliper.

An electronic caliper (see Fig. 2-9) is more expensive (\$100 – \$200) than a dial caliper. However, electronic calipers can do a great deal more and have greater precision (± 0.013 mm (± 0.0005 in)). A measurement can be locked in by pressing a button; so, a measurement will not change even if the sliding part moves. Electronic calipers can alternate between metric and English measurements, can be zeroed anywhere so "difference" measurements can be read directly rather than subtracting or adding, and they can be set to flash their numbers on and off when an object is out of maximum or minimum tolerance. A tape readout of measurements can be made with extra equipment.

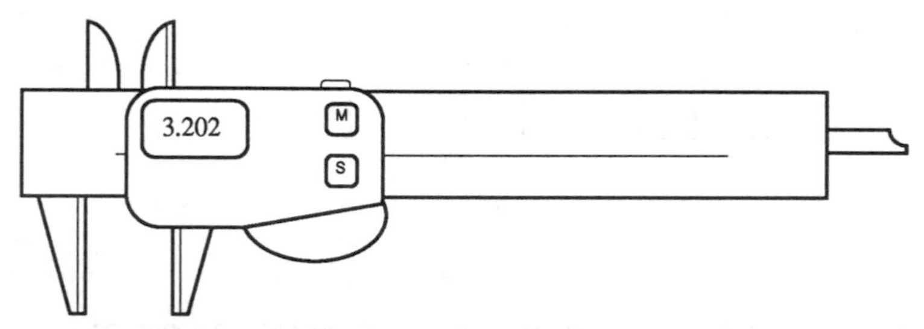

Fig. 2-9 The electronic caliper.

2.2.4 The Micrometer

Another common tool for length measurement is the micrometer. Like the vernier caliper, in the U. S., it is typically made with English measurements. Whereas the caliper can provide precision measurement (to ± 0.002 in), you need a micrometer for greater precision (± 0.0001 in).

The micrometer is a very accurately machined machine bolt and nut. If you rotate a machine bolt into a nut a specific number of times, you'll notice that the bolt will have traveled a specific distance (see Fig. 2-10). On the micrometer, the rotating handle is the nut, and the plunger is the machine bolt. Fig. 2-11 demonstrates how to read a micrometer.

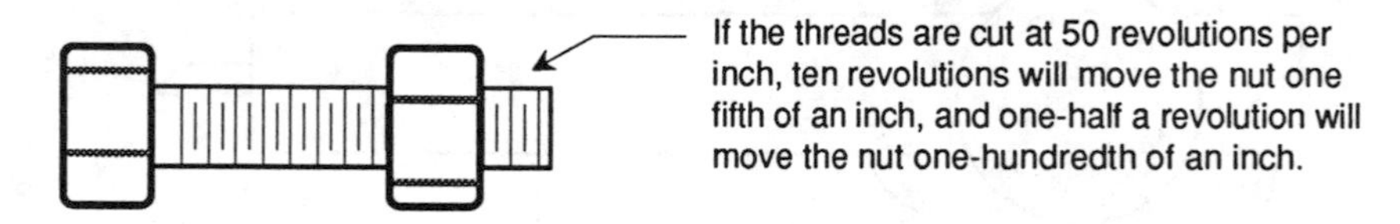

If the threads are cut at 50 revolutions per inch, ten revolutions will move the nut one fifth of an inch, and one-half a revolution will move the nut one-hundredth of an inch.

Fig. 2-10 The micrometer is a highly machined nut and bolt.

A good-quality micrometer is not as expensive as a caliper of equal quality. However, each design of a micrometer is capable of only one type of measurement and can measure only a limited range of that measurement (typically one inch). By comparison, the caliper can read inside, outside, and depth measurements. The caliper, as well, can read from zero to (typically) six inches. Therefore, to have an equivalent set of micrometers to do what one caliper can do would require up to eighteen micrometers. Despite this shortcoming, a good machinist would not consider using a caliper for anything but a rough measurement, and would insist upon using a micrometer for all final measurements.

To make a micrometer measurement, place the item to be measured against the *anvil* of the micrometer. Then, rotate the *handle* or *ratchet* until contact is made. All final contact of the *spindle* against the measured item should be made with the ratchet. The ratchet provides limited control over the handle. Once it encounters resistance, it will spin with a clicking noise but will offer no more pressure against the handle. This feature guarantees consistent spindle pressure against the anvil for all readings.

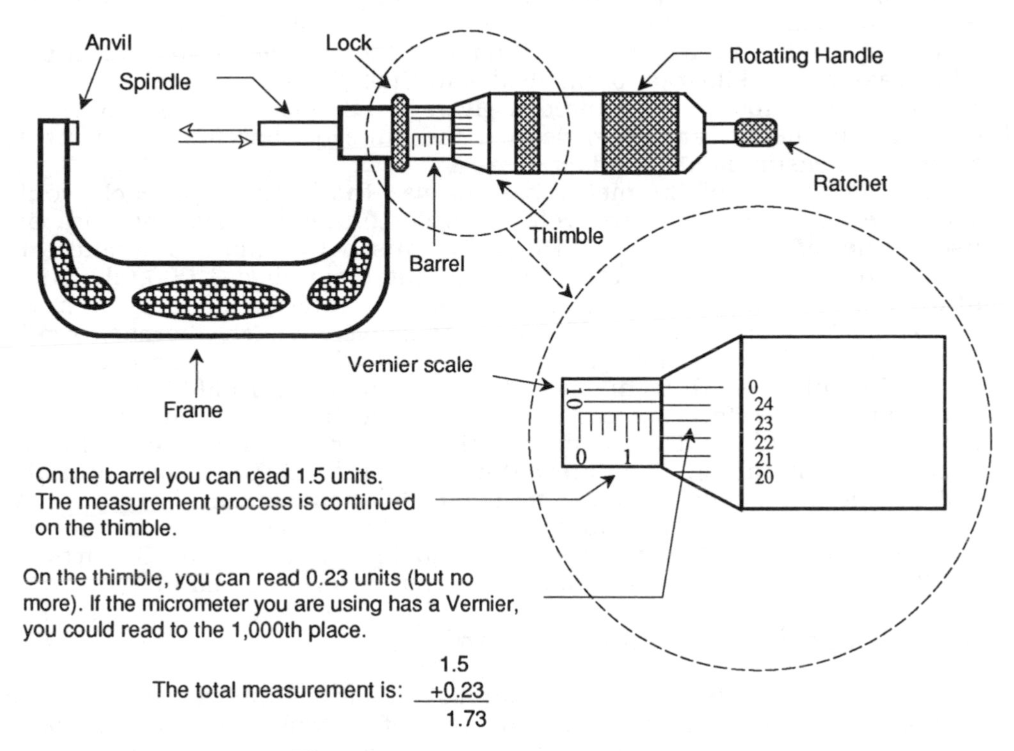

Anvil
Spindle
Lock
Rotating Handle
Ratchet
Thimble
Barrel
Frame
Vernier scale
10
0
1
0
24
23
22
21
20
On the barrel you can read 1.5 units.
The measurement process is continued
on the thimble.
On the thimble, you can read 0.23 units (but no
more). If the micrometer you are using has a Vernier,
you could read to the 1,000th place.
The total measurement is:
1.5
+0.23
1.73

Fig. 2-11 How to read a micrometer.

A micrometer can also be used as a go/no-go gauge. The *lock* can be tightened, preventing the handle from rotating. Thus, by setting a measurement and by tightening the lock, the measurement is set. Then anything that fits (or does not fit) between the spindle and anvil is accepted or not accepted depending on your needs.

For accuracy greater than $1/_{10,000}$ in; an air gauge is used. These gauges have accuracy to $1/_{1,000,000}$ in, and are used in the most demanding situations. The air gauge is specialized beyond the scope of this book and no further explanation is included.

2.3 Volume

2.3.1 The Concepts of Volume Measurement

It does not seem that it should be difficult to calculate the volume of any given container. First you establish a unit of volume, then you base everything on that unit. Despite the apparent simplicity of such a process, two widely divergent approaches to calculating a volume unit have developed.

One approach established that a liter was the volume of space occupied by the mass of one kilogram of distilled water (at 4°C and 30 inches of mercury). The other approach required a given length measurement to be defined (one decimeter), and then defined the cube of that measurement as the volume measurement (one liter).

The original idea of the metric system was that either approach would provide the same unit of metric volume. Unfortunately, it did not work because of the subtle differences in density caused by subtle differences in temperature. Thus, the kilogram-based milliliter equalled 1.000,027 cubic centimeters. Because of the discrepancy, the International System for Weights and Measures had to make a choice between which approach would be accepted to obtain volume measurements, and the nod was eventually given to the cubic length technique. The use of liters and milliliters in volumetric ware is therefore misleading because the unit of volume measurement should be cubic meters (cubic centimeters are used as a convenience for smaller containers). The International System of Units (SI) and the ASTM accept the use of liters and milliliters in their reports providing that the precision of the material does not warrant cubic centimeters. Because the actual difference in one cubic centimeter is less than 3 parts in 100,000, for most work it is safe to assume that 1 cm^3 is equal to 1 mL.

2.3.2 Background of Volume Standards

Rigorous standards have been established for volumetric ware. These standards control not only specific standards of allowable error for volumetric ware, but the size of the containers, the materials of construction, their bases, shapes, sizes, the length and width of index lines, and how those lines are placed on the glass or plastic. The painstaking work to establish these guidelines was done by agencies such as the NIST (National Institute of Standards and Technology), the ASTM (American Society for Testing and Materials), and the ISO (International Standards Organization).

Occasionally you may see references to Federal Specification numbers. For example, the Federal Specification number, 'NNN - C 940-C', is for graduated cylinders. All Federal Specification numbers are no longer being updated and these specifications are being superseded by those established by the ASTM. The preceding number for graduated cylinders is now under the specifications of ASTM 1272-89. All ASTM documents are identified by a designation number. If there are updates to any document, the number will be followed by the acceptance year of the update. For example, the preceding example ASTM document cited shows that it was accepted by the ASTM for publication in 1989. If there is a conflict in ASTM guidelines, the document with the later publication date takes precedence.

All manufacturers abide by the standards set by these organizations. Therefore with the exception of quality and control, one manufacturer's volumetric ware (of comparable type) should not be more accurate than another manufacturer's. It is important to keep this similarity in mind so that ASTM standards are not used to imply more than they are meant to.

Manufacturers refer to these standards in their catalogs, both to let you know what you are buying, as well as to enhance importance that may not otherwise be there. For instance, only glassware made to specific established tolerances can have the symbol of "A" or "Class A" on their sides signifying highest production quality. For example, for a standard tolerance graduated cylinder, you may see statements in catalogs like '...conforms to ASTM Type I, Style 1 specs for volumetric ware'. This description translates to mean '*the graduated cylinder is made out of borosilicate glass* (Type I) *and has a beaded lip with a pour spout* (Style 1)'. In reality, this enhanced description is harmless, and much safer than statements saying 'the most accurate graduated cylinder in town'.

Do not let the designation Class A mean more than it was meant to. Class A can only mean that it is the best tolerance readily available for that specific type of volumetric ware. For example, a Class A volumetric pipette does not have the same degree of tolerance as a Class A measuring pipette. Equally, a Class A graduated cylinder does not have the same degree of tolerance as a Class A volumetric flask. See Table 2-6 for a representative cross comparison of Class A tolerances.

Table 2-6‡

Cross Comparison of Class A Volumetric Ware (25 mL)

Item	Class A	Item	Class A
Grad. Cylin.	±0.17 mL	Volumetric Pipette	±0.03 mL
Vol. Flask	±0.03 mL	Measuring Pipette	±0.05 mL
Burette	±0.03 mL	Serological Pipette	(only Class B & lower)

‡ Based on ASTM guidelines.

In addition, do not be misled by ASTM designations. The words "Class," "Style," and "Type" are constantly used to describe different attributes to different types of variables in ASTM literature. They seldom refer to the same attribute. Thus it is important to know what the identifying word is attributed to before assuming that you know what it is signifying. An example of these differences are shown in Table 2-7.

Table 2-7‡

An Example of How the Term "Style"is Used for Different Meanings[a]

	Graduated Cylinder Here the term "Style" refers to the physical structure of the opened end.	**Pharmaceutical Cylinder** Here the term "Style" refers to the calibrations.
Style 1	Beaded lip with pouring spout	Metric calibration
Style 2	Top with ground glass stopper	English (in-lb) calibrations
Style 3	Beaded lip with pouring spout and reinforced rim	Both English and metric calibrations.

‡ From ASTM specifications E 1094-86 *(Standard Specification for Pharmaceutical Glass Graduates)* and E 1272-89 *(Standard Specification for Graduated Cylinders)*.

The ASTM always refers to itself in its own specifications for equipment when describing one of its own procedures for a given test. Thus, when you see manufacturer statements in catalogs describing (for example) a graduated cylinder that '... meets specifications of ASTM E133 for use in tests D86, D216, D285, D447, and D 850 ...', all the manufacturer is saying is that if you are performing any of these particular ASTM tests, this graduated cylinder satisfies ASTM requirements for use. These descriptive entries may be confusing if you are unfamiliar with the numbers.

2.3.3 Categories, Markings, and Tolerances of Volumetric Ware

There are four categories of containers used to measure volume: *volumetric flasks*, *graduated cylinders*, *burettes*, and *pipettes*. Of the four, volumetric flasks are used exclusively to measure how much has been put into them. This use is known as '*to contain*.' Graduated cylinders and a few pipettes are used to measure how much has been put into them as well as how much they can dispense. The latter measurement is known as '*to deliver*.' Burettes and most pipettes are used solely *to deliver*.

The term *to deliver* is based on the concept that when you pour a liquid out of a glass container, some of that liquid will remain on the walls of that container. Because not all of the measured liquid is completely transferred, the material left behind should not be considered part of the delivered sample. Pipettes have two different types of *to deliver*: one which requires you to "blow out" the remaining liquid, and one that does not. Some volumetric containers are made out of plastic which does not 'wet' like glass. Because these containers drain completely, the *to contain* is the same as the *to deliver*. Because some materials (i.e., mercury) do not 'wet' the walls of any container, they should be used with only *to contain* measuring devices.

The abbreviations TC and TD are commonly used to denote *to contain*, and *to deliver*, respectively in the U. S. Old glassware might be labeled with a simple 'C' or 'D'. The ISO (International Standards Organization) uses the abbreviations IN for *to contain* and EX for *to deliver*. These abbreviations come from the Latin root found in *INTRA* (within) and *EXTRA* (out of). This classification has been accepted for use in the U. S. by the ASTM and is slowly being introduced by U. S. manufacturers. Because there have been no formal deadlines set, manufacturers are waiting for their current silkscreen printing setups to wear out by attrition and then replacing them using the new classification. Although some volumetric ware may be labeled with both English and metric calibrations, glassware is seldom labeled with both TD and TC calibrations. If you are ever using a volumetric container that is double-labeled, be careful that you are reading the scale relevant to what you are doing.

Other common laboratory containers such as beakers, round bottom flasks, and Erlenmeyer flasks often have a limited graduated volume designated on their sides. These markings provide an approximate volume and cannot be used for quantitative work. The required accuracy of these containers is only 5% of volume. When there are no calibration lines on a flask, it still is possible to obtain an approximate volume measurement based on the stated volume: in general, the stated volume will approximately fill any given non-volumetric container to the junction of the neck and container (see Fig. 2-12). Thus, if you need about 500-mL of water, it is safe to fill a 500 mL flask up to the neck and you will have approximately the needed volume.

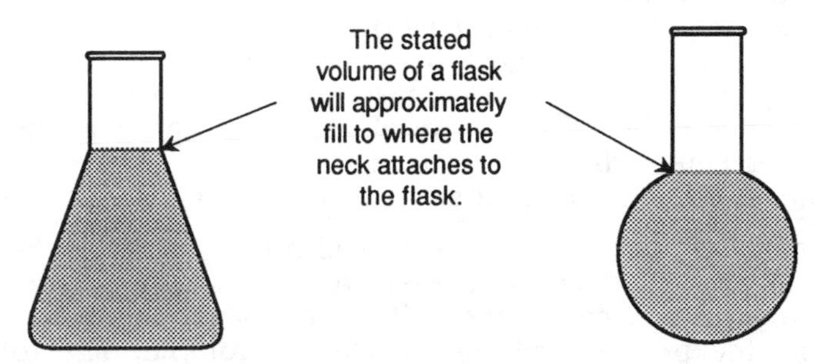

The stated
volume of a flask
will approximately
fill to where the
neck attaches to
the flask.

Fig. 2-12 The volume of flasks.

The quality of any given volumetric ware is based on how accurate any given calibration line will deliver the amount it claims. For instance, say that a one-liter flask is accurate to ±5%, meaning that the flask is likely to contain anywhere from 950 mL to 1050 mL of liquid. For comparison, a one-liter Class B volumetric flask is accurate to ± 0.60 mL, or ±0.06% accuracy, and a one-liter Class A is accurate to ± 0.30 mL, or ±0.03% accuracy. Needless to say, it costs more for greater accuracy.

Precision volumetric glassware must be made following the standards set by ASTM Standards E542 and E694. These standards establish not only the degree of tolerance, but how the container is to be made, how the lines are engraved, the width and length of calibration lines, the type of glass used, the design and type of base (if any), the flow rate of liquids through tips, and many other limitations. In reality, if the manufacturers are properly following the ASTM guidelines, there is little room for any volumetric ware to be poorly made.

ASTM guidelines also establish that volumetric ware is to be calibrated at 20°C. This (or any) constant is required because gases, liquids, and solids vary in size as temperature changes. Thus, any individual item is accurate only at one temperature. If manufacturers did not adhere to one temperature calibration, constant mathematical adjustments would have to be made as one switched from one brand of glassware to another.

There are four basic grades of volumetric ware:

1) ***Special Tolerance***, also called ***Calibration Tolerance*** and ***Certified Ware***. This glassware is calibrated not only to Class A tolerances, but to each mark on the glass. This calibration ensures high accuracy for every measurement. It needs to be specially ordered and calibrated.

2) ***Class A***. This code is for the highest-class glassware that is production made. Some manufacturers refer to this grade as *Precision grade*. High-quality work can be performed using glassware that has this designation. The tolerance is the same

as *Special Tolerance*, but in Class A it is based on a container's total capacity, not for each measurement division. Containers of Class A tolerance are always designated as such in writing on the side.

3) ***Class B***. This classification is average grade glassware and can be used for general quality work. Some manufacturers refer to this quality as *standard grade* or *standard purpose*. By ASTM designation, Class B tolerances are twice those of Class A. The tolerance is based on a container's total capacity, not for each measurement division. Containers of Class B tolerance are seldom, if ever, designated as such in writing on their side.

4) ***General Purpose***. This glassware is not accurate for any level of quality work. It is often referred to as *student grade* or *economy grade*. Because the calibration is never verified, it is the most-inexpensive volumetric glass available. The tolerance is limited to plus or minus the smallest subdivision of the container. Containers of General Purpose tolerance are never specifically identified as such and therefore may be confused as Class B volumetric glassware. Such glassware made by Kimble is identified as "Tekk." Corning does not have a generic name for its student ware. General purpose glassware is often made out of soda-lime glass rather than borosilicate glass.

2.3.4 Materials of Volumetric Construction #1 Plastic

There are four types of plastic that are commonly used in volumetric ware: polypropylene (PP), polymethylpentene (PMP or TPX), polycarbonate (PC), and polystyrene (PS). Plastic can be less expensive than glass, more difficult to break, and can be used equally well for *to contain* and *to deliver* measurements. However, it cannot tolerate temperatures over 130° to 170°C (depending on the type of plastic), can be dissolved in some solvents, and the best quality available is Class B.

General Character of Each Type of Plastic. *Polypropylene* is translucent and cannot be dissolved by most solvents at room temperature. It can crack or break if dropped from a desk.

Polymethylpentene is as transparent as glass and is almost as resistant to solvents as polypropylene. It can withstand temperatures as high as 150°C on a temporary basis and 175°C on an intermittent basis. It can crack or break if dropped from a desk.

Polycarbonate is very strong and as transparent as glass. It is subject to reactions from bases and strong acids and can be dissolved in some solvents.

Polystyrene is as transparent as glass. It can be soluble to some solvents and because it is inexpensive, it is often used for disposable ware. It can crack or break if dropped from a desk.

Resistance to Chemicals. For a comprehensive list of plastic resistance to chemical attack, see Appendix B.

Cleaning. Volumetric ware must be cleaned both to prevent contamination of other materials and to ensure accurate measurements. Because volumetric ware is often used as a temporary carrier for chemicals, it comes in contact with a variety of different materials which may be inter-reactive. Also, any particulate or greasy material left behind can alter a subsequent measurement.

Never use abrasive or scouring materials on plastic ware. Even though the bristles of a bottle brush are not likely to scratch plastic, the metal wire that they are wrapped on can! Scratches can alter volumetric quality, are more difficult to clean, increase the surface area open to attack by chemicals (that would otherwise be safe for short exposures), and can prevent complete draining from a *to deliver* container.

The following are some general cleaning guidelines:

1. Generally, mild detergents are safe with all plastic ware.
2. Do not use strong alkaline agents with polycarbonate.
3. Never leave plastic ware in oxidizing agents for an extended time as they will age the plasticisers and weaken the plastic. Chromic acid solutions are okay to use, but never soak longer than four hours.
4. Polystyrene should not go through laboratory dishwashers because of the high temperatures.
5. Polypropylene and polymethylpentene can be boiled in dilute sodium bicarbonate ($NaHCO_3$), but not polycarbonate or polystyrene.
6. Sodium hypochlorite solutions can be used at room temperature.
7. Organic solvents can be used cautiously *only* with polypropylene. However, test each solvent before use for possible negative effects on the outside of the volumetric ware, away from a calibrated area.

Autoclaving. Polypropylene can be autoclaved, polycarbonate can be autoclaved for a limited time (<20 minutes) but can be weakened from repeated autoclaving. The quality of volumetric measurements is likely to be eroded with repeated autoclaving. Polystyrene should never be autoclaved.

2.3.5 Materials of Volumetric Construction #2 Glass

Although glass is more likely to chip, crack, or break than plastic ware, it is safe to use with almost all chemicals. Type I glass can safely be put into a drying oven, and Type I and II glass can be autoclaved (glass types are explained in the next section). No volumetric ware should ever be placed over a direct flame. Occasionally, tips and/or ends of burettes or pipettes are tempered to provide additional strength. The tempering process typically used is heating followed by rapid cooling.

General Characteristics of Each Glass Type. Volumetric ware made of glass provides different properties depending on the type of glass used. The ASTM refers to the different glasses used in volumetric ware as *Types*. The use of the terms 'Class A' and 'Class B' in reference to types of glass have no relationship to volumetric quality.

1. ***Type I, Class A*** glass is usually Pyrex or Kimex glass. It can be used for all classifications of volumetric ware. This glass is chemically inert, repairable (in non-volumetric regions away from calibration), can safely be placed in a drying oven, and can be fused onto other similar borosilicate laboratory glassware.
2. ***Type I, Class B*** glass is an alumino-silicate glass. It can be used for all classifications of volumetric ware. The trade names used are Corning's 'Corex' glass or Kimble's 'Kimex-51.' They are more chemically resistant than standard borosilicate. However, they have larger coefficients of expansion, and therefore cannot directly be fused onto other lab glass items made out of Pyrex or Kimex. An intermediary glass with a coefficient of expansion of about 40 – 42 x 10^{-6} is required for fusing Type I, Class B glass to Type I, Class A glass.
3. ***Type II*** glass is often identified as *soda-lime*, *flint*, or *soft glass*. Corning does not have a specific trade name, but Kimble uses the code number R-6. It is not as chemically resistant as Type I glass, but short duration containment is acceptable (never leave chemicals in soda-lime volumetric ware).

Type II volumetric ware cannot be repaired if cracked or chipped nor fused onto other laboratory ware. On the other hand, it is inexpensive and therefore commonly used for disposable ware and lower standard calibration ware. One type of Type II glass, Exax (from Kimble), has anti-static properties, which make it particularly suitable for powders.

Resistance to Chemicals. Exposure to all alkalines should be kept to a minimum (minutes). Exposure to hydrofluoric and perchloric acid should be limited to seconds. Volumetric ware used once for these materials should be downgraded (glassware that was Class A should now be considered Class B) or not used at all. If measurements of these acids or alkalines are required, use plastic ware because it is resistant to these chemicals.

Cleaning. All standard cleaning techniques used on glass are safe on volumetric ware. However, base baths and HF should never be used as they are glass-stripping agents and during the cleaning process remove glass. This stripping would alter any calibrations, making the volumetric ware useless.

Contrary to common belief, it is safe to place borosilicate volumetric ware in a drying oven. While there has not been a study indicating the effects of heating on Type I, Class B or Type II glass, there was a study on Type I, Class A volumetric ware. The study was done by Burfield and Hefter[6] and the results (which are shown in Table 2-8) clearly indicate that any variations from the original volume are within tolerances even for Class A glassware.

The results of Table 2-8 notwithstanding, there still is the question of how to dry and/or store other types of volumetric glassware as well as plastic ware. First, plastic ware does not need oven drying because plastic does not absorb water. Secondly, Type II glassware measurements will not be significantly affected by remaining distilled water (from the final rinse), so again there is no reason to dry this glassware. There is also no reason to dry *to contain* glassware between repeated measurements of the same solution. However,, all other volumetric glassware should be dried prior to use.

Because of contamination concerns, it is not recommended that you place volumetric ware on a standard drying rack because the pegs may introduce foreign materials into the container. It is also not safe to balance any volumetric ware on its end to drain or dry as it might be tipped over. Williams and Graves[7] came up with an ingenious approach to drying and/or storage similar to how wine glasses are often stored. Although they limited their explanation to volumetric flasks, the idea can be expanded to any volumetric container with a bulge or protuberance at one end. Such volumetric

Table 2-8‡

Effect of Heat Treatment on Volumetric Glassware

Apparatus	Nominal Volume (mL)	Nature of Heat Treatment	Initial[b] Volume (mL)	Final[b] Volume (mL)
Pipette	10	24 h @ 320°C	9.970 ± 0.009	9.977 ± 0.004
Pipette	10	Cycled to 320°C[a]	9.954 ± 0.030	9.954 ± 0.020
Standard Flask	100	Cycled to 320°C[a]	99.95 ± 0.01	99.94 ± 0.01
Standard Flask	100	168 h @ 320°C	99.81 ± 0.02	99.82 ± 0.02
Standard Flask	25	168 h @ 320°C	24.909 ± 0.010	24.917 ± 0.010

a Consists of eight heat-cool cycles from RT to 320°C with 15 min at 320°C.
b Mean and standard deviation of eight determinations at 26°C.
‡ From the Journal of Chemical Education, 64 (1987), 1054. With permission

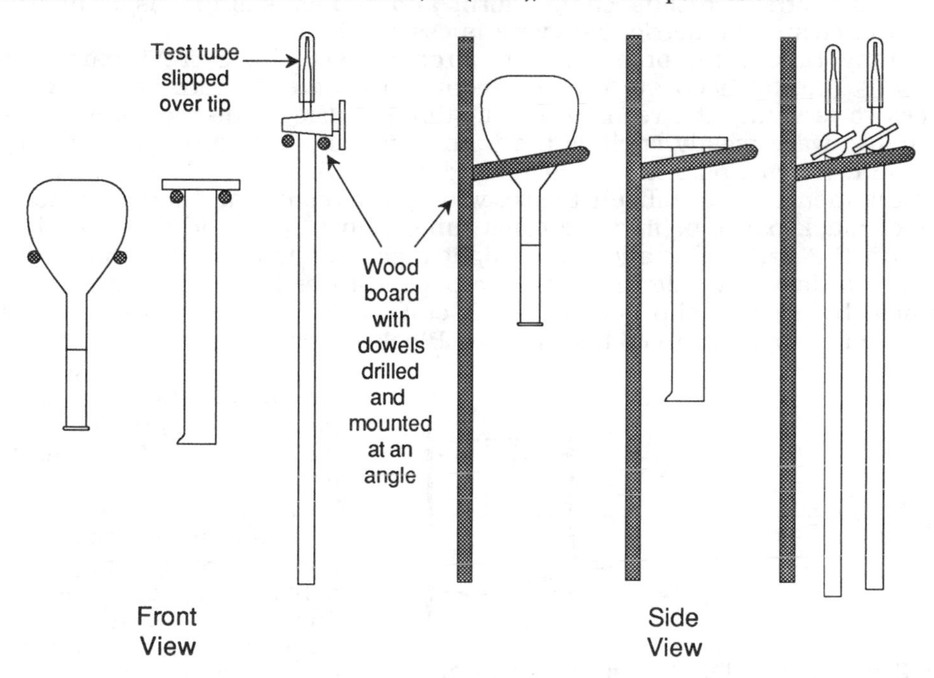

Fig. 2-13 Storage or drying of volumetric ware.

ware items would include the base of a graduated cylinder or the stopcocks of a burette (see Fig. 2-13). This technique obviously will not work with pipettes.

2.3.6 Reading Volumetric Ware

The parallax problems of linear measurement are compounded with volumetric ware because there are two distinct lines to read. One line is where the liquid makes contact with the walls of the volumetric container, and the other is in the center of the volumetric tube (see Fig. 2-14).

The distortion of the liquid is caused by the surface tension of that liquid and the walls of the container. This distorted line is called the meniscus. When liquid wets the walls of a container, you read the bottom of the meniscus. When liquid does not wet a containers walls (such as mercury or any liquid in a plastic container), you read the top of the meniscus. Incidentally, if the glassware is dirty, a smooth meniscus cannot form and proper sighting is impossible; therefore, clean volumetric glassware is essential.

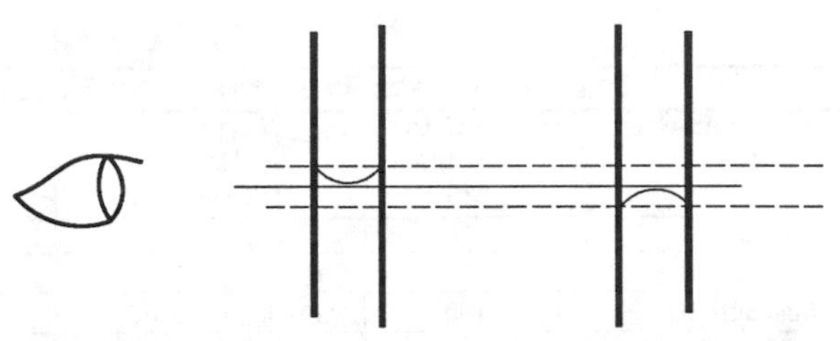

Fig. 2-14 When viewing fluid volumes, observe the center of the meniscus at eye level.

It may take a bit of practice to properly see the correct part of the meniscus line for accurate measurement. Fortunately, there are tricks and devices to facilitate the reading. For instance, if the graduation lines on the volumetric ware mostly encircle the tube, it is easy to line up your vision so that you can avoid parallax problems (see Fig. 2-15).

If the meniscus is difficult to see, you can make it stand out by placing a piece of black paper behind the glass tube, below the liquid line. The liquid picks up the dark color and, like a light pipe, the end of the liquid column will have a dark line. Alternatively, the paper can be folded around the tube, and held by a paper clip (like a French cuff). This 'paper cuff' can be raised or lowered easily, and frees both hands. Black paper is inexpensive and easy to use, but if the paper gets wet it will need to be replaced. This problem is not major, but if it occurs mid-experiment, it can be an inconvenience.

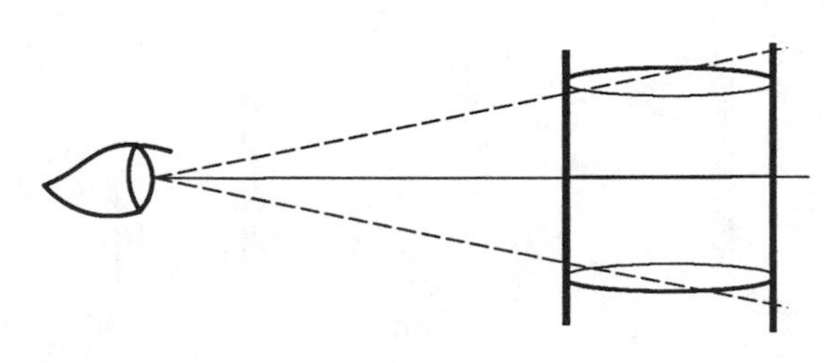

Fig. 2-15 By using the lines above and below the one you are sighting on, you can avoid parallax.

Another technique that does not require paper uses a black ring cut from flexible tubing. These rings can be purchased to fit a variety of tube sizes, but if you have a hose of the right size

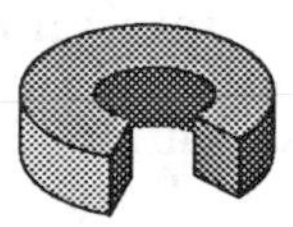

Fig. 2-16 A properly-cut piece of black flexible tube that facilitates seeing the meniscus.

available, they are easy to make by cutting out a disk from black tubing or by slicing a disk from a black rubber stopper (and a hole), and cutting a wedge out of either disk (see Fig. 2-16). By sliding this ring up or down so that it is below the top of the liquid column, the meniscus can be easily seen.

2.3.7 General Practices of Volumetric Ware Use

After using volumetric ware, wash if necessary and always rinse thoroughly, first with water and then with distilled water. Although it is not necessary to dry *to deliver* volumetric ware between measurements, it is necessary to dry the insides of *to contain* containers before use. Reagent grade acetone or methyl alcohol may be used to facilitate drying.

If using volumetric ware for gravimetric calibration, unintended weight (i.e., dirt or fingerprint oils) may alter the results of any weighing. Therefore, be very careful where you place volumetric ware on a bench and how you hold it. You may choose to use cotton gloves to prevent fingerprint marks or handle all glassware with tweezers or tongs.

Excess solution on the walls (inside or out) of volumetric ware adds weight to measurement, but are not considered part of the volume. Therefore, be careful not to splash your solution into the container. Such splashing can cause drops to settle on the side walls. To prevent splashing, place the tip of a burette or pipette against the walls of the receiving container so the liquid slides down the walls. This procedure should always be done above the calibration line. Wait a few minutes for any residual fluid along the walls to settle before making any final volume determination. Because fluids may cling to the walls of the container, it may take a few minutes for everything to drain to its lowest level (for an exaggeration, think of honey). During the settling period, cap the container to limit the amount of fluid that would otherwise evaporate.

Do not leave alkaline materials in glass volumetric ware. Aside from the damage they may do to the volumetric ware, dissolved glass can affect the pH of your solution. In addition, a glass plug or stopcock may freeze in place if left in contact with an alkaline solution. This freezing prevents the stopcock or plug from being turned or removed. A long-standing alkaline solution can roughen a surface of the glass barrel even in a Teflon stopcock. The rough surface can later scratch the surface of the Teflon plug when it is rotated.

2.3.8 Calibrations, Calibration, and Accuracy

The volumetric marks on volumetric ware are called calibrations. How they were located on the volumetric ware is called calibration. All volumetric ware is calibrated to provide its stated volume at 20°C. The International Standards Organization has recommended that the standard volumetric temperature should be changed to 27°C. However, so far there has not been any significant movement toward this goal. The ASTM recommends that those labs in temperate climates that are unable to maintain an environment at 20°C, maintain one at 27°C.

There are many labs that not only are unable to maintain a 20°C temperature, but cannot maintain *any* temperature consistently. Temperature variations can create problems if you are attempting to do accurate work. The more accurate the work, the greater the need to maintain room temperature at 20°C. Alternatively, include the temperature of the liquid at the time of measurement in your experiment log. Then, at a later time, apply corrections to all measurements made to conform to ASTM 20°C measurements.

However, do not waste your time doing unnecessary work. The correction for any given measurement may be irrelevant if either: 1) its result is smaller than the tolerances of your equipment: 2) it involves more significant figures than can be supported by other aspects of your work:* or, 3) you do not require that level of accuracy.

Another concern for accuracy is based on how accurately the user can read calibrations on the volumetric ware. The reproducibility of an individual user will be more consistent than the reading made by a variety of users. Therefore, if there will be a variety of users on any given apparatus, all who are likely to use it should make a series of measurements. This way, the individual errors can be calculated. The ASTM has analyzed the range of errors made by trained personnel and the reproducibility of these results are shown in Table 2-9.

Table 2-9‡

Precision Data		
Vessel	**Nominal Size** cm^3	**Reproducibility** $cm^{3\,A}$
Transfer Pipette	1	0.002
	2	0.002
	5	0.002
	10	0.003
	15	0.005
	25	0.005
	50	0.007
	100	0.010
Flasks	10	0.005
	25	0.005
	50	0.007
	100	0.011
	200	0.014
	250	0.017
	500	0.021
	1000	0.042
Burettes	10	0.003
	25	0.005
	50	0.007
	100	0.012

A The term "reproducibility" refers to the maximum difference expected between two independent determinations of volume.

‡ From the ASTM document E 542, Table 4. With permission.

* For example, suppose you are calculating your car's 'miles per gallon.' It is a waste of time to measure the gas to hundredths of a gallon if your odometer can only read tenths of a mile.

2.3.9 Correcting Volumetric Readings

Volumetric readings can be made two ways. The easiest and most common is simply reading the volume directly from a piece of volumetric ware. Alternatively, you can weigh a sample and, if you know the molecular weight of the material, you can calculate the volume. Each approach can be affected by barometric pressure, humidity, and temperature. The calculations and tables needed to obtain true volume from observed volume or calculated weight are not difficult to use but should only be used when necessary, that is, when accuracy or precision demand their use.

There are two different approaches for properly correcting volumetric readings caused by environmental variations because there are two approaches to making volumetric readings: those done by reading volume directly from volumetric ware, and those made indirectly by weight.

The simplest corrections are made when reading directly from volumetric ware. As the volumetric container (and the liquid contained) expands and contracts by temperature variations from 20°C, *volumetric* corrections are required. These corrections can be found on Table 2-10.

For instance, say you had a 100-mL borosilicate volumetric pipette whose liquid was measured at 24°C. Table 2-10 shows "-0.09" for these conditions, which means that you must subtract 0.09 mL from the stated volume to compensate for water and glass expansion. Thus you actually delivered only 99.91 mL of liquid.

You may occasionally see a soda-lime glass equivalent for Table 2-10, but I have omitted it from this book. Although soda-lime glass volumetric ware was common many years ago, it is not used for any accurate volumetric purposes today. It is unnecessary to provide highly accurate corrections for non-accurate glassware.

The corrections shown in Table 2-10 are only valid for distilled water. Different liquids will have different coefficients of expansions and therefore will require different corrections. Table 2-11 provides a few representative solutions at different normalities and a factor to correct Table 2-10 for volumetric discrepancies.

Corrections required when weighing volumetric flasks are somewhat different than straight volumetric readings. Both single- and double-pan balances have four common parameters which can affect the accurate weighing of liquids, but the single-pan balance has one separate parameter of its own. The common parameters are *water density*, *glass expansion*, and the *buoyancy effect*.

Water density varies because as materials get hot, they expand and take up more space. Despite their taking up more space, they still have the same mass and are thus less dense. One liter of hot water would therefore weigh less than one liter of cold water.

Glass expands as it gets hot. The effects of expansion of solid materials are well demonstrated with the ring and ball demonstration. In this demonstration, a ring is unable to get past a ball on the end of a rod. If the ring is heated just a small amount, it expands sufficiently to easily slide past the ball. In a similar fashion, a warm glass container holds more liquid than a cool glass container.

Table 2-10

Temperature Corrections for Water in Borosilicate Glass													
	Capacity of apparatus in mL at 20°C												
Measurement Temperature	2000	1000	500	400	300	250	200	150	100	50	25	10	5
	Correction in mL to give volume of water at 20°C												
15	1.69	0.85	0.42	0.34	0.25	0.21	0.17	0.13	0.08	0.04	0.02	0.01	0.00
16	1.39	0.70	0.35	0.28	0.21	0.17	0.14	0.10	0.07	0.03	0.02	0.01	0.00
17	1.08	0.54	0.27	0.22	0.16	0.14	0.11	0.08	0.05	0.03	0.01	0.01	0.00
18	0.74	0.37	0.19	0.15	0.11	0.09	0.07	0.06	0.04	0.02	0.01	0.00	0.00
19	0.38	0.19	0.10	0.08	0.06	0.05	0.04	0.03	0.02	0.01	0.00	0.00	0.00
20	0.00	0.00	0.00	0.00	0.00	0.00	0.00	0.00	0.00	0.00	0.00	0.00	0.00
21	-0.40	-0.20	-0.10	-0.08	-0.06	-0.05	-0.04	-0.03	-0.02	-0.01	-0.01	0.00	0.00
22	-0.83	-0.42	-0.21	-0.17	-0.12	-0.10	-0.08	-0.06	-0.04	-0.02	-0.01	0.00	0.00
23	-1.27	-0.64	-0.32	-0.25	-0.19	-0.16	-0.13	-0.10	-0.06	-0.03	-0.02	-0.01	0.00
24	-1.73	-0.87	-0.43	-0.35	-0.26	-0.22	-0.17	-0.13	-0.09	-0.04	-0.02	-0.01	0.00
25	-2.22	-1.11	-0.56	-0.44	-0.33	-0.28	-0.22	-0.17	-0.11	-0.06	-0.03	-0.01	-0.01
26	-2.72	-1.36	-0.68	-0.54	-0.41	-0.34	-0.27	-0.20	-0.14	-0.07	-0.03	-0.01	-0.01
27	-3.24	-1.62	-0.81	-0.65	-0.49	-0.41	-0.32	-0.24	-0.16	-0.08	-0.04	-0.02	-0.01
28	-3.79	-1.90	-0.95	-0.76	-0.57	-0.47	-0.38	-0.28	-0.19	-0.09	-0.05	-0.02	-0.01
29	-4.34	-2.17	-1.09	-0.87	-0.65	-0.54	-0.43	-0.33	-0.22	-0.11	-0.05	-0.02	-0.01
30	-4.92	-2.46	-1.23	-0.98	-0.74	-0.62	-0.49	-0.37	-0.25	-0.12	-0.06	-0.02	-0.01

Table 2-11‡

% Volume Corrections for Various Solutions			
	Normality		
Solution	N	N/2	N/10
HNO_3	50	25	6
H_2SO_4	45	25	5
NaOH	40	25	5
KOH	40	20	4

‡ From the ASTM document E 542, Table 6. With permission.

The differences in the *buoyancy effect* (based on Archimedes principle) of materials in air at different barometric pressures is not as great as the buoyancy differences in water versus air, but it still exists and can affect accurate weighings.

The actual effect of these parameters on any measurement can be calculated by volumetric measurements. Then, their relative significance can be properly considered. The ASTM calculated the values shown in Table 2-12.

As Table 2-12 shows, for most laboratory work, the parameter that is likely to have the greatest effect on measurements when using a two-pan balance is water temperature. Any weight measurements made when the liquid temperature is not 20°C can be corrected by using Table 2-13. This table is used when the volumetric flask is Type I, Class A (borosilicate glass). It may also be used for Type I, Class B (aluminosilicate glass) by adding

0.0006 degree for every degree below 20°C. Likewise, subtract 0.0006 degree for every degree above 20°C.

Table 2-12‡

Relative Significance of Environmental Parameters on Volume Measurements		
Parameter	Parametric Tolerance	Volumetric Tolerance
Relative Humidity	± 10 %	1 in 10^6
Air Temperature	± 2.5°C	1 in 10^5
Air Pressure	± 6 mm	1 in 10^5
Water Temperature	± 0.5°C	1 in 10^4

‡ From the ASTM document E 542, Sec. 14.2.1. With permission

Table 2-13

Corrections for Weight Determinations for Borosilicate Glass Using a Two Pan Balance (nominal capacity - 100 ml)

Temp. in °C	tenths of degrees 0	1	2	3	4	5	6	7	8	9
15	0.200	0.201	0.202	0.204	0.205	0.207	0.208	0.210	0.211	0.212
16	.214	.215	.217	.218	.220	.222	.223	.225	.226	.228
17	.229	.231	.232	.234	.236	.237	.239	.241	.242	.244
18	.246	.247	.249	.251	.253	.254	.256	.258	.260	.261
19	.263	.265	.267	.269	.271	.272	.274	.276	.278	.280
20	.282	.284	.286	.288	.290	.292	.294	.296	.298	.300
21	.302	.304	.306	.308	.310	.312	.314	.316	.318	.320
22	.322	.324	.327	.329	.331	.333	.335	.338	.340	.342
23	.344	.346	.349	.351	.353	.355	.358	.360	.362	.365
24	.367	.369	.372	.374	.376	.379	.381	.383	.386	.388
25	.391	.393	.396	.398	.400	.403	.405	.408	.410	.413
26	.415	.418	.420	.423	.426	.428	.431	.433	.436	.438
27	.441	.444	.446	.449	.452	.454	.457	.460	.462	.465
28	.468	.470	.473	.476	.479	.481	.484	.487	.490	.492
29	.495	.498	.501	.504	.506	.509	.512	.515	.518	.521
30	.524	.526	.529	.532	.535	.538	.541	.544	.547	.550
31	.553	.556	.559	.562	.565	.568	.571	.574	.577	.580
32	.583									

A sample calculation (using Table 2-13) for Type I, Class A glass is as follows:

Nominal capacity of vessel	= 25 mL
Temperature of weighing	= 22.5°C
Weight on pan before filling receiver	= 24.964 gms
Weight on pan after filling receiver	= 0.044 gms
Apparent weight of water at 22.5°C	= 24.920 gms
Correction for 25 mL at 22.5°C (0.25 times* value in Table 2-13)	= 0.083
Volume of vessel at 20°C	= 25.003 mL

As can be seen from this example, the correction is ten times smaller than the tolerances capable from the flask itself (0.03 mL). Thus, to make the time spent on any such correction worthwhile, you need to see that any changes caused by temperature are at least equal to, or greater than inaccuracies of the container. Otherwise, do not bother with any correction.

Table 2-14

Corrections for Water Weight Determinations for Borosilicate Glass Using a One Pan Balance (nominal capacity - 100 ml)

Temp.	Barometric Pressure (mm of Hg)									
in °C	620	640	660	680	700	720	740	760	780	800
18.5	0.2390	0.2418	0.2446	0.2473	0.2501	0.2529	0.2557	0.2585	0.2613	0.2641
19.0	.2480	.2508	.2536	.2564	.2592	.2620	.2647	.2675	.2703	.2731
19.5	.2573	.2601	.2629	.2657	.2685	.2713	.2740	.2768	.2796	.2824
20.0	.2669	.2697	.2725	.2753	.2780	.2808	.2836	.2864	.2892	.2919
20.5	.2768	.2795	.2823	.2851	.2879	.2906	.2934	.2962	.2990	.3017
21.0	.2869	.2897	.2924	.2952	.2990	.3007	.3035	.3063	.3091	.3118
21.5	.2973	.3000	.3028	.3056	.3083	.3111	.3139	.3166	.3194	.3222
22.0	.3079	.3107	.3134	.3162	.3190	.3217	.3245	.3272	.3300	.3328
22.5	.3188	.3216	.3243	.3271	.3298	.3326	.3353	.3381	.3409	.3436
23.0	.3299	.3327	.3354	.3382	.3410	.3437	.3465	.3492	.3520	.3547
23.5	.3413	.3441	.3468	.3496	.3523	.3551	.3578	.3606	.3633	.3661
24.0	.3530	.3557	.3585	.3612	.3640	.3667	.3695	.3722	.3750	.3777
24.5	.3649	.3676	.3704	.3731	.3758	.3786	.3813	.3841	.3868	.3896
25.0	.3770	.3797	.3825	.3852	.3880	.3907	.3934	.3962	.3989	.4017
25.5	.3894	.3921	.3949	.3976	.4003	.4031	.4058	.4085	.4113	.4140
26.0	.4020	.4047	.4075	.4102	.4129	.4157	.4184	.4211	.4239	.4266
26.5	.4149	.4176	.4203	.4231	.4258	.4285	.4312	.4340	.4367	.4394
27.0	.4280	.4307	.4334	.4361	.4389	.4416	.4443	.4470	.4498	.4525
27.5	.4413	.4440	.4467	.4495	.4522	.4549	.4576	.4603	.4631	.4658
28.0	.4549	.4576	.4603	.4630	.4657	.4685	.4712	.4739	.4766	.4793

It is interesting to note that for both Table 2-13 and 2-14, there are corrections for 20°C and 760 mm of atmospheric pressure. Because the glassware is calibrated for this temperature and pressure, one may wonder why a correction is necessary. The reason goes back to Archimedes principle which cannot be accounted for when the glass is calibrated. The only way to avoid use of these tables at STP is to do all weighing in a vacuum to avoid the effects of air's weight.

Corrections for single-pan balances include the preceding parameters, plus a fourth one: the *apparent mass* of the built-in weights. The *apparent mass* of weights can vary somewhat from their true mass because the weights in a single-pan balance were calibrated at a specific temperature and barometric pressure. However, in a lab of different temperature and barometric pressure, you are weighing against their apparent mass.* Thus, when making accurate weighings with a single-pan balance, you must keep track of the temperature and barometric pressure of the room in which you are working.

* The figures in the tables are based on 100-ml sample sizes. The sample calculation is made with a 25-ml flask, therefore the calculations need to be multiplied by 0.25.

Calibration correction tables for single-pan balances are provided in Table 2-14, which is used in the same manner as the previous example. Like Table 2-13, this table is used when a volumetric flask is Type I, Class A (borosilicate glass). It may also be used for Type I, Class B (aluminosilicate glass) by adding 0.0006 for every degree below 20°C. Likewise, subtract 0.0006 for every degree above 20°C.

2.3.10 Volumetric Flasks

Volumetric flasks are used to measure very precise volumes of liquids for making standard solutions or weighing for density calibrations. Each flask has one measurement line** for the specific volume of the flask (see Fig. 2-16). Volumetric flasks are of either Class A or Class B quality; the tolerances for Class B are twice those for Class A. There are no general purpose volumetric flasks. The tolerances for the standard sizes of volumetric flasks are listed in Table 2-15. Volume calibrations were made at 20°C.

All volumetric flasks can be capped, thus preventing material in the flask from evaporating as well as maintaining the material's purity. There are three ways to cap the top of a volumetric flask. Each type of cap can only be used with the proper corresponding volumetric flask. For example, a volumetric flask designed for a screw cap cannot use a snap-on, or standard taper stopper. Each following type of cap is available for Class A and Class B flasks:

1) Plastic caps that snap on
2) Plastic caps that screw on
3) Standard Taper stoppers made out of plastic or glass.

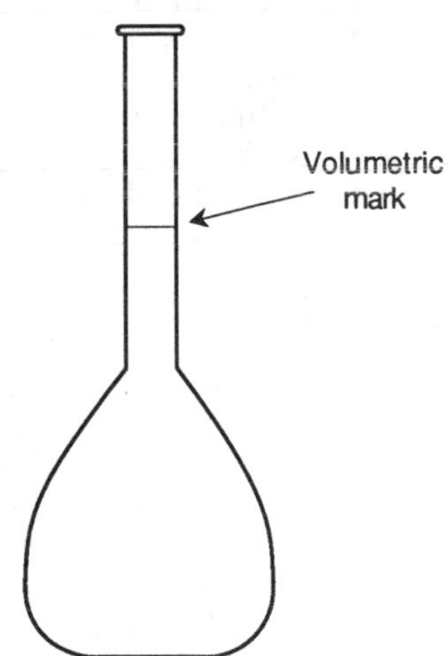

Fig. 2-17 The volumetric flask.

The advantage of snap-on caps is that they are inexpensively replaced. They can also provide a small amount of protection against leakage if a flask is knocked over or is being agitated to mix the contents. The major disadvantage of a snap-on cap is that you need both hands to remove the cap. Screw-on plastic caps provide the best protection against spillage, but are more expensive. They also require both hands for removal.

Standard taper stoppers will prevent spillage only so long as the stopper is firmly in place, but they can be removed with only one hand. Plastic standard taper stoppers do not require grease, but may be chemically attacked

* Old single pan balances used brass weights with a specified density (at 20°C) of 8.3909 g/cm^3. Now, stainless steel weights are used with a specified density of 8.0 g/cm^3. Fortunately, the differences of this change are too small for most purposes to be concerned about requiring any further calibration between the two.

** There are specialized volumetric flasks with calibration lines up the neck.

by materials within the flask. Glass plugs placed over alkaline solutions without grease are likely to be frozen in place. On the other hand, the contents of the flask may dissolve any grease used.

Table 2-15‡

Capacity Tolerances of Volumetric Flasks					
Capacity mL	Tolerances, ± mL		Capacity mL	Tolerances, ± mL	
	Class A	Class B		Class A	Class B
5	0.02	0.04	200	0.10	0.20
10	0.02	0.04	250	0.12	0.24
25	0.03	0.06	500	0.20	0.40
50	0.05	0.10	1000	0.30	0.60
100	0.08	0.16	2000	0.50	1.00

‡ From ASTM document E 288, Table 1. Omitted were columns entitled "*Inside Diameter of Neck at Capacity Line (mm)*," all information on the "*Position of Capacity Line*," and "*Stopper Size*." With permission

The proper procedure for filling a volumetric flask is as follows:

1) Carefully place the liquid to be measured into the flask using a pipette, burette, or any other device that can supply a steady stream against a wall, about a centimeter above the calibration line. Be careful not to splash. The tip may touch the wall.
2) When the level is just below the calibration line, cap the container and let it sit for a few minutes to let the liquid drain from the walls of the container.
3) Complete the filling process using Steps 1 and 2 again. If you overshoot the calibration line, a pasture pipette can be used to draw out excess fluid. The surface tension in the tip of the pipette should be sufficient to draw out the fluid. Once completed, cap the container.

The proper procedure for using a volumetric flask for gravimetric purposes is as follows:

1) Clean and dry (if necessary).
2) Weigh the flask, its cap, and any other materials that may be included in the final weighing.
3) If accurate measurements are required, note the liquid's temperature and the room's barometric pressure and humidity.
4) Carefully place the liquid to be measured into the flask using a pipette, burette, or any other device that can supply a steady stream against a wall, about a centimeter above the calibration line. Be careful not to splash. The tip may touch the wall.
5) When the level is just below the calibration line, cap the container and let it sit for a few minutes to let the liquid drain from the walls of the container.
6) Complete the filling process using Steps 4 and 5 again. If you overshoot the calibration line, a pasture pipette can be used to draw out excess fluid. The surface tension in the tip of the pipette should be sufficient to draw out the fluid. Once completed, cap the container.

7) Weigh the filled flask.
8) If Step 3 was performed, repeat it to verify the readings.
9) Subtract the weighed flask by the empty flask weight.
10) If Steps 3 and 8 were performed, use the appropriate table and make any necessary corrections.

The procedure for using a volumetric flask for making a solution of a definite strength is:

1) Clean and dry (if necessary).
2) Carefully place the pre-calculated concentrated liquid into the flask using a pipette, burette, or any other device that can supply a steady stream against a wall about a centimeter above the calibration line. Be careful not to splash. The tip may touch the wall.
3) Follow this by adding to the flask distilled water up to the calibration line using a pipette, burette, or any other device that can supply a steady stream against a wall about a centimeter above the calibration line trying to rinse any of the concentrated liquid remaining on the flask's neck. Be careful not to splash. The tip may touch the wall. When the level is just below the calibration line, cap the container and stop for a few minutes to let the liquid drain from the walls of the container.
4) Complete the filling process using Step 3 again. Try not to overshoot because any excess liquid will dilute the intended concentration.

For emptying a volumetric flask, the following procedure should be followed:

1) Slowly incline the flask to provide a steady stream of liquid from the spout. Be careful not to splash.
2) Continue inclining the flask until it is vertical and hold for about half a minute.
3) Touch the drop at the tip of the neck to the wall of the receiving container.
4) Remove the flask horizontally from the wall of the receiving container with no vertical motion.

2.3.11 Graduated Cylinders

Graduated cylinders are generally not used for high-quality volumetric work. Although they are available in Class A, Class B, and Student Grade, the acceptable tolerance of graduated cylinders is considerably greater than volumetric flasks. Following standard practice, Class B tolerances are twice those of Class A. For specific capacities, graduations, and tolerances, see Table 2-16. Earlier federal specifications required more accurate tolerances on *to contain* than *to deliver* containers, but this case no longer applies and now both have the same tolerances.

* To receive standard taper stoppers.

There are three styles of graduated cylinders: *Style 1* has a beaded lip and a pour spout, *Style 2* has a ground standard taper joint on the top,* and *Style 3* is just like Style 1 but has a heavy glass reinforcing bead just below the top for strength. Graduated cylinders of *Style 1* are manufactured in both *to contain* and *to deliver* designs. Graduated cylinders of *Styles 2* and *3* are manufactured in *to contain* designs only. Often, graduated cylinders come with plastic or foam rubber cylinder guards to protect their tops if they are accidentally tipped. Use of these guards is highly encouraged.

Because liquids cannot wet plastic walls, plastic cylinders can provide constant volume for both *to deliver* and *to contain*. However, they are only accurate to Class B tolerances and susceptible to chemical attack from organic solvents.

Table 2-16‡

Dimensions and Tolerances of Graduated Cylinders

Capacity, mL	Main Graduations	Graduations Intermediate	Least[c] Graduations	Standard Taper Stopper Number[d]	Tolerances to Contain or to deliver ±mL	
					Class A	Class B
5	1.0	0.5	0.1	9	0.05	0.10
10	1.0	0.5	0.1	9	0.10	0.20
10	1.0		0.2	13	0.10	0.20
25	2.0	1.0	0.2	13	0.17	0.34
25	5.0	1.0	0.5	13	0.17	0.34
50[b]	5.0 or 10.0	5.0	1.0	16	0.25	0.50
100	10.0	5.0	1.0	16 or 22	0.50	1.00
250	20.0	10.0	2.0	22 or 27	1.00	2.00
500	50.0	25.0	5.0	27 or 32	2.00	4.00
1000	100.0	50.0	10.0	32	3.00	6.00
2000	200.0	100.0	20.0	38	6.00	12.00
4000	500.0	250.0	50.0	–	14.50	29.00

[a] Main graduations may be each 5 or 10 mL. If main graduations are 5 mL each 5 mL, intermediate graduations do not apply.
[b] Lines below the first numbered line may be omitted.
[c] Applies to Style 2 only.
‡ From the ASTM document E 1272-89, Table 1. Omitted were columns titled "*Distance From Scale to Top, mm (Min and Max)*," "*Minimum Wall Thickness (mm)*," and "*Maximum Inner Diameter (mm)*." With permission.

Regardless of the quality, or construction design, graduated cylinders have no calibration lines at the section closest to the base because when the body is fused onto the base, the overall shape of the tube is distorted and accurate calibration on a production basis is not possible. This feature is equally applicable to graduated cylinders that use a plastic press-on foot.

Pharmaceutical graduated cylinders, which can be identified by their non-vertical walls,* have very poor accuracy. Their accuracy varies at any given inside wall diameter, even on the same flask. For instance, at regions where the inside diameter is between 21–25 mm, the accuracy is ±0.4-mL. However, at regions where the inside diameter is between 46–50 mm, the accuracy is ±1.8 mL.

*Those smaller than 15 mL have vertical walls.

To fill a *to contain* cylinder (Style 1, 2, or 3), the following procedure should be followed:

1) Clean and dry (if necessary).
2) Carefully place the liquid to be measured into the flask using a pipette, burette, or any other device that can supply a steady stream against a wall about a centimeter above the calibration line, being careful not to splash. The tip may touch the wall.
3) When the level is just below the calibration line, stop for a few minutes to let the liquid drain from the walls of the container.
4) Finish the fill using Steps 2 and 3 again. If you overshoot the calibration line, a pasture pipette can be used to draw out excess fluid. The surface tension in the tip of the pipette should be sufficient to draw out the fluid. If you are using a Style 2 cylinder, be sure not to get the ground joint wet with fluid. If the cylinder is a Style 2, cap the container to prevent evaporation or contamination.

For filling a *to deliver* cylinder (Style 1 only), the following procedure should be followed:

1) Clean (if necessary) for the first filling. If there is a water film, you may wish to rinse the cylinder with the liquid about to be measured to stabilize the molarity. Subsequent fillings do not require the rinse nor will they need a dry container.
2) Carefully place the liquid to be measured into the flask using a pipette, burette, or any other device that can supply a steady stream against a wall about a centimeter above the calibration line, being careful not to splash. The tip may touch the wall.
3) When the level is just below the calibration line, stop for a few minutes to let the liquid drain from the walls of the container.
4) Finish the fill using Steps 2 and 3 again. If you overshoot the calibration line, a pasture pipette can be used to draw out excess fluid. The surface tension in the tip of the pipette should be sufficient to draw out the fluid.

For emptying a *to deliver* cylinder (Style 1 only), the following procedure should be followed:

1) Slowly incline the cylinder to provide a steady stream of liquid from the spout. Be careful not to splash.
2) Continue inclining the cylinder until it is vertical and hold for about half a minute
3) Touch the drop at the tip of the spout to the wall of the receiving container.
4) Remove the cylinder horizontally from the wall of the receiving container with no vertical motion.

2.3.12 Pipettes

There are three types of volumetric pipettes: *Volumetric Transfer*, *Measuring*, and *Serological*. The differences are based on whether the volume within the pipette is subdivided and if the volume in the tip is included in the calibration (see Fig. 2-18).

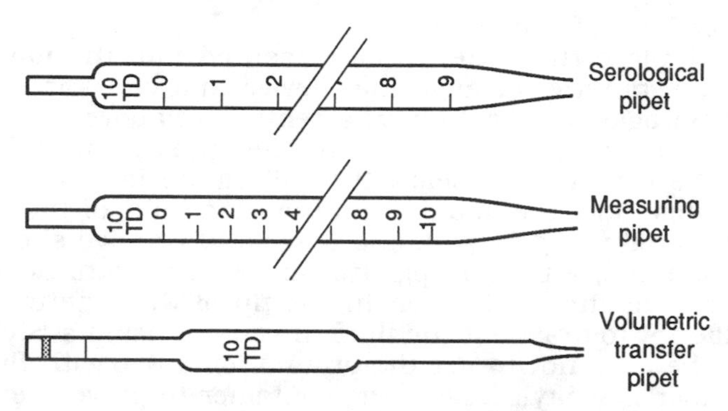

Fig. 2-18 The three types of pipettes.

The serological pipette is used to dispense varying volumes of liquid similar to the burette. Like a burette, the serological pipette is calibrated volume does not include the tip region. Volumetric pipettes are designed to dispense one volume of liquid, whereas the measuring pipette is calibrated to dispense varying volumes. Both of these pipettes are designed to include the tip region in their entire dispensed volumes.

There are two different methods for completely draining a volumetric transfer or measuring pipette, and each method requires a different pipette design. One method is to leave the tip against the side of the receiving vessel and let the material drain into the vessel. This method leaves a small amount of solution remaining at the tip. The difference between this remaining portion and what has been transferred has been accounted for in the pipette's calibration. The other method of pipette draining requires the user to *blow out* the remaining liquid in the pipette; after all the liquid has drained out naturally, an extra burst of air is applied to the end of the pipette to expel any remaining fluid out of the tip. All blow-out pipettes are identified by either an opaque band one-quarter of an inch wide, or two bands one-quarter inch apart at the end of the pipette.

There are pros and cons to both types of pipettes. The decision to use one or the other depends on the reliability, and repeatability, of your laboratory technique as well as the nature of the liquid you are pipetting. All non-blow-out pipettes are calibrated with water. Thus if your liquid has different surface tension or viscosity characteristics than water, your measurements will not be accurate. On the other hand, not everyone will exert an equal amount of blow-out force. Thus different people may deliver different volumes using the same equipment.

Pipettes often have color coding bands on their ends to help identify them. Although the colors are designated by the ASTM, they do not necessarily specify a specific volume. Their volumes do not necessarily correspond to any other pipette design. In fact, a given color may identify two volumes for any given type of pipette: however, pipettes with the same color band are always distinctively different in volume.

Traditionally, pipettes were filled by sucking fluid into the tube with the mouth. This procedure, however, carries the risk of getting chemicals in the mouth. To protect the user, some pipette ends are designed to receive a cotton plug (see Fig. 2-19). The cotton plug allows gases past, but inhibits the flow of liquids. Thus, if a user sucked too hard, the liquid would clog the cotton plug and prevent the liquid from reaching the user's mouth. Although people still suck liquids *into* pipettes, the use of *pipette fillers*, or even rubber bulbs, are strongly recommended.

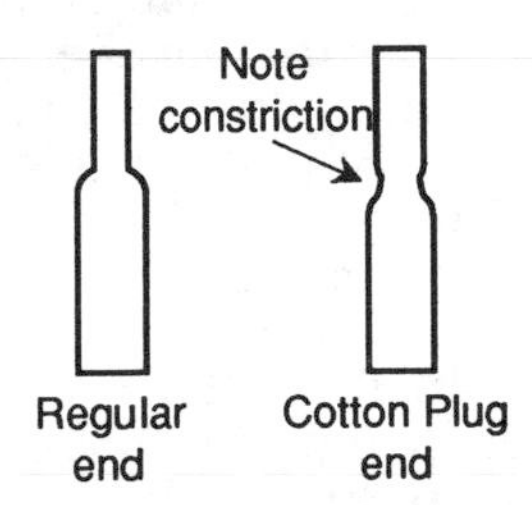

Fig. 2-19 Regular and cotton plug ends for pipettes.

Volumetric Transfer. These pipettes are used to deliver a (single) specified volume. They are characterized by a bulb mid-span on the tube (see Fig. 2-18). The bulb is used to achieve greater capacity in the pipette and maintain the general length throughout the different volumes.

Volumetric pipettes are used solely *to deliver* in both drain and blow-out models. They are calibrated in both Class A and Class B tolerances based on guidelines established in the ASTM Designation E 969 - 83. Tolerances are provided in Table 2-17.

The color-coding band(s) on the end of the volumetric pipette are used for quick identification purposes only. The repetition of the colors is sufficiently separated by the size of the pipette to not confuse (for instance) the 10 mL and 50 mL sizes.

To fill a volumetric measuring pipette, draw the solution to just above the volumetric level, then let the solution fall to the calibration mark. Remove the pipette from where you drew the fluid, and wipe the tip of the pipette with a laboratory tissue to remove any solution on the outside of the tip so it is not included with the dispensed liquid from the inside of the pipette.

When draining a volumetric pipette, let the tip touch the side of the receiving container and let the fluid flow. After emptying the pipette, count to two (to allow for any remaining fluid to flow to the bottom), and remove the tip sideways away from the receiving wall. Do not remove the tip with an upward or downward motion.

If the end of the pipette indicates the pipette is of blow-out design, gently (but firmly) provide (by lips or by the pipette filler) extra air pressure to 'blow out' the last drop of liquid. Do not maintain a long continuous blow, especially if you are using the mouth to blow out the excess because you may contaminate your solution.

Measuring and Serological Pipettes. These pipettes have graduations along their sides. The difference is that the graduations on *measuring* pipettes stop before reaching the taper of the tip while the stated volumes of *serological* pipettes include the contents of the tip (see Fig. 2-18). Measuring pipettes should never be drained or blown out when delivering solution because the extra volume in the tip is not part of the pipette's calculated volume. Some serological pipettes are blow-outs, and others are not. You need to examine the end for the one-quarter inch opaque mark.

Measuring pipettes come in both Class A and Class B and are available in two styles: *Style 1* is a standard taper tip, and *Style 2* is a long taper tip (Class B only). Serological pipettes are only made to Class B tolerances and have no special styles. Volumes, tolerances, and other data of measuring pipettes are provided in Table 2-18. Information for serological pipettes is provided in Table 2-19. All calibrations were made at 20°C.

Class A pipettes have slower outflow times than Class B pipettes. This slowness provides the user with more reaction time to control liquid flow and thereby achieve better accuracy. It also allows the fluid draining from the walls to keep up with the fluid being dispensed from the tip. Otherwise you may dispense what you thought was 5 mL of liquid from a pipette, but after the liquid in the pipette settled, you discover only 4.9 mL of solution had been dispensed.

To fill a measuring or serological pipette, draw the solution to just above the volumetric level, then let the solution fall to the calibration mark. Remove the pipette from where you drew the fluid, and wipe the tip with a laboratory tissue to remove any excess solution from the outside of the pipette so that it is not included with the calibrated liquid.

When dispensing fluid from a measuring pipette, let the tip touch the side of the receiving container and let the fluid flow. If you are dispensing the liquid by hand, you need to control the flow rate by placing your thumb on the end of the pipette. However, never let your thumb wander away from the end of the pipette because you will need it to stop the fluid flow. If you completely drain the pipette, count to two (to allow for any remaining fluid to flow to the bottom), and remove the tip sideways away from the receiving wall. Do not remove the tip with an upward or downward motion.

Table 2-17‡

Requirements for Volumetric Transfer Pipettes

Nominal Capacity, mL	Class A Capacity Tolerance, mL	Class A Minimum Outflow Time, sec[b]	Class B Capacity Tolerance, mL	Class B Minimum Outflow Time, sec[a]	Color-Coding Band
0.5	±0.006	5	±0.012	3	black (2)
1	±0.006	10	±0.012	3	blue
2	±0.006	10	±0.012	3	orange
3	±0.01	10	±0.02	5	black
4	±0.01	10	±0.02	5	red (2)
5	±0.01	10	±0.02	8	white
10	±0.02	15	±0.04	8	red
15	±0.03	25	±0.06	10	green
20	±0.03	25	±0.06	10	yellow
25	±0.03	25	±0.06	15	blue
50	±0.05	25	±0.10	15	red
100	±0.08	30	±0.16	20	yellow

a Maximum outflow time for A and B shall be 60 seconds.

‡ From the ASTM document E 969-83, Table 1. Omitted were columns titled "*Length of Delivery Tube, mm (Min and Max),*" "*Inside Diameter at Capacity Mark, mm (Min and Max),*" and "*Maximum Distance Between Bulb and Graduation Mark, mm.*" With permission.

Table 2-18‡

Requirements for Measuring Pipettes

Capacity mL	Capacity accuracy ± tolerance, mL		Graduations, mL		Outflow time, seconds				Color of coding bands
			Least graduation value	Main graduation (numbered)	Class A		Class B		
	Class A	Class B			Min.	Max.	Min.	Max.	
0.1	–	0.005	0.01	0.01	–	–	0.5	10	White
0.1[A]	–	0.005	0.001	0.1	–	–	0.5	10	2-Green
0.2	–	0.008	0.01	0.02	–	–	0.5	8	Black
0.2[A]	–	0.008	0.001	0.02	–	–	0.5	8	2-Blue
0.5[A]	–	0.01	0.01	0.10	–	–	2	6	2-Yellow
1.0	0.01	0.02	0.10	0.10	20	60	3	22	Red
1.0	0.01	0.02	0.01	0.10	20	60	3	22	Yellow
2.0	0.01	0.02	0.10	0.20	20	60	3	22	Green
2.0[A]	–	0.02	0.01	0.20	–	–	3	22	2-White
5.0	0.02	0.04	0.10	1.0	30	60	8	20	Blue
10.0	0.03	0.06	0.10	1.0	30	60	8	30	Orange
25.0[B]	0.05	0.10	0.10	1.0	40	70	15	35	White
50.0[B]	–	0.16	0.20	2.0	–	–	20	40	Black

A Style 2 only.
B Style 1 only.
‡ From the ASTM document E 1293-89, Table 1. Omitted was the column titled "*Outside Diameter of Grad. Portion min. mm.*" With permission.

Table 2-19‡

Requirements for Serological Pipettes

Capacity mL	Capacity Tolerance ± mL	Graduations, mL			Style I & II Outflow time		Style III Nominal Tip Opening mm	Color Coding Band
		Least Value	Main Numbered	Interval Graduated 0.0 to at least	min.	max.		
0.1	0.005	0.01	0.01	0.09	0.5	3		White
0.1[a]	0.005	0.01	0.01	0.09	0.5	3		2 Green
0.2	0.008	0.001	0.02	0.19	0.5	3		Black
0.2[a]	0.008	0.001	0.01	0.19	0.5	3		2 Blue
0.25[b]	0.008	0.0125	0.05	0.25	0.5	3		. . .
0.5	0.01	0.01	0.05	0.45	0.5	3		2 Black
0.5	0.01	0.05	0.05	0.45	0.5	3		2 Yellow
0.60[a]	0.01	0.15	0.15	0.45	0.5	3		. . .
1.0	0.02	0.01	0.1	0.95	1	5		Yellow
1.0	0.02	0.1	0.1	0.9	1	5	2.0	Red
2.0	0.02	0.01	0.02	1.9	1	5	2.5	2 White
2.0	0.02	0.1	0.2	1.9	1	5	2.5	Green
5.0	0.04	0.1	1.0	4.5	3	10	3.0	Blue
10.0	0.06	0.1	1.0	9.5	5	15	3.0	Orange
25.0	0.10	0.1	1.0	23.0	5	15	3.0	White

a Kahn serological pipettes, calibrated to tip.
b Kahn serological pipe, calibrated to base.
‡ From the ASTM document E 1044-85, Table 1. Omitted was the column titled "*Outside Diameter of Graduated Portion, min., mm.*" With permission.

There are three different styles of serological pipettes. Style I has a standard end piece. Style II has an end that can receive a cotton plug (see Fig. 2-19). Style III has the same type of end as Style II, but also has a larger tip opening to speed the emptying process.

Table 2-20‡

Tolerances of Disposable (Serological) Pipettes[a]

Capacity cm^3	Glass	Plastic
0.1	±7%	—
0.2	±6%	—
0.5	±3%	+3%
1.0	±3%	+3%
2.0	±3%	+3%
5.0	±3%	+3%
10.0	±3%	+3%
25.0	—	+3%

[a] Applies to the stated capacity, not individual calibrations.

‡ From ASTM Tables E 714-86 *(Standard Specification for Disposable Glass Serological Pipettes)*, Table 2 (Omitted was the column titled "*Coefficient of Variation*"), and E 934-88 *(Standard Specification for Serological Pipette, Disposable Plastic)*, Table 2 (Omitted was the column titled "*Coefficient of Variation (Equal to or less than 1.5%)*"). With permission.

Pipettes are also made as *disposable serological pipettes*. They are just as easy to use as regular pipettes and are typically made out of plastics or soda-lime glass. They have very low-quality standards, and their calibrations are not required to have the same permanency as regular pipettes. Thus, if you try to wash them, the calibrations may wash off with the dirt. The tolerances of disposable pipettes are shown in Table 2-20.

The biggest problem with disposable pipettes (or any disposable labware) is that they are wasteful and not environmentally sound. The concept of a disposable lab is not practical, in the long run, for the individual, or the earth.

An alternate approach to measurement is to use a syringe rather than a pipette. Although tolerances of syringes are not very good, syringes are particularly useful when adding materials to labware through a stopcock (see Fig. 2-20), and essential when delivering material past a septum.

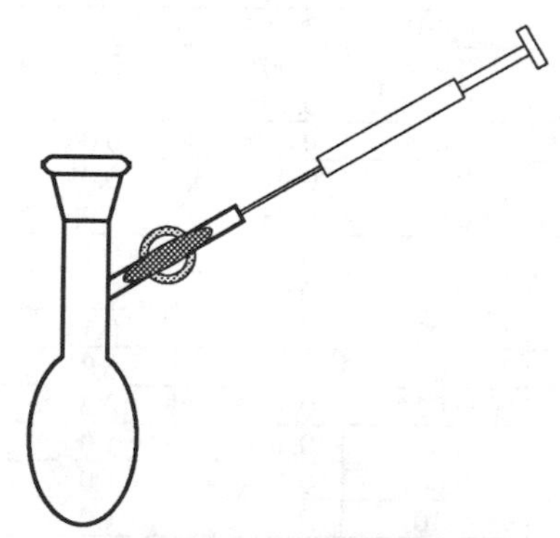

Fig. 2-20 If the plug is opened, a syringe may be inserted into the stopcock to dispense liquid.

Syringes are made of plastic or glass. Like other plastic ware, you must be on-guard that solvents do not dissolve the item. Glass, on the other hand, is more likely to break, have the piston stick, or have the fluid leak past the piston plunger. The latter problem can be a greater danger than just a nuisance if the material is toxic and/or is likely to dissolve any protective gloves you should be wearing.

2.3.13 Burettes

Burettes are a specialized form of measuring pipette. To control the rate and amount of material flow from the tip, a pinch clamp, stopcock, or valve is attached to the bottom of a burette, thus providing control on the liquid outflow and allowing accurate dispensing. Burettes are made with Class A, Class B, and Student Grade tolerances and are calibrated at 20°C. Class A are marked as such, and other grades are not required to identify their volumetric tolerances. The tolerances and general characteristics of burettes are listed in Table 2-21.

Like pipettes, Class A burettes have slower outflows than Class B burettes. They allow the user to better control the dispensing fluid. The slower speed also allows the liquid in the burette ample time to flow from the walls at the rate of the dispensing liquid. Thus the user does not need to wait before making a reading.

A burette should be mounted securely in a vertical position with a burette clamp or several three-fingered clamps. A single three-fingered clamp is likely to wobble and swing off vertically. If the accuracy of your work is critical, a plumb is required to achieve a true vertical orientation.

A burette can be filled from a side tube or three-way stopcock on the bottom. If there is no bottom-filling capability, the liquid can be poured into the top of the burette with a funnel. If any liquid spills on the outside of a burette, wipe the burette dry with a laboratory tissue.

Unless a burette is automatic, overfill the burette about 10 mm past the zero line. Let the liquid settle a minute, then release some of the liquid into a beaker or some other receptacle by slightly opening the stopcock. Let the fluid lower to the zero line. Wait another minute to allow the fluid to settle to the new level, and re-check the level of the meniscus at the zero line. Release or add more liquid as necessary.

To make a burette reading, first read and record the liquid volume in the burette, dispense the required amount, and reread the liquid volume in the burette. Then subtract the first reading from the second reading to calculate the amount of fluid delivered. Self-zeroing burettes do not require a first and second reading, because all readings start from 0.00 ml.

Table 2-21‡

Requirements for Burettes				
Capacity mL	**Subdivision, mL Class A & B**	**Class A & B Number each mL**	**Volumetric Tolerances ± mL Class A**	**Class B**
10	0.05	0.5	0.02	0.04
25	0.1	1	0.03	0.06
50	0.1	1	0.05	0.10
100	0.2	2	0.10	0.20

‡ From the ASTM document E 287, Table 1. Omitted were columns titled "*Class A Scale Length, mm, 0 to Capacity*," "*Class B Scale Length, mm, 0 to Capacity*," and "*Distance, Top of Burette to "0" Graduation, mm.*" With permission.

2.3.14 Types of Burettes

There are four types of burette designs. Not every type of burette design is made in all tolerances, and some burettes have limited use.

Some burettes have a tube for easier filling attached between the calibration lines and above the tip. Flexible tubing is attached to the tube used for filling the burette. There are two methods to stop liquid flow. One technique uses a pinch clamp on the flexible tubing, and the other has a stopcock on the side tube. The pinch clamp can lead to inherent errors as the pressure from the weight of the liquid in the burette causes a small expansion of the flexible tube. As the burette is emptied, this pressure decreases, and the amount of error decreases as well. If your work requires limited accuracy, these changes are well within tolerance. Alternatively, a stopcock can be used instead of a pinch clamp. There are no hydrostatic complications with the stopcock.

The least accurate burettes are *Mohr* burettes (see Fig. 2-21). Mohr burettes do not have stopcocks at their tips and therefore require flexible tubes with pinch clamps to control dispensing liquid.

The standard lab burette is the *Geissler*, which can be identified by the stopcock at the bottom. It can be refilled by pouring liquid in at the top or by a filling tube from the side. Some Geissler burettes provide three-way stopcocks* for easier filling (see Fig. 2-21). If the stopcock is turned one way, the burette fills, if it is turned 180°, the burette empties.

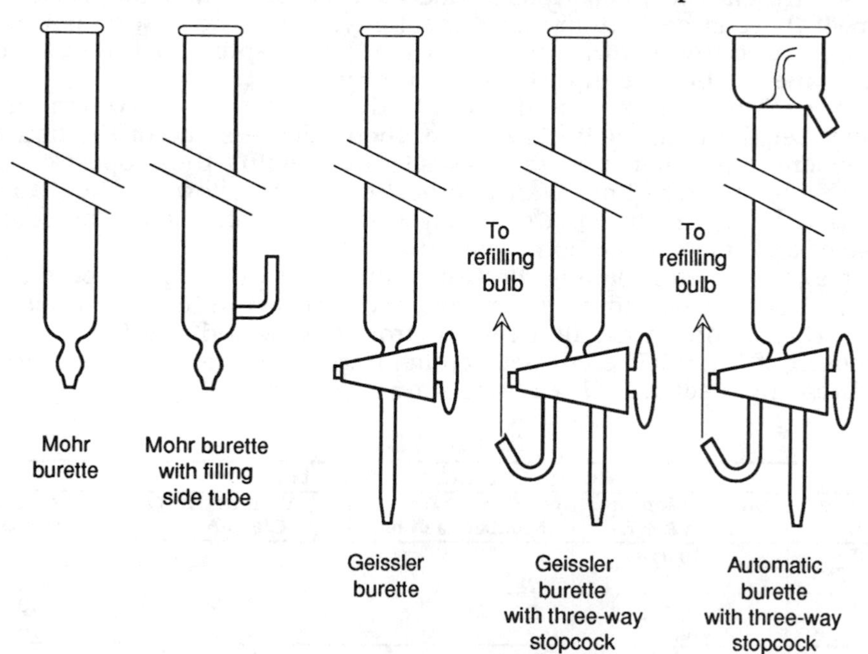

Fig. 2-21 Different types of burettes.

* The three-way stopcock receives its name because it has three arms leading to the barrel. A two-way stopcock has two arms leading to the barrel.

A third type of burette is the *automatic*, or *self-zeroing*, burette. It can repeatedly be filled quickly and easily to exactly 0.00 ml, meaning that to make a measurement, no arithmetic is required to determine the amount dispersed. This method of dispersing liquid is not only a valuable time-saver, it can help avoid errors due to poor arithmetic. The self-zeroing burette is filled by intentionally over-filling the top which is enclosed and has a drainage for collecting the over-flow (see Fig. 2-21). The top portion of a self-zeroing burette is not calibrated, so if you dispense less liquid than is contained in this region, there is no way to determine the amount removed.

The last type of common burette is the *dispensing* burette. It is easy to recognize by its size. It can carry up to one liter of liquid and is capable of fast, effective liquid dispensing. Its accuracy is about ±0.5% of total volume.

2.3.15 Care and Use of Burettes

Burettes are very seldom tip- or end-heat strengthened as are pipettes, and are therefore more prone to chipping or cracking. Burettes with removable tips and/or stopcocks can be useful for salvaging burettes that otherwise would be thrown away. Because the burette's calibration is exclusively on the column, removable tips and stopcocks have no effect on the buret's tolerance quality, nor should they imply the level of quality.

In addition to tip care, the care of stopcocks on a burette is equally important. They should not jam, leak, or provide inconsistent release of fluids.

Glass stopcocks on burettes are prone to more wear than stopcocks on other laboratory apparatus due to the conditions under which burettes are used. Therefore, it is important to periodically remove the glass plug of a burette's stopcock, clean the plug *and* the barrel, and re-grease it. The removal of the old grease is critical because it may carry particulate matter that can scratch a stopcock plug.

It is common for an old burette (that was working well) to leak if the original plug is lost or broken and is replaced with a new plug. This leakage is because the old plug and barrel wore together and evenly. The new plug is not worn with depressions that match the depressions on the old plug. Regrettably, the best recourse may be to replace the entire stopcock or to discard the entire burette. Re-grinding can be done by a trained person, but unfortunately, you once again will have a matched plug and barrel.

It is not always possible to successfully take the plug out of one glass stopcock and place it in the barrel of another burette made by the same company. However, it is just about impossible to do such a switch if the brands are different. Although either plug will fit in the barrel, it is unlikely that the holes will line up and fluid will be unable to pass by the plug. Anytime you place a plug into the stopcock barrel of a burette, sight down the burette tube and see if you can see the hole of the plug. If the hole is offset, or not in sight, try a different plug until a good match is found.

If you suspect that a stopcock is leaking, remove all grease (if present) from the stopcock, and replace the plug (which should be wet with water) ungreased into the barrel in the closed position. Push the plug in firmly, but do not rotate the plug (if the plug is glass, rotation without grease may jam the plug in the barrel). Next, fill the burette with water and see the rate of water loss over a period of about a half-hour. If the water drops more than 3 mm in length, the stopcock can be considered poor. If there is no significant loss, pull the stopcock out, rotate 180°, replace, and re-test. Any loss of water less than 3 mm will easily be assisted by the application of grease. However, any attempts to repair a leaky stopcock by extra grease will be doomed to failure. Adding excess grease to repair imperfections in a stopcock is like adding extra lacquer to fill up the rough surface of unsanded wood: it just does not work. Regardless, the excess grease is likely to fill the holes in the plug resolving the leaking problem in an unacceptable manner.

How often you need to clean and re-grease a stopcock depends on the quality and type of grease you are using, how often the burette is being used, and the nature of the fluids within the burette. It is safe to say that an inexpensive grease will require cleaning and replacement more often than a grease of higher quality.

Do not use silicon-based stopcock grease on burettes unless the nature of your chemicals absolutely requires it. The reason is that silicon-based greases require constant cleaning and replacement to maintain their slippery nature. Also, the most effective way to clean silicon grease is with a base bath. However, the base bath is considered one of the two worst cleaning methods for use on volumetric ware. Silicon grease may be inexpensive in the short run, but it can prove to have many hidden costs. If the chemicals in your work require you to use silicon grease, it would be better to use a Teflon stopcock or rotary valve as a substitute. They are a much better choice as they do not require any grease and require very little maintenance.

The plugs from Teflon stopcocks and rotary valves must be protected from scratches. So, when a stopcock is disassembled for cleaning, lay the plug down where it is not likely to be scratched or pick up particulate matter. Wipe the plug with some methanol (or acetone) on a Kimwipe before reinserting it into the stopcock or valve barrel. Be sure to follow the correct sequence of end pieces when reassembling the stopcock: The sequence is (white) washer, (black) O-ring lock washer, and (colored) plastic nut. If the washer and lock washer are reversed on a Teflon stopcock, the plug will tighten as you rotate it clockwise (CW), and loosen (and probably leak) as you rotate it counterclockwise (CCW).

When storing burettes with stopcocks, leave the locking nut on a Teflon stopcock loose. Then, if the stopcock gets warm, the swelling Teflon plug is less likely to swell and stick, or split the walls of the barrel.

Do not store solutions in a burette, and never store alkaline solutions in a burette. Alkaline solutions will react with the glass and cause a glass stopcock to freeze. Also, an alkaline solution will react with the glass creating a rough surface which can scratch a Teflon plug.

The procedure for cleaning burettes, is similar to general glass cleaning already discussed. Because burettes are only used *to deliver*, it is never necessary to dry a burette before use. However, a wet burette can change the concentration of the solution being placed within it. To avoid changing concentrations, shake the burette dry and pour a small amount of the solution you will be working with into the burette. Swirl and/or rotate it around the burette, and pour it out. If you have a very low molarity solution, you may want to repeat this process again. The pre-rinse (with the solution to be used) will reduce, and should eliminate, any change in concentration caused by standing water in the burette.

When first filling a burette, any air bubbles in the tip region (below the stopcock) should be removed. If a bubble remained where it originally was, there would be no problem. However, if a bubble comes out while making a measurement, it takes the place of fluid that was recorded, but never left the burette. The standard practice for removing bubbles is to overfill the burette, and open the stopcock fully to try and force the bubble out. This technique often works, but it can also be wasteful of material.

An alternate method of tip bubble removal was reported by Austin.[8] First, pour about 1 mL of fluid into the burette, open the stopcock to let the fluid into the tip region and close the stopcock. Now observe the tip and see if a bubble is trapped within the liquid of the tip. If there is no bubble, fill the burette to the top and proceed with your work. If there is a bubble, turn the burette so the stopcock is on top and quickly rotate the stopcock plug 180°. This procedure will lower the liquid in the tip of the burette, but if done quickly enough, it should not empty the tip. It should however, remove the bubble from the tip region. The advantage of this approach is that it uses less solution.

2.4 Weight and Mass

2.4.1 Tools for Weighing

We weigh things by comparing the unknown weight (or force) of an object with a known weight (or force). The device used to weigh things is called a balance. The word comes from the Latin bi-lancis, or *two dishes*. The word *balance* is still used despite the fact that two-dish (or two-pan) balances are seldom (if ever) used nowadays. In lieu of a second pan, the counter (opposing) force now may be springs, built-in calibrated weights, or a magnetic device called a servomotor.

There are four classes of balances, each based on their ability to split hairs as it were, or more specifically, the number of intervals used within the scale capacity. For example, if a laboratory balance has a capacity of 200.00 grams and it reads to two decimals, it would have 20,000 scale intervals. The formal identification of four classes was devised by the *International Organization of Legal Metrology* (OIML) and they are shown in Table 2-22.

Table 2-22[9, 10]

Classifications of Weighing Equipment

Class	Class Name	Scale Intervals
I	Special Accuracy (fine)	50,000 < = n[a]
II	High Accuracy (precision)	5,000 < = n < = 100,000
III	Medium Accuracy (commercial)	500 < = n < = 10,000
IIII	Ordinary Accuracy (course)	100 < = n < = 1,000

[a] n = number of intervals.

Balances can be calibrated or verified for accuracy with special weights, called *calibrated weights*, whose specific mass is known. Calibrated weights* vary in quality and tolerance. They are classified by type, grade, and tolerance.** All calibrated weights are compared directly or indirectly to the international prototype *one kilogram mass* to verify their accuracy.

2.4.2 Weight versus Mass versus Density

It is fair to say that an object weighing a ton is heavy, and that few, if any, people could move or lift it. However, if on the moon, the same object would weigh only a bit over 300 lbs—and although most people still could not move or lift it—there are some Earthlings who could. If we were on the space shuttle in a free-fall environment, anyone could move the object around with relative ease. When we weigh an object, we are measuring its inertia to Earth's gravitational pull. That measurement is its weight, not its mass. The weight of an object, not its mass, will change depending upon its inertia.

Unlike weight, which varies relative to its inertia (such as gravity), mass is an inherent and constant characteristic of any object. In any given gravitational environment, an object with a lot of material (mass) will weigh more than an object of the same type, but less material. Because of this quality, we can make calculations of, and about, an object's mass from its gravitational weighing.

Density is not directly related to mass or weight, but is calculated from an object's weight divided by its volume g/m^3. For example, a large object of little mass (such as foam rubber) is considered to have little density. On the other hand, a small object of tremendous mass (such as a neutron star) has tremendous density. *Density* refers to the amount of space ("volume") a given amount of mass occupies. *Mass* refers to the amount of material in an object, not to the amount of space it occupies.

In everyday parlance, we imply an object's mass when we speak of its weight. However, because we weigh an object based on its attraction to Earth, we are, in effect, measuring its force. In a nutshell, we measure an object's *force* to obtain its *weight*, from which we can calculate its *mass*. The validity of this approach holds despite the fact that the force of gravity varies over the earth's surface by over 5% in addition to changes in elevation.

2.4.3 Air Buoyancy

The fact that weights occupy space creates an interesting problem. The space occupied by a weight is normally occupied by air, and because air has weight, it provides a buoyancy effect (known as Archimedes' Principle) against the real weight of the object. This effect influences the measured weight of an object.

The problem is more easily explained by examining what happens when you place something in water (because water weighs more than air and provides a greater buoyancy effect, its effects are more dramatic). If you put a

* Calibrated weights are verified to weigh what they say they do within a given tolerance. The smaller the tolerance of a calibrated weight, the better the quality and the more expensive it will be.

** These parameters are discussed further in the section on calibrated weights (see Sec. 2.4.13).

cube of metal in water, it sinks to the bottom of the container. That cube weighs less in the water than it did in air, by an amount equal to the weight of the water it displaced. On the other hand, if you put a similar sized block of wood in the water, it would float because the amount of water that the wood displaces weighs more than the wood, preventing the wood from sinking. The same phenomena occurs in air which is why helium balloons rise and wood balloons fall.

Even if an object's density and size change, its mass does not. Thus, it is possible for two objects to have the same mass but weigh different amounts due to the effects of air buoyancy caused by the weight of air. This principle can be demonstrated by taking a calibrated weight whose density is 8.0 g/cm^3 and weight (in air) is 100.000 g, then using it to weigh an equal weight of pure water whose density is 1.0 g/cm^3. Because their densities are different, they will occupy different volumes of air: their volumes are 12.5 cm^3 and 100 cm^3, respectively. If you were to weigh each object in a vacuum to eliminate the air buoyancy factor,* the calibrated weight would now weigh 100.015 g and the pure water would now weigh 100.120 g.

Factors that affect buoyancy are the density of a sample, ambient air pressure, and relative humidity. Thus, a barometer and humidity indicator should be located within any balance room where highly accurate readings are required. The counterbalance weights within single-pan balances are also affected by air buoyancy, but in equal amounts to the sample, so they should cancel each other out. The exact amount of the buoyancy effect varies depending on the density of the material being weighed and the density of the air at the time of weighing. This phenomenon was studied in detail by Schoonover and Jones[11], and by Kupper.[12]

The formula for converting a weighing result to true mass is given in Equation 2.1:

$$m = M \cdot \frac{(1 - \frac{d_a}{D})}{(1 - \frac{d_a}{d})} = M \cdot k \qquad \text{Eq. (2.1)}$$

where
m = mass of the sample
M = weighing result, i.e., the counterbalancing steel mass
d_a = air density at time of weighing**
D = density of mass standard, normally 8.0 g/cm^3
d = density of sample
k = ratio of mass to weighing result for the sample.

To calculate the density of the air, use the formula in Equation 2.2:

$$d_a = (0.0012929\ g/cm^3) \cdot (\frac{273.15}{T}) \cdot (\frac{B}{760}) \qquad \text{Eq. (2.2)}$$

where
T = Temperature in °K (Centigrade plus 273.15)
B = Barometer reading in millimeters of mercury.

* Instead of making the weighings in a vacuum, you can also make use of Eqs. 2.1 and 2.2).
** For the density of moist air, see the *Handbook of Chemistry and Physics*, The Chemical Rubber Company, published annually.

There are only two occasions where measured weight equals true mass, when k = 1. This occasion occurs when measurements are made in a vacuum or the density of a sample is equal to the density of the mass standard. Fortunately, the greatest differences only occur when an object's density is particularly low (0.1% for density ≈ 1.0 g/cm³ and about 0.3% for density ≈ 0.4 g/cm³). In most situations, the effect of air buoyancy is significantly smaller than the tolerance of the analytical balance. The effects of varying densities (of objects being weighed) and varying air densities are shown in Table 2-23.

Table 2-23‡

Variation in Weight with Atmospheric Density

Object[a]	Density (g/cm³)	Weight (kg) in Denver, CO (d_a = 0.00098 g/cm³)	Weight (kg) in vacuum (True Mass)
Oven dried eastern white pine	0.373	1.000564	1.003077
Live oak wood	0.977	1.000198	1.001080
Aluminum	2.7	1.000054	1.000295
Calibration weight in Mettler AT balance	7.97	1.000000	1.000001
Stainless steel as used in mass standards	8.0	1.000000	1.000000
Brass	8.4	0.999999	0.999993
Gold	19.3	0.999984	0.999912
Platinum-iridium (kilogram prototype)	21.5	0.999983	0.999906

‡ From Tables 1 and 2 from "Honest Weight—Limits of Accuracy and Practicality" by W.E. Kupper from the *Proceedings of the Measurement Science Conference 1990*. With permission.

[a] 1 kg of the following materials, as weighed in sea level atmosphere of density d_a = 0.0012 g/cm³ against steel weights of density d = 8.0 g/cm³.

One approach to avoiding the problem of air buoyancy is to weigh an object in a vacuum (known as "weight in vacuo"). Such readings provide an object's true mass as opposed to its apparent mass. There are a variety of vacuum balances made precisely for this purpose. However, vacuum balances are expensive, require expensive peripheral equipment (such as vacuum systems), and are neither fast nor efficient to use.

2.4.4 Accuracy, Precision, and Other Balance Limitations

The amount of accuracy required in a balance, like most things in the lab, depends the balancing of your needs versus the capacities of your pocketbook. You do not need great accuracy if all you are doing is weighing letters. On the other hand, weighing volumetric flasks to calibrate volume requires tremendous accuracy. Not only is the cost greater for a more accurate balance, but the support equipment and personnel for the maintenance of such equipment are also substantially greater. When analyzing the different attributes and characteristics that help to define the quality and capabilities of balances, many different terms are used. The following terms (and concepts) are used to describe various features of all balances:

The greater the accuracy of a balance, the closer the balance will read the *nominal weight** of a calibration weight. If the calibration weight reads

* The nominal weight of a calibrated mass is the calibration that is printed on, or stamped into, its surface.

10 mg, the balance should read 10 mg. If, for example, the balance reads 10.5 mg, it has less than desirable accuracy, and if it reads 12 mg, it has poor accuracy.

The *precision* of a balance is related to how well it can repeatedly indicate the same weight over a series of identical weighings under similar environmental conditions. Precision is not a measure of how accurately a scale can make a single weight reading. The *index of precision* is the standard deviation for a collection of readings. If the index of precision is greater than the readability of the balance, accuracy is significantly jeopardized.

Readability refers to the smallest measurement that a balance can indicate and that can be read by an operator when the balance is being used as intended. Generally, triple-beam balances have a readability of ±0.1 g, centigram balances have a readability of ±0.01 g, and analytical balances have a readability of ±0.0001 g. Neglect or abuse can damage a balance resulting in no deflection for the original smallest units. The result is a change in the readability for the damaged balance to a new, and larger, value.

Linearity is the ability of a balance to accurately read the entire range of weights that it was designed to weigh. A balance which accurately weighs 10 mg, but poorly weighs 100 mg, but again accurately weighs 200 mg on up to its full scale calibration, is said to have poor linearity.

Off-center errors are problems specifically associated with top-loading balances. Placing a balance pan above the fulcrum places different torques and friction on balance pieces that do not exist when a balance pan hangs. The problem is exhibited if an object has different weight readings when moved to various locations across the surface of a top-loading balance pan.

Accuracy is the ability of a balance to precisely and repeatedly read a weight. The ability for a balance to read a particular weight is based not only on the above attributes and characteristics, but also three other factors:

1) The quality of the machine
2) The quality of the weighing process conditions (typically dependent upon the balance's location (see Sec. 2.4.5))
3) The skill of the person operating the balance.

Each of these factors is dependent the other two, and the failure of any one of them can affect the accuracy of a weighing. For example, do not expect great accuracy from a balance that is located above a radiator. Likewise, do not expect accuracy from a balance which has just demonstrated poor precision. Finding the source of errors in weighing is a step-by-step process. You must rule out each problem before moving on to the next level.

2.4.5 Balance Location

By their nature, balances are fragile pieces of equipment. The more sensitive a balance is, the more susceptible it is to environmental influences. The following should be considered before placing permanently in any location:

1) Balances should be away from sources of vibration. Rooms used for balance work should be away from elevators and ventilation motors. They should be located near support walls (as opposed

to walls that just separate rooms), which can help dampen vibration. To observe the amount of vibration your balance is receiving, lay small dishes of water at several locations on the supporting table you plan to use and float microscope cover slips on the surface of the water in each dish. Shine a light on each cover slip in turn, such that the light is reflected onto an adjacent wall. Make observations over several times during the day. Have someone walk around the room, close or open the door, and perform other activities that may cause vibrations. The more stable the light's reflection, the better the table (and location) will be for weighing purposes.

2) Balances should be away from winds and drafts. Many buildings have sealed windows, but vents can generate drafts as well. Although fume hoods generally do not create a significant draft, opening and closing doors can.
3) Balances should be placed away from sources of varying temperature, including direct sunlight, windows, heaters, vents, drying ovens, refrigerators, doors, and rooms on the south side of buildings.
4) Balances should be kept away from rapid changes in humidity such as those that occur when steam heaters are used. Not only will steam heaters affect temperature, the change in humidity can affect the workings of the balance.
5) Balances should be kept away from stored chemicals, especially those with low vapor pressures. Because most of the internal workings of modern balances are not in plain sight, you will not be aware of any corrosion until the the balance begins to malfunction. At that point, it may be too late to remedy the situation.
6) Hanging balance pans should be removed before balances are moved from one location to another to take the strain off the beam balance points as well as to prevent a swinging pan from damaging the balance.
7) Balances should be re-calibrated after they have been moved.
8) Balance rooms should be maintained free of dust. Dust can affect both mechanical and electronic balances.
9) Balances should be placed on nonmagnetic, nonferrous surfaces. Cement, stone, and wood are all acceptable.

If, after reading these rules, you conclude that the best place for a balance is in a special room by itself, you are right. Such a room ideally should be windowless, with one separate, shielded entry and filtered, baffled vents. The room should be small, with support beam walls and very heavy benchtops, and should be maintained consistently at 20°C.

In lieu of a perfect room, the specific problems your room presents should be known to protect the validity of your weighing as much as possible. For example, if you are having problems limiting your vibration sources, make your weighings at odd hours when elevators are not in use or when people are unlikely to enter the room. For drafts, baffle vents, place screens around the balance, or use other similar stop-gap measures.

2.4.6 Balance Reading

Consider the following scenario: Many people each make one reading on the same balance, using the same weight for each reading. All readings are within one-hundredth of a gram except that of one user, whose reading was one-tenth of a gram off. It is safe to assume that the user does not know how to use, or read, that particular balance, or how to make an accurate weighing on any balance. Either way, this user needs some help in balance use.

To best examine a person's ability to use a balance, have him or her make multiple weighings (e.g., ten times) using the same balance and weight. If he or she consistently makes the same type of error, the readings will have very little variation. If he or she makes a variety of random errors, the readings will vary with no consistent pattern or be caused by not preventing the effects of electrostatic forces (see Rule 11 below).

It is impossible to provide instruction for all types of balances in a book such as this one. However, it is possible to provide simple, generic, general rules for the operation of a balance. These general guidelines for balance operation are the same regardless of the specific type of balance:

1) Never force a lever, pan, door, or other part of a balance. If balance parts do not move smoothly and easily, there is probably a reason, and brute force will not be the cure.
2) Never drop a weight or a sample, or let one fall on any part of the balance (for that matter, do not drop a weight or sample at any time).
3) Never handle a calibration weight with your fingers. Even if the tolerances you are working with are well above the limits for weighing a fingerprint, the oils from your skin could corrode the weights. Besides, if you get in the habit of handling low-quality weights (see Sec. 2.4.12) with your fingers you are likely to continue the habit with high-quality weights.
4) Never handle an object to be weighed with your fingers when making analytical weighings, because your fingerprints will alter the weight. Use tongs or tweezers.
5) Never let a spill sit. Keep some paper towels within reach of a balance to help mop up any spill immediately. Such a spill may pose a health danger or could damage the balance.
6) Never place or remove a weight on a balance without first verifying that the pans or trays are in the arrested position, that is, the pan(s) are securely supported and cannot move. Otherwise the action could cause the blades to slip from their pivot points and/or damage the blades.
7) Never move, twist, or turn any part of the balance with a quick jerky motion. The balance is a delicate machine and such actions could cause the blades to slip from their pivot points and/or damage the blades.
8) Never place a powder or liquid directly on any part of the balance. Aside from the difficulty in *completely* removing the entire weighed sample, there is the danger of the material being corrosive to the balance pan, or the powder drifting off, or the

liquid dripping off into the balance's mechanical sections. There should always be glass, a weighing pan, or weighing paper between the object and the balance.

9) Never rush. Allow time for the object to reach the balance room's ambient temperature. One way to verify such a temperature alignment is to put a comparable material and thermometer into a dummy vessel. When the material in the dummy vessel reaches ambient temperature, the weighing sample is also likely to be at an acceptable temperature.
10) Never leave a single-pan balance without first setting the pan(s) to the arrested position, and then turning the weights back to zero. Otherwise the constant pressure prematurely ages the balance's mechanical parts.
11) Never weigh objects (particularly powders or granules) if the air is dry (<40% relative humidity), and never use plastic containers.
12) Never take the calibration of a balance for granted. Likewise, the balance should be calibrated each time it is moved regardless of whether it was across the country, to a different building, or to a different table.

There are some interesting quirks as to why temperature (Rule # 9) is important to weighings made on an analytical balance. Heat has some seemingly strange effects on the weight of an object. An object that is hot will weigh less than it does at room temperature due to hot air rising, and pulling the weight and pan up with the current of air. It is also because hot materials have less adsorbed and absorbed water on them. Thus when you weigh a hot object, you are weighing the weight without any surface water. As an object cools off, water begins to attach to the surface of the object and 'weigh it down.' The actual significance of this *water weight* is generally ignored as it is too small for most purposes. Likewise, if a cool object is placed in a weighing chamber, the reverse occurs and the object will appear heavier than it really is. Because cold objects are likely to have water condense on the container's surfaces as well as on the object itself, the amount of collected water is very likely to have a significant effect on an object's measured weight.

Electrostatic problems (Rule # 11) are likely to occur when atmospheric (relative) humidity is below 40%. If a container becomes positively (or negatively) charged and the weighing apparatus is likewise charged, they will repel each other and the object will appear heavier than it really is. Conversely, if the container and balance are oppositely charged, they will attract and the object will appear lighter than it really is. There are several ways to prevent this problem:[13]

1) Use a humidifier to increase the humidity to between 45% to 60%.
2) Place weighing vessels in a metal container to screen electrostatic forces.
3) Use an appropriate weighing vessel (plastic is poor, glass is good, and metal is best) to avoid electrostatic forces.

4) Antistatic guns are commercially available, but they should not be relied on.
5) The balance is likely to be grounded through the power cord's third wire. Therefore do not bypass this wire if using an extension cord. Be sure to use a three-wire extension cord and an appropriate wall socket.

Although the need to calibrate (see Rule #12) a balance that was moved to a different table is just to verify that nothing got misaligned, the changes in gravity forces of a balance moved cross-country can be significant. Likewise, a balance move in a vertical motion some 10 meters can show signs of mis-calibration. Aside from the need to re-calibrate a balance to counter the effects of changing gravity, the wear and tear of moving balances should be enough reason to not move a balance any more than necessary.

2.4.7 The Spring Balance

A *spring balance* compares two different forces (gravitational attraction vs. tensile forces of the spring) to weigh an object. Assuming that the tensile forces of a spring balance remain constant, as any inertial forces (i.e., gravity) change, weight readings will change. A spring balance therefore is typically used to illustrate differences in weight when the gravitational effects of celestial bodies are compared. A lever arm (pan) balance compares the same force and thereby cannot reflect changes in gravity.

Unfortunately, metal spring balances suffer from a variety of problems. For instance, the spring in a balance can suffer from hysteresis (extended beyond the point of resiliency, the spring cannot return back to its original position). In addition, very accurate spring balances need to be calibrated for changes in gravity depending on where on the Earth they are being used.

Perhaps the biggest problem with spring balances is that the linear motion of the spring is typically converted to some type of circular motion for reading the weight. Such gear reductions introduce friction which decreases readability, precision, and accuracy. Linear spring balances, such as those used for weighing fish on a boat, also have built-in friction from the parts that rub on the scale face.

Despite their general lack of accuracy, their popularity is understandable. Their limited accuracy and reliability are more than sufficient however for the demands of grocery stores or bathroom scales. To use anything more complicated or expensive, when the need is not justified, is a waste of time and money.

There is an interesting story, from the gold rush days, about people taking advantage of the inaccuracies peculiar to the spring balance. As you go from the north or south pole toward the equator, the centrifugal force caused by the earth spinning increases, which counter-effects the force of gravity. In addition, the earth is somewhat oblate (due to centrifugal force) and the effects of gravity are somewhat less at the equator because the earth's surface is farther from the earth's center. Thus, using a spring balance, an object weighs less at the equator than it does at a pole. During the California gold rush, some people would purchase gold in California, using a spring balance for weight measurement, and sell the same gold in Alaska using the same balance. Because the gold weighed more in Alaska, it was worth more. Thus, a (good) living was made by buying and selling gold at the same price per ounce. Such a practice would not be possible with a two-pan balance.

Some highly accurate spring balances are typically made out of fused silica (quartz glass) and used in vacuum systems. The use of fused silica has two advantages. For one, the material is extremely nonreactive. Thus it is unlikely to corrode as a metal might when in contact with acids and/or oxidizers. In addition, glass is considered perfectly elastic until the point of failure. Thus, there are no hysteresis problems.

These balances simply are a hanging coil of fused silica with a glass basket on the bottom. An optical device with a calibrated vertical measurement system sights on a point of the glass basket. A calibrated weight is placed within the basket and the amount of deflection is noted. Several more weight measurements are made to verify linearity. By observing the amount of spring travel for a given amount of weight, very accurate measurements can be made. Unfortunately, the amount and type of supporting equipment for this type of spring balance are expensive.

2.4.8 The Lever Arm Balance

The lever arm balance compares the mass of an object to the mass of calibrated weights. There are two major types of lever arm balances. The *equal arm balance* (see Fig. 2-22(a)) works by directly comparing a known weight to an unknown weight, both of which are placed an equal distance from the fulcrum. Once in balance, there is a one-to-one ratio (of the weights) between the unknown and known weights. A well-made balance of this type is capable of high accuracy and precision. The disadvantage of this type of balance is that many standard weights are needed to provide the balance with its full range of weighing capabilities.

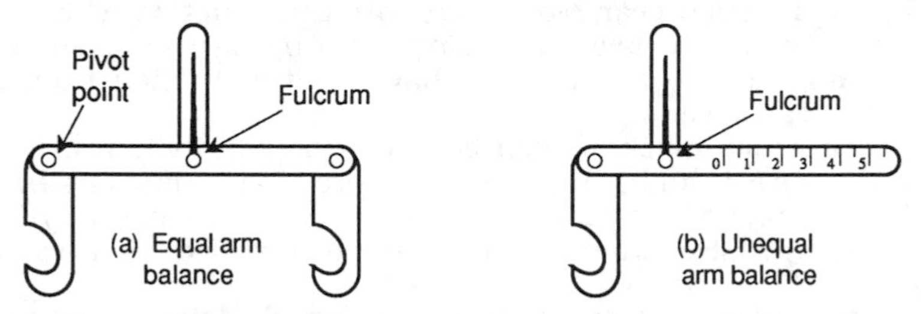

Fig. 2-22 Two basic lever arm designs.

The *unequal arm balance* (see Fig. 2-22(b)) compares an unknown weight to the distance a known weight is to the fulcrum. When weighing an object, the further the balance weight is moved from the fulcrum, the heavier the object. The primary advantage of the unequal arm balance is that it needs only a limited number of calibrated weights. The disadvantage of this type of balance is that its accuracy is limited by how many divisions can accurately be defined between the units on the balance arm.

Another problem of the equal arm balance is ensuring that the weight is directly under its pivot point. This alignment is important because the pivot point helps define how far the weight is from the fulcrum. As the unequal arm balance demonstrates, variations in the distance a weight is from the fulcrum can change its effective weight. In the 17th century, a mathematician named Roberval devised a balance arm that guaranteed that weights would act as if they were under the pivot point, regardless of how far they were placed from the fulcrum. The basic configuration of a balance utilizing the *Roberval Principle* is shown in Fig. 2-23, and the Roberval Principle used in equal and unequal style top-loading balance designs is shown in Fig. 2-24.

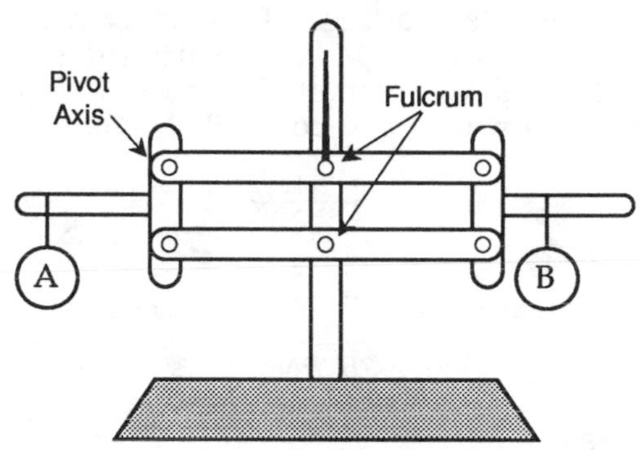

Fig. 2-23 The Roberval Principle balance.‡

‡ Based on Figure 5, from *The National Bureau of Standards Handbook* 94, "The Examination of Weighing Equipment" by M.W. Jensen and R.W. Smith. U.S. Printing Office, 1965., p.155. With permission.

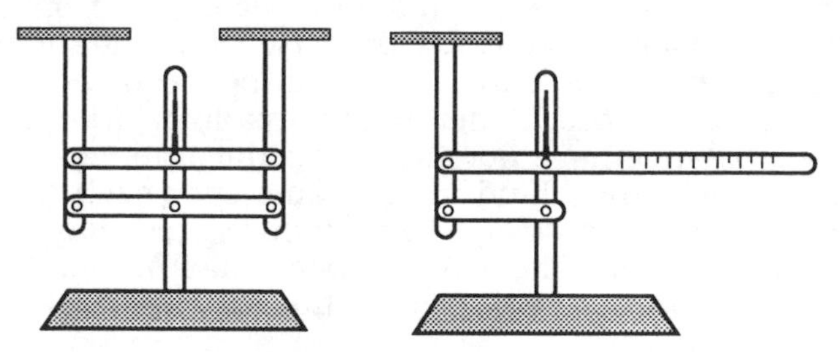

Fig. 2-24 Applications of the Roberval Principle used with top loading balances.‡

‡ Based on Figure 6, from *The National Bureau of Standards Handbook* 94, "The Examination of Weighing Equipment" by M.W. Jensen and R.W. Smith. U.S. Printing Office, 1965., p. 155. With permission.

The advantage of the Roberval Principle balance is that the placement of the weights has no effect on weight readings because the momentum of the weights is transferred to (and located at) the pivot axis. As far as the functions of the balance are concerned, the effective location of the weight is at the pivot axis regardless of its real location. The Roberval Principle is used on all top-loading balances and many pan balances. The major drawback to this system is that it doubles the axis points and therefore, doubles the friction on the system, which can reduce the balance's sensitivity.

The pivot axes of laboratory balances are not drawn correctly in either Figs. 2-23 or 2-24, but instead are either a *knife edge* or *flexural pivot* as shown in Fig. 2-25. The *knife edge* has the advantage of freedom of movement and no inherent restrictions. Its disadvantage is that its edge can wear or collect dust at its edge contact point. Either of these conditions causes friction, decreasing the accuracy and reliability of the balance. Static electricity can also reduce the freedom of movement of a knife edge.

The *flexural pivot* is free of the problems associated with the knife edge. However, it can act as a spring and therefore may add an extra force to the weighing process. This force may artificially subtract from an object's true mass and provide inaccurate readings. This effect may arise as a result of time or as temperature changes.* Manufacturing design may decrease this effect, and it may also be compensated for in the electronics, especially when used in conjunction with servo-controlled balances.

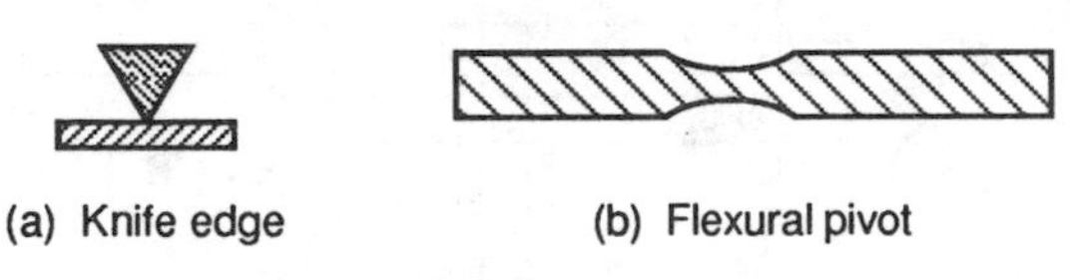

Fig. 2-25 Pivot axes.

2.4.9 Beam Balances

Beam balances fall into the unequal arm balance category. Generally, they are not capable of weighing very light objects (<0.1 g). They obtain greater precision by having a separate beam for each decade of weight capability (by splitting up the measurements on different beams, you obtain greater distance for each subdivision of weight. Otherwise, the gradations are closer together, making it more difficult to distinguish between them).

As seen in Fig. 2-26, the higher weight readings (the 100 and 10 unit beams) of a beam balance have notches at regular intervals, which forces the sliding weight to consistently land in the same spot each time the weight is moved to that location. This design helps insure precision (repeatability) for those weight ranges as well. The 1.0 unit beam has no notches which allows the user to slide the counter weight along the beam until the balance arm is in balance.

Before using a balance, be sure that it is level with the surface of the table. Some beam balances have a built-in level for this purpose. Screw the legs of the balance until the balance is level. Balances that do not have built-in levels are not expected to have the accuracy of those requiring precision-leveling. Regardless, you should always level the balance as best you can by eye.

Before an object is weighed, the balance should first be *nulled*, or zeroed. This zeroing procedure is done with nothing on the balance pan and by placing all of the weights (on each beam) to their zero points. If the balance

* The flexural pivot may become more or less stiff due to time and/or temperature.

does not indicate 'centered,' rotate the *balancing screw* (on the pan side of the balance) as necessary to zero, or 'balance' the balance. Balances that require you to complete the balance process with the beam in the same balanced (null) position at which you started are called *null-type balances.*

To weigh an object, place the unknown object on the center of the balance pan. Move the heaviest weight over, a few notches at a time, until the beam side is heavier than the pan side. Move that weight back one notch, then take the next heaviest weight and slide it over, a few notches at a time, until the beam side is again heavier than the pan side. Move that weight back one notch. Repeat this process for all of the weights on notched beams.

The lightest weight beam has no notches. Slide the counter weight until the beam and pan are balanced. Reading the weights on the balance involves adding up the weights on the various beams. Fig. 2-26 shows the counter weights on the three beams at "70," "8," and "0.76." Thus, the reading of this triple-beam balance is "78.76."

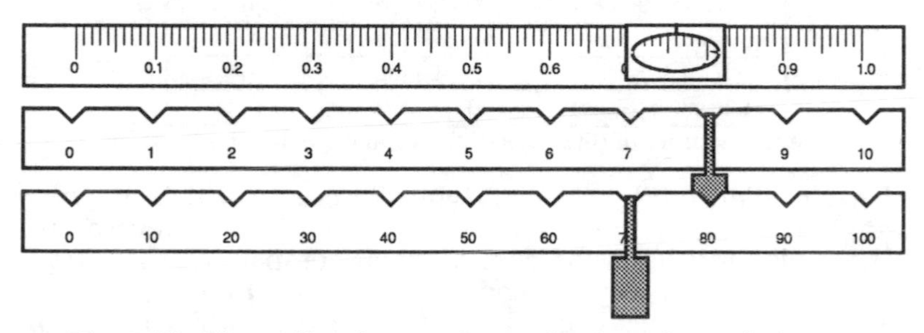

Fig. 2-26 The calibration readings of a three-beam balance.

After you complete a weighing and record the value in your log book, remove the object and return the counter weights to their zero value.

2.4.10 Analytical Balances

Analytical balances are those of Class I and Class II (see Table 2-22), and require a division of the scale's capacity that is typically smaller than 10^{-5} of the total capacity with a readability of between 0.1 µg to 0.1 mg (10^{-7} to 10^{-4} g). Analytical balances have the greatest precision for the most demanding work. For many years, only a properly-made two-pan balance* could achieve the precision required to obtain the required level of accuracy. Despite their accuracy, weighings made on two-pan balances were slow and inefficient. Eventually, technology improvements in single-pan balances brought them to meet and eventually exceed the capabilities of two-pan balances, yet maintain their speed and efficiency.

* It is unlikely that you will find the classic two-pan balance in use in laboratories. In a way this omission is a pity because they are typically beautifully-made instruments. For those of you who may still use them, I cannot fault you for your romantic actions. For those of you who are considering getting rid of one to make room, please do not. You do not need to use it, just keep it and let its beautiful construction remind you of the art that is possible in technology. Because of the rarity of two-pan balances in contemporary labs, there will be no discussion of their use or operation in this book.

As can be seen in Table 2-22, a balance that can read 5 grams to 4 decimal points is in the same class as a balance that can read 500 grams to 1 decimal point. Thus, an analytical balance is not one that can read only small amounts to very small intervals, it is practically any balance that can read its capacity to very small intervals. The terms micro balance or semimicro balance do not refer to the size of the calibrations, rather they refer to the capacity of the scale. In fact, an analytical balance can have a capacity as great as 50 kg so long as the scale divisions are smaller than 10^{-5} of the total capacity. The names of the standard analytical balances used in the laboratory are shown in Table 2-24.

Table 2-24[14]

Laboratory Analytical Balances		
Balance name	**Capacity**	**Scale Divisions**[a]
Macro balance	100 - 200 grms	10^{-5}
Semimicro balance	30 - 100 grms	10^{-5}
Micro balance	3 - 50 grms	10^{-5}
Ultramicro balance	3 - 5 grms	10^{-6}

[a] Scale divisions are the number of divisions of the total capacity of a balance.

A *single-pan balance* (Fig. 2-27) is typically one in which all that the user sees is a single pan in a container with several knobs and levers on the front or sides. There also will likely be some optical component that the user may see as a frosted screen or digital readout. There are four major types of single-pan balances: mechanical, mechanical with an optical deflector (for the smallest measurements), servomotor, and hybrids of mechanical (for larger weights) and servomotor (for smaller weights). Although you will still find the former two types of balances in laboratories, most of the balances currently in production today are of the latter two types.

Most of these balances will have either a magnetic or air damper, which arrests the free movement of the balance beam. This damper also allows the balance to come to a rest position much faster than otherwise possible. In addition, hybrid mechanical single-pan balances have a lever arm that can prevent, partially free, and completely free the balance beam's movement. This three-part varying restriction is important, for it protects the knife edge on which the balance beam rocks during gross, or rough, movement actions, yet it allows free and easy movement when final measurements are made. When objects are being placed on the pan, the lever should be in the fully arrested position to prevent any movement of the beam and protect the balance's mechanical parts. Partial movement of the balance

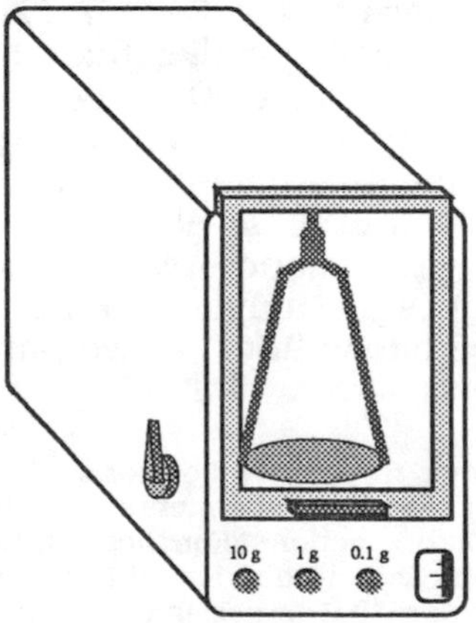

Fig. 2-27 The single pan balance.

beam allows safe addition and subtraction of heavier weights. Full motion of the balance beam is necessary when light (mg) weights or optics are used for final weighing.

The mechanical aspect of these balances typically employs a crankshaft or rocker arm assembly that supports the weights. The weights are engaged or disengaged on the balance beam as the shaft is turned. Typically, only four weights per decade are required. An example of the weights necessary to go from 0 to 10 would be 1, 1, 3, and 5 (the 0 to 10 range as shown in Table 2-25). The selection of weights is handled internally by the crankshaft. The user simply rotates the dial until the proper deflection is noted on the balance's viewing screen.

The counterbalance weights within a single-pan balance are described as having an *apparent mass*. The term *apparent mass* is used because what you are reading is different from true mass due to the effects of air buoyancy. If corrections are made for air buoyancy (see Sec. 2.4.3), you have *true mass*. In more recently manufactured single-pan balances,* the apparent mass has been changed from 8.3909 g/cm^3 (brass) to the current 8.0 g/cm^3 (stainless steel). This change was made to better approximate the density of grade S weights (see Sec. 2.4.13 and Table 2-27). The density of the apparent mass used within a balance should be identified somewhere on the balance and/or in its literature. If you are using a balance with the older apparent mass and switch to a balance with the newer apparent mass, you would need to add a 0.0007% correction to the total weight. This correction is equivalent to 7 mg in 1 kilogram and so significantly lower than the tolerance of most balances that the correction is not really necessary.

Table 2-25

The Addition of Weights From 0 to 10

nothing									=	0
1									=	1
1	+	1							=	2
3									=	3
1	+	3							=	4
5									=	5
1	+	5							=	6
1	+	1	+	5					=	7
3	+	5							=	8
1	+	3	+	5					=	9
1	+	1	+	3	+	5			=	10

Regardless of whether a single-pan balance is completely mechanical, or a hybrid, there are several advantages of the single-pan balance over the two-pan balance:

1) Because the weights are inaccessible to the user, there is no chance for the weights to be accidentally touched. Analytic balances are so sensitive that fingerprints, or corrosion from fingerprints, can alter the weight of analytic weights.

* This was changed in the mid 1970's.

2) The single-pan balance can make use of optics and/or servomotors (see next section) to obtain the smallest weight measurements. By eliminating mechanical deflection of weights for the final part of the measurement process (such as a gold chain* or rider**) repeatability is substantially improved.
3) Single-pan balances typically have damping devices that help to minimize the free swinging of the lever arm. This device speeds the weighing process immensely over the time spent 'counting swings' for two-pan balances.

There are many different brands and models of single-pan balances. Each has its own mechanism for weighing an object. As in all other lab equipment, the original manual, or a copy of the same for information on how to use the balance should be readily available. Regardless, the general procedures are typically similar: The larger weights are invariably manipulated by dials for each decade of weight. These dials are rotated until the appropriate weight is obtained or until the final weighing is complete.

Weighing an object on a single-pan balance typically involves first making sure that the balance is *nulled*. This procedure is done by setting all dials to "0," rotating the lever arm to the free arm position, and manually setting the servomotor or optical window to "0.00" to verify its 'null' starting point. Once the balance is set, return the lever arm to the arrest position before placing the object to be weighed on the center of the pan, inside the door of the balance. For very sensitive weighing, do not handle the object with your bare hands. Select tongs that are sufficiently long so that your hands do not get into the weighing chamber area. Close the door and wait a few minutes to allow the internal sections of the balance to regain equilibrium for both drafts and temperature.

The weighing procedure is made by rotating the lever arm to the semi-release position. Slowly rotate the weight knobs from lighter to heavier. When the scale deflects, back up one weight calibration (for instance, if the scale deflects when the gram dial is rotated to '7,' rotate the dial back to '6'). Continue the process for each weight dial until the final weight measurement is attained.

It is not possible to specifically explain how to make the smallest weight division weighing for single-pan balances in a book like this one because there are so many different ways it can be done depending on the brand or model of balance. Regardless, the general procedure for a final measurement is made by rotating the arresting arm from the *partial freedom position* to the *full freedom position*. Then dial, turn, or observe the smallest weight division for the completed weight. Once the weight has been determined, return the arresting arm back to the full arrest position. Open the door and remove the object that was weighed. If there are any liquid or powder spills, clean them up immediately.

* A gold chain was used on one design of the single-pan balance. The gold chain was unwound onto the balance pan until the sample and internal weights balanced. The length of chain on the pan was directly equal to its proportional weight and therefore the length of chain was used to make the final measurement digit.

** The rider is a sliding counterbalance that is used to make the final weight calibration on two-pan balances.

2.4.11 The Top-loading Balance

The advent of the servomotor brought new levels of accuracy to the *top-loading balance*. The servomotor gave top-loading balances the ability to weigh very small amounts quickly. Top-loading balances can generally measure as little as one-hundredth of a gram; more sensitive (and therefore more expensive) models can measure one-ten thousandth of a gram.

The operation of the servomotor is completely different from balance-beam balances because it does not require the use of counterbalancing weights. By their simplicity, servomotors have done to weighing what quartz crystals have done to time-keeping. This revolutionary device has allowed the removal of sluggish counterbalance weights and replaced them with electronics. Basically, the servomotor works by transferring linear motion to an electromagnetic force. The pan is established in a null position with an electronic light sensor. Any weight placed on the pan deflects the light sensor off its established position and an electromagnetic current is initiated to return the pan to its original position. Because a greater weight requires a greater electric current to accomplish this task, the current can be directly read as a weight.

Because the neutral (or null) position can easily be established with a weight on the pan, re-calibration and *taring** to the *null* position (before actual weighing) are accomplished by pressing a button.

Servomotors may be found in both single-pan balances, and in *top-loading balances*. The operation of top-loading balances typically requires turning on the balance, pressing the null, or taring, button, and placing an object on the balance. The weight is calibrated and displayed on a screen within moments. Because the entire operation is electronic, the information can be sent to a printer for a permanent record, or to a computer for automatic processing. Computational capabilities (by software) can also be included to process such things as counting or statistical information about objects being weighed.

Lest you think that the laboratory balance has been made perfect by the servomotor, realize that the servomotor is electronic and therefore is susceptible to various types of interference. Sources of interference include:

1) Weighing magnetic materials or placing the balance on a magnetic or ferrous table or surface. Magnetic objects cannot be weighed on a servomotor. To verify whether the magnetic or ferrous surface is affecting the readings, you can place the balance at various locations around the table and note any differences. In most circumstances, you do not need to bother trying to weigh with a servomotor on such a surface.
2) Electromagnetic interference. This interference can come from any electromagnetic-emitting field or source such as a CRT (from computer screens), RF generator, and radio transmitter. Using a hand-held radio transmitter, one can test the effects of electromagnetic interference on a balance. Erratic behavior

* Taring is the act of setting the scale to zero when a container is on the balance. Consequently, further weighings do not require subtracting the already-weighed container.

of the balance's display may also be caused by interference on a floor above or below the balance location.

3) Dust contamination. Although it is easy to associate problems with dust on mechanical balances, it is less apparent why dust would affect an electronic apparatus. The answer is that because there is movement within the servomotor itself, dust collections between the magnet and electric coils is likely to cause erratic measurements. Additionally, if dust-sized ferrous particles find their way to the electromagnet, the servomotor could be shorted and rendered useless.

Analytical top-loading balances (those that can measure one-thousandth of a gram or smaller) have covers or doors to isolate the balances from drafts. These covers also provide limited protection from accidental spills. Typically, general-use, top-loading balances do not have covers, and are therefore subject to damage from accidents. Plexiglass covers may be obtained for many models of top-loading balances to protect them against such problems. Even a cardboard box placed over a balance will help reduce dust and limit accidental spills on the balance.

One problem inherent in top-loading balances is that the weight of objects to be weighed can vary when placed at different locations on the weighing pan.* Although this inconsistency should not apply, the problem is often complex and has to do with the geometry of how a top-loading balance is made and where the weight is distributed. Fortunately, testing for this problem is easy: make several weighings of the same object at various locations on the balance pan. You may want to pre-mark the pan with some numbered geometric pattern (such as a star) to readily identify the location of any weight changes.

2.4.12 Balance Verification

If you have determined that your balance is making inaccurate measurements, and you have eliminated human error, you not only cannot trust any future weighings, you must question all past weighings to the point of the last balance verification. By maintaining *written records* of balance accuracy tests on a routine basis, the reliability of past measurements can be verified. Otherwise, every weighing made between the last verification and the first appearance of faulty readings is in doubt.** If you find errors during equipment testing, you need to track their source and correct the problem. Otherwise, all future data will also be in doubt.

The tolerance of a given balance is based on the level of accuracy that the balance is designed to provide. The greater the tolerance, the less the precision. The less the tolerance, the greater the precision. The tolerance of a balance is a percentage of its last significant figure (in fact, tolerance is often defined by the last significant figure). If you have a balance which is accurate to ± 0.1 gram, you should not report a reading of 0.02 grams.

* Hanging pan balances, by their design, cannot have this problem.

** If the amount of inaccuracy is less than your experimental limits, there is no reason to throw out any measurement.

When we discuss a balance's quality, we generally are referring to its reliability and accuracy. A balance, no matter how sensitive, is not a quality balance unless it is reliable and accurate in its measurements. Because the accuracy of a balance can decrease from wear, dirt, or contamination, routine periodic verification is required. The manufacturer can provide suggested verification schedules that may have to be increased or decreased depending on the conditions in your lab.

All balances should be checked for:

1) **Precision**. Does the balance read the same weight over a series of measurements for the same object?
2) **Accuracy and linearity**. Does the balance read the same weight as that given for a calibration's nominal weight, and does the balance provide the same accurate weighings over the full weight range of the balance?
3) **Readability**. Is there accurate, repeatable deflection at the smallest unit of measurement that the balance is supposed to read, including any vernier or micrometer calibration (if present)?
4) **Settling time**. Does the balance take the same amount of time to settle at the final weighing as it did to null?
5) **Response to temperature**. Does the balance provide the same reading at 20°C as it does at 25°C?
6) **Responses to other environmental disturbances**. What are the effects on the balance of drafts, vibrations, electromagnetic fields, magnetic fields, and other conditions?

Tests 4,5, and 6 help identify under what conditions you should not bother making weighings. They should be re-evaluated each time balance verification is made, because general wear and tear may exacerbate any environmental influence.

All electronic balances should also be checked for:

7) **Warm-up variations**. Some balances may indicate different weight values for the same object depending on how long the balance has remained in operation. See if any of the six previous tests is affected if the balance has just been turned on, left on for one-half hour, or left on for several hours.

Finally, top-loading balances should also be checked for:

8) **Off-center errors**. Does the balance make consistent weight measurements when an object is placed at different locations on the balance pan?

2.4.13 Calibration Weights

Calibration weights should only be used to calibrate, or verify the accuracy of a balance. Calibration weights should never be used to make weight determinations. They should never be handled directly with hands, and should be stored in safe locations away from environmental dangers.

A balance should be verified using the same calibrated weight each time. Because calibrated weights have expected variations in tolerance, using different weights may yield varying test results and could lead you to believe your balance requires constant (minor) re-calibration when, in fact, no such calibration is required.

The ASTM has categorized laboratory weights into the following divisions:

1) ***Types* (I and II)**. *Type* refers to how the weights were constructed. Type I is of better quality than Type II. See Table 2-26.
2) ***Grades* (S, O, P, and Q)**. *Grade* refers to how the surfaces of the weights are finished. S is better in quality than O and, likewise in turn, P and Q. See Table 2-27.
3) ***Classes* (1, 1.1, 2, 3, 4, 5, 6)**. *Class* refers to the amount of weight tolerance. The lower the Class number, the smaller the tolerance. Class 1.1 is a specialized class for calibrating low-capacity, high-sensitivity balances. See Table 2-28 & 2-29.

Table 2-26‡

Laboratory Weight Types	
Type I	**Type II**
These are made from one piece of material and have no compensating added materials. They are required when a precise measurement of density must be made.	These can be made of multiple materials for purposes of correcting the weight. This can be done by adding material or adding rings or hooks. The added material must not be able to separate from the weight.

‡ From ASTM Designation E 617-81 (Reapproved 1985)ε1 "Standard Specification for Laboratory Weights And Precision Mass Standards." With permission.

Table 2-27‡

Laboratory Weight Grades				
Grade	**Density**	**Surface Area**	**Surface Finish**	**Surface Protection**
S	7.7 to 8.1 (50 mg and larger)	Should not be greater than twice the area of a cylinder of equal height and diameter.	Highly polished and free of pits or markings except for identification markings.	None, must be pure.
O	7.7 to 9.1 (1 g and larger)	(Same as Grade S)	(Same as Grade S)	May be plated with platinum, rhodium, or other suitable material that will meet specification for corrosion resistance, magnetic properties, and hardness.
P	7.2 to 10 (1 g and larger)	No restrictions but those made out of sheet metal should not be overly thin.	Smooth and free of irregularities that could retain foreign matter.	May be plated or lacquered. Coating material should resist handling or tarnishing.
Q	7.2 to 10 (1 g and larger)	(Same as Grade P)	(Same as Grade P)	May be plated, lacquered, or painted to resist tarnishing and handling. Weights 50 kg or larger may have opaque paint.

‡ From ASTM Designation E 617-81 (Reapproved 1985)ε1 "Standard Specification for Laboratory Weights And Precision Mass Standards." With permission.

Table 2-28

Class 1[‡]			Class 2[‡‡]			Class 3[‡‡‡]		
Weight (kg)	Tolerance (mg) Individual	Tolerance (mg) Group	Weight (kg)	Tolerance (mg) Individual	Tolerance (mg) Group	Weight (kg)	Tolerance (mg) Individual	Tolerance (mg) Group
50	125		50	250		50	500	
30	75		30	150		30	300	
25	62	135	25	125	270	25	250	625
20	50		20	100		20	200	
10	25		10	50		10	100	
5	12		5	25		5	50	
3	7.5		3	15		3	30	
2	5.0	13	2	10	27	2	20	62.5
1	2.5		1	5.0		1	10	
(g)			(g)			(g)		
500	1.2		500	2.5		500	5.0	
300	0.75		300	1.5		300	3.0	
200	0.50	1.35	200	1.0	2.7	200	2.0	6.3
100	0.25		100	0.50		100	1.0	
50	0.12		50	0.25		50	0.60	
30	0.074		30	0.15		30	0.45	
20	0.074	0.16	20	0.10	0.29	20	0.35	2.00
10	0.050		10	0.074		10	0.25	
5	0.034		5	0.054		5	0.18	
3	0.034		3	0.054		3	0.15	
2	0.034	0.065	2	0.054	0.105	2	0.13	0.70
1	0.034		1	0.054		1	0.10	
(mg)			(mg)			(mg)		
500	0.01		500	0.025		500	0.080	
300	0.01		300	0.025		300	0.070	
200	0.01		200	0.025	0.055	200	0.060	0.325
100	0.01		100	0.025		100	0.050	
50	0.01		50	0.014		50	0.042	
30	0.01	0.020	30	0.014		30	0.038	
20	0.01		20	0.014		20	0.035	0.183
10	0.01		10	0.014	0.034	10	0.030	
5	0.01		5	0.014		5	0.028	
3	0.01		3	0.014		3	0.026	
2	0.01		2	0.014		2	0.025	0.128
1	0.01		1	0.014		1	0.025	

‡ From the ASTM document E 617-81 (Reapproved 1985), Table X3.1, Class 1 Metric. With permission.

‡‡ From the ASTM document E 617-81 (Reapproved 1985), Table X3.3, Class 2 Metric. With permission.

‡‡‡ From the ASTM document E 617-81 (Reapproved 1985), Table X4.1, Class 3 Metric. With permission.

Table 2-29

Class 4‡

Weight (kg)	Tolerance (g) Individual	Group
5,000	100	
3,000	60	
2,000	40	250
1,000	20	
500	10	
300	6.0	
200	4.0	25
100	2.0	
	(mg)	
50	1,000	
30	600	
25	500	1,250
20	400	
10	200	
5	100	
3	60	
2	40	250
1	20	
(g)		
500	10	
300	6.0	
200	4.0	25.0
100	2.0	
50	1.2	
30	0.90	
20	0.70	4.00
10	0.50	
5	0.36	
3	0.30	0.75
2	0.26	
1	0.20	
(mg)		
500	0.16	
300	0.14	
200	0.12	0.65
100	0.10	
50	0.085	
30	0.075	
20	0.070	0.363
10	0.060	
5	0.055	
3	0.052	
2	0.050	0.255
1	0.050	

Class 5‡‡

Weight (kg)	Tolerance (g) Individual	Group
5,000	250	
3,000	150	
2,000	100	625
1,000	50	
500	25	
300	15	
200	10	63
100	5.0	
50	2.5	
30	1.5	
25	1.2	3.0
20	1.0	
10	0.50	
	(mg)	
5	250	
3	150	
2	100	625
1	50	
(g)		
500	30	
300	20	
200	15	88
100	9	
50	5.6	
30	4.0	
20	3.0	17.5
10	2.0	
5	1.3	
3	0.95	
2	0.75	4.25
1	0.50	
(mg)		
500	0.38	
300	0.30	
200	0.26	1.40
100	0.20	
50	0.16	
30	0.14	
20	0.12	0.65
10	0.10	
5	0.080	
3	0.070	
2	0.060	0.325
1	0.050	

Class 6‡‡‡

Weight (kg)	Tolerance (g) Individual	Group
500	50.0	
300	30.0	no
200	20.0	
100	10.0	group
50	5.00	
30	3.00	tolerance
20	2.00	
10	1.00	
	(mg)	
5	500	
3	300	
2	200	
1	100	
(g)		
500	50	
300	30	
200	20	no
100	10	
50	7	
30	5	group
20	2	
10	2	
5	2	tolerance
3	2	
2	2	
1	2	
(mg)		
500	1	
300	1	
200	1	
100	1	

‡ From the ASTM document E 617-81 (Reapproved 1985), Table X5.1, Class 4 Metric. With permission.

‡‡ From the ASTM document E 617-81 (Reapproved 1985), Table X6.1, Class 5 Metric. With permission.

‡‡‡ From the ASTM document E 617-81 (Reapproved 1985), Table X7.1, Class 6 Metric. With permission.

2.5 Temperature

2.5.1 The Nature of Temperature Measurement

Most of the measurements discussed in this chapter deal with *physical* properties, such as length, volume, or weight. Measurement of these properties can be made directly. Temperature is different because it is an *energy* property, and energy cannot be measured directly. However, we can quantify the effect that one body's energy (in this case heat) has on the physical properties of another body, and measure that physical effect.

Unfortunately, heat energy does not have the same percentage of effect on all materials in the same way. For example, heat makes most materials expand, but few materials, if any, expand the same amount for an equal amount of heat. Thus, the size increase for one material (for a given amount of heat change) is unlikely to equal the size increase for another material (with the same amount of heat change).

On the other hand, it is possible to obtain the same temperature from two different materials if they are *calibrated* the same. This operation is done as follows: take two different materials and heat them to a specific (and repeatable) temperature. Place a mark on some reference material that has not expanded (or contracted). Then heat the materials to another specific and repeatable temperature and place a new mark as before. Now, if equal divisions are made between those two points, the specific temperature readings along the calibrated region should be the same even if the actual changes in lengths of the materials are different.

An interesting aspect about temperature measurement is that calibration is consistent across different types of physical phenomena. Thus, once you have calibrated two or more established points for specific temperatures, the various physical phenomena of expansion, resistance, emf, and other variable physical properties of temperature will give the same temperature reading.*

The establishing, or *fixing*, of points for temperature scales is done so that anyone, anywhere can replicate a specific temperature to create or verify a thermometer. The specific temperature points become (in effect) the International Prototypes for heat. The General Conference of Weights and Measures accepted the new International Practical Temperature Scale of 1968 (IPTS 1968) with thirteen fixed points (see Table 2-30). The new (IPTS 1968) scale was a revision from the IPTS 1948 (which had been amended in 1960).

There are two reasons for having many points with which to fix a temperature scale. One is that, as mentioned before, few materials affected by heat change length equally or linearly. Having many points allows scales to be calibrated in short ranges, where nonlinearity is less likely to have a pronounced effect. The second is that few, if any, thermometers can read all temperatures. Most thermometers are calibrated to read a small range of

* Although not all of these temperature measurement techniques provide a uniform linear measurement, the variations are known and can be calibrated and accounted for.

temperatures. Many 'fixing' points allows for a robust system of calibration. Unfortunately, most of these points require expensive equipment, and even then they are not easy to obtain and/or verify.

Table 2-30‡

The International Practical Temperature Scale of 1968 for K and °C

Material and Condition of Fixing Point[a]	K	°C
Hydrogen, solid-liquid-gas equilibrium	13.81	-259.34
Hydrogen, liquid gas equilibrium at 33,330.6 Pa (25/76 standard atmosphere)	17.042	-256.108
Hydrogen, liquid-gas equilibrium	20.28	-252.87
Neon, liquid-gas equilibrium	27.102	-246.048
Oxygen, solid-liquid-gas equilibrium	54.361	-218.789
Argon, solid-liquid-gas equilibrium	83.798	-189.352
Oxygen, liquid-gas equilibrium	90.188	-182.962
Water, solid-liquid-gas, equilibrium	273.16	0.01
Water, liquid-gas equilibrium	373.15	100.00
Tin, solid-liquid equilibrium	505.1181	231.9681
Zinc, solid-liquid equilibrium	692.73	419.58
Silver, solid-liquid equilibrium	1235.08	961.93
Gold, solid-liquid equilibrium	1337.58	1064.43

[a] Except for the triple points and one equilibrium hydrogen point (17.042 K), the assigned values of temperature are for equilibrium states at a pressure:

p_0 = 1 Standard atmosphere (101,325 Pa)

‡ From the ASTM document E 77-84, Section 3.2.2. With permission.

However, in addition to these primary reference points, a secondary series of reference points was established by the IPTS-68 (see Table 2-31). These secondary points can more easily be used (than the primary temperature points) for testing temperature equipment such as liquid-in-glass thermometers. They are useful because they require less equipment and are therefore easier to obtain. Remember that these points are secondary standards and should not be considered primary standards.

Note that Table 2-30 refers to the temperature 'K,' or the temperature "°C.' Both of these measurement scales are *temperature measurement units.* There is another scale, also known as 'K,' which is the unit of measurement for the Kelvin thermodynamic scale. Because heat is a thermodynamic property, temperature measurements should be capable of easy referral to the Kelvin temperature 'K,' more properly known as the Thermodynamic Kelvin Temperature Scale (TKTS). One degree K is exactly equal to the Thermodynamic unit of K, and likewise is exactly equal to one degree C. Otherwise, any specific temperature in one scale can easily be converted to another by the relation: K = C + 273.15.

Values given in Kelvin temperature (TKTS) are designated as '*T*.' Values given in degrees centigrade (°C) are designated as '*t*'. When either *T* or *t* is used to express temperature from the International Practical Temperature Scale, it should be designated as T_{68} or t_{68}.* For example, if you were to refer to the freezing point of mercury, you would write -38.862°t_{68}.

* The subscripted '68' is used to indicate that these temperatures conform to the IPTS-68 guidelines.

Table 2-31‡

Secondary Reference Points of the IPTS-68 in °C	
Temperature of	**°C**
Equilibrium between solid carbon dioxide and its vapor	-78.476
Freezing mercury	-38.862
Freezing water	0
Triple point of phenoxy benzene (diphenyl ether)	26.87
Triple point of benzoic acid	122.37
Freezing indium	156.634
Freezing cadmium	321.108
Freezing lead	327.502
Boiling mercury	356.66
Freezing aluminum	660.46

‡ From the ASTM document E 77-84, Section 3.2.2. With permission.

2.5.2 The Physics of Temperature-taking

When we measure the temperature of a body, we are depending on the heat of the body to be transferred to (or from) our measuring device. Once the heat has been taken to (or from) our measuring device, any physical changes in that device are interpreted as a temperature change. The process where we analyze the effects caused by a property to determine the amount of that property is known as *inferred measurement*. For temperature we have a variety of physical properties from which to infer the amount of energy (heat) that a given object has.

Measurement of temperature (or any energy property) has one major difference from measurement of physical materials: it is not cumulative. To measure the length of a room, you can lay several meter sticks end to end. The sum of the number of meter sticks will be the length of the room. Temperature, however is not cumulative. If you have a liquid that is hotter than the range of temperatures measurable on one thermometer, you cannot use a second thermometer to obtain the remaining temperature.

Because the temperature ranges and conditions of a system can vary, a variety of materials have been incorporated into different types of thermometers. The following is a list of common thermometer types and the property measured in each to obtain a heat measurement.

1) **Liquid-in-glass thermometer**. Volume of liquid increases as heat increases.

2) **Gas or vapor at constant volume**. Pressure of gas increases as heat increases.

3) **Dilatometer (or bimetal coil)**. As heat rises, the length of one metal expands more than the length of the other metal.

4) **Platinum wire**. Electrical resistance of the wire increases as heat increases.

5) **Thermocouple**. Thermal emf* goes up as heat increases.

* emf stands for electromotive force, which is further discussed in Subsection 2.5.11.

Table 2-32 displays some common temperature ranges of a variety of thermometers.

Table 2-32‡

Ranges of Common Temperature Measuring Devices			
Measuring Device (method)	**Approximate Ranges**		
Mercury in glass thermometer	-38°C	to	400°C
Alcohol in glass thermometer	-80°C	to	100°C
Constant volume gas thermometer	4K	to	1850K
Bimetallic thermometer	-40°C	to	500°C
Thermocouple	-250°C	to	1600°C
Resistance thermometer	0.8K	to	1600°C
Optical pyrometer	600°C	to	up
Total radiation pyrometer	100°C	to	up
Speed of sound	no limits		
Thermodynamic	no limits		

‡ From Robert L. Weber, *Heat and Temperature Measurement*, ©1950, Englewood Cliffts, NJ: Prentice Hall, p. 7. With permission

In addition to desired temperature ranges, variables affecting the selection of a thermometer may include:

1) **The material to be studied.** Is it acidic, alkaline, oxidizing, flame, plasma, hydrocarbon solvent, or conducting? One environment may affect one type of thermometer, but have no effect on a different type of thermometer.

2) **The amount of material to be studied.** The smaller the sample, the more you need to be concerned about the heat capacity of the thermometer. More specifically, a large thermometer with a large heat capacity can change the temperature of a small sample, whereas a small thermometer with a small heat capacity will read the temperature of a small sample without changing the temperature.

2) **The environment.** Is is cramped, dusty, wet, hot, cold? Some thermometers require very controlled environmental conditions to properly operate.

3) **The cost.** Platinum resistance thermometers are incredibly accurate over a wide temperature range. Unless you need the accuracy they can provide, you might be wasting your money.

4) **The amount of support equipment required.** A thermocouple is not very expensive, but the controller can be. Similarly, constant volume gas thermometers can be very bulky and cumbersome.

The reason for the variety of thermometers is that no one type is economical, practical, accurate, or capable of measuring all temperatures in all conditions. Selecting the right thermometer is a matter of analyzing your needs for a given job and identifying what thermometer best satisfies those needs.

As stated before, temperature (or heat) is a form of energy. Heat always travels from hot to colder bodies until thermal equilibrium is achieved. Thus, ice in a drink does not cool the drink. Rather, the heat in the drink is transferred to the cold ice, causing the ice to warm and melt. What remains is a 'colder' drink because heat energy was lost melting the ice.

The significance of this 'ice story' leads us to an important principle of temperature measurement: The act of taking an object's temperature changes the object's temperature because we depend on some amount of heat being given off or absorbed by the object to affect our measuring device. If the object being studied is very large in comparison to the temperature measuring device, this effect is negligible. If the material being studied is in a dynamic system (with heat constantly being introduced), the effect is irrelevant. However, if a small, static amount of material is being studied, the thermometer may have a significant effect. You need to select a temperature recording device that will have the least amount of effect on the sample's temperature. Optimally, you want the heat capacity of the temperature measuring device to be so much smaller than the heat capacity of the material being studied that it provides an insignificant change.

Temperature (heat) is transferred from one body (or region) to another by three thermal processes, conduction, convection, and radiation-adsorption, or some combination of the three.

1) ***Conduction*** is the transfer of heat from one body to another by molecules (or atoms) in direct contact with other molecules (or atoms). This mechanism explains how a coffee cup is heated by hot coffee.

2) ***Convection*** is the physical motion of material. Examples of this process would include hot air (or water) rising and cold air (or water) sinking. Rigid materials cannot have convection.

3) ***Radiation and Absorption*** are the results of heat energy being transformed into radiant energy (the energy, for example, that gives us a suntan).

One excellent example of an energy-efficient vessel is the Dewar. The Dewar effectively limits heat transfer by limiting all three heat transport mechanisms. Because its glass is a poor conductor of heat, the liquids on the inside of the Dewar do not readily conduct their heat to the outside surfaces. The vacuum in the space between the two glass layers prevents heat transfer both by conduction and convection. Finally, the walls of the Dewar are silvered to prevent loss of heat by radiation and absorption.

2.5.3 Expansion-based Thermometers

Because most materials expand as heat increases, the measurement of such expansion is used as a basis of heat measurement. Because the expansion of most materials is reasonably constant across a given range of temperatures, the amount of expansion can be quantified by a *coefficient of expansion* formula.

A *linear coefficient of expansion* is based on the following formula:

$$\alpha = \frac{L_t - L_0}{L_0 \Delta t} \qquad \text{Eq. (2-3)}$$

where α = linear expansion
L_t = final length
L_0 = original length
Δt = change in temperature.

A *volumetric coefficient of expansion* is based on the following formula:

$$\beta = \frac{V_t - V_0}{V_0 \Delta t} \qquad \text{Eq. (2-4)}$$

where β = volume expansion
V_t = final volume
V_0 = original volume
Δt = change in temperature.

A *pressure coefficient of expansion* is based on the following formula:

$$\beta_v = \frac{P_t - P_0}{P_0 \Delta t} \qquad \text{Eq. (2-5)}$$

where β_v = volume expansion
P_t = final volume
P_0 = original volume
Δt = change in temperature.

As can be seen, all coefficients of expansion are based on the amount of size change divided by the product of the original size and the change of temperature that occurred. The result of this type of equation can be calculated for any material. Because a coefficient of expansion is not necessarily consistent across a range of temperatures, coefficient of expansion tables (or listings) will be an average across a given temperature range. For example, to state that the coefficient of expansion of glass is 0.0000033 means nothing unless you specify that you are talking about laboratory borosilicate glass in the temperature range of 0 to 300°C.

It is critical to be precise about the composition and/or nature of the material being analyzed. By changing the composition of any material, even a small amount, the coefficient of expansion can be altered significantly.

2.5.4 Linear Expansion Thermometers

Linear expansion is most commonly used in bimetal spiral thermometers, which use two metals with different coefficients of expansion (see Fig. 2-28). The two metals can be welded, soldered, or even riveted together. As the metals are heated, the metal with the greater expansion will cause the spiral to flex open or close depending on which side the metal with the greater coefficient of expansion is on. A reverse in temperature will cause a commensurate reversal in the flexing.

Spiral thermometers are easily recognized as part of most room thermostats. They also are used in meat and oven thermometers.

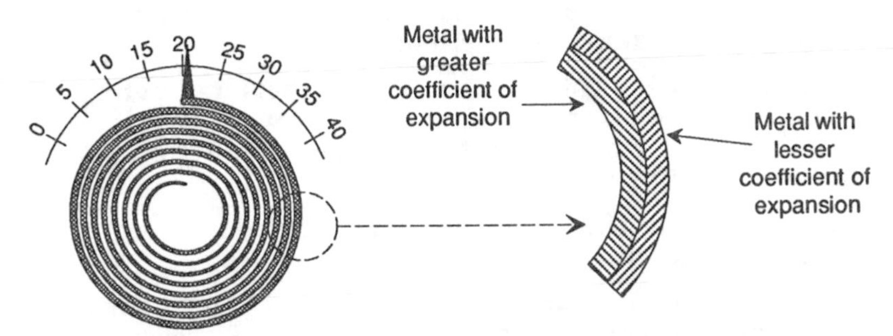

Fig. 2-28 Bimetal thermometers use two metals of different expansion to create spiral thermometers.

2.5.5 Volumetric Expansion Thermometers

When you mention thermometers, *volumetric expansion thermometers* are what typically come to mind (see Fig. 2-29). The material that expands within a volumetric expansion thermometer is typically mercury or (ethyl) alcohol. Another name for a volumetric expansion thermometer is a liquid-in-glass thermometer.

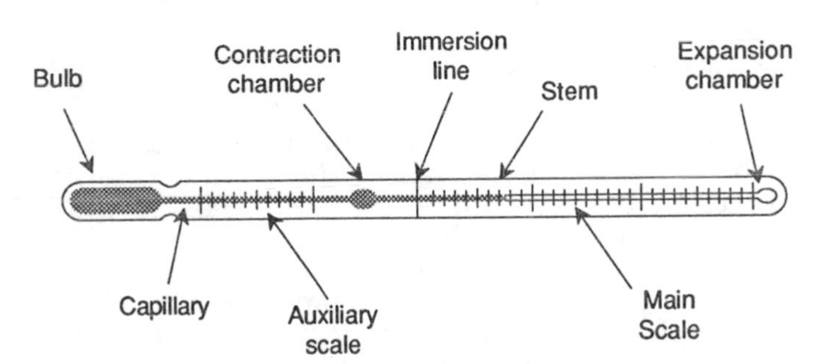

Fig. 2-29 The principle features of the solid-stem liquid-in-glass thermometer.‡

‡ From Figure 3 from the NBS Monograph 90, "Calibration of Liquid-in-Glass Thermometers," by James F. Swindells. With permission.

The parts of a standard liquid-in-glass laboratory thermometer are as follows:*

1) **The *bulb*.** The storage area for the liquid. The size of the bulb is based on the size of the thermometer.
2) **The *stem*.** The main shaft of the thermometer.
3) **The *capillary*.** The channel that carries the liquid up the stem. The narrower the capillary, the greater the accuracy that can be achieved. However, at a certain point, temperature readings are affected by surface tension of the liquid and the glass of

* Not all liquid-in-glass thermometers have all these parts.

the thermometer, so manufacturers are limited as to how accurate a liquid-in-glass thermometer can be.

4) **The *main scale*.** This scale is where the temperature is read. Some thermometers are designed to read a specific temperature range for a specific test. A doctor's thermometer is one example of this type of scale.

5) **The *immersion line*.** Sets the placement depth for partial-immersion thermometers.

6) **The *expansion chamber*.** An expanded region at the top of the capillary designed to prevent a buildup of excessive pressure from the expanding liquid.

7) **The *contraction chamber*.** Used to reduce the necessary length of a thermometer when the desired temperature range would otherwise require a very long thermometer.

8) **The *auxiliary scale*.** Required on thermometers whose calibrated region does not include an IPTS (International Practical Temperature Scale) calibration point. For example, say you have a thermometer with the range of 20° to 80°C. The auxiliary scale would include the range of -5° to 5°C so that the thermometer could be verified against the triple point of water.

Along the shaft of the thermometer, above the liquid in the thermometer capillary, is an air space typically filled with nitrogen. The nitrogen is under pressure to prevent condensation of the liquid in the upper portions of the thermometer. The pressure of the gas in the confined space will vary according to changes in temperature. Therefore, exposing the air space of a thermometer to unusually hot or cold temperatures can affect readings.

Within the bulb is a large repository of the expansion liquid. However, be aware that you cannot obtain an accurate temperature reading by placing just the thermometer bulb in the test material. When only the thermometer's bulb is under the heat's influence, the amount of expansion (or contraction) of the liquid beyond the bulb region is unknown. Any liquid not immersed in the sample being measured is not under the same influence as the liquid that is immersed. For example, if the bulb were placed in a boiling solution while the stem was in an arctic frost, the liquid in the stem would be contracted more than it would be if the stem was in a warm room.

It is possible to compensate on the calibration lines for these limitations to a certain degree. To make this compensation, three different types of liquid-in-glass thermometers have been designed with three different immersion requirements. They are:

1) ***Total-immersion thermometers*.** Thermometers that require the liquid in the stem to be completely immersed in the measured liquid. The placement of the thermometer must be adjusted during use so that the liquid in the bulb and stem are always immersed in the sample. These thermometers are the most accurate.

2) ***Complete-immersion thermometers.*** Thermometers that require the entire thermometer to be immersed in the measured liquid.
3) ***Partial-immersion thermometers.*** Thermometers that require only the bulb and a specified portion of the stem to be immersed in the measured liquid. There will be a mark or a line on the thermometer stem designating how far into the material the thermometer must be placed. The standard partial-immersion thermometer has a line 76 mm (3 inches) from the end of the bulb.

There are many specialized thermometers available. Some are used to obtain maximum and minimum temperatures, others for specific tests. The ASTM has defined a series of special partial-immersion thermometers for specific tests. These thermometers are identified as *ASTM thermometers* and are marked with a number followed by a 'C' (for centigrade) or an 'F' (for Fahrenheit). The number is strictly an identifying number with no relation to the temperature range the thermometer can read. Because these thermometers are specialized, they have immersion lines at unique locations on their stems.

Some thermometers have standard taper joints or ridges on the body of the glass to fit specific equipment such as distillation or melting point apparatus. These thermometers provide two types of position control. They set the bulb at just the right height within specialized equipment and insure the liquid column is sufficiently immersed in the heated sample. However, most thermometers do not have built-in controls and the user must not only select the right thermometer, but also adjust the thermometer to its proper level within the equipment.

The use of complete-immersion thermometers is fairly obvious. However it is not always possible, or practical, to completely immerse a thermometer. For example, if a sample solution is not transparent, it is not possible to see the temperature.

There are tables that provide correction values for readings made when total-immersion thermometers are not sufficiently immersed. In the absence of such tables, use the formula for calculating *stem correction* given in the following equation.

$$\text{Stem correction} = Kn(t_b - t_s) \qquad \text{Eq. (2.6)}$$

where K = the differential expansion of the liquid
n = number of units (in degrees) beyond the immersed stem section
t_b = is the temperature of the bath
t_s = is the temperature of the liquid column (a second thermometer is required for this reading).

Some general values for K* are:
K = 0.00016 for centigrade mercurial thermometers; and
K = 0.001 for centigrade organic liquid thermometers.

* Although the specific value of K varies as the mean temperature of the thermometer liquid varies, these values are sufficient for most work.

To place stem corrections into the proper light, consider the following example:

Thermometer reading: 105°C
Temperature of thermometer stem: 37°C
Number of units (in degrees) of stem beyond immersed liquid: 43

stem correction = 0.00016 x 43(105°C - 37°C)
stem correction = 0.47°C

Final thermometer reading: 105.47°C

Although this stem correction is relatively small, stem corrections of 10 to 20 degrees are not out of the question. Stem correction may be unnecessary depending on the difference between the sample and room temperature, the temperature ranges you are working with, or your tolerance requirements. It is a good practice to see what the stem correction would be to see if it is significant or not before assuming that it is not necessary.

Table 2-33‡

Tolerance (±) and Accuracy for Mercury Thermometers

	Total-immersion			**Partial-immersion**		
Temperature range in degrees	**Graduation interval in degrees**	**±°C**	**Accuracy in degrees**	**Graduation interval in degrees**	**±°C**	**Accuracy in degrees**
Thermometers graduated under 150°C						
0 up to 150°	1.0 or 0.5	0.5	0.1 to 0.2	1.0 or 0.5	1.0	0.1 to 0.3
0 up to 150°	0.2	0.4	0.02 to 0.05	1.0 or 0.5	1.0	0.1 to 0.5
0 up to 100°	0.1	0.3	0.01 to 0.03			
Thermometers graduated under 300°C						
0 up to 100	1.0 or 0.5	0.5	0.1 to 0.2	1.0	1.0	0.1 to 0.3
Above 100 up to 300	1.0 or 0.5	1.0	0.2 to 0.3	1.0	1.5	0.5 to 1.0
0 up to 100	0.2	0.4	0.02 to 0.05			
Above 100 up to 200	0.2	0.5	0.05 to 0.1			
Thermometers graduated above 300°C						
0 up to 300	2.0	2.0	0.2 to 0.5	2.0 or 1.0	2.5	0.5 to 1.0
Above 300 up to 500	2.0	4.0	0.5 to 1.0	2.0 or 1.0	5.0	1.0 to 2.0
0 up to 300	1.0 or 0.5	2.0	0.1 to 0.5			
Above 300 up to 500	1.0 or 0.5	4.0	0.2 to 0.5			

‡ From Tables 5 and 7 from *Liquid-in-Glass Thermometry*, NBS Monograph #150 by Jacquelyn A. Wise. Printed by the U.S. Government Printing Office. With permission.

If you make stem corrections, be sure to indicate this fact in any work you publish. Likewise, when temperature measurements are cited in literature, and no stem correction is mentioned, it is safe to assume that no stem correction was made.

Partial-immersion thermometers have a greater tolerance (and therefore less precision) than total immersion thermometers.* Interestingly enough,

* Because partial-immersion thermometers are designed for a specific test, uniformity of procedure is more important than overall accuracy.

when a total immersion thermometer is only partially immersed and no stem correction is made, the accuracy is likely to be less than a partial-immersion thermometer.

The tolerance ranges for all thermometer designs are quite different from tolerance ranges for other calibrated laboratory equipment such as volumetric ware (see Sec. 2.3). Tolerance varies mostly with graduation ranges and secondly with whether the thermometer is of total- or partial-immersion design. Table 2-33 shows NIST tolerance and accuracy limitations. Remember that tolerance is a measure of error (±), or how different a measurement is from the real value. A large tolerance indicates less accuracy and a small tolerance indicates greater accuracy. Accuracy is the agreement of the thermometer reading to the actual temperature after any correction is applied. Conditions that can affect a thermometer's accuracy include variations in capillary diameter and external pressure variations on the bulb.

2.5.6 Short- and Long-term Temperature Variations

Thermometers do not maintain their accuracy over time. Depending on how they are used, they are subject to short-term or permanent changes in measurement. The problem stems from the fact that the density of glass and the volume of the bulb change as temperature changes. If a thermometer is brought to a high temperature and allowed to cool very fast, its density may not return to the original density. Under these conditions the glass may 'set' to a new density while cooling and may thereafter be too viscous to return to the originally manufactured density. By selecting special types of glass, thermometer designs, and manufacturing techniques, manufacturers limit the amount of temperature-caused errors as much as possible.*

Generally, using a thermometer only within the scale range for which it was designed should limit changes. Overheating a thermometer (not designed to be used for high temperatures) above 260°C should be avoided. When change occurs, it typically results in a low reading that is often called an *ice-point depression.* During temperature-caused changes, the bulb increases in size as the temperature increases, but it does not contract to its original size as the temperature returns to normal. Thereafter, once re-calibrated at the ice-point, the thermometer temperature will always read less than the real temperature.

These changes may either be temporary or permanent, often depending on whether the thermometer was cooled slowly through the higher temperature regions, or simply removed and haphazardly laid on a table. Depending on a thermometer's glass quality, the hysteresis** effect can cause from 0.01 to 0.001 "of a degree per 10 degrees difference between the temperature being measured and the higher temperature to which the thermometer has recently been exposed."[15]

A thermometer will indicate short-term changes when used below 100°C, and will recover an error of 0.01 to 0.02 degrees in several days. Admittedly, these are small amounts of error. Greater amounts of temporary

* There is likely to be changes in the stem volume as well, but the amount of those changes is likely to be negligible.

** Hysteresis is when a material is stretched or distorted to a new position or shape, and does not return to its original position or shape.

error can be exhibited when a thermometer is used for relatively high temperature readings and then is immediately used for relatively low temperature readings. Thus, it may be advisable to separate thermometers not only by their temperature ranges, but also by the temperature ranges for which they are used. This practice is especially important if your work requires accurate thermometric readings.

Long-term changes are generally caused when a thermometer is maintained at high temperatures for extended periods of time (> several hundred hours).[16] Such conditions can cause errors as great as 12°C. It is also possible to cause permanent changes from repeated cycling at lower temperatures between -30°C and room temperature.[17]

It is uncommon to observe an alteration due to a temperature-induced change in the readings along selective parts of the thermometer scale. Typically they are more of a dropping of the entire scale reading. Thus, periodically checking the zero-point with an ice bath, or water boiling-point with a steam bath, can ensure the accuracy of your thermometer. It is not necessary to re-calibrate the entire stem of the thermometer.

2.5.7 Thermometer Calibration

Steam bath calibration for the liquid-steam point of H_2O is not easy for many labs to use because of its expense and the equipment is difficult to use. In addition, the user needs to account for atmospheric pressure variations and the effects of `local variations of gravity on the barometer. Fortunately, the ice-point calibration is easy to set up and use.

Any thermometer with 0.0°C on either its main or auxiliary scale can be calibrated with an *ice-bath calibration apparatus* that can be assembled with the following material:

1) A 1- to 1-$^1/_2$ liter Dewar flask, cleaned and rinsed with distilled water
2) Crushed ice made from distilled water
3) A (clean) siphon to remove excess melted ice.

Fill the Dewar approximately one-third full with distilled water. Place the end of the siphon tube to this level, and then fill the Dewar to the top with crushed ice.

The thermometer should be inserted so that the thermometer is immersed to the 0.0°C level (if calibrating a complete-immersion thermometer) or inserted to the immersion line (if calibrating a partial-immersion thermometer). The thermometer is likely to require support to maintain its proper position. Let the entire apparatus sit for fifteen to thirty minutes, to reach equilibrium. Periodically add more ice, as needed, and remove any excess water with the siphon. If the ice is kept clean and tightly packed around the thermometer bulb and stem, it is possible to achieve accuracy to within 0.01°C.

If you are concerned about accurately calibrating the entire scale of a particular thermometer, it may be sent to the NIST for calibration for a fee. You will need to contact the NIST for more information on that service.

Once calibrated, any variations in the zero point should not significantly alter the corrected calibrations along the scale.

NIST-calibrated thermometers are expensive, but very accurate tools. Unfortunately, they require special use and maintenance to maintain their integrity. Not only can abuse alter their calibration, but general use can as well. For example, if you are using an NIST-calibrated liquid-in-glass thermometer on a regular basis, an ice-point re-calibration should be taken after each measurement. These variations should be added to the adjustments made to the corrected scale temperatures.

2.5.8 Thermometer Lag

Liquid-in-glass thermometers require a finite amount of time to achieve a final, equilibrium temperature. The time required can vary for individual thermometer types depending on the diameter of the thermometer, the size and volume of the bulb, the heat conductivity of the material into which the thermometer is placed, and the circulation rate of that material.

For taking the temperature of a material that maintains a constant-temperature, thermometer lag is only a nuisance. Most laboratory thermometers placed in a 75°C bath will come within 0.01°C of the final temperature within 19 to 35 seconds of first contact. The time range is dependent on the type and design of the thermometer.

When attempting to measure a changing temperature, the problem is more critical because you are limited to knowing what the temperature was, not is. Because of the thermometer's lag time, you will be unable to know any specific temperature instantaneously. Thus, the greater the rate of temperature increase, the less a thermometer is able to keep up with the temperature change. Interestingly enough, if the temperature change varies uniformly, White[18] found that any thermometer lag is likely to be canceled because any rate-of-change will not have been altered.

2.5.9 Air Bubbles in Liquid Columns

Sometimes, because of shipping, general use, or sloppy handling, the liquid column of a thermometer will separate, leaving trapped air bubbles. Because of the capillary size, it is difficult for the liquid to pass by the air space and rejoin. Sometimes such separations are glaringly obvious. Other times, the amount of liquid separation is small and is difficult to see.

One technique for monitoring the quality of a thermometer is to use a second thermometer for all measurements. Any disagreement between the temperatures from the two thermometers may be caused by a problem as air breaks in the liquid columns. If both thermometers have breaks in their liquid columns, it is almost impossible for them to agree across a range of temperatures. Detection of liquid breaks is virtually guaranteed. Unfortunately, this technique will not identify which thermometer is at fault.

There are two techniques for rejoining separated liquid ends: heating and cooling. The option chosen will depend on the location of the break, and the existence and location of contraction or expansion chambers.

Generally it is better (and safer) to cool the liquid in thermometer than it is to heat it. First, try to cool the thermometer with a (table) salt-and-ice slush bath. This method should bring the liquid into the contraction chamber or bulb. Once the liquid is in the chamber or bulb it should rejoin, leaving the air bubble on top. If there is not a clean separation of the air bubble, it may be necessary to softly tap the end of the thermometer. This tapping should be done on a soft surface such as a rubber mat, stopper, or even a pad of paper. Alternatively, you may try swinging the thermometer in an arc (such as a nurse does before placing it in your mouth).* Once joined, the liquid in the thermometer can slowly be reheated.

If the scale or design of the thermometer is such that these methods will not work, try touching the bulb end to some dry ice. Because dry ice is 38°C colder than the freezing point of mercury, you must pay attention and try to prevent the mercury from freezing. As a precaution, when reheating, warm from the top (of the bulb) down to help prevent a solid mercury ice plug from blocking the path of the expanding mercury into the capillary. With no path to expand into, the expanding liquid may otherwise cause the bulb to explode.

An alternate technique to rejoin broken liquid columns is to expand the liquid into the contraction or expansion chamber by heat. Be careful to avoid filling the expansion chamber more than two-thirds full, as extra pressure may cause the top of the thermometer to burst. Never use an open flame to intentionally heat any part of a thermometer as the temperature from such a source is too great and generally uncontrollable.

If the thermometer has no expansion or contraction chamber, heating the bulb should not be attempted because there is no place into which the liquid may expand to remove bubbles. Occasionally it may be possible to heat the upper region of the liquid along the stem (do not use a direct flame; use a hot air gun or steam). Observe carefully, and look for the separated portion to break into small balls on the walls of the capillary. These balls may be rejoined to the rest of the liquid by slowly warming the bulb of the thermometer so that the micro-droplets are gathered by the expanding mercury. Once collected, the mercury may be slowly cooled.

In addition to air bubbles in the stem, close examination of thermometer bulb may reveal small air bubbles in the mercury. Carefully cool the mercury in the bulb until the liquid is below the capillary. While holding the thermometer horizontal, tap the end of the thermometer against your hand (you are trying to form a bubble within the bulb). Next, rotate the thermometer, allowing the large air bubble to contact the internal surface of the bulb, capturing the small air bubbles. Once the small air bubbles are 'caught,' the thermometer can be reheated as before.

IMPORTANT: Many states are limiting the use of mercury in the laboratory. Such laws severely limit the use of mercury-in-glass thermometers despite the fact that the mercury is isolated from the environment in normal

* Medical thermometers have a small constriction just above the bulb that serves as a gate valve. The force of the mercury expanding is great enough to force the mercury past the constriction. However, the force of contraction is not great enough to draw the mercury back into the bulb. The centrifugal force caused by shaking the thermometer is great enough to draw the mercury back into the bulb.

use. Unfortunately, the laws apply because a broken thermometer can easily release its mercury into the lab. Fortunately, thermocouples (see Sec. 2.5.11) are relatively inexpensive, and the price of controllers has gone down significantly—while the accuracy of thermocouples and the ease of their use have gone up significantly. Thermocouples should be considered as an alternative even if you are not receiving pressure to reduce mercury use because of ease of use and robust design.

2.5.10 Pressure Expansion Thermometers

Gas law theory maintains that pressure, volume, and temperature have an interdependent relationship. If one of these factors is held constant and one changes, the third has to change to maintain an equilibrium. At low temperatures and pressures, gases follow the standard gas law equation much better than they do at other conditions (see Eq. 2-7).

$$PV = nRT \qquad \text{Eq. (2-7)}$$

where P = pressure (in torr)
and V = Volume (in Liters)
n = number of molecules
R = Gas constant (62.4)
T = temperature (in K).

The typical *pressure expansion thermometer* is a volume of gas that maintains the same number of molecules throughout a test. Because the volume, number of molecules, and gas constant are all constant, any drop in temperature will subsequently cause a drop in pressure. Likewise any rise in temperature will cause a rise in pressure.

These thermometers are difficult to use, sensitive, and require a great deal of supporting equipment. They are typically used for the calibration of other (easier to use) thermometers. In addition, they are used to determine the temperatures of melting, boiling, and transformation points of various materials. Because of their limited use in the average laboratory, further discussion of pressure expansion thermometers is beyond the scope of this book.

2.5.11 Thermocouples

Thermocouples are robust and inexpensive. Reading the temperature from a thermocouple is invariably as easy as reading a controller's dial, liquid crystal, or LED.* The following commentary is intended to provide a basic understanding of the operation of thermocouples and their limitations. If you have any questions about the final selection of a thermocouple, controller, lead, and extension for any specific type of job or desired use, contact a thermocouple manufacturing company as it will likely provide you with the best information on how to make your system work.

* LED stands for Light Emitting Diode. These devices are the lighted numerals typically seen on digital clocks.

Table 2-34

Thermocouple Types and Characteristics			
ANSI Symbol	**Composition** (pos. – neg. lead)	**Temperature Range (°C)** (Thermocouple) (Extension)	**Comments**
J	Iron - Constantan (Cu-Ni)	0 to 750 0 to 200	Recommended for reducing atmosphere. Displays poor conformance characteristics due to poor iron purity. Above 538°C, oxidation of Fe in air is rapid so heavy-gauge wire is recommended for extended use. Should not be used bare in sulfurous atmospheres above 538°C.
T	Copper – Constantan (Cu-Ni)	-200 to 350 -60 to 100	Recommended for mildly oxidizing and reducing atmospheres as well as moist environments. Performs very well at low temperatures and limits of error are guaranteed in the sub-zero range although type E may be preferred due to higher emf. The copper lead makes secondary compensation unnecessary.
K	Chromel – Alumel	-200 to 1250 0 to 200	Recommended for clean oxidizing atmospheres. They can be used in reducing atmospheres, but they should not be cycled from reducing to oxidizing and back repeatedly. The higher temperature ranges can only be achieved with heavy-gauge wires. They should not be used in vacuums or bare in sulfurous atmospheres.
E	Chromel – Constantan (Cu-Ni)	-200 to 900 0 to 200	Recommended for vacuum use or inert, mildly oxidizing or reducing atmospheres. These thermocouples produce the greatest amount of emf and therefore can detect small temperature changes.
S	Platinum – Platinum 10%, Rhodium	0 to 1450 0 to 150	Recommended for use in oxidizing or inert atmospheres to 1400°C although may be brought to 1480°C for a short time. Most stable, especially at high temperatures. Type S is used for standard calibration between the antimony point (630.74°C) and gold point (1064.43°C). Should only be used inside non-metallic sheaths such as alumina due to metallic diffusion (from metal sheaths) contaminating the Pt.
R	Platinum – Platinum 13%, Rhodium	0 to 1450 0 to 150	Recommended for use in oxidizing or inert atmospheres to 1400°C although may be brought to 1480°C for a short time. Should only be used inside non-metallic sheaths such as alumina.
B	Platinum – Platinum 30%, Rhodium	50 to 1700 0 to 100	These thermocouples are excellent for vacuum use. The output between 0 and 50°C is virtually flat and therefore is useless at these temperatures. The reference junction can be between 0° to 40°C. Should only be used inside non-metallic sheaths such as alumina.
N[a]	Omega-P™ "Nicrosil" (Ni-Cr-Si) – Omega-N™ "Nisil" (Ni-Si-Mg)	-270 to 1300 0 to 200	Similar to type K, but with improved oxidation resistance at higher temperatures.
G[a]	Tungsten – Tungsten 26%, Rhodium	0 to 2320 0 to 260	Useful in reducing or vacuum environments at high temperature, but not for oxidizing atmospheres.
C[a]	Tungsten 3% Rhodium – Tungsten 26%, Rhodium	0 to 2320 0 to 870	Useful in reducing or vacuum environments at high temperature, but not for oxidizing atmospheres.
D[a]	Tungsten 3% Rhodium – Tungsten 25%, Rhodium	0 to 2320 0 to 260	Useful in reducing or vacuum environments at high temperature, but not for oxidizing atmospheres.

[a] Not ANSI designations.

In 1821 Thomas Seebeck discovered that when two different types of metal wires were joined at both ends and one of the ends was heated or cooled, a current was created within the closed loop (this current is now called the Seebeck Effect). Specifically, heat energy was transformed into measurable electrical energy.

Unfortunately, there is not enough current produced to do any work, but there is enough to measure (1 to 7 millivolts). This rise in potential energy is called the *emf* or *electromotive force.*

By hooking the other end of joined dissimilar wires up to a voltmeter and measuring the emf (output), it is possible to determine temperature. Such a temperature-measuring device is called a thermocouple.

There are seven common thermocouple types as identified by the American National Standards Institute (ANSI). They are identified by letter designation and are described in Table 2-34. There are four other thermocouple types that have letter designations; however, these four are not official ANSI code designations as one or both of their paired leads are proprietary alloys. They are included at the end of Table 2-34. Although Table 2-34 lists the standard commercially-available thermocouples, there are technically countless other potential thermocouples because all that is required for a thermocouple is two dissimilar wires.

At a basic level, it seems that one should be able to take any thermocouple, attache it to a voltmeter, and determine the amount of electricity generated from the heat applied to the dissimilar junction. Then the user could look in some predetermined thermocouple calibration table* to determine the amount of heat the that amount of electricity from that specific type of joined wires produces.

One of the negative complications of thermocouples is that they do not have a linear response to heat. In addition, as temperature changes, thermocouples do not produce a consistent emf change. Therefore, there must be individual thermocouple calibration tables for each type of thermocouple.

If you look at a thermocouple calibration table, you will see that it has a *reference junction at 0°C.* This reference point is used because of an interesting complication that arises when a thermocouple is hooked up to a voltmeter. To explain this phenomenon, first look at one specific type of thermocouple, a type T (copper-constantan design). Also, assume that the wires in the voltmeter are all copper. Once the thermocouple is hooked to the voltmeter, we end up having a total of three junctions (see Fig. 2-30):

Junction 1) The original thermocouple junction of copper and constantan.
Junction 2) The connection of the constantan thermocouple wire to the copper wire of the volt meter.
Junction 3) The connection of the copper thermocouple wire to the copper wire of the volt meter.

We are hoping to find the temperature of Junction 1, but it seems we now have two more junctions with which to be concerned. Fortunately,

* Thermocouple calibration tables have been compiled by the NIST and can be found in a variety of sources such as the *Handbook of Chemistry and Physics* by the Chemical Rubber Company , published yearly.

Junction 3 can be ignored because this connection is of similar, not dissimilar metals. Therefore, no emf will be produced at this junction. Junction 2 presents a problem because it will produce an emf, but of an unknown temperature; meaning that, we have two unknowns in the same circuit, and we are unable to differentiate between the two. Any voltmeter reading taken at this point will be proportional to the temperature difference between the first and third junction.

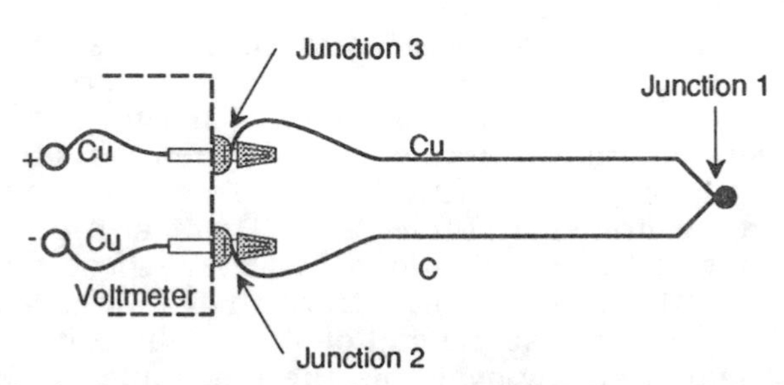

Fig. 2-30 A simple diagram of a copper-constantan thermocouple.‡

‡ This illustration is from *The Temperature Handbook* ©1989 by Omega Engineering, Inc., and is reproduced with the permission of Omega Engineering Inc.

If we take Junction 2 and place it in a known temperature, we can eliminate one of the two unknowns. This procedure is done by placing Junction 2 (now known as the reference junction) in an ice bath of 0°C (known as the reference temperature). Because the temperature (voltmeter) reading is based on an ice-bath reference temperature, the recording temperature is referenced to 0°C (see Fig. 2-31). References for voltmeter/thermocouple readings (found in such books as the CRC Handbook) will usually indicate that they are referenced to 0°C. Because Junctions 3 and 4 are of similar metals, they have no effect on the emf.

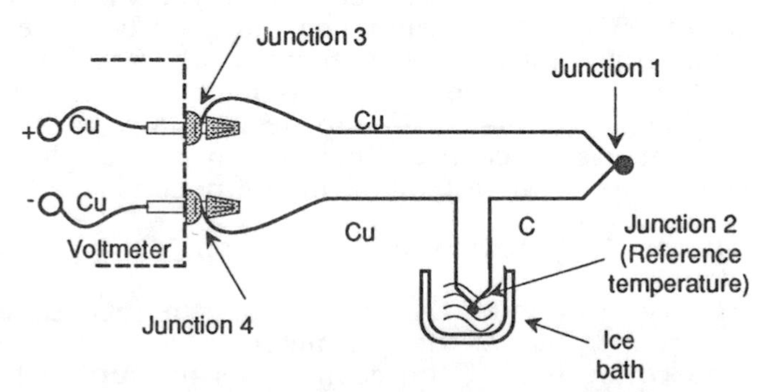

Fig. 2-31 A thermocouple using an ice bath for a reference temperature.‡

‡ This illustration is from *The Temperature Handbook* ©1989 by Omega Engineering, Inc., and is reproduced with the permission of Omega Engineering Inc.

This example is limited because it is representative only of thermocouples with copper leads. In all other thermocouples, there are likely to be four dissimilar metal junctions and therefore up to four Seebeck Effects.

However, by extending the copper wires from the voltmeter and attaching them to an isothermal block, placing the same type of wire from one of the thermocouple leads to the part between the ice-bath and the isothermal block, and then attaching the thermocouple wires to the isothermal block (to eliminate thermal differences at these two junctions), it is possible to cancel out all but the desired Seebeck Effect (see Fig. 2-32).

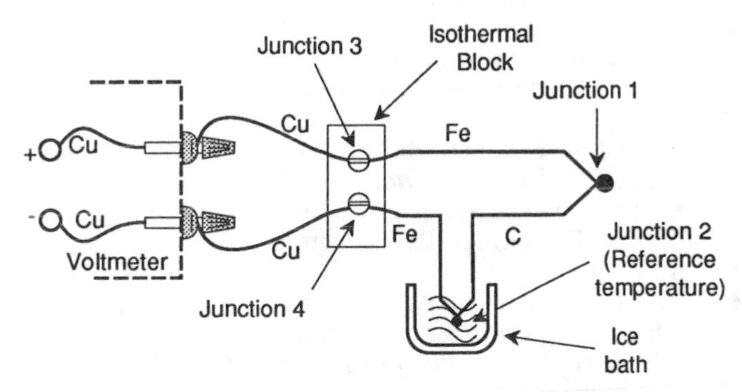

Fig. 2-32 An iron-constantan thermocouple using an isothermal block and an ice bath.‡

‡ This illustration is from The Temperature Handbook ©1989 by Omega Engineering, Inc., and is reproduced with the permission of Omega Engineering Inc.

It is obvious that the use of an ice bath is inconvenient and impractical. Maintaining a constant 0.0°C temperature can be difficult, and there always is the possibility of the ice-bath tipping over. Fortunately there are two convenient techniques to circumvent the need for an actual ice bath for thermocouple measurements.

One approach requires measuring the equivalent voltage of the ice-bath reference junction and having a computer compensate for the equivalent effect. This technique, called *software compensation*, is the most robust and easiest to use. However, depending on the equipment used, the lag time involved for the compensation may be unacceptable.

Alternatively, it is possible to have *hardware compensation* by inserting an electric current (typically from a battery) to provide a voltage which electronically offsets the reference junction. This approach is called an *electronic ice point reference*. This method has the simplicity of an ice bath, but is far more convenient to use. The major disadvantage is that a different electric circuit is required for each type of thermocouple.

Ironically, the sophistication of modern controllers typically makes most of the concerns about how temperature is determined irrelevant. The greatest concern for the user is to select the right type and size of thermocouple for the specific job and environment, then to select the proper controller for that type of thermocouple.

Although there are many overlapping temperature ranges among various thermocouple types, all types of thermocouples perform better for some jobs than others. For example, Type K is significantly less expensive than Type R, although both thermocouples can read into the 1000°C range. This similarity would lead one to believe that either choice is adequate, however Type K is preferred because of its lower cost. If all you need are occasional readings of such temperatures, you could get by with Type K. However, if you expect to do repeated and constant cycling up to temperatures, in the 1000°C range, Type K would soon fail and would need to be replaced. Unfortunately, you are not likely to notice failure developing until something has obviously gone wrong. Because the potential costs of failure are likely to be greater than the additional cost of the more expensive thermocouple, it may be more economical to obtain the more expensive type R at the outset.

Sample size should also be considered when selecting thermocouple size. If a large thermocouple (with a high heat capacitance) is used on a small sample (with a low heat capacitance), the sample's temperature could be changed. As stated before: *the act of taking an object's temperature can change its temperature.* Because thermocouples come with varying wire diameters (see Table 2-35), select the thermocouple wire size best suited to measure your sample.

Be advised that the upper range of temperatures cited (within thermocouple catalogs) for a given thermocouple are related to the larger wire sizes. Thus, a small-wire thermocouple is likely to fail at the upper temperature ranges for that particular thermocouple type.

Table 2-35

American Wire Gauge Size Comparison Chart

AWG[a]	Dia. mils	Dia. mm
8	128	3.3
10	102	2.6
12	81	2.1
14	64	1.6
16	51	1.3
18	40	1.0
20	32	0.8
22	25	0.6
24	20	0.5
26	16	0.4
28	13	0.3

[a] American Wire Gauge.

Environment also influences the selection of thermocouples. The condition of the region where the thermocouple will be placed can be oxidizing, reducing, moist, acidic, alkaline, or present some other condition that could cause premature failure of the thermocouple. Selection of the right type of thermocouple can help avoid premature failure. Fortunately, there are sleeves and covers available for thermocouples that prevent direct contact with their various environments. These covers are made of a variety of materials, from metals to ceramics, making selection of the right material easy. On the other hand, covers add to the heat capacitance of the entire probe, and therefore can slow thermocouple response time and, because of a greater heat capacity, they are more likely to affect the temperature of the material being studied.

2.5.12 Resistance Thermometers

In 1821, Sir Humphrey Davy discovered that as temperature changed, the resistance of metals changed as well. By 1887 H.L. Callendar completed studies showing that purified platinum wires exhibited sufficient stability and reproducibility for use as thermometer standards. Further studies brought the Comité International des Poids et Measures in 1927 to accept the *Standard Platinum Resistance Thermometer* (SPRT) as a calibration tool for the newly-adopted practical temperature scale.

Platinum resistance thermometers are currently used by the NIST for calibration verification of other thermometer types for the temperature range 13.8 to 904 K. In addition, they are one of the easiest types of thermometers to interface with a computer for data input. On the other hand, platinum resistance thermometers are very expensive, extremely sensitive to physical changes and shock, have a slow response time and therefore can take a long time to equilibrate to a given temperature. Thus, resistance thermometers are often used only for calibration purposes in many labs.

Platinum turned out to be an excellent choice of materials because it can withstand high heat and is very resistant to corrosion. In addition, platinum offers a reasonable amount of resistivity (as opposed to gold or silver), yet it is very stable and its resistance is less likely to drift with time. However, because it is a good conductor of electricity, the SPRT requires a sufficiently long enough piece of platinum wire* to record any resistance.

One of the easiest ways to get a long piece of material in a small (convenient) area is to wrap or wind the material around a mandrel. The typical mandrel used on SPRTs is made of either mica or alumina. The sheaths covering the wrapped thermometer may be borosilicate, silica glass, or a ceramic (see Fig. 2-33). Note that both examples in Fig. 2-33 show four leads where, logically, there should be only two. This is done to reduce any unwanted resistance from the region beyond the thermometer.

An alternative approach to "loose winding" the wires is to form the platinum wire and then flow melted glass around the wire to 'lock' the wire in place. This method protects the wire from shock and vibration. These resistance thermometers are, unfortunately, limited to temperatures where the expanding platinum will not crack the containment glass. To limit this occurrence, the expansion of the platinum must closely match the glass around which it is wrapped. Resistance thermometers of this design are more rugged, and therefore, are more likely to be used in the laboratory.

There are many challenges in the construction and use of resistance thermometers, including:

1) The wire itself must be very pure; any impurities will affect the linearity and reliability of the resistance change. The wire itself must also be uniformly stressed, meaning that after construction, the wire must be temperature-annealed to achieve uniform density. Uniform density provides uniform resistance.
2) When the wire is wound on the support device, it must be left in a strain-free condition, as any strain can also affect the resis-

* Currently, about 61 cm of 0.075 mm platinum wire are typically used on resistance thermometers.

tance of metals. Because the platinum will expand and contract as the temperature changes, it must be wound in such a manner that there is no hold-up or strain. Even this strain could affect the resistance of the metal.

3) Although it is doubtful that a lab will make its own SPRTs, I mention construction demands to impress upon the user the importance of maintaining a strain-free platinum wire. Strain on the wire can be introduced not only during construction, but also in use. The most likely opportunities for wire strain are through vibration and bumping.

4) The SPRT must not receive any sharp motions or vibrations. Such actions can affect the resistance of the metal by creating strain. To place this challenge in the proper perspective, consider a SPRT rapped against a solid surface loud enough to be heard (but not hard enough to fracture the glass sheath). The temperature readings could be affected by as much as 0.001°C. Although this variation may not seem like much, over a year's time, such poor use could cause as much as 0.1°C error.[19]

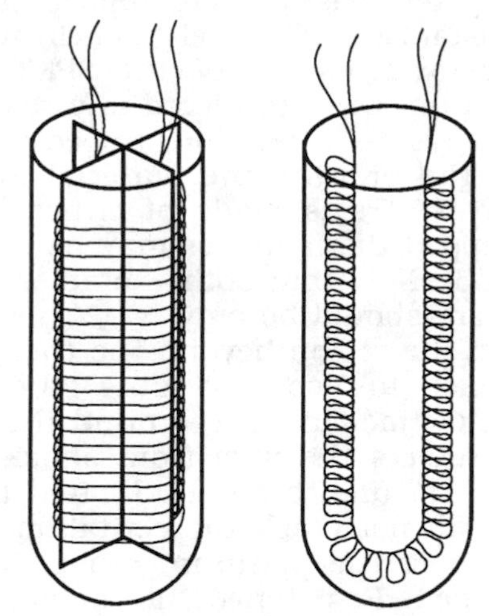

Fig. 2-33 Two different SPRT wrapping designs.

Transportation of SPRTs should be as limited as possible. If shipping, special shock-absorbing boxes are recommended. If you have an SPRT shipped to you, keep the box for any future shipping needs and storage. Calibrated SPRTs should be hand carried whenever possible, to minimize shock or vibration. For instance, you should not lay an SPRT on a lab cart for transportation down the hall. The vibration of the cart may cause changes in subsequent temperature readings.

Other precautions that should be taken with SPRTs include:

1) When SPRTs are placed into an apparatus, they should be inserted carefully, to avoid bumps and shocks.

2) Try to avoid rapid dramatic changes in temperature. A cold-to-hot change can cause strains as the wire expands within the thermometer. A hot-to-cold change can cause fracture of the glass envelope encasing the thermometer, or of any glass-to-metal seals (which are structurally weak). A hot-to-cold heat change can also cause a calibration shift of the thermometer.
3) Thermometers with covers of borosilicate glass should not be used in temperatures over 450-500°C without some internal support to prevent deformation.
4) Notable grain growth has been observed in thermometers maintained at 420°C for several hundred hours.[20] Such grain growth causes the thermometer to be more susceptible to calibration changes from physical shock and therefore, inherently unstable.

There are several inherent complications in the use of SPRTs. One involves the fact that an SPRT is not a passive responding device. What the SPRT records is the change in resistance of an electric current going through the thermometer. The mere fact that you are creating resistance means that you are creating heat. Thus, the device that is designed to measure heat also creates heat. The resolution of this situation is to 1) use as low a current as possible to create as little heat as possible and 2) use as large an SPRT as possible (the larger the SPRT, the less heat generated).

Although an SPRT is not a thermocouple, an emf is created at the junction of the SPRT's platinum wires and the controller's copper wires. Fortunately, this *emf* is automatically dealt with electronically by the controller with the *offset-compensation ohms technique* and can be ignored by the user.

References

[1] Verney Stott, *Volumetric Glassware* (London: H.F. & G. Witherby, 1928), pp. 13-14.

[2] R.B. Lindsay, "The Temperature Concept for Systems in Equilibrium" in *Temperature; Its Measurement and Control in Science and Industry*, vol. 3, F.G. Brickwedde, ed., Part 1. (New York: Reinhold Publishing Corporation, 1962), pp. 5-6.

[3] W.E. Knowles Middleton, *A History of the Thermometer and Its Use in Meteorology*, (Baltimore, Maryland: The John Hopkins Press, 1966), pp. 58-61.

[4] A.V. Astin, "Standards of Measurement," *Scientific American*, 218: (1968), pp. 50-62.

[5] E. Ehrlich, et al, "*Oxford American Dictionary*" Oxford University Press, 1980.

[6] D.R. Burfield and G. Hefter, "Oven Drying of Volumetric Glassware," *J. of Chem. Ed.*, 64 (1987), p. 1054.

[7] H.P. Williams and F.B. Graves, "A Novel Drying/Storage Rack for Volumetric Glassware," *J. of Chem. Ed.*, 66 (1989), p. 771.

[8] D.J. Austin, "Simple Removal of Buret Bubbles," *J. of Chem. Ed.*, 66 (1989), p. 514.

[9] Dr. L. Biétry, *Mettler, Dictionary of Weighing Terms*, (Switzerland: Mettler Instrumente AG, 1983), p. 12.

[10] W.E. Kupper, "Validation of High Accuracy Weighing Equipment," *Proceedings of Measurement Science Conference 1991*, Anaheim, CA.

[11] R.M. Schoonover and F.E. Jones, "Air Buoyancy Correction in High-Accuracy Weighing on Analytical Balance," *Analytic Chemistry*, 53 (1981), pp. 900-902.

[12] W.E. Kupper, "Honest Weight — Limits of Accuracy and Practicality," *Proceedings of Measurement Science Conference 1990*, Anaheim, CA.

[13] from "Weighing the Right Way with METTLER," © 1989 by Mettler Instrumente AG, printed in Switzerland.

---

[14] Dr. L. Biétry, *Mettler, Dictionary of Weighing Terms*, (Switzerland: Mettler Instrumente AG, 1983), pp. 69. 74, 100, and 118.

[15] Jacquelyn A. Wise, *Liquid-in-Glass Thermometry*, (Washington D.C.: U.S. Government Printing Office, 1976), p. 23.

[16] E.L. Ruh and G.E. Conklin, "Thermal stability in ASTM Thermometers," *ASTM Bul.*, No. 233, 35 (Oct., 1958).

[17] W.I. Martin and S.S. Grossman, "Calibration Drift with Thermometers Repeatedly Cooled to -30° C," *ASTM Bul.*, No. 231, 62 (July, 1958).

[18] W.P. White, "Lag Effects and Other Errors in Calorimetry," *Physical Review*, 31 (1910), p. 562-582.

[19] J.L. Riddle, G.T. Furukawa, and H.H. Plumb, "*Platinum Resistance Thermometry*," National Bureau of Standards Monograph No. 126, U.S. Government Printing Office, April 1972, p. 9.

[20] *ibid*, p. 11.

Chapter 3

Joints, Stopcocks, and Glass Tubing

3.1 Joints and Connections

The ability to assemble and disassemble apparatus components has been a long standing requirement in laboratories. The use of ground joints for this purpose is not new, but standardizing the taper so that one piece from one lab can be attached to another section of apparatus in another lab has only been in common use since just after World War II. Currently, there are many standardized ways to join and separate laboratory glass apparatus.

3.1.1 Standard Taper Joints

Standard Taper Ground Joints. These joints have the same taper throughout the world. This taper, a 1:10 ratio, provides for the ground section to narrow one unit in width for every ten units of length. These units are always in "mm" (see Fig. 3-1).

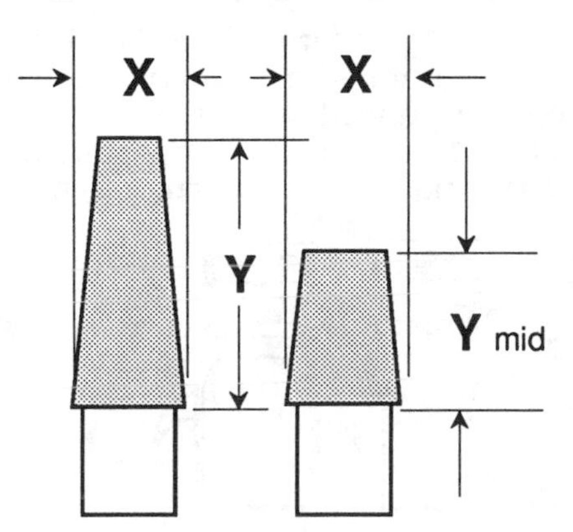

These male joints show the two areas that are measured and which thereby identify any Standard Taper "ST" joint. These pairs of numbers are always given in X, Y coordinates. The first number (X) is the widest part of the ground area. The second number (Y) is the vertical length of the ground section.

Mid-length joints differ from full-length joints only in their "Y" dimension. It is as if the short end had been cut off. The example to the left shows a regular and mid-length joint.

Fig. 3-1 Full- and mid-length standard taper joints.

Standard taper joints are used in matching inner and outer (also called male and female) members. They are made in a variety of standardized sizes and shapes. Figures 32 and 3-3 illustrate the shape variations available. In general, how a joint will be used determines which joint design should be selected.

Inner Joints. Common varieties of *inner joints* are shown in Fig. 3-2. Figure 3-2(a) is a standard full-length joint, the one most commonly seen in the lab. Figure 3-2(b) has a small 'drip tip' extended from the small end of the joint. This joint is commonly placed on condensers to prevent distillate from coming in contact with, and/or dissolving, stopcock grease. Figure 3-2(c) has an extended drip tip that a glassblower can bend or seal to other pieces of glass. The manufacturer adds these extensions during initial fabrication, before the joint is ground to the proper taper. If a glassblower were to try to seal such an extension onto a Type (a) joint, the distortion caused by the sealing action would alter the taper of the joint, necessitating regrinding. Manufacturers also provide joints that already have hooks for springs on them, as in (d).

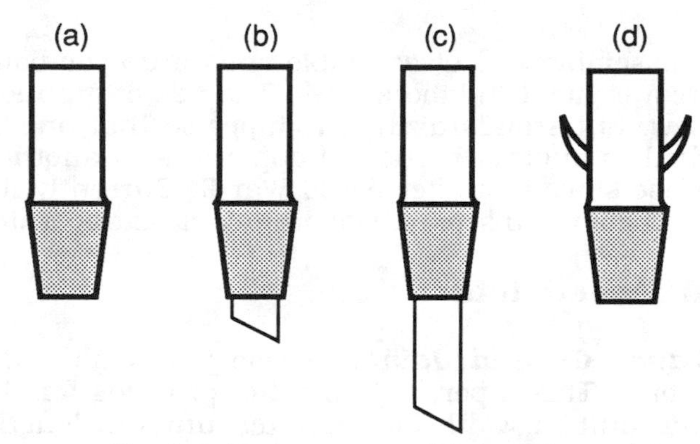

Fig. 3-2 Various types of inner joints.

Outer Joints. These joints also have a variety of forms (see Fig. 3-3). Both (a) examples have heavy walls and can receive a moderate amount of physical abuse. Example (b) has thinner walls on the ground joint section. Ribs are placed around the joint for extra strength. Thin walls are important for experiments with large temperature fluctuations that, in turn, might cause

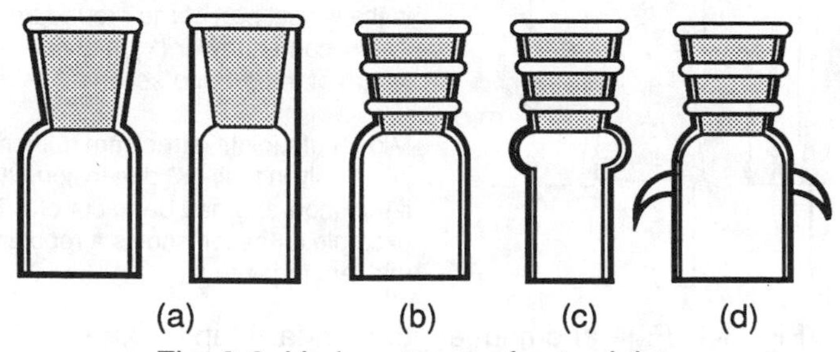

Fig. 3-3 Various types of outer joints.

glass expansion or contraction. Changes in glass size could cause separation of the inner and outer joints, thus causing leaks. Example (c) is often referred to as an "old style" joint. It can easily be identified by the small bulb just below the ground section. This bulb provides a resting place for the joint on a two-finger clamp without requiring extra pressure (see Fig. 3-4). If the joint is not held vertically, the bulb can also cause a minor hold-up of fluids. Example (d) shows outer joints with pre-attached hooks.

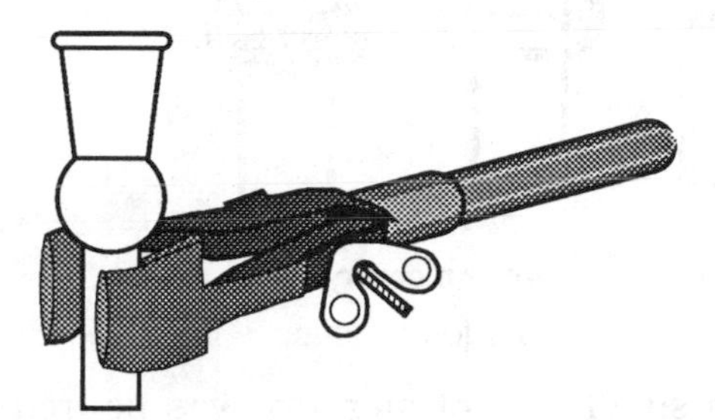

The "old" style female joint is very good to use with two-finger clamps. It is not neccessary to squeeze the wing nut tightly to securely hold the joint. Rather, the bulb of the joint rests easily on the clamp.

Fig. 3-4 Old style outer joint in a two-finger clamp.

Standard taper joints are formed by a process called "tooling," which establishes the physical form from a simple glass tube. After the joint is made, it is ground using various grades of Carborundum grinding compounds to achieve the proper tapered smooth finish. In addition to the standard ground surface joint, Wheaton Scientific Co. also makes a joint with a smooth polished glass surface. These joints are known as *Clear-Seal®* joints. Instead of being ground, the joint is heated until soft and then suctioned over a mandrel. The suctioning causes the glass to collapse onto and form to the mandrel.

Wheaton Clear-Seal® joints have the unique characteristic that they do not need any stopcock grease. However, they require very special handling and storage, because any scratch can destroy their sealing abilities. Additionally, when joint members are joined, special care must be taken to ensure that no particulate matter lodges between the pieces. Such matter can prevent a proper seal, scratch the sides of the members, and/or cause the members to jam or stick. Because no stopcock grease is used with Clear-Seal® joints, it is more difficult to separate stuck members.

Clear-Seal® joints are mandatory when using diazomethane (CH_2N_2) as a methylating reagent for carboxylic acids. Diazomethane is a toxic, irritating gas which has the tendency to self-detonate. This self-detonation can occur when the diazomethane is in either a gaseous or liquid state. It has been proven[1] that rough surfaces (such as ground joint surfaces) can initiate such detonations. Thus, only Clear-Seal® joints should be used in diazomethane chemistry.

Mercury Seal Joint. This joints is no longer used for its intended purpose because of improvements in joint manufacturing as well as safety concerns about mercury in the lab. The mercury seal joint (see Fig. 3.5(a) and (b) was used to prevent gases from passing by ground joint seals. After the

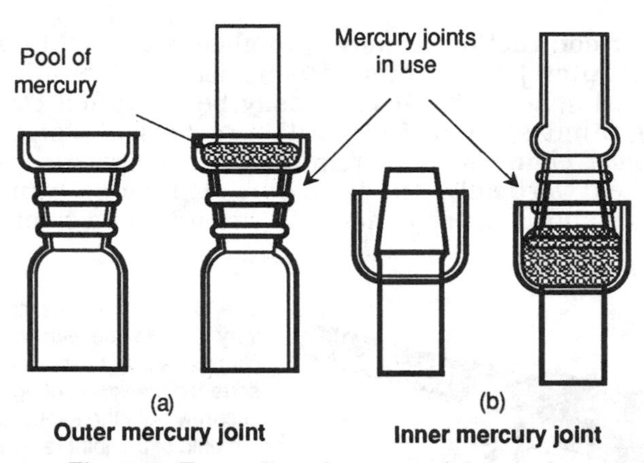

Fig. 3-5 Examples of mercury joint seals.

ground glass joints were in place, a small pool of mercury was poured in the trap to complete the seal.

The outer mercury joint is still commonly used by glassblowers who need to seal tubing onto the ground end of an outer joint (for example, on a water-cooled condenser where the joint also needs to be water cooled, as in Fig. 3-6). Without the extension of glass, a seal of this type would distort the walls of the joint, making it useless. This outer joint, therefore, provides the same advantage afforded by extended drip tip inner joints (Fig. 3-2(c)).

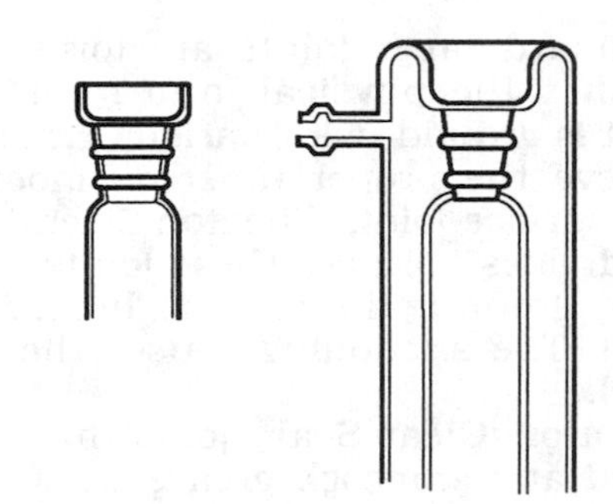
Fig. 3-6 Sample of a water-cooled condenser using a mercury seal joint.

The Stopper Plug. There are two types of *ground stoppers*: *Reagent bottle stoppers* and *flask stoppers*. They have the same 1:10 taper as the joints previously mentioned, however, they are listed by only one number based on their width measurement (see Fig. 3-7). Both stopper types have a flattened disk above the ground section for easy holding. Because the term 'penny-head stoppers' can refer to either reagent or flask stoppers, a complete description of the stopper you want is important for complete identification.

Reagent bottle stoppers are essentially the same size as mid-length joints, although their lengths are somewhat shorter. A size 14 stopper is the same size as a 14/20 joint, and can fit into a 14/20 female joint (See Table 3-1). Do not expect a stopper plug to maintain a vacuum, however, unless the stopper is hollow. The thinner walls of a hollow stopper will adjust to temperature changes better than a solid plug so there is less chance for the members to become separated due to the expansion or contraction of the glass. In addition, only hollow stoppers are ground to the standards required for vacuum components.

Flask stoppers come in widths (sizes) that are different from other joints and stoppers; therefore, flask stoppers are not interchangeable with joints or stoppers (see Table 3-1).

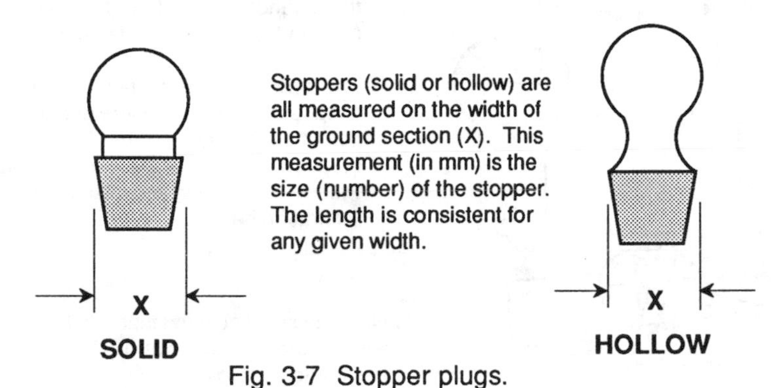

Stoppers (solid or hollow) are all measured on the width of the ground section (X). This measurement (in mm) is the size (number) of the stopper. The length is consistent for any given width.
X
SOLID
X
HOLLOW

Fig. 3-7 Stopper plugs.

3.1.2 Ball-and-Socket Joints

Ball-and-socket joints may be required when the union between apparatus pieces is not linear. Ball-and-socket joints have a different type of measurement system than standard taper joints (see Fig. 3-8). They cannot be used for significant (e.g., > 10^{-3} torr) vacuums.

Ball-and-socket joints must be held together with either a metal spring type of clamp (like a clothespin), or with a semi-rigid plastic clamp. Because ball-and-socket joints cannot be used without one or the other type of clamp, be sure to include the right size clamp when ordering apparatus that have ball-and-socket parts, if they are not otherwise supplied.

The clamp size required for ball-and-socket joints is identified from the outer diameter measurement. Thus, a size 18 clamp is used for either an 18/7 or an 18/9 size O-ring joint.

Table 3-1

Comparable Sizes of Standard Taper Joints and Stoppers

Mid - length joint size width / length	**Reagent bottle stopper no.** width (≈length)	**Flask stopper no.** width (≈length)
7/15		
		8 (≈10)
		9 (≈14)
10/18		
		13 (≈14)
14/20	14 (≈20)	
	16 (≈15)	
19/22	19 (≈22)	19 (≈17)
24/25	24 (≈30)	22 (≈20)
29/26	29 (≈35)	27 (≈22)
34/28	34 (≈40)	32 (≈22)
40/35		38 (≈30)
	45 (≈47)	

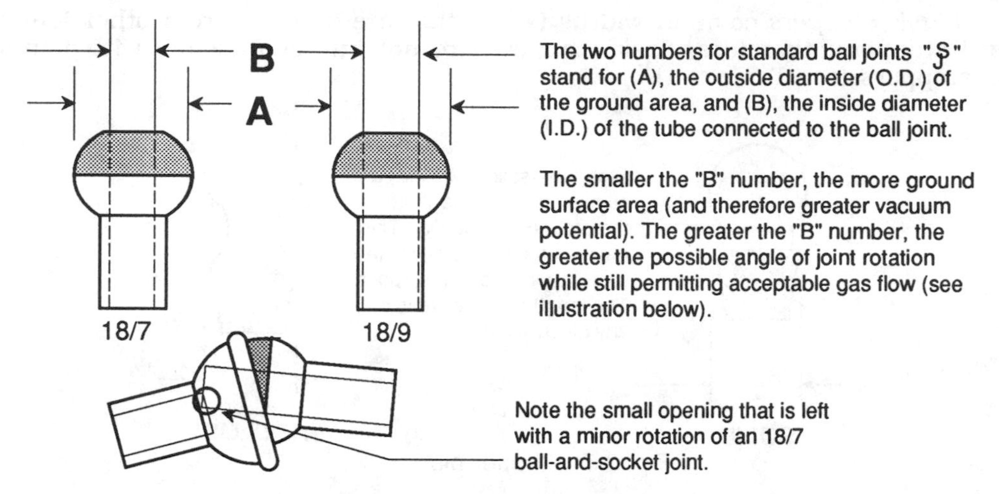

B
A
The two numbers for standard ball joints "Ş" stand for (A), the outside diameter (O.D.) of the ground area, and (B), the inside diameter (I.D.) of the tube connected to the ball joint.
The smaller the "B" number, the more ground surface area (and therefore greater vacuum potential). The greater the "B" number, the greater the possible angle of joint rotation while still permitting acceptable gas flow (see illustration below).
18/7
18/9
Note the small opening that is left with a minor rotation of an 18/7 ball-and-socket joint.

Fig. 3-8 Ball and socket joints.

3.1.3 The O-ring Joint

The *O-ring joint* (see Fig. 3-9) has no ground sections, and as opposed to other joint types, both members of an O-ring joint are identical. For sealing, O-ring joints require placing an O-ring, made from some polymer material (see Sec. 1.4), between the joint members, which are held together by the same type of "clothespin" clamp used with ball-and-socket joints.*

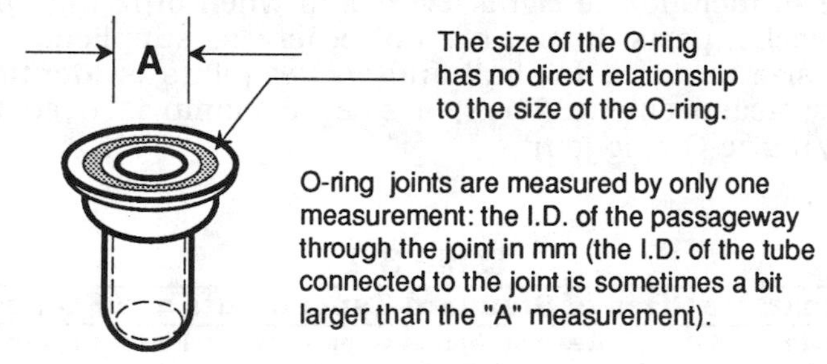

A
The size of the O-ring has no direct relationship to the size of the O-ring.
O-ring joints are measured by only one measurement: the I.D. of the passageway through the joint in mm (the I.D. of the tube connected to the joint is sometimes a bit larger than the "A" measurement).

Fig 3-9 The O-ring joint.

Because clamp identification numbers are derived from their relationship with ball-and-socket joints, there is no relationship between which size clamp should be used with which O-ring joint. There is no one-to-one relationship with O-ring joints and ball-and-socket joints; one clamp size is likely to fit several different size O-rings joints. Table 3-2 provides the recommended O-ring and clamp sizes for respective O-ring joints. Like ball-and-socket joints, be sure to order the appropriate size clamp when ordering apparatus with O-ring connections.

* The one-piece plastic clamp used with ball-and-socket joints will not work with O-ring joints.

O-ring joints can easily be used with vacuum systems ($\leq 10^{-7}$ torr) and are particularly useful for quick-release connection. Occasionally it is necessary to leave a thin film of stopcock grease on an O-ring's surface to either help the vacuum capabilities of the joint, or to protect the O-ring's surface from the effects of a solvent.

Table 3-2

O-ring and Clamp Sizes For O-ring Joints

O-ring joint	O-ring size	Clamp size
5	110	12
7	111	18
9	112	18
15	116	28
20	214	35
25	217	35
41	226	65
52	229	75
115	341	102

3.1.4 Hybrids and Alternative Joints

A hybrid *O-ring ball joint* (see Fig. 3-10) is available. This joint has the vacuum capabilities of the O-ring joint along with the variable angle abilities of a ball-and-socket joint. Typically, the *socket* side of such a union is ground and an O-ring cannot achieve as good a vacuum as it could against a non-ground surface. Because of this, some manufacturers are supplying un-ground sockets for use with O-ring ball joints. Regardless of the type of socket used, because this design does not require stopcock grease it can be used in conditions where a connection would otherwise be impossible. Unfortunately, there may be liquid hold-up within such a joint.

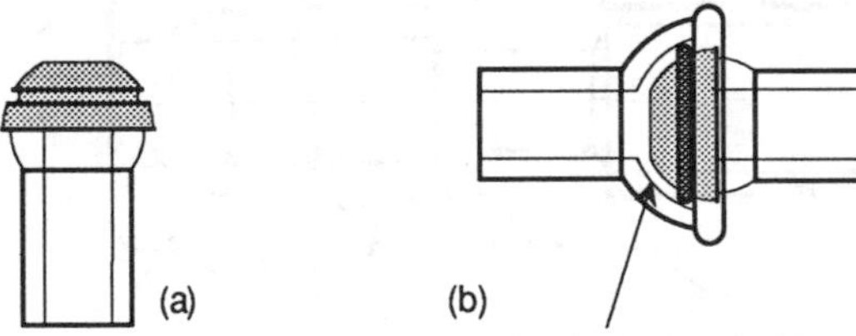

Fig. 3-10 The O-ring ball joint, without the O-ring in place (a), and the hold-up it can create (b).

There are occasions when a solvent must remain in contact with a connection for several days and the connection must maintain a vacuum. Depending on the nature of the solvent, standard grease and most O-rings cannot be used in these situations.

An alternate solution to complications with long-term solvent contact with O-ring polymers is the use of Solv-Seal joints by Lab Crest Scientific (a Subsidiary of Fischer & Porter Co.). These joints seal together with a double-ringed Teflon seal (see Fig. 3-11) within a specially designed glass joint. Although there is a Viton O-ring within the Teflon sealing unit, the O-ring is used as backup protection. The primary containment for the material within the system is the Teflon.

Fig. 3-11 Solv-Seal joints.‡

‡ The Solv-Seal joint, from the Lab Crest Scientific Product line, is made by Andrews Glass Co., a division of Fischer & Porter Co. With permission.

3.1.5 Special Connectors

There are several special connectors that provide 'demountable' connections between glass to glass, glass to metal, or metal to metal. These connectors provide many advantages over standard connections.

Swagelok® fittings (see Fig. 3-12) are used to connect metal tubes to metal tubes. These fittings, which provide permanent connections on metal tubing, are not meant for use on glass. These connectors have a pair of small metal ferrules (like a collar), one of which tightens around a metal tube as the Swagelok® nut is tightened, the other is a washer. The tightened ferrule is permanently attached to the metal tube and cannot be removed. It provides a leak-tight fit that can be used in high-vacuum or high-pressure conditions.

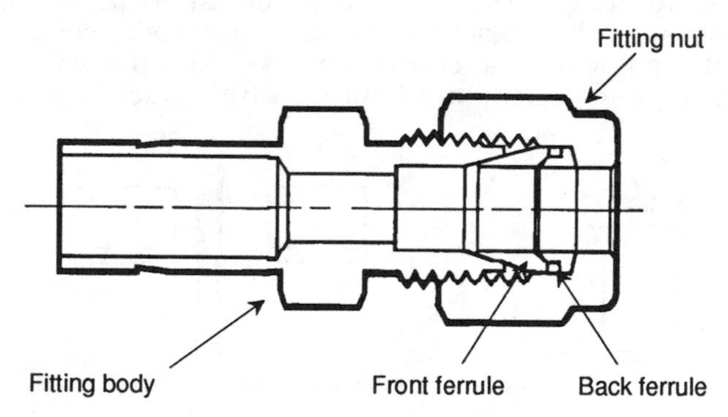

Fig. 3-12 The Swagelok® fitting.‡

‡ The illustration of a Swagelok® Fitting is used with permission of Swagelok Co., Solon, Ohio 44139. The illustration was supplied by the Swagelok Co.

Swagelok® fittings are made in a wide variety of styles, connections, and sizes. They are especially useful when connecting copper tubing for delivering nitrogen gas to various systems around a lab, or when making connections for extension tubes on metal high-vacuum systems.

Cajon® Ultra-Torr® fittings are similar to Swagelok® fittings except they are reusable and can be used on metal as well as glass and plastic (see Fig. 3-13). Instead of a metal ferrule, there is an O-ring that imparts a leak-tight seal. They are made in a wide variety of styles and sizes. Unlike Swagelok®

fittings, Cajon® Ultra-Torr® fittings cannot be baked out for ultra-high vacuum work.

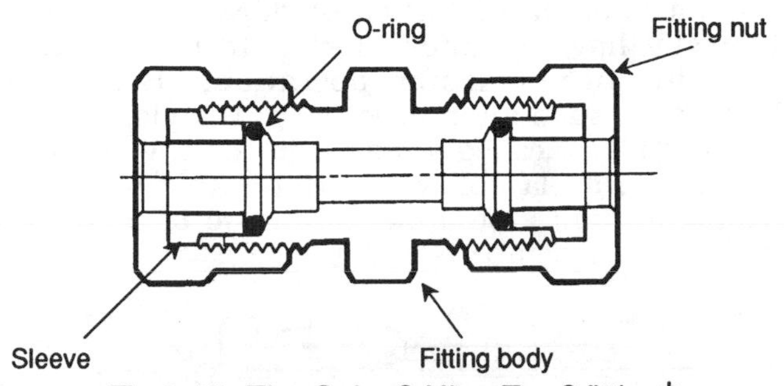

Fig.3-13 The Cajon® Ultra-Torr® fitting‡.

‡ This illustration of a Cajon® Ultra-Torr® Fitting is used with permission of Swagelok Co., Solon, Ohio 44139. The illustration was supplied by the Swagelok Co.

If you are going to connect a Cajon® Ultra-Torr® fitting to glass, first verify whether the fitting is metric or not. If the fitting is non-metric, use the appropriate type of tubing. Medium and heavy wall tubing are both made in English sizes, but are identified in metric numbers (see tables in Sec. 3.4.3). Because glass tubing has variations in O.D. tolerance (see Sec. 3.4.2), pre-test the fitting before you commit yourself to any construction. This pre-testing can be done by simply ensuring that the Ultra-Torr® tightening nut can slip over the tube. Because glass has no macroscopic compressional abilities, do not force a Cajon® fitting onto glass. If it does not fit, select another piece of glass that fits.

There is a glass variation of the Cajon® fitting design available from several glass industry sources. These variations go under a variety of names (depending on the manufacturer), but are typically under the classification of glass-threaded connections (see Fig. 3-14). There are adapters made with a threaded connection on one end and either a glass tube ready to be attached to some glass apparatus, or a standard tapered joint on the other, made out of glass or Teflon. These connectors can be used (for example) to support thermometers within a flask.

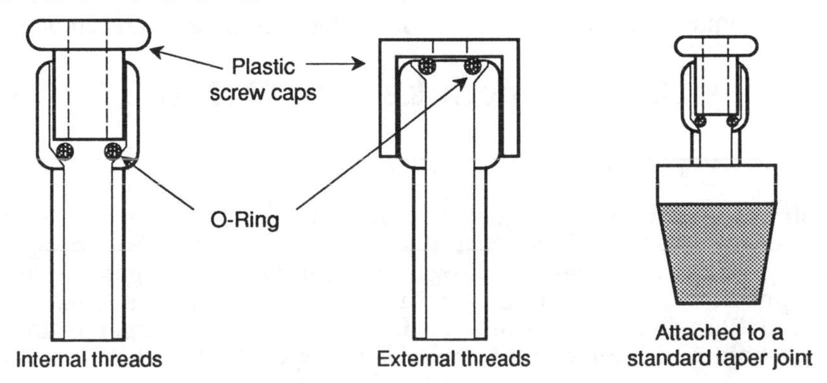

Fig. 3-14 Several generic designs of glass-threaded connections.

Glass-threaded connectors can be used for limited vacuum operations. However, there often is nothing to stop the inserted tube from being sucked in unless there is some step in the fitting or ridge on the tube. Cajon® Ultra-Torr® fittings typically have an internal step to prevent the inserted tube from being sucked in during vacuum operations. Because glass-threaded connections seldom have such an internal step, the glassware needs to be altered by adding a bulge (also called a maria) to the tube by a glassblower (see Fig. 3-15). Alternatively, a glassblower may be able to form (depending on the apparatus configuration) a step on the inside of a connector to prevent inward motion of an inserted tube (see Fig. 3-16).

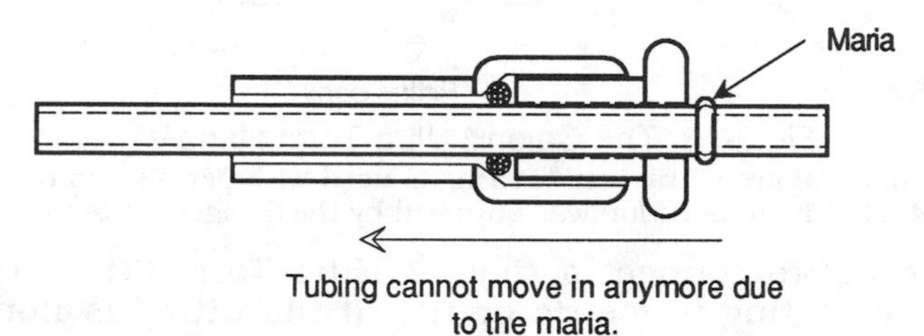
Maria
Tubing cannot move in anymore due
to the maria.

Fig. 3-15 Internal motion of the inserted tube is blocked by the addition of a maria.

The advantages of glass-threaded connections are that they allow easy connections to be made and, disassembled, require no grease.* Additionally, there is little if any chance of two pieces seizing. The disadvantages are that they may not be ideal for high-vacuum work, pieces can wiggle around, and you need to be careful about the materials that come in contact with the particular O-rings being used.

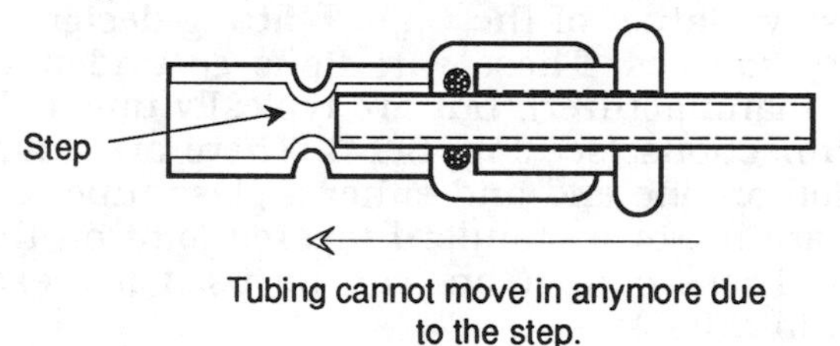
Step
Tubing cannot move in anymore due
to the step.

Fig. 3-16 Internal motion of the inserted tube is blocked by the addition of a maria.

3.2 Stopcocks and Valves

3.2.1 Glass Stopcocks

The stopcock is more than just an on/off valve as it may also be used to direct liquid or gas flow through a system. In addition, depending on their design, stopcocks have limited to excellent ability to vary gas or liquid flow rates. Stopcock size is identified by the size of the hole through the plug. The internal diameter of each arm is likely to be much larger than the size of the plug hole and should not be used for stopcock size identification.

* A small film of grease on the O-ring can ease sliding the O-ring onto glassware and help provide a better vacuum seal.

Like the ground joint, glass stopcocks are made in a 1:10 taper. Interchangeability of new plugs and barrels (a plug from one stopcock placed in a similar sized barrel from the same manufacturer) used to be a standard feature by all manufacturers. Currently however, not all manufacturers guarantee interchangeability of new plugs and barrels. Talk with a glassblower, or your supplier, about the interchangeability of any particular brand. Older, used stopcocks are far less likely to provide interchangeability because of uneven wear during use. Therefore, look for leaks when placing a new plug within an old barrel.

To verify if a stopcock plug and barrel are a good match, assemble the pieces, then sight down the inside of a side arm to see how well the holes align. Misalignment may be minor, causing a slight decrease in potential flow, or be so great that nothing can pass through (see Fig. 3-17).

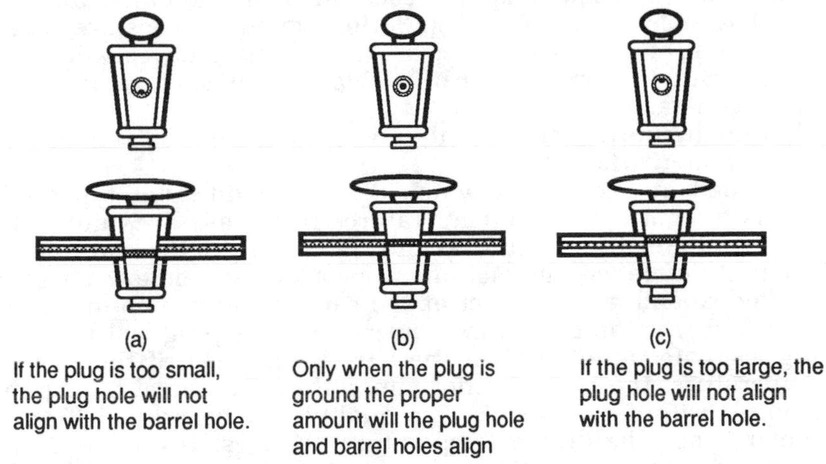

(a)
If the plug is too small, the plug hole will not align with the barrel hole.
(b)
Only when the plug is ground the proper amount will the plug hole and barrel holes align
(c)
If the plug is too large, the plug hole will not align with the barrel hole.

Fig. 3-17 Plugs of the wrong size can prevent flow in a stopcock.

The primary difference between a regular and a high-vacuum stopcock is the final grind. Stopcocks and joints are ground (lapped) with abrasive compounds just as rocks are ground in lapidary work. The finer the grade of grinding compound, the smoother the finish on the ground (lapped) surface. The smoother the finish on the glass parts, the easier they will slide past each other when rotated. Joints can be used for high-vacuum purposes without the same degree of grinding because they are not rotated.

Vacuum stopcocks are *not* meant to be interchangeable. Each plug and barrel receive their final grind together, and therefore are mated for life. To prevent mismatching a plug and barrel set, numbers and/or letters are inscribed for each stopcock on the handles of the plug and on the side of the barrels (see Sec. 3.3). Aside from alignment problems, cross-matched plugs and barrels of high-vacuum stopcocks may leak or jam when rotated.

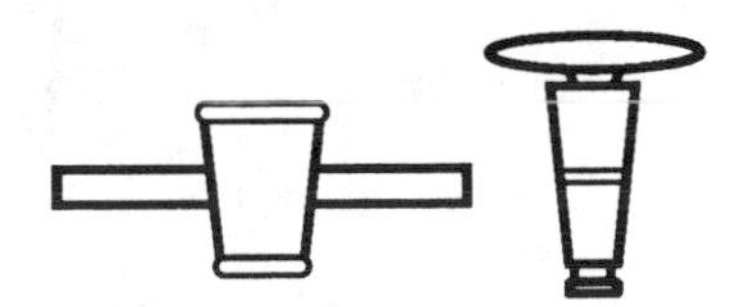

Fig. 3-18 A standard two way stopcock.

A simple, basic two-way glass stopcock (shown in Fig. 3-18) will have a single hole drilled through a solid plug. The arms are straight and placed 180° from each other. The plug is typically held in place by a rubber washer or metal clip at the small end. This stopcock design is easy to make and is inexpensive. The usual problems that arise from this type of stopcock are:

1) Dirt can collect on the grease and work its way into the plug's hole. From there the dirt can eventually wear a groove on the inside of the barrel. This groove, leading from one arm to another, can eventually become great enough to prevent a complete seal when the plug is rotated to the "off" position.
2) As stopcock grease becomes old and/or cold, it becomes brittle, and rotation can cause a "shear" type break, allowing the plug to be free from the barrel.

By setting the two arms oblique to each other on the barrel and having a hole drilled diagonally in the plug (Fig. 3-19), two advantages are achieved:

1) If the first of the above problems arises, the grooves will not line up with the opposite arm and potential leakage is reduced considerably.
2) This configuration allows only one position (in 360° of rotation) for material flow.

As mentioned in Sec. 1.1.3, which discussed different types of glass, glass expands or contracts with temperature. Thin glass does not expand (or contract) the same relative amount as thick glass under the same conditions. Because of this difference, all vacuum stopcocks have hollow plugs so that the plugs may expand and contract at the same relative amount as the barrels. Thus, when there is a drop in temperature, the plug will not loosen by contracting a greater amount than the barrel. Likewise, if there is a sharp rise in temperature, the plug will not jam within the barrel by expanding a greater amount within the barrel. This quality can be critical if an experiment or environment has rapid temperature changes. Figure 3-19(a) shows a solid plug. The holes in solid plugs are drilled. Figure 3-19(b) shows a hollow plug. The hole in a hollow stopcock plug is a glass tube sealed obliquely within and later ground to fit the barrel. This inner tube provides the passageway for gases* through the hollow plug.

Oblique-arm stopcocks are not immune from 'shear' breaks caused by cold or old stopcock grease. Because an accidental rush of air into a vacuum system can cause damage to the system or destroy oxygen-sensitive materials within the system, vacuum stopcocks have a design feature to prevent accidental removal of the plug from the barrel. The stopcock design shown

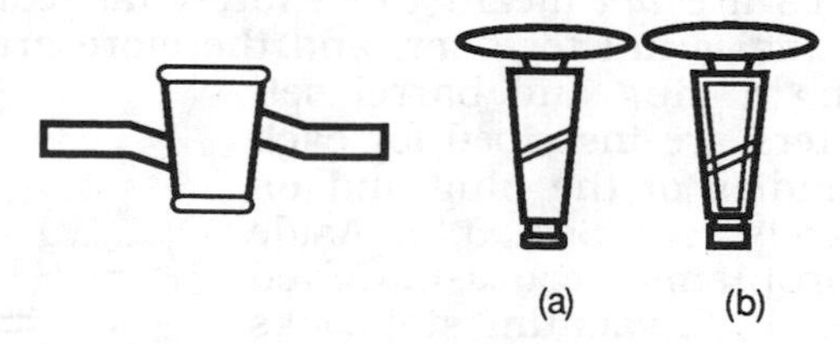

Fig. 3-19 Several different designs of an oblique 2-way stopcock.

* Vacuum stopcocks are not intended for liquid transport use.

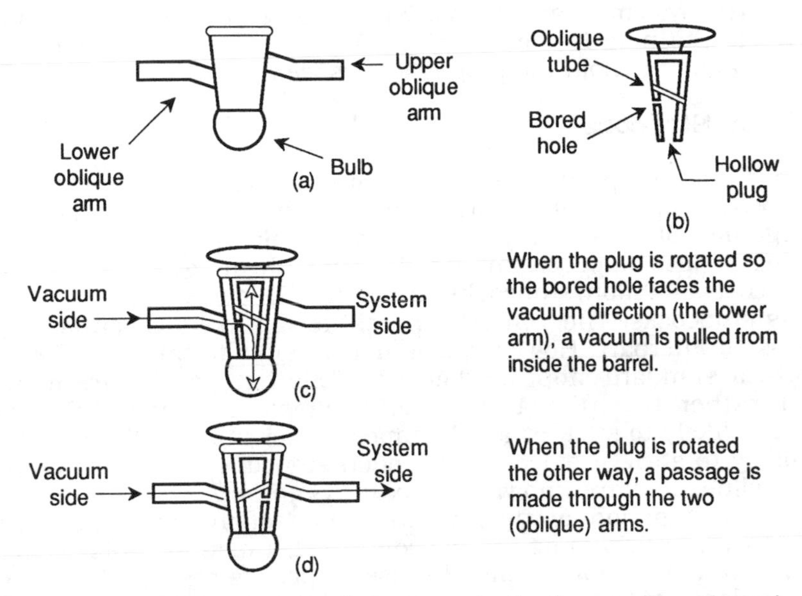
Upper
oblique
arm
Lower
oblique
arm
Bulb
(a)
Oblique
tube
Bored
hole
Hollow
plug
(b)
When the plug is rotated so the bored hole faces the vacuum direction (the lower arm), a vacuum is pulled from inside the barrel.
Vacuum
side
System
side
(c)
Vacuum
side
System
side
(d)
When the plug is rotated the other way, a passage is made through the two (oblique) arms.

Fig. 3-20 How to evacuate the vacuum bulb of a vacuum stopcock.

in Fig. 3-20 demonstrates how a vacuum stopcock uses a vacuum to secure its plug from slipping out of the barrel. The bottom of the barrel is closed with a bulb (see Fig. 3-20(a)). A hole is bored opposite the low opening of the oblique tube (see Fig. 3-20(b)). Aligning the hole with the lower oblique arm of the stopcock (see Fig. 3-20(c)) creates a passage for the vacuum that holds the plug securely in the barrel. A 180° rotation from this position is the "open" position for this stopcock (see Fig. 3-20(d)). The only way to separate the plug from the barrel is to release the vacuum from inside the stopcock by rotating the bored hole toward the lower oblique arm when no vacuum is present.

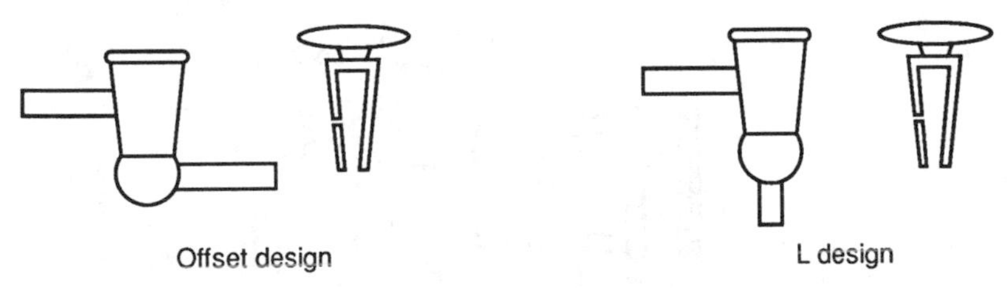
Offset design
L design

Fig. 3-21 The offset design of a two-way vacuum stopcock.

Fig. 3-22 The L design of a two-way vacuum stopcock.

Figures 3-21 and 3-22 show two different alignments of the same type of stopcock. This plug design is simpler, and therefore is somewhat less expensive. There is no advantage of the "offset" design (Fig. 3-21) to the "L" design (Fig. 3-22). The choice depends on how your system is built, that is, whether the "offset" design or the "L" design stopcock physically fits better into your given system or apparatus. With these stopcocks, the vacuum of the system holds the plug in continuously. However, once the system loses its vacuum, its ability to hold the plug is lost as well.

3.2.2 Teflon Stopcocks

Teflon stopcocks provide excellent alternatives to standard glass stopcocks because no grease is required. They can be used in distillation systems where organic solvents, UV radiation, or oxidizing gases would normally make using a glass stopcock impossible. This advantage does not come without some cost as Teflon stopcocks are generally more expensive than standard glass stopcocks. The Teflon stopcock design looks essentially the same as its glass counterpart. However, it is not possible to take a Teflon plug and use it with a standard stopcock because Teflon stopcocks are made with a 1:5 taper rather than the 1:10 taper of glass. Because Teflon flows, or 'creeps,' it is likely to stick at a 1:10 taper by flowing into the arm holes, but far less likely to stick at a 1:5 taper. Teflon stopcocks with a 1:5 taper follow ASTM guidelines and are known as *Product Standard.*

There was a period of time when some Teflon stopcocks were made with 1:7 tapers, but this line was discontinued. I mention this point because some of you may have found one of these stopcocks and are having difficulty finding a replacement plug or are frustrated by trying to fit an old 1:7 plug in a new 1:5 barrel. The taper variations are small and subtle, so it is easy to confuse the two. Because the 1:7 taper design is no longer made, it is recommended that you phase any pieces with that design out of your lab.

The Teflon stopcock has more pieces (see Fig. 3-23) than the glass stopcock, and often, the pieces are assembled in the wrong order. Inserting the plug into the barrel provides no challenges, but the white and black washers invariably get inverted. Each piece serves a specific function for the successful operation of the Teflon stopcock: the white washer helps everything on the plug shaft rotate together when the plug is turned; the colored locking nut helps maintain a certain amount of tension on the plug; and finally, the

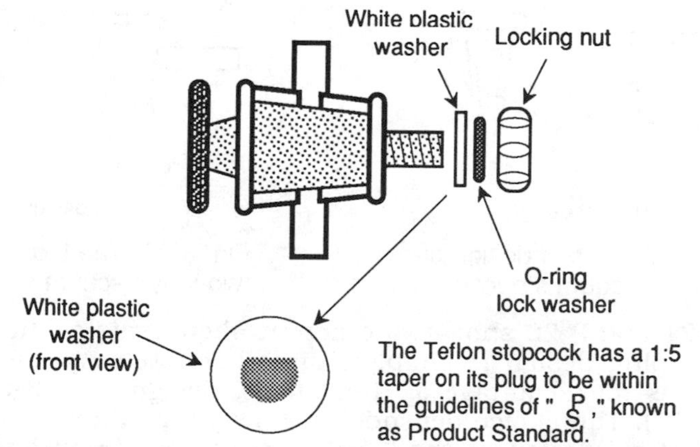

Fig. 3-23 Proper placement of the parts of a Teflon stopcock.

black washer (O-ring) serves as a lock washer to prevent the locking nut from rotating itself off the threaded section of the plug.

If the lock washer is placed on the plug shaft before the white washer, the washer's friction will grab the surface of the barrel, and tend to make everything past the black washer not rotate while the plug is turned, causing the plug to tighten when rotated clockwise and loosen when rotated counterclockwise.

The white washer has a flat side on the hole through the center to maintain its non-slip position on the plug. Occasionally, the flat spot on the white washer becomes worn round, or a groove is worn into the threaded section of the plug. Either of these effects causes the washer to not rotate with the plug and is the same as switching the positions of the black and white washer. Replace the washer or plug if either are worn.

3.2.3 Rotary Valves

Standard Teflon stopcocks cannot be used for any vacuum work. A completely different design from those discussed thus far is required for a Teflon vacuum stopcock. This design is shown in Fig. 3-24 and is called a *rotary Teflon valve.* It is called a valve instead of a stopcock because its operation and design is similar to a metal-designed valve.

There are many designs of rotary valves: some have externally threaded glass barrels, others have internally threaded barrels; some have exposed O-rings, others have covered O-rings, and some have no replaceable O-rings. Regardless, they all work with the same operational technique. Rotating cap motion translates into vertical motion on a central shaft, which rides up and down to open or close a pathway.

When closing a metal valve, it is customary to rotate the valve until it cannot be rotated anymore. Then you must give it a final twist to make sure it is closed tight. This *extra twist* should never be done with glass rotary valves. These valves cannot withstand the same forces as metal valves, nor is

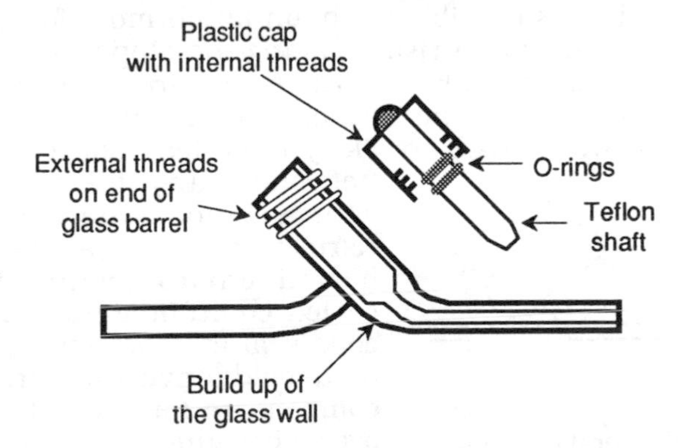

Fig. 3-24 The rotary valve operates by rotating a plastic cap. The rotational motion of the cap is translated into linear motion on the Teflon shaft. When the shaft is inserted all the way into the glass barrel, the tip of the shaft pushes against a build-up of glass and stops fluid motion. The O-rings on the shaft prevent fluid leaks past the cap.

it necessary to use such force on a glass rotary valve. Glass is likely to break if too much torque is applied to the cap. Fortunately, because you can see through glass, it is easily possible to see if the shaft is closed against the barrel. This closed position is shown in Fig. 3-25 on the walls of the barrel with a 'wet' mark.

Both Figs. 3-24 and 3-25 show O-rings on the Teflon shaft. Some manufacturers cover these O-rings with Teflon so that the rotary valve can be used with solvents that might otherwise destroy the O-ring. Others have 'fins' that sweep in front of the O-ring to prevent solvents from contacting the O-ring.

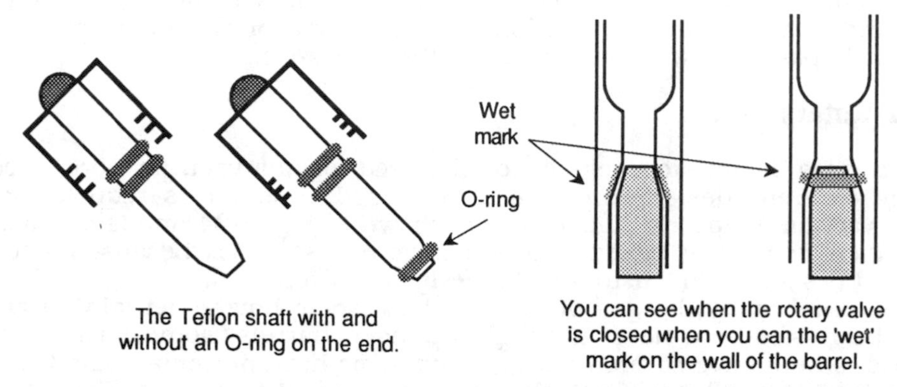

Wet
mark
O-ring
The Teflon shaft with and
without an O-ring on the end.
You can see when the rotary valve
is closed when you can the 'wet'
mark on the wall of the barrel.

Fig. 3-25 The rotary valve plug with and without an O-ring on the end on the left, and the shaft showing 'wet' marks on the right.

Although all vacuum rotary valves have O-rings along the side of the shaft, some have O-rings at the end as shown in Fig. 3-25. This design has advantages as it provides a better quality seal without over-tightening. The reason is simple: the same force applied over a small area will have a greater pounds-per-square-inch pressure than that applied over a large area.* With an O-ring at the end, it is possible to obtain much more force without having to bear down on the barrel and risk breaking the stopcock.

All Teflon stopcocks and valves must remain free of particulate contamination so the surface will not be scratched or damaged. Any abrasions, or distortions, of the central shaft are likely to be leak sources. In addition, do not place any filled container closed with a Teflon stopcock (or valve) in a refrigerator. Just as Teflon expands in heated environments, it contracts in cooled environments. Thus, what was a good seal at room temperature may leak in a cold environment. Do not try to compensate for the cold temperature by overtightening the shaft for two reasons: One, it probably will not prevent a leak,

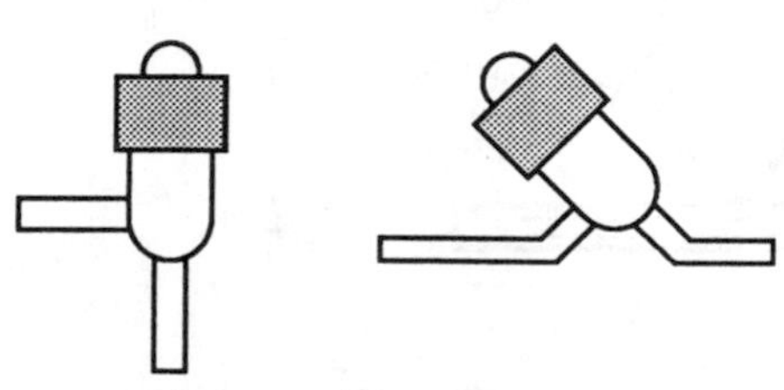

Fig. 3-26 General designs for rotary valves.

* This amount is analogous to the pressure exerted by spiked heels vs. running shoes on a sidewalk.

and two, when it is removed from the cold environment, its expansion may break the apparatus.

Although the specific design of a valve varies between manufacturers, rotary Teflon valves are almost always a variation of the in-line and right-angle designs shown in Fig. 3-26. Although the rotary design may be restrictive, it has provided new possibilities for valve design such as the Inlet Valve Chem Cap™ (shown in Fig 3-27), which is designed for medium vacuum system use (up to approximately 10^{-4} torr). The Chem Cap™ does not have a second or third arm as is customary with just about all other stopcocks because the transmission of gas is through the Teflon shaft itself. This stopcock can be used in place of a standard glass vacuum stopcock for grease-free operations where a standard Teflon stopcock could not be used because of its inability to maintain a vacuum.

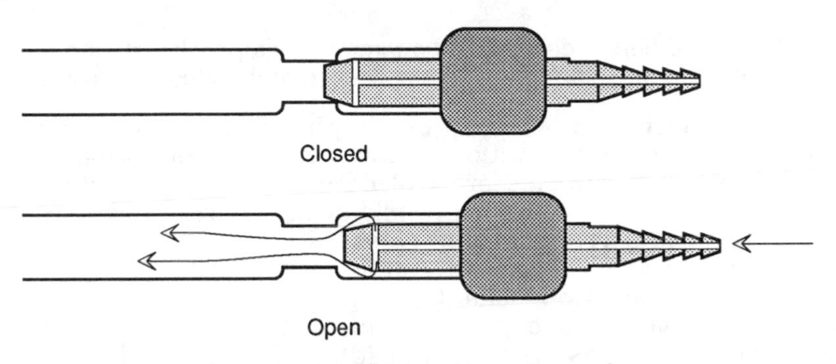

Fig 3-27 Inlet Valve Chem Cap™.‡

‡ The Inlet Valve Chem Cap™ is from Chemglass Co. Inc., Vineland, N.J. 08360. With permission.

Another advantage of the rotary valve design is apparent when an extension from the end of the shaft is added and the valve becomes a needle valve. It can then permit fine control for the release or addition of a gas or liquid (see Fig 3-28).

Rotary Teflon valves have greater initial costs than comparable vacuum glass stopcocks. However, due to their lack of maintenance needs, as well as not requiring any stopcock grease, their long-term expense can be considerably less than standard vacuum stopcocks. However, because Teflon stopcocks or valves require a limited-temperature environment, all-glass stopcocks will always remain in demand.

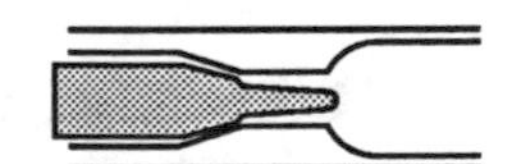

Fig. 3-28 A close-up of the end of the Teflon shaft with a needle valve extension.

3.2.4 Stopcock Design Variations

All of the above designs are called *two-way stopcocks* because there are two arms, or connections, to the stopcock. Two-way stopcocks typically are open/close valves with no proper orientation. The exception for this statement are the stopcock designs shown in Figures 3-20, 3-21, and 3-22

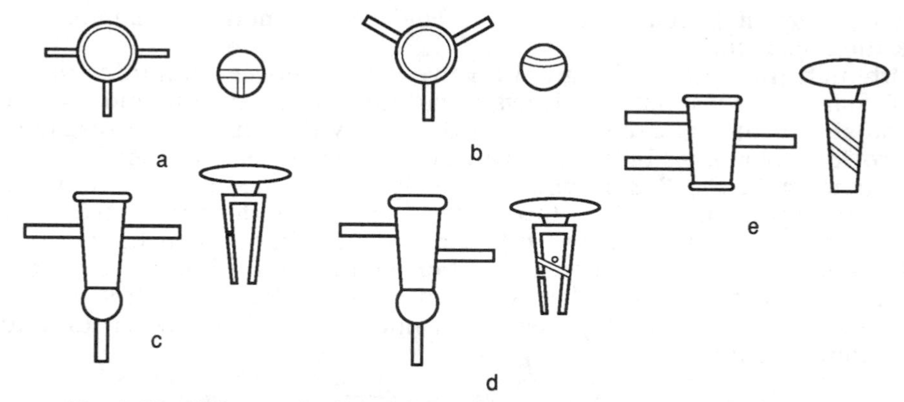

Fig. 3-29 Different designs of the three-way stopcock, with the barrel on the left and a side view or cross-section of the plug on the right.

which require a specific orientation of the plug to the stopcock to evacuate the inside. There are no orientation requirements for rotary valves.

Stopcocks with three arms are called *three-way stopcocks*. Five of the most common three-way glass stopcock designs are shown in Fig. 3-29. Represented are the barrel and plug cross-sections (or end views) of three-way stopcocks. The three-way stopcocks (a), (b), and (e) are available in solid plug, Teflon, and high-vacuum designs. The (c) and (d) designs are exclusively for vacuum systems. Because of their design, rotary valves can only have two arms. It is possible to creatively place two of them together in a three-way alignment, however. An example of a three-way rotary Teflon valve is shown in Fig. 3-30

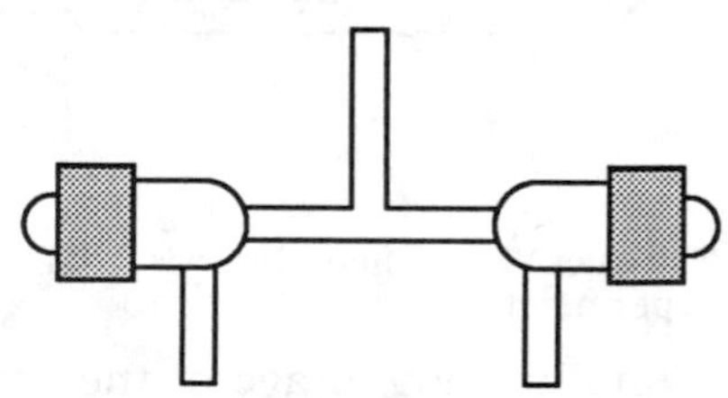
Fig. 3-30 A three-way Teflon rotary valve.

3.3 Maintenance and Care of Joints, Stopcocks, and Glassware

3.3.1 Storage and Use of Stopcocks and Joints

So much depends on the proper performance of stopcocks and joints that they should be treated with the same care, and consideration, as an optical lens. If your chemicals are stored in a volumetric flask, or held within a storage container by a stopcock, the material becomes worthless if you lose access to it because of a stuck stopper or a frozen stopcock.

Most occurrences of stuck stopcocks and joints can be prevented by proper storage, preparation (of the item to be used), application of grease (if required), and proper use. All of these points may seem obvious, but regrettably they are seldom executed. The greatest disasters I have seen from stuck stopcocks and joints occurred not from the original jammed piece, but rather from inexperienced attempts at separating the members, which often make it impossible to follow through on other separation approaches. For information on how to prevent joints and stopcocks from sticking, see Sec. 3.3.6. For a comprehensive list of stuck stopcock and joint separation techniques, see Sec. 3.3.7.

When putting items away for storage, the apparatus should be cleaned of all grease and then thoroughly cleaned so that all items are ready for use at all times. Glassware should never be stored dirty with the intent to clean a piece before use.* In addition, dried grease and dirt often cause stopcocks and joints to stick. In addition, apparatus should be stored with all pieces assembled (note precautions in Secs. 3.3.2 and 3.3.3).

Round Bottom Flasks, Pennyhead Stoppers, Plugs, and Other Loose Items. Many items are stored in drawers and (as the drawer is opened and closed) can bang around. This situation applies especially to round bottom flasks, which are hard to prevent from rolling around. When these flasks bang around, star cracks, or flaws for future fracture, are created. For round bottom flasks, Erlenmeyer flasks (on their sides) and other glass items that can roll, line the drawer with soft, irregular-surface materials (like bubble pack) and if possible, place cardboard dividers between the pieces. Finally, identify on the outside of the drawer what size flasks are stored within limit the amount of opening and closing required.

Small, loose items, like pennyhead stoppers, should be stored by *size* and *type* in small boxes in a drawer. Occasionally you may have extra glass plugs from glass stopcocks (i.e., from broken burettes). These plugs also should be stored by *size* and *type* in small boxes in a drawer. Do not store Teflon stopcock plugs with glass stopcock plugs as the glass may scratch the Teflon leaving it worthless. Wrap Teflon plugs in a tissue and then mark the tissue with its contents to help protect the surface of the plug as well as let people know the tissue's contents.

Stopcocks. After a stopcock has cleaned, a small slip of paper (measuring about 6 mm by 50 mm) should be inserted between the plug and barrel to prevent jamming (see Fig. 3-31). You may remove solid plugs and store them separately, but this storage is generally not recommended as it may cause future problems. For example, many manufacturers guarantee that their plugs are interchangeable. If the plugs are not interchangeable, the hole of the plug will not align with the holes of the stopcock. However, not all manufacturers are as good as their claims. Regardless, plugs from one manufacturer seldom fit the barrels of another. If you have stopcocks from Kimble, Kontes, Corning, and Ace (to name a few) and do not wish to deal with trying to match plugs with barrels (and then try to match alignment, if necessary) store matched sets with a paper slip between the barrel and plug.

Vacuum Stopcocks. *Hollow plugs always indicate vacuum and should always be kept with their respective barrels. Vacuum stopcocks should never be stored with their plugs separated from their mated barrels.* On the side of a plug and barrel is a numeric or an alpha/numeric code that is used to

* It is always more difficult to clean glassware that has sat then it is to clean glassware that is recently soiled.

match the two units. All high-vacuum stopcocks receive the final grind of the plug and barrel together, making them matched and individual. You should never cross-match high vacuum stopcock parts, even if they *seem* to fit. If one or the other parts should be lost or broken, replace the entire unit. For storage, slip a piece of paper (about 6mm X 50 mm) between the plug and barrel, then support the two parts together with a rubber band as shown in Fig. 3-31.

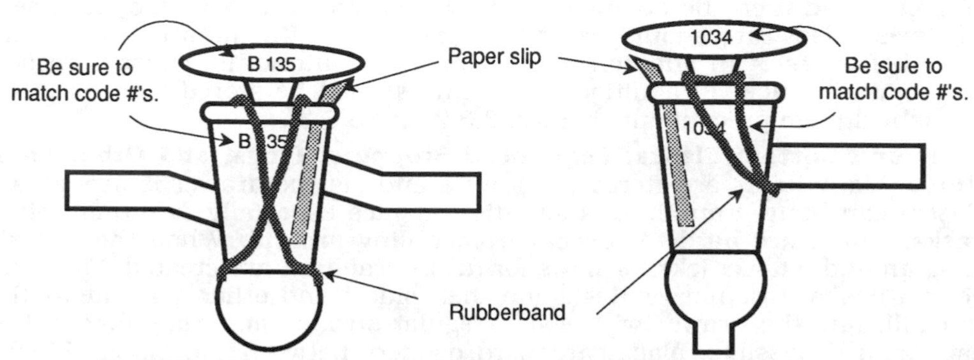

Fig. 3-31 When storing a high-vacuum stopcock, use a rubber band to secure the plug onto its respective barrel. Be sure to match the plug and barrel with their respective code numbers.

Cold Traps and Apparatus. Items like cold traps should be stored as complete units to protect the inner tubes. A small slip of paper (measuring about 6 mm by 50 mm) should be inserted between the inner and outer joints to prevent jamming. If there are hooks, use a rubber band or springs to hold the parts together. If not, use masking tape to hold the parts together (see Fig. 3-32). Unfortunately, masking tape ages and loses its sticking abilities. The remains of old masking tape can be removed with acetone.

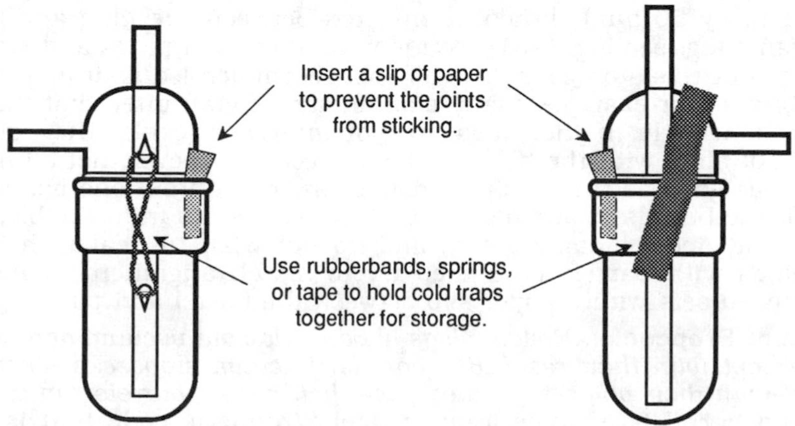

Fig. 3-32 Proper storage of apparatus will not only protect the joints, but will also protect any internal glassware that could otherwise be broken.

Teflon Stopcocks. Teflon stopcocks should never be stored with the locking nut tightened. Teflon flows under pressure, and such pressure could squeeze the plug up and into the holes of the barrel (see Fig. 3-33). Teflon plugs are tapered at a 1:5 ratio rather than the 1:10 ratio of glass stopcocks to help relieve some of these pressures.

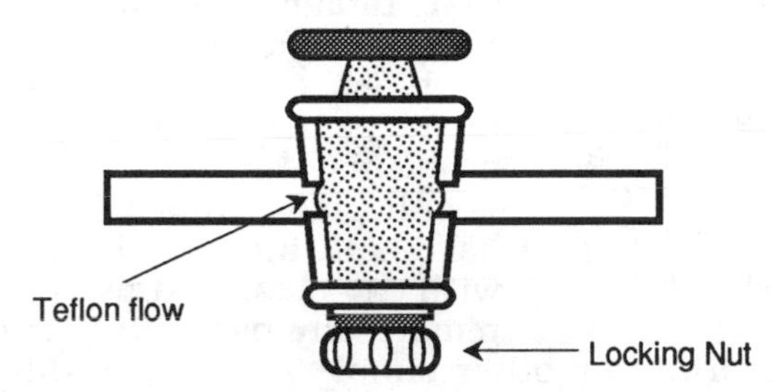

Fig. 3-33 If the locking nut is left too tight, the Teflon will flow (over time) into the holes of the barrel.

If the barrel of a Teflon stopcock was designed too thin, and the locking nut is left on too tight, any increase in temperature causing the Teflon to expand may cause the barrel to crack open into pieces.

The order of the washers on Teflon plug is important. As described in the previous section, after the plug is inserted in the barrel, the first piece to go on is the plastic washer, followed by the O-ring, which acts as a lock washer. The final piece to go onto the plug is the locking nut.

The hole through the plastic washer is not round; rather, it has a flat part on the hole (some manufactures use two parallel flat sections). This flat section matches a flat section on the threaded part of the Teflon plug and keeps the washer, O-ring, and locking nut as a unit to rotate with the plug when it is rotated. If this flat part wears round, the washer, O-ring, and locking nut do not rotate with the plug as it is turned. The result of this condition is that when the plug is rotated clockwise (CW), the locking nut (which is not rotating) tightens onto the plug. When the plug is rotated counterclockwise (CCW), the locking nut loosens from the plug. The occurrence of this phenomenon could either be just annoying or disastrous.

Teflon Rotary Valves. Teflon rotary valves should also be stored with the plug in place. The primary danger is to the plug, which — if scratched — could be worthless. The plug should be left screwed into the barrel, but never to the closed position. If the plug is screwed too tightly into the closed position it is likely to place too much pressure on the threaded part of the Teflon, which may in turn cause the threads to wear off.

3.3.2 Preparation for Use

Any dust, fibers, or particulate material between the inner and outer members of a joint or stopcock can cause jamming and/or excessive wear on glass or scratch Teflon.

If you examine the cleaned plug from an old burette's stopcock (especially one from a freshman chemistry lab), you will see a lot of scratching

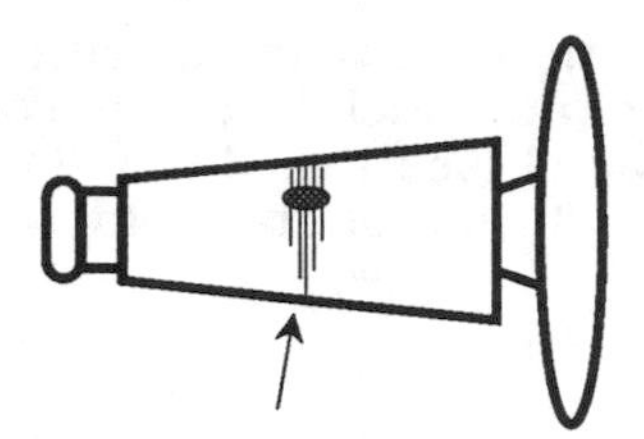

Fig. 3-34 Scratches on glass (or Teflon) plugs caused by particulate matter collec-ted on grease and never cleaned can cause wear, scratches, or grooves on a plug.

and wear on either side of the hole. This wear is caused by particulate matter collected in the hole (see Fig. 3-34). In time, this wear circumscribes the entire plug causing a leak that nothing can stop save replacement of the entire stopcock. Frequent cleaning (often not provided to freshmen chemistry burettes) can prevent unnecessary wear and subsequent problems.

All glass stopcocks and joints should have both inner and outer members wiped with a clean Kimwipe® before use to remove any particulate matter. In addition, to remove any fingerprints or other simple greases, a simple glass cleaner can be used to clean the glass. Rinse with water followed by a distilled water rinse. Ground sections of high-vacuum pieces should not be handled after cleaning because perspiration can degas slowly within the system.

Teflon items (Teflon stopcock plugs, rotary valve plugs, and Teflon sleeves) should be wiped off with some acetone (using a Kimwipe®) before being assembled for use.

3.3.3 Types of Greases

When selecting a grease, keep in mind that a grease can only lubricate and/or seal. Stopcock grease cannot repair leaks or imperfections of damaged or poorly-made stopcocks or joints. A grease's lubrication abilities are necessary for stopcocks to rotate, as well as to facilitate the easy separation of joint members. A grease's sealing abilities are necessary to help the joint or stopcock maintain the separation between a vacuum and atmospheric pressure, or between one (or more) compound(s).

A sealant, such as Apiezon W is a good example of a material that was designed for an extremely limited application. Apiezon W is a hard black wax that needs to be heated to 80°–90°C before it is soft enough to apply to the members you wish to join. At room temperatures it is hard and has no lubrication abilities whatsoever and is therefore not usable for a stopcock. It has a relatively high vapor pressure (10^{-3} torr), so it cannot be used for most vacuum work. However, at temperatures of about 100° to 150°C, it becomes very thin like hot honey and can easily be applied to joint members. Therefore, if you have a standard taper joint that will not be in a high-heat vacuum environment, needs to separate often, and could possibly separate accidentally, Apiezon W is the type of sealant you need.

There are three types of stopcock greases: hydrocarbon-based, silicon-based, and fluorinated hydrocarbon-based. The general characteristics and attributes of the various hydrocarbon- and silicon-based greases are listed in Tables 3-3 and 3-4 respectively (the third category, the fluorinated greases, requires no table because there are only two fluorinated greases, both made by the same manufacturer and no cross-distinctions are necessary). Table 3-5 lists the available technical information on all of the greases. These tables are by no means inclusive of all the greases available, nor is it complete. Regrettably, much technical information is not available. Occasionally, the technical information that was available was contradictory, and when this case applied I listed only the less grandiose of the claims.

Hydrocarbon-based stopcock greases are made (usually) from a refined hydrocarbon base. They offer efficient sealing and good to excellent vapor pressures. Some are refined specifically for stopcock use, whereas most others are made for the simpler requirements of joints. Most stopcock greases can be easily cleaned with hydrocarbon solvents, which is also part of their problem: most solvents can easily strip the grease from a stopcock or joint. Working with solvents, solvent vapors, high temperatures, or UV wavelengths can break down organic stopcock greases (such as the Apiezon

Table 3-3

Hydrocarbon-based Stopcock Greases		
Type	**Best used for**	**Watch out for**
Apiezon H	Wide temperatures (-15° to 250°C).	Its price (≈$80/25 gm). Not intended for stop-cock use.
Apiezon L	Joints where it's used will be in applica-tions of very low vapor pressure (10^{-10} to 10^{-11} torr).	Not intended for stop-cock use. Moderately expensive (≈$55/25 gm).
Apiezon M	Similar to Ap. L for general applica-tions, but cheaper (≈$35/25 gm).	Higher vapor pressure (10^{-7} to 10^{-8} torr) than Ap. L. Not intended for stopcock use.
Apiezon N	Stopcocks because of special 'slippery' properties.	Its price (≈$85/25 gm).
Apiezon T	Joints used in high temperature.	Its price (≈$80/25 gm). Not intended for stop-cock use.
Apiezon Q	Semi-permanent sealing of joints.	High vapor pressure (10^{-4} torr) or when evacuation must be maintained.
Apiezon W	Permanent sealing of joints.	Must be heated both to be applied as well as to separate inner and outer members.
Cello-Grease Lubricant	Very thick, max. operating temp of 119°C.	Unknown vapor pressure.
Cello-Seal	Maintains consistency up to 100°C. May be mixed with Cello-Grease for thicker consistency.	Unknown.
Thomas "Lubriseal"	Very cheap (≈$14.00/ 75 gm).	Because of its thinness, it tends to squeeze out of joints too easily.
Thomas "Lubriseal" High-Vacuum	Very cheap (≈$2.95/ 25 gm).	Because of its heaviness it is not recommended for general use.
Podbielniak Phynal Stopcock Lubricant	Designed for stopcocks, insoluble in hy-drocarbons and most organic solvents, but soluble in water for easy clean-up.	Soluble in water.

brand of stopcock grease). One option is to use a silicon-based stopcock grease.

Silicon-based stopcock grease maintains constant viscosity over a wide range of temperatures (-40°C to 300°C). These greases are water-insoluble and inert to most chemicals and gases. Silicone grease is inexpensive compared to many other stopcock greases.

There are, however, several drawbacks to using silicone grease: silicone grease is not impervious to all solvents and ethers. Silicone grease ages rapidly and it breaks down under UV irradiation. Some brands of silicone grease begin to age once they leave the factory and are only good for about eighteen months once you receive the tube. Old silicone grease becomes very hard, which makes rotation of a stopcock very difficult. When hard, silicone grease will not flow as a joint is being rotated, thus causing the grease to 'crack,' leaving a network of air lines (called a spider web) that will leak to either the environment or to the other sections of the stopcock. Despite its low vapor pressure, silicone grease 'creeps' within a vacuum system, which can lead to poor visibility, make glassblowing repair difficult to impossible, and increase cleaning complications.

Both hydrocarbon and silicon-based greases show signs of aging, which are typically increases in viscosity. When hydrocarbon-based greases age, lighter factions of the grease evaporate leaving heavier factions of the same material remaining. Softening this material can be done by heat in the same manner that heat is used to soften cold honey.

When silicon-based greases age, light factions evaporate leaving behind a different material that does not have light factions. Heat cannot soften this material. Thus, silicone grease requires more maintenance than hydrocarbon-based stopcock grease. For best results, old silicone grease can be (sufficiently) removed with a chlorinated hydrocarbon and new grease applied every two to three months. Reapplication of new silicone grease should be done regardless of whether a joint was used or not.

When silicone grease becomes hard, the material remaining is not the same material that has volatilized off. Thus, hot air technique used on hydrocarbon grease for softening the remaining material cannot be used on

Table 3-4

Stopcock Greases Containing Silicone		
Type	**Best used for**	**Watch Out for**
Dow-Corning Silicone Compound	Plastics, rubber, metals, and adhesives.	Shelf life is 18 months from date of shipping.
Dow-Corning High-Vacuum grease	Non-melting so it can be placed on hot (140° to 150°C). Maintains consistency from -40° to 260°C.	Needs to be cleaned and reapplied every several weeks wether in constant use or not.
Pldbielniak Silicone Stopcock	Contains silicone and carbon particles as lubricants, which may provide a more slippery lubricant.	Because of the carbon, it should not be used with Teflon because there is a potential of scratching the Teflon. It should also not be used with compressed oxygen as it could detonate.

silicone grease. You must replace the old grease immediately. Blowing a hot air gun on silicone grease with the intention to soften it up will in fact hasten its aging by removing the remaining volatile materials.

When all volatile compounds are removed from silicone grease, all that remains is silica. This material can be very difficult to remove from glass, and can severely hinder any repair of glass items by a glassblower. Therefore, silicone grease must be completely removed before any intense heat (>400°C) is applied. Silicone grease can be removed from small systems by letting them soak in a base bath for a half-hour (see Sec. 4.1.6).

The last grease type is *fluorinated greases*. These greases go under the brand name of Krytox® (registered and manufactured by Du Pont). They are all derived from the base oil of Krytox 143. To this oil are added different amounts of a thickening agent and other materials that impart a variety of characteristics. The thickening agent for Krytox greases is a component of a compound related to Teflon called VYDAX. VYDAX is the lubricating agent used on all stainless steel razor blades. Krytox comes in two forms for common lab use: one for general use, *Krytox® GPL*, and one with low vapor pressure, *Krytox® LPV* (10^{-12} torr at 20°C).

Krytox seems like a product from heaven. In fact, many of its attributes are spectacular. It is a fluorinated grease and is therefore impervious to hydrocarbon- and water-based solvents. It cannot burn, so it has no flash point and can be exposed directly to oxygen, even at high pressures. It can be used over the widest temperature range of all the greases and works equally well for stopcocks as with joints. It has remarkable characteristics for stability and can withstand long-term abuse. And, Krytox greases are nontoxic.

There are, however, two problems with the Krytox greases. Although they do not burn, they break down in high-heat environments (>260°C) into lethal fluorine compounds. Thus, a habit such as smoking can truly become a real killer: if any Krytox grease remains on your hands when you handle a cigarette, burned fumes from the grease left on the cigarette can kill you. Because of the dangerous fumes generated by highly heated Krytox grease, it is obvious that any glassware contaminated with Krytox must be completely cleaned before any repairs can be made.

The cleaning of Krytox grease presents the second problem. To remove the grease from glass apparatus or your hands, you must scrub it off (simple wiping is not sufficient) with a chlorofluorocarbon solvent such as Freon® 113 or Freon® TF.* Both of these compounds are directly implicated in the destruction of the world's protective ozone layer. Of course it can be argued that the amount needed is small compared to a leaky air conditioner or many industrial uses, but it also can be argued that every bit hurts.

However, supposed you used just a tiny bit of Freon solvent to hurt the environment as little as possible. You wipe off the grease with a small amount of Freon 113 or Freon TF and assume this cleaning will be safe. As mentioned before, the thickening agent for Krytox grease is called VYDAX whose fumes are also toxic. VYDAX can smear into the rough surfaces of a ground joint or stopcock adding to the lubrication capabilities. To remove Krytox from these surfaces requires rough scrubbing with a toothbrush or fingernail brush, and sufficient solvent to remove the material.

* The production of both are due to be phased out.

Table 3-5

Technical Data Chart for Various Greases

Type	Melting Point (°C)	Vapor Pressure (mm Hg)	Recommended Temperature Range (°C): Stopcocks	Recommended Temperature Range (°C): General	Comments
Apiezon H	n/a	10^{-9}	5 to 150	-15 to 250	a, b
Apiezon L	47	10^{-11}	10 to 30	10 to 30	b, c
Apiezon M	44	10^{-8}	10 to 30	10 to 30	b, d
Apiezon N	43	10^{-9}	10 to 30	10 to 30	b, e
Apiezon T	125	10^{-8}	10 to 80	0 to 120	b, f
AP 100	30	10^{-11}	? ?	? to ?	b, g, h
AP 101	185	10^{-6}	? ?	? ?	b, h, i
Cello-Grease	130	?	? ?	? to ≈120	j, k
Cello-Seal	?	10^{-6}	? ?	? to ≈100	j, l
DC Silicone	?	$<10^{-5}$	? ?	-40 to 204	m, n
DC High-Vacuum	?	$<5\times10^{-6}$	? ?	-40 to 260	m, n
Gen. Purp. Comp.	?	?	? ?	? ?	o, p
Insulgreases	?	?	? ?	-70 to 200	o, q
Kel-F Grease	?	$<10^{-3}$	? ?	-18 to 180	r
Krytox GPL	—	10^{-7}	-35 to 250	-35 to 250	s, t
Krytox LVP	—	10^{-12}	-35 to 250	-35 to 250	s, t
Lubriseal	40	?	? ?	? ?	e
Lub. Hi-Vac	50	3×10^{-6}	? ?	? ?	u
Nonaq	?	?	? ?	? ?	j, v
Stpck. Lub. Sil.	?	10^{-6}	? ?	? 250	w
Stpck. L. Phynal	?	?	? ?	? 50	w, x
Versilubes	?	?	? ?	-73 to 230	o

a Softens at 250°C, but does not melt to a free-flowing liquid.
b Apiezon greases are made in England, and are sold by many suppliers.
c Designed for ultra-high vacuum.
d Good for general use.
e Designed for stopcocks.
f Designed for high temperature uses.
g Ultra-high vacuum use.
h Not leached by aggressive solvents, yet are removable with aqueous detergents.
i Moderate vacuum use.
j Fisher Scientific product.
k Heavier and darker than Cello-Seal, high vacuum use.
l Gas-tight seals between glass and rubber tubing; can be coated outside of tubing to prevent diffusion.
m Dow Corning product.
n Constant viscosity in that temperature range.
o General Electric product (Silicone Products, Waterford N.Y.).
p Corrosion protection; heat sink.
q Non-lubrication; dielectric grease.
r Made by 3M and M.W. Kellog Co.
s Made by E.I. Du Pont de Nemours & Co., Inc.
t Fluoronated grease.
u More refined than Lubriseal.
v Contains silicone and graphite.
w Made by Podbielniak.
x Insoluble with most organic and hydrocarbon materials, but soluble in water.

Krytox, Teflon, and other similar products can be removed by incineration, and the amount of fluorine released from several stopcocks or joints is a relatively small amount. However, if the oven used is not properly vented, in the confined space of the oven's room, the fumes could be toxic. At the time of the writing of this book, there are no environmentally-safe solutions. Accordingly, those concerned about the environment may be inclined to conclude that there is no justification for using Krytox grease. However,

Krytox is the only grease for stopcocks that can be used for very low temperatures (-35°C) to very high temperatures (220°C), has a very high level of chemical inertness, and will not affect elastomers or metals. Thus, there are conditions when Krytox is the only viable alternative and must be used.

As a final note, a recent (December 1990) call to Du Pont elicited the information that they do have several materials that can clean Krytox and may be both environmentally safe and nontoxic. However, until they have completed tests and decide which of these products must go into full-scale production, they still recommend using Freon 113. They hope by the end of 1991 to have a safe cleaning product in production.

3.3.4 The Teflon Sleeve

There is one other substitute for stopcock grease on standard taper joints: the Teflon sleeve. These sleeves are like socks for your joints. Because they are made of Teflon, they are not attacked by solvents, alkalines, and most other chemicals. Thus, they are wonderful for items like solvent flasks which are under constant fume or chemical contact. However, they are not capable of maintaining a static vacuum and should not be used for vacuum work. Thin Teflon sleeves are less expensive, but cannot take physical abuse. Heavier Teflon sleeves have a greater initial cost, but can be used over and over.

3.3.5 Applying Grease to Stopcocks and Joints

The concept of "if a little is good, then a lot is better, and too much is just right" is incorrect for the application of grease on stopcocks and joints. Excess grease can spread throughout an apparatus creating a mess that can affect the work in progress, decrease the potential vacuum and the speed of obtaining a vacuum, and worse, plug up holes in stopcocks, making them inoperative.

Stopcock grease is applied to the inner member of a joint or stopcock, inserted within the outer member, then twisted and/or rotated to spread the grease throughout the rest of the ground surface. Applying a mild amount of pressure will facilitate an even application throughout a joint. The colder a joint, the more difficult the grease will be to spread. Lightly heating members prior to assembly will help create an even distribution of grease.

When applying stopcock grease to pieces used on vacuum systems, do not use your fingers because perspiration from fingers can result in slow degassing within the system, limiting your potential vacuum. If the grease is supplied in a flat metal tin or glass jar, use a wooden spatula to spread the grease on the inner piece. Grease that is supplied in a tube (similar to toothpaste), is easy to apply by laying the grease directly on the inner member in lengthwise strips (see Fig. 3-35). Regardless of the method of application, one to two strips is all that is necessary for very small joints and stopcocks (14/35 joints or 2 mm stopcocks), and up to four strips for large joints or stopcocks (45/50 joints or 10 mm stopcocks).

Once the grease is applied, insert the inner member into the outer member, twist and/or rotate the inner piece until there is a thin, smooth film of grease between the two pieces. If the hole in the stopcock plug becomes filled with grease, remove the plug and, using the wooden stem of a cotton swab, push out the excess grease.

If you are re-greasing a stopcock or joint, be sure to remove all old grease (from both members) before applying new grease. If a solvent is used to facilitate grease removal, let the solvent dry before applying the new grease or it may be prematurely aged by the solvent.

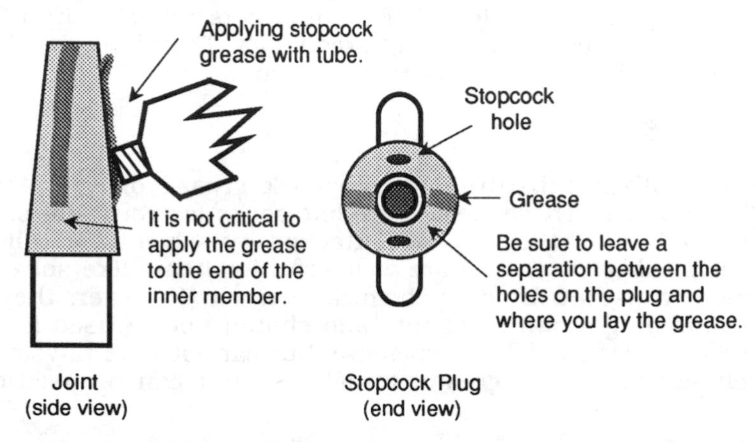

Fig. 3-35 The application of grease on a stopcock or joint.

3.3.6 Preventing Glass Stopcocks and Joints From Sticking or Breaking on a Working System

Few things can be more frustrating, or economically devastating, than a frozen joint or stopcock preventing access to a compound. Part of experimental time should be spent making sure your equipment is ready to provide you safe and reliable use. Time for such preparation is part and parcel of experimentation.

Periodic *removal* and *reapplication* of stopcock grease is a must. The frequency depends on the size of the stopcock or joint, how often it is used, under what conditions it is used, and under what type of conditions it will be used in the future. Conditions that can increase the need for cleaning and reapplying stopcock grease are age, heat, the continued use of hydrocarbons within a system, and constant exposure to UV radiation. Silicone grease should be changed as often as once every two months to every week, depending on the type and nature of use. Even if an apparatus is not being used, silicone grease ages and must be changed.

Because of the constant heating that large stopcocks often receive (to aid their rotation), and because heating helps to age and drive out the higher volatile components of stopcock grease, large stopcocks should be cleaned and new grease reapplied fairly often. Specific frequency for changing stopcock grease is difficult to provide because it depends on the amount of abuse and frequency of use. However, with constant use, hydrocarbon-based stopcock grease should be replaced about every six months. Some general tips for joint and stopcock maintenance are:

1) Never force an item. The glass might not be as strong as you, and you do not want to find out the hard way. Also, forcing a stopcock or joint is more likely to permanently jam an item than to free it.
2) Never soak a joint or stopcock in a joined position with its mate in an alkali solution (i.e., a base bath). Alkali solutions should not be stored in apparatus that have ground sections (joints or stopcocks) as they are likely to jam them.
3) Never leave a piece of apparatus with a stopcock or ground joint with its mate in a drying oven.
4) Never leave a Teflon plug or stopcock in a drying oven. Fumes from a Teflon item in a drying oven can be toxic. Also, the Teflon item will expand with the heat at a greater rate than the glass causing the glass item to crack or break.
5) Always hold the barrel or free-standing arm of a stopcock when rotating the plug. The torque applied to rotate the plug may be greater than the torque necessary to snap the stopcock off the apparatus.

3.3.7 Unsticking Joints and Stopcocks

Even if you follow all the rules of storage, preparation, and use, it is still possible to have a joint or stopcock stick. The separation of these items, like cleaning glassware, is more an empirical art than a specific science. Perhaps it is pessimistic to say, but you should approach the unsticking of joints or stopcocks as if you had already lost the apparatus in question. If you are able to free them, you have gained. Otherwise, you cannot lose something that you have already lost. The general rules for unsticking joints and stopcocks are as follows:

Your objective is to loosen, not remove any grease between the mated members. Never perform cleaning operations with solvents on mated stopcocks or joints. Never let mated stopcocks or joints bake in a drying oven or soak in solvents that can dissolve the grease. If there is no grease, there is little hope in separating the pieces.

Flame Heat. You can heat the entire joint or stopcock with a large bushy flame from a gas–oxygen torch, if 1) no Teflon is involved (a sleeve on a joint or Teflon stopcock), 2) the piece is open to air on both sides*, and 3) the piece does NOT contain a flammable mixture. First, be sure to wear heavy, heat-resistant gloves (such as gloves made with Kevlar®), and safety glasses (didymium glasses** are recommended). Be sure to *rotate* the object, while in the flame, so that *all sides* of it are equally heated. The length of heating

* For instance, do not heat a round bottom flask with a single joint opening closed off. Excess pressure can cause the item to blow up.

** Didymium glasses are special glasses used by glassblowers. They are obtainable from most of the manufacturers listed in Appendix C, Chapter 8. Didymium has special filtering properties which eliminate the sodium spectra, allowing the wearer to see through the normally bright light emitted by glass. Sunglasses and welder's glasses cannot be used for substitution.

time is only for half a minute at most. After heating, *remove* the item from the flame and try to separate the pieces immediately. If this method fails, you may let the piece cool and try again. However, your chance of success goes down rapidly if the first attempt is a failure.

Electrical Conduction Heat. If the item is soft glass, such as the ground stopper section on a storage container (or even a perfume bottle), a softer heat is required. This heating can be done by wrapping bare copper or nichrome wire (18 gauge) tightly around the joint, leaving an inch or so of tail. Heat the wire tail with a flame and conductance will heat the joint.[2] If you wish to use a voltage regulator (such as a Variac®) to heat the wire, use nichrome wire that is insulated with nonflammable covering. Do not use copper because it does not have enough resistance (while trying to get the wire hot, you will draw too much current and likely blow a fuse), and make sure that the wires do not overlap. In addition, *be sure to turn the Variac off before handling.*

Chemical Soaking Solutions. If it is not possible to use the heat from a flame, there are a few soaking solutions that may help. Their chemical mixtures are as follows:[3]

Solution #1

Alcohol (2 parts)
Glycerin (1 part)
Sodium Chloride (1 part)

Solution #2

Chloral hydrate (10 parts)
Glycerin (5 parts)
Water (5 parts)
25% Hydrochloric Acid (3 parts)

Soak the frozen joint or stopcock in either solution overnight or longer. Try to free the item with a moderate amount of force, never an overt amount of force. If either solution fails, try the other solution or either of the methods below.

Soda Pop. Surprisingly, soda pop has been used to successfully separate stuck stopcocks. Some people say you need to use stale cola, others say fresh cola. I've tried both with equally mixed success. Butler,[4] using the fresh cola approach, suggested that the mechanism at work is the decomposition of carbonic acid as follows:

$$H_2CO_3 \rightarrow H_2O + CO_2$$

Butler also suggested that it is the resulting CO_2 pressure that de-freezes the stuck stopcocks. Additionally, he suggested that there may be some interaction between the carbonic acid and various sites on the glass surface. However, his suggestions do not explain why *stale* cola should work. Regardless of the freshness of the cola, the joint needs to be soaked anywhere from three to forty-eight hours. If you choose to use the fresh cola approach, it should be replenished periodically to allow for fresh bubbles.

Beeswax. Heat the joint to just above the melting temperature of beeswax (about 65°C), then apply the wax at the lip of the two parts. The wax will melt and surface tension will cause the wax to penetrate between the male and female sections. Let cool, then reheat and the joints may separate.[5]

The Vibrating Etching Tool.[6] First, remove the carbide tip from an engraving tool (to prevent damaging the glassware), and then place the vibrating end to the ground section while pulling on one of the pieces. You may also try to preheat the joint as mentioned above.

The Crab Mallet. Even if glassware with stopcocks or joints is properly stored in drawers, occasionally the members may stick. The following de-sticking technique will work on dry (ungreased) items only. Use a wooden stick, or crab mallet (see Fig. 3-36), and lightly tap the stuck member in the same alignment as the joint itself (see Fig. 3-37). Be sure to lay the item on a table so that if you are successful, the item will skid on the tabletop a few inches rather than fall to the ground. The use of wood for this technique is mandatory as anything harder is likely to chip, crack, or demolish the glass.

Fig. 3-36 A crab mallet.

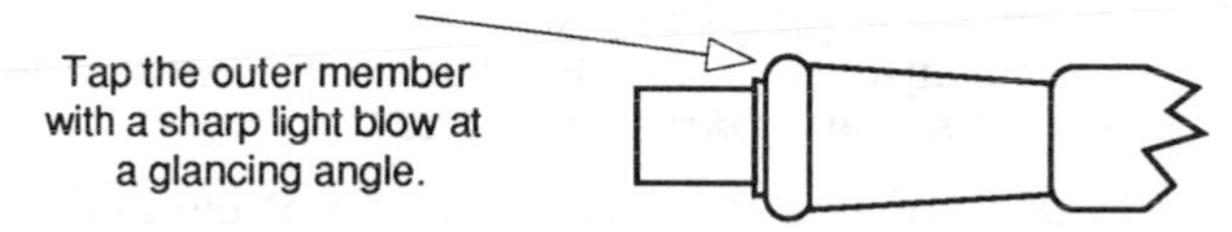

Tap the outer member
with a sharp light blow at
a glancing angle.

Fig. 3-37 Lightly tapping stuck (ungreased) joint or stopcock members with wood can sometimes separate the pieces.

The False Handle. Finally, there may be times when part of a stopcock or plug has broken off leaving little left to hold onto. This situation can make removal very difficult. Trying to grab onto remaining glass with a pair of pliers often is futile as the glass breaks and chips off. If there is access to the opposite end, it may be possible to tap, or push, the piece out. Alternatively, if the plug or joint is hollow and is closed like a cup, it may be possible to pour melted lead, or melted electrical solder into the item like a screwdriver tip. Then when the melted metal hardens, you have a good handle with with to remove the remaining portion of the plug. The solder may then be removed and reused.[7]

WARNING: Although the materials used in these unsticking techniques are, for the most part, inert or nonreactive, practice safe chemistry. If you have a chemical (that is trapped by a stuck stopcock or joint) that may react with one of the materials or solutions mentioned in this subsection, select a different approach.

3.3.8 Leaking Stopcocks and Joints

There are three reasons that a joint or stopcock might leak: either the item is not being used in the conditions, or manner, for which it was designed, the stopcock or joint is worn, or the item was poorly manufactured. Although extra grease may be used for short-term emergency situations, it is not a solution because extra grease is not fixing the problem. Similarly, adding extra stopcock grease to a regular stopcock will neither make it act like a high-vacuum stopcock nor help a defective stopcock or joint.

Surprisingly, many assumed defects of a joint or stopcock are caused by using an item improperly. By trying to "make do" with what you have, you may either jeopardize the quality of your work, or subject yourself and others around you to potential dangers. Thus, adding extra grease onto a standard stopcock to that it can be used for vacuum work will be a losing battle.

Assuming that the joint or stopcock selected is proper for the application, and there is evidence of leakage, you should look for potential human errors:

1) Did you apply the stopcock grease properly, or is the current grease old?
2) Did you properly clean the items before assembly? Particulate matter left in a joint or stopcock cannot only cause a leak, but can also cause the stopcock to jam.
3) Are the pieces of a Teflon stopcock assembled in the correct order? For example, on a Teflon stopcock, if the black O-ring is placed next to the stopcock barrel, counterclockwise rotation may cause the plug to loosen.
4) If you have a vacuum system with glass stopcocks, do the numbers on the plugs and barrels match?

Even if failure to prevent the above errors has not caused a leak outright, prolonged misuse eventually causes wear on a joint or stopcock, which can then cause permanent damage. Not all defects caused by wear can be distinguished from manufacturing defects by appearance alone. However, if a new item leaks from structural defects, it probably is a manufacturing imperfection.

Although the quality of joint and stopcock construction from the glass manufacturing industry is generally excellent, mistakes do happen. It is possible to have received hundreds, or thousands, of excellent joints and stopcocks from the same manufacturer, plus one that leaks. Fortunately, manufacturers will promptly replace defective materials.

The Rock or Jiggle Test. It is possible for a joint or stopcock to be so badly made that a mated pair can rock, or jiggle, when a little lateral motion is applied. This test should be done without any stopcock grease on the pair to prevent artificial damping of the motion. Do not force or twist any joint without grease as it is likely to jam. The biggest problem with this test is that it is impossible to know which of the pair is the bad one. To use a pair with this problem will only make both pieces bad as the bad one will wear the good one unevenly, thus making it unusable.

The Pencil Test. Before performing the pencil test, completely clean the entire apparatus in question. Remove all traces of any oil or grease (if the grease is hydrocarbon-based, you can heat the glass item in an oven to approximately 400°C to burn off any remaining grease). Next, take a standard no. 2 pencil and scribble a line all over the inside surface of the outer piece and the outside surface of the inner piece, as shown at the left of Fig. 3-38. Then, wet the joint (or stopcock) members with water (for lubrication) and rotate back and forth five to ten times (do not bear down, let the water be a lubricant). If the ground sections of both faces of the joint (or stopcock) are perfectly flat, the pencil will have been removed while the two faces ground against each other. If there are any depressions or raised areas on either surface, pencil lines will remain or be selectively ground away as seen at the right of Fig. 3-38.

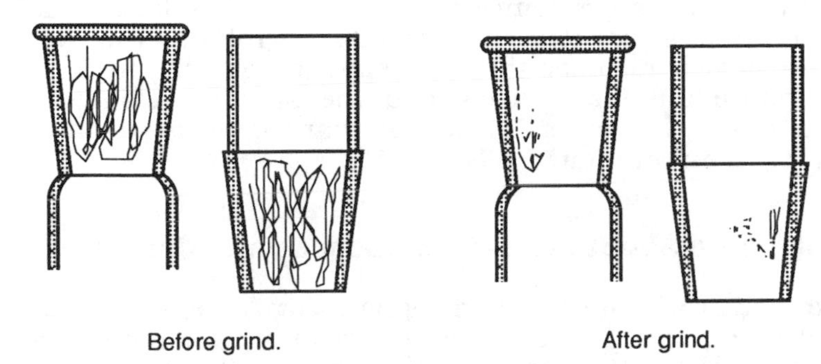

Fig. 3-38 Sample of using the pencil test to examine the quality of a joint.

This technique works well with both regular stopcocks and joints, but the surfaces on vacuum stopcocks are ground so smooth that they are unable to grind off the pencil. However, by adding a *very small* amount of an optical finishing powder (i.e., aluminum oxide #50) you will be adding sufficient roughness without doing any damage to the stopcock. If there is too much of the optical finishing compound, the slurry will grind all the pencil off regardless of the quality of the stopcock. Use plenty of water for lubrication; a joint that becomes stuck with grinding compound is almost impossible to separate. Once you have finished using the grinding compound, be sure to thoroughly rinse it out. Grinding compound left on a joint or stopcock is likely to destroy the item over time.

As previously mentioned, it is important that all oil and grease be removed before trying the pencil test. This point becomes even more critical if any optical finishing compound or grinding compound is used because it is likely to stick to any remaining oil or grease. If the stopcock is used with any particulate matter mixed in with the grease, it will grind into the glass and destroy the stopcock or joint.

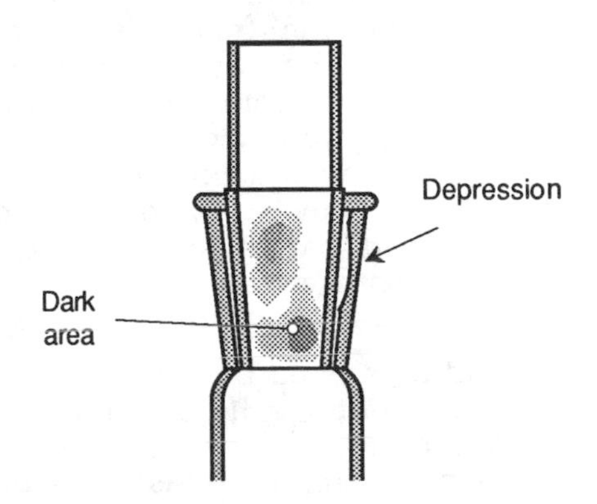

Fig. 3-39 Sample of using the prussian blue(or merthiolate) test to examine the quality of a joint.

The Merthiolate or Prussian Blue Test. The pencil test will not work on a Teflon stopcock because there is no rough surface with which to rub a pencil mark off. Grinding compound cannot be used because it will destroy the surface of the Teflon, thus destroying the plug.

Rather than using a rough material to grind off a pencil line, this test uses either *Prussian Blue* or *Merthiolate* to collect in the depressed surface of the member. The item is cleaned as before, but this time several drops of Prussian Blue or Merthiolate are placed on the inner piece before assembly with the outer member. As seen in Fig. 3-39, any dark spots signify a thicker film of the indicating solution, which implies a concave impression on one or the other piece. A perfect joint shows clear with no dark areas. This test can also be used on glass stopcocks or joints. By rotating the two pieces you can note with which member the dark spot travels. If you hold the outer piece still, rotate the inner piece, and the spot travels, you know the depression is on the inner piece. Similar logic is used to identify depressions on the non-rotated piece.

3.3.9 What To Do About Leaks in Stopcocks and Joints

Unfortunately, there are not many options available when you have a defective stopcock or joint. The best solution is to replace the defective part of the apparatus, or the entire apparatus if necessary. If the item is new, return it to your supplier and have it replaced.

If the defective part is the plug of a Teflon stopcock, you can try to simply replace the plug. If the plug was damaged by abrasion (or other abuse), replacement offers a simple resolution. The possibility that a distorted Teflon plug is causing a leak is very unlikely. If the glass barrel of a Teflon stopcock is distorted (showing the problems seen in Fig. 3-39) you have no recourse but replacement.

If the item is all glass, it may be reground. However, do not regrind a glass stopcock if :

1) You have any other recourse (see Point 5).
2) You are not experienced in grinding stopcocks or joints.
3) You are unable to completely remove <u>all</u> traces of oil or grease from the item. Any oil or grease will retain grinding compound onto the part and subsequent use of the item with the grinding compound on it will cause it to jam or wear unevenly leaving you back where you started. The best technique for removing organic greases is to burn them off in a high-heat oven.
4) If the item is a stopcock and the overlap of the holes on the plug and barrel are such that any grinding will close the hole passageway.
5) You can accept that the piece in question is expendable.

Grinding should be done with 120-grit Carborundum and plenty of available water. Adding a little dishwater soap will help lubricate of the pieces as they are ground.

After removing all grease, wet both glass members with water and place a small bit of grinding compound on one of the members. Insert the inner member into the outer member and, while applying a light amount of pressure, rotate the members around and around, *not* back and forth (this point is very important because back and forth rotation with grinding compound will cause uneven wear). Remember, incomplete rotation will mean you are making a depression in only one region of the rotation. Never let the members become dry while rotating, otherwise they are likely to freeze. Members that freeze while grinding cannot be separated.

If you chose to regrind, and you use a glass part with which to grind the errant member, the parts are "married" to each other and should be numbered like vacuum stopcocks. These pieces are likely to leak with any other similar sized member.

3.3.10 General Tips

To obtain a high vacuum, heat the walls of a system to help drive off moisture. On a glass system, this heating is typically done with a hot air gun. It is important that any heat source is not aimed at stopcocks or joints on the system. However, it is often necessary for very large stopcocks or joints to be heated before they can be rotated or separated. Therefore, pay attention and do not heat any more than necessary. These stopcocks or joints require frequent grease replacement.

Keep in mind that if you heat a vacuum stopcock or joint in a vacuum influence, any grease on the member will be squeezed out because the pressure of the vacuum will pull the plug deeper into the barrel causing the (now less viscous) grease to be squeezed out. After repeated heating, grease under these conditions will age prematurely and become so thin that it will be useless.

Very high heat (> 300°C) can cause organic greases to volatilize. Once the grease is gone, rotation or separation of the joint or stopcocks becomes very difficult to impossible.

Using high heat (> 350°C) with silicone grease can cause parts of the grease to volatilize, while leaving the non-organic remains to form a hard, crusty deposit that makes separation or rotation of stopcocks, or separation of joints, impossible. The material remaining from overheated silicone grease can only be removed with hydrofluoric acid. Silicone grease at room temperature ages in two to three months. At higher temperatures, it ages significantly faster and needs to be replaced even more frequently.

3.4 Glass Tubing

3.4.1 The Basics of Glass Tubing

The glass tubing often used in beginning chemistry labs is soft glass that can be worked with the flame of a Bunsen burner. For more information on this type of glass, see Sec. 1.1.3. The vast majority of the remaining glass within the laboratory is borosilicate glass. For more information on borosilicate glass, see Sec. 1.1.3.

Tables 3-7 through 3-13 list most of the different types of tubing manufactured in the U.S. and Germany. These tables list the various sizes of tubes available as well as their inside diameters (I.D.), volume in $^{ml}/_{cm}$ (with the tolerance of that volume), and volume (in ml) of one half-sphere (representing the rounded end of a tube). The next two columns indicate the number of pieces shipped per case, and the price per length of tube (based on the price of one tube when purchasing one case of that size tube at 1990 retail prices). The final column provides the maximum tolerable pressure which that size tubing can withstand, assuming a maximum bursting pressure of 750 psi. The reader should closely follow the information on *pressure in glass tubing* in Sec. 1.1.9.

Standard Wall Tubing (Table 3-7). This tubing is made in metric sizes (unless otherwise stated, all measurements on these charts are metric). The vast majority of all tubing seen in the laboratory is Standard Wall Tubing.

Medium and Heavy Wall Tubing (Table 3-8 and 3-9). This tubing is made in English sizes, but with the push toward the metric system such tubing is now listed in catalogs in its metric equivalent (therefore, 1" tubing is listed as 25.4 mm). Both English and metric measurements are listed in Tables 3-8 and 3-9.

Special Wall Tubing (Table 3-10). This tubing comes in medium and heavy wall types, but there are fewer size selections. In addition, these tubings are not sold as "medium" or "heavy" wall. Rather, when ordering, you must specify the desired *Outside Diameter* (O.D.) and wall thickness.

Capillary Tubing (Table 3-11). This tubing is selected on the basis of its *Inside Diameter* (I.D.), also called its *bore*. Capillary tubing is sold in a continuous (and sometimes overlapping) range of bore sizes. The O.D. is seldom the deciding factor in selection.

Schott Glass Tubing (Table 3-12). This tubing manufacturer uses a different system for determining sizes. For one, this tubing is 1.5 meters long (approximately five feet). Secondly, it does not come in medium and heavy wall thicknesses based on the English measurement system. Rather, for each outside diameter, one, two, or three different wall thicknesses are available. Because of these differences, all calculations for volume and pressure previously made would not be relevant. Table 3-12 provides all the above information, but the tubing is not separated by wall thickness.

Schott Glass Capillary Tubing (Table 3-13). This tubing is similar to regular Schott Glass Tubing because several I.D.'s are available for each size O.D.

3.4.2 Calculating the Inside Diameter (I.D.)

In manufacturing tubing, two out of three parameters—the outside diameter and wall thickness—can be monitored; the inside diameter (I.D.) is maintained through the manufacturer's control of the other two parameters. Therefore, the I.D. on all charts (except Capillary Tubing) is calculated by inference using Equation 3.1:

$$\text{O.D.} - [2 \times (\text{wall thickness})] = \text{I.D.} \qquad \text{Eq. (3.1)}$$

Note that there are tolerances for each of these measurements, and each of these tolerances can cause variation in I.D. Fig. 3-40 shows the potential variation.

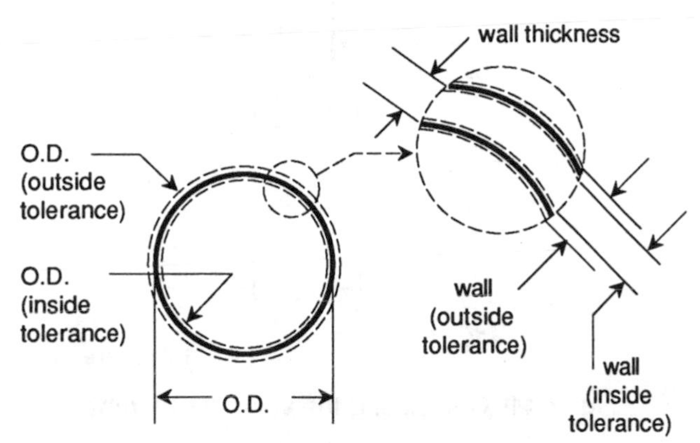

Fig. 3-40 The tolerances (±) of the wall thickness and outside diameter of glass tubing.

Variations of the O.D. include any possible distortion of the tube diameter. This distortion is referred to as "out of round" or API (All Points In). For example, 20 mm tubing has a potential variation of ±0.6 mm. The wall thickness for this tubing is 1.2 mm with a potential variation of ±0.1 mm. The range of variations can be seen in Table 3-6. Although there can be a range of 2.2 mm (±1.1) on the inside diameter of a 20 mm tube, most manufacturers maintain their tolerances within half of these amounts.

Table 3-6

Ranges of variations of Outside Diameter Due to Out of Round						
1)	Max O. D.	–	Min. wall thickness	=	I.D.	
	20.6	–	**1.1**	=	**19.6**	**mm**
2)	Max O. D.	–	Max wall thickness	=	I.D.	
	20.6	–	**1.3**	=	**18.0**	**mm**
3)	Min. O. D.	–	Min. wall thickness	=	I.D.	
	19.4	–	**1.1**	=	**17.2**	**mm**
4)	Min. O. D.	–	Max wall thickness	=	I.D.	
	19.4	–	**1.3**	=	**16.8**	**mm**

3.4.3 Sample Volume Calculations

With the use of Tables 3-7 through 3-13, you can determine what size tubing may be required to satisfy a specific liquid volume amount, or whether the apparatus you have will contain a given liquid volume, by calculating the volume based on the O.D. of the container and its length.

For example, we can make a sample volume calculation using the information provided in Fig. 3-41 to determine the volume of "A." If the diameter of "A" is 20 mm tubing, we can see on Table 3-7 that the volume of the half-sphere is 1.4 ml and the volume of each centimeter is 2.4 ml; therefore the total volume is: (2 x 1.4) + (3 x 2.4) = 10 ml. The centimeter of the 10 mm tube on the top can contain 0.50 ml.

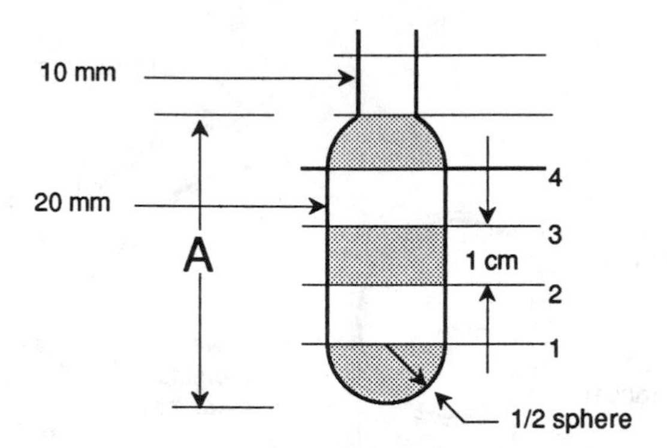

Fig. 3-41 Calculating the volume of a tube.

Although tolerance is given for the volume per centimeter of tubing, no tolerance is given for the half-sphere. Because half-spheres cannot be formed with the same precision as with tubing, the volume measurement for half-spheres should be used only as a working tool and not as an accurate measure.

Table 3-7

Standard Wall Tubing

O.D. (mm)	I. D. (mm)	Vol (ml)/ HT (cm)	± ml	Vol (ml)/ half-sphere	Pieces/ case	Cost ($)/ piece[a]	Max int pres (psi)[b]
2	1.0	0.01	0.00	0.00	357	$0.24	450.0
3	1.8	0.03	0.01	0.00	551	$0.39	352.9
4	2.4	0.05	0.01	0.00	520	$0.43	352.9
5	3.4	0.1	0.0	0.01	390	$0.48	275.7
6	4.0	0.1	0.0	0.02	271	$0.70	288.5
7	5.0	0.2	0.0	0.03	223	$0.85	243.2
8	6.0	0.3	0.0	0.06	189	$1.00	210.0
9	7.0	0.4	0.0	0.09	164	$1.15	184.6
10	8.0	0.5	0.0	0.13	148	$1.27	164.6
11	9.0	0.6	0.0	0.2	132	$1.43	148.5
12	10.0	0.8	0.0	0.3	120	$1.57	135.2
13	10.6	0.9	0.1	0.3	93	$2.03	151.0
14	11.6	1.1	0.1	0.4	86	$2.19	139.4
15	12.6	1.2	0.1	0.5	80	$2.36	129.5
16	13.6	1.5	0.1	0.7	75	$2.51	120.8
17	14.6	1.7	0.1	0.8	70	$2.69	113.3
18	15.6	1.9	0.1	1.0	65	$2.90	106.6
19	16.6	2.2	0.1	1.2	62	$3.04	100.7
20	17.6	2.4	0.1	1.4	58	$3.25	95.4
22	19.0	2.8	0.2	1.8	53	$3.56	109.2
25	22.0	3.8	0.2	2.8	37	$5.10	95.4
28	25.0	4.9	0.2	4.1	33	$5.72	84.6
30	26.4	5.5	0.2	4.8	26	$7.25	95.4
32	28.4	6.3	0.3	6.0	24	$7.86	89.1
35	31.0	7.5	0.3	7.8	20	$9.57	90.6
38	34.0	9.1	0.4	10.3	18	$11.54	83.1
41	37.0	10.8	0.4	13.3	16	$12.98	76.7
45	41.0	13.2	0.5	18.0	15	$13.85	69.6
48	44.0	15.2	0.5	22.3	14	$14.84	65.1
51	47.0	17.3	0.5	27.2	13	$15.98	61.1
54	49.2	19.0	0.6	31.2	4	$20.35	69.6
57	52.2	21.4	0.7	37.2	4	$23.15	65.8
60	55.2	23.9	0.7	44.0	4	$23.15	62.4
64	59.2	27.5	0.8	54.3	4	$28.55	58.4
70	65.2	33.4	0.9	72.6	4	$31.13	53.2
75	70.2	38.7	1.1	90.6	4	$33.75	49.5
80	75.2	44.4	1.1	111.3	4	$36.33	46.3
85	80.2	50.5	1.2	135.0	4	$39.10	43.5
90	85.2	57.0	1.3	161.9	4	$41.53	41.1
95	90.2	63.9	1.4	192.1	4	$44.13	38.9

[a] 1990 prices.

[b] Maximum internal pressure is a calculation based on a variety of safety factors. One of these factors is surface abrasion, which can significantly lower the potential pressure that a given tubing can withstand. The use of safety shields and fiber-tape wrapping of tubing is strongly recommended. For more information see Sec. 1.1.9.

Table 3-7

Standard Wall Tubing (cont.)							
O.D. (mm)	I. D. (mm)	Vol (ml)/ HT (cm)	± ml	Vol (ml)/ half-sphere	Pieces/ case	Cost ($)/ piece[a]	Max int pres (psi)[b]
100	95.2	71.2	1.5	225.9	4	$46.75	36.9
110	104.8	86.3	2.3	301.3	4	$54.53	36.3
120	114.0	102.1	2.7	387.9	4	$70.08	38.4
125	119.0	111.2	2.9	441.2	4	$72.70	36.9
130	124.0	120.8	3.2	499.2	4	$75.48	35.4
140	133.0	138.9	3.8	615.9	2	$93.45	38.4
150	143.0	160.6	4.3	765.6	2	$103.85	35.8
178	171.0	229.7	6.3	1309.1	2	$133.20	30.1

[a] 1990 prices.
[b] Maximum internal pressure is a calculation based on a variety of safety factors. One of these factors is surface abrasion, which can significantly lower the potential pressure that a given tubing can withstand. The use of safety shields and fiber-tape wrapping of tubing is strongly recommended. For more information see Sec. 1.1.9.

Table 3-8

Medium Wall Tubing								
O.D. inches	O.D. (mm)	I. D. (mm)	Vol (ml)/ (cm)	± ml	Vol (ml)/ half-sphere	Pieces/ case	Cost ($)/ piece[a]	Max int pres (psi)[b]
1/4	6.4	4.0	0.1	0.0	0.0	276	$0.88	328.7
1/2	12.7	9.5	0.7	0.1	0.5	96	$2.52	210.0
5/8	15.9	12.7	1.3	0.2	1.1	75	$3.23	164.6
3/4	19.1	15.9	2.0	0.2	2.1	61	$3.97	135.2
1	25.4	20.6	3.3	0.4	4.6	31	$7.82	153.5
1 1/4	31.8	27.0	5.7	0.6	10.3	24	$10.28	120.8
1 1/2	38.1	33.3	8.7	0.8	19.4	18	$14.60	99.6
1 3/4	44.5	39.7	12.4	1.0	32.7	15	$16.93	84.6
2	50.8	44.5	15.5	1.4	46.0	4	$26.28	99.6
2 1/4	57.2	50.8	20.3	1.7	68.6	4	$28.48	87.9
2 1/2	63.5	57.2	25.7	2.0	97.7	4	$38.58	78.7
2 3/4	69.9	63.5	31.7	2.3	134.1	4	$41.13	71.3
3	76.2	69.9	38.3	2.6	178.4	4	$47.30	65.1
3 1/4	82.6	76.2	45.6	3.0	231.7	4	$49.93	59.9
3 1/2	88.9	82.6	53.5	3.3	294.5	4	$56.95	55.5
4	101.6	92.1	66.6	4.3	408.7	4	$94.93	73.6
4 1/2	114.3	104.8	86.2	5.6	602.2	2	$108.50	65.1

[a] 1990 prices.
[b] Maximum internal pressure is a calculation based on a variety of safety factors. One of these factors is surface abrasion, which can significantly lower the potential pressure that a given tubing can withstand. The use of safety shields and fiber-tape wrapping of tubing is strongly recommended. For more information see Sec. 1.1.9.

Table 3-9

Heavy Wall Tubing								
O.D. inches	**O.D. (mm)**	**I. D. (mm)**	**Vol (ml)/ HT (cm)**	**± ml**	**Vol (ml)/ half-sphere**	**Pieces/ case**	**Cost ($)/ piece[a]**	**Max int pres (psi)[b]**
3/8	9.5	4.8	0.2	0.1	0.1	96	$2.13	450.0
1/2	12.7	7.9	0.5	0.1	0.3	61	$3.35	328.7
5/8	15.9	11.1	1.0	0.2	0.7	45	$4.54	256.7
3/4	19.1	12.7	1.3	0.3	1.1	28	$7.30	288.5
7/8	22.2	15.9	2.0	0.4	2.1	30	$8.58	243.2
1	25.4	17.5	2.4	0.5	2.8	24	$11.99	268.6
1 1/4	31.8	23.8	4.5	0.7	7.1	18	$15.99	210.0
1 1/2	38.1	30.2	7.1	1.0	14.4	12	$21.16	172.1
1 3/4	44.5	36.5	10.5	1.4	25.5	10	$24.59	145.7
2	50.8	41.3	13.4	1.7	36.8	4	$36.45	153.5
2 1/4	57.2	47.6	17.8	2.1	56.6	4	$41.48	135.2
2 1/2	63.5	54.0	22.9	2.4	82.3	4	$51.40	120.8
2 3/4	69.9	60.3	28.6	2.8	114.9	4	$58.73	109.2
3	76.2	66.7	34.9	3.3	155.2	4	$63.65	99.6
3 1/4	82.6	73.0	41.9	3.8	203.9	4	$70.98	91.5
3 1/2	88.9	79.4	49.5	4.2	261.8	4	$82.10	84.6
4	101.6	88.9	62.1	5.5	367.9	2	$125.90	99.6
4 1/2	114.3	101.6	81.1	6.4	549.1	2	$142.35	87.9
5	127.0	114.3	102.6	7.8	781.9	2	$158.70	78.7
5 1/2	139.7	127.0	126.7	9.3	1072.5	2	$125.15	71.3
6	152.4	136.5	146.4	11.8	1332.4	2	$235.10	82.2
6 1/2	165.1	149.2	174.9	13.8	1739.9	1	$262.80	75.6
7	177.8	158.8	197.9	16.5	2094.8	1	$328.40	84.6

[a] 1990 prices.

[b] Maximum internal pressure is a calculation based on a variety of safety factors. One of these factors is surface abrasion, which can significantly lower the potential pressure that a given tubing can withstand. The use of safety shields and fiber-tape wrapping of tubing is strongly recommended. For more information see Sec. 1.1.9.

Table 3-10

Special Wall Tubing

O.D.	I.D.	Vol (ml)/ Ht (cm)	± ml	Vol (ml)/ half-sphere	Pieces/ case	Cost ($)/ piece[a]	Max int pres (psi)[b]
8	5.0	0.2	0.1	0.3	136	$1.93	328.7
9	6.0	0.3	0.1	0.6	117	$2.25	288.5
10	7.0	0.4	0.1	0.9	104	$2.53	256.7
11	8.0	0.5	0.1	1.3	93	$2.83	231.1
14	10.8	0.9	3.0	3.3	67	$3.93	190.4
8	4.0	0.1	0.1	0.2	109	$2.41	450.0
9	5.0	0.2	0.1	0.3	94	$2.80	396.2
10	6.0	0.3	0.1	0.6	82	$3.21	352.9
11	7.0	0.4	0.1	0.9	73	$3.60	317.6
19	14.5	1.7	0.3	8.0	35	$7.52	197.9
22	17.5	2.4	0.3	14.0	30	$8.77	168.7

[a] 1990 prices.

[b] Maximum internal pressure is a calculation based on a variety of safety factors. One of these factors is surface abrasion, which can significantly lower the potential pressure that a given tubing can withstand. The use of safety shields and fiber-tape wrapping of tubing is strongly recommended. For more information see Sec. 1.1.9.

Table 3-11

Capillary Wall Tubing

O.D. (mm)	Bore (mm)	Vol (ml)/ length (cm)	± (ml)	Vol (ml)/ half-sphere	Pieces/ ctn	Cost ($)/ piece[a]	Max int pres (psi)[b] min wall[c]	Max int pres (psi)[b] max wall[c]
5.5	0.5	0.004	0.001	0.001	107	$1.71	718.2	747.3
6.5	1.0	0.008	0.002	0.004	78	$2.35	696.5	730.3
6.5	1.3	0.010	0.002	0.008	57	$3.22	674.2	715.3
7.0	2.8	0.022	0.002	0.087	76	$2.41	661.8	703.6
8.0	3.4	0.027	0.002	0.162	59	$3.11	640.0	681.5
7.5	1.5	0.012	0.002	0.014	58	$3.16	543.1	626.1
7.5	1.8	0.014	0.002	0.022	61	$3.00	543.1	600.0
9.0	2.4	0.019	0.002	0.056	43	$4.26	528.1	600.0

[a] 1990 prices.

[b] Maximum internal pressure is a calculation based on a variety of safety factors. One of these factors is surface abrasion, which can significantly lower the potential pressure that a given tubing can withstand. The use of safety shields and fiber-tape wrapping of tubing is strongly recommended. For more information see Sec. 1.1.9

[c] Because of the ranges in O.D. and Bore dimensions of capillary tubing, these two columns represent the extreme minimum and maximum bursting pressures based on the resultant variations of wall thickness.

Table 3-12

Duran Tubing

O.D. (mm)	I. D. (mm)	Vol (ml)/ HT (cm)	± ml	Vol (ml)/ half-sphere	Pieces/ case	Cost ($)/ piece[a]	Max int pres (psi)[b]
3	1.6	0.02	0.01	0.00	941	$0.35	417.8
4	2.4	0.05	0.01	0.00	555	$0.55	352.9
5	3.4	0.09	0.01	0.01	343	$0.71	275.7
6	4.0	0.13	0.02	0.02	245	$0.72	288.5
6	3.0	0.07	0.02	0.01	211	$0.97	450.0
7	5.0	0.20	0.02	0.03	190	$0.86	243.2
7	4.0	0.13	0.02	0.02	172	$1.19	380.8
8	6.0	0.28	0.02	0.06	149	$1.01	210.0
8	5.0	0.20	0.03	0.03	147	$1.39	328.7
9	7.0	0.38	0.03	0.09	119	$1.14	184.6
9	6.0	0.28	0.03	0.06	119	$1.60	288.5
10	8.0	0.50	0.03	0.13	95	$1.29	164.6
10	7.0	0.38	0.04	0.09	90	$1.82	256.7
10	5.6	0.25	0.03	0.05	56	$2.42	391.9
11	9.0	0.64	0.04	0.19	86	$1.42	148.5
11	8.0	0.50	0.05	0.13	74	$2.02	231.1
11	6.6	0.34	0.04	0.08	42	$2.75	352.9
12	10.0	0.79	0.04	0.26	130	$1.57	135.2
12	9.0	0.64	0.05	0.19	67	$2.24	210.0
12	7.6	0.45	0.04	0.11	42	$3.07	320.6
13	11.0	0.95	0.04	0.35	119	$1.72	124.1
13	10.0	0.79	0.06	0.26	56	$2.42	192.4
13	8.6	0.58	0.05	0.17	36	$3.39	293.4
14	12.0	1.13	0.05	0.45	110	$1.86	114.7
14	11.0	0.95	0.06	0.35	46	$2.65	177.4
14	9.6	0.72	0.05	0.23	30	$3.72	270.3
15	12.6	1.25	0.05	0.52	86	$2.37	129.5
15	11.4	1.02	0.06	0.39	56	$3.40	200.8
15	10.0	0.79	0.06	0.26	25	$4.46	288.5
16	13.6	1.45	0.05	0.66	81	$2.52	120.8
16	12.4	1.21	0.07	0.50	49	$3.64	187.1
16	11.0	0.95	0.06	0.35	25	$4.79	268.6
17	14.6	1.67	0.06	0.81	76	$2.69	113.3
17	13.4	1.41	0.08	0.63	49	$3.89	175.2
17	12.0	1.13	0.07	0.45	25	$5.15	251.2
18	15.6	1.91	0.06	0.99	66	$2.89	106.6
18	14.4	1.63	0.08	0.78	49	$4.17	164.6
18	13.0	1.33	0.07	0.58	20	$5.52	235.8
19	16.6	2.16	0.07	1.20	63	$3.02	100.7
19	15.4	1.86	0.09	0.96	42	$4.43	155.3

[a] 1990 prices.

[b] Maximum internal pressure is a calculation based on a variety of safety factors. One of these factors is surface abrasion, which can significantly lower the potential pressure that a given tubing can withstand. The use of safety shields and fiber-tape wrapping of tubing is strongly recommended. For more information see Sec. 1.1.9.

Table 3-12

Duran Tubing (cont.)

O.D. (mm)	I. D. (mm)	Vol (ml)/ HT (cm)	± ml	Vol (ml)/ half-sphere	Pieces/ case	Cost ($)/ piece[a]	Max int pres (psi)[b]
19	14.0	1.54	0.08	0.72	36	$5.89	222.2
20	17.6	2.43	0.08	1.43	55	$3.22	95.4
20	16.4	2.11	0.11	1.15	36	$4.68	146.9
20	15.0	1.77	0.12	0.88	20	$6.26	210.0
22	19.6	3.02	0.09	1.97	42	$3.57	86.3
22	19.6	3.02	0.09	1.97	42	$3.57	86.3
22	18.4	2.66	0.12	1.63	30	$5.22	132.6
22	17.0	2.27	0.14	1.29	30	$6.99	189.2
24	21.6	3.66	0.10	2.64	38	$3.94	78.7
24	20.4	3.27	0.13	2.22	25	$5.70	120.8
24	19.0	2.84	0.15	1.80	25	$7.62	172.1
26	23.2	4.23	0.15	3.27	33	$4.95	85.1
26	22.0	3.80	0.14	2.79	25	$6.86	124.1
26	20.4	3.27	0.16	2.22	12	$9.30	178.4
28	25.2	4.99	0.16	4.19	25	$5.33	78.7
28	24.0	4.52	0.15	3.62	20	$7.49	114.7
28	22.4	3.94	0.18	2.94	20	$10.05	164.6
30	27.2	5.81	0.24	5.27	36	$6.13	73.3
30	26.0	5.31	0.23	4.60	16	$8.52	106.6
30	24.4	4.68	0.26	3.80	16	$11.61	152.8
32	29.2	6.70	0.26	6.52	25	$6.56	68.5
32	28.0	6.16	0.25	5.75	16	$9.18	99.6
32	26.4	5.47	0.28	4.82	16	$12.51	142.5
34	31.2	7.65	0.27	7.95	25	$7.03	64.3
34	30.0	7.07	0.26	7.07	16	$9.79	93.4
34	28.4	6.33	0.30	6.00	16	$13.42	133.5
36	33.2	8.66	0.29	9.58	25	$7.32	60.6
36	32.0	8.04	0.28	8.58	25	$10.46	87.9
36	30.4	7.26	0.32	7.36	12	$14.16	125.6
38	35.2	9.73	0.31	11.42	20	$7.84	57.3
38	34.0	9.08	0.30	10.29	20	$10.90	83.1
38	32.4	8.24	0.34	8.90	9	$15.14	118.6
40	36.8	10.64	0.41	13.05	16	$9.26	62.4
40	35.4	9.84	0.45	11.61	16	$13.25	91.2
40	33.6	8.87	0.49	9.93	9	$18.07	129.5
40	30.0	7.07	0.49	7.07	9	$26.63	210.0
42	38.8	11.82	0.43	15.29	16	$9.88	59.3
42	37.4	10.99	0.48	13.70	16	$13.93	86.6
42	35.6	9.95	0.52	11.81	9	$18.88	122.9
44	40.8	13.07	0.46	17.78	16	$10.33	56.5

[a] 1990 prices.

[b] Maximum internal pressure is a calculation based on a variety of safety factors. One of these factors is surface abrasion, which can significantly lower the potential pressure that a given tubing can withstand. The use of safety shields and fiber-tape wrapping of tubing is strongly recommended. For more information see Sec. 1.1.9.

Table 3-12

Duran Tubing (cont.)

O.D. (mm)	I. D. (mm)	Vol (ml)/ HT (cm)	± ml	Vol (ml)/ half-sphere	Pieces/ case	Cost ($)/ piece[a]	Max int pres (psi)[b]
44	39.4	12.19	0.51	16.01	16	$14.53	82.5
44	37.6	11.10	0.54	13.92	9	$19.97	116.9
45	35.0	9.62	0.57	11.22	9	$30.51	184.6
46	42.8	14.39	0.48	20.53	16	$10.78	54.0
46	41.4	13.46	0.53	18.58	9	$15.29	78.7
46	39.6	12.32	0.57	16.26	9	$21.00	111.5
48	44.8	15.76	0.50	23.54	16	$11.24	51.7
48	43.4	14.79	0.56	21.40	16	$15.97	75.3
48	41.6	13.59	0.60	18.85	6	$21.73	106.6
50	46.4	16.91	0.52	26.15	9	$15.21	55.9
50	45.0	15.90	0.58	23.86	9	$20.75	78.7
50	43.0	14.52	0.62	20.81	9	$28.56	112.3
50	40.0	12.57	0.64	16.76	6	$39.19	164.6
50	36.0	10.18	0.64	12.21	6	$52.80	237.9
50	32.0	8.04	0.63	8.58	6	$64.40	314.1
52	48.4	18.40	0.54	29.68	9	$15.71	53.7
52	47.0	17.35	0.60	27.18	9	$21.68	75.6
52	45.0	15.90	0.65	23.86	9	$29.66	107.7
54	50.4	19.95	0.56	33.52	9	$16.47	51.7
54	49.0	18.86	0.63	30.80	9	$22.60	72.6
54	47.0	17.35	0.68	27.18	9	$30.92	103.5
55	45.0	15.90	0.72	23.86	4	$43.67	148.5
56	52.4	21.57	0.58	37.67	9	$17.05	49.8
56	51.0	20.43	0.65	34.73	9	$23.36	69.9
56	49.0	18.86	0.71	30.80	9	$32.43	99.6
58	54.4	23.24	0.61	42.15	9	$17.81	48.0
58	53.0	22.06	0.68	38.98	9	$24.28	67.4
58	51.0	20.43	0.73	34.73	9	$33.35	95.9
60	55.6	24.28	0.93	45.00	9	$22.26	57.0
60	53.6	22.56	0.99	40.31	9	$31.84	84.2
60	51.6	20.91	1.04	35.97	4	$40.83	112.3
60	50.0	19.63	1.01	32.72	4	$48.02	135.2
60	46.0	16.62	1.00	25.48	4	$65.03	194.7
60	42.0	13.85	0.99	19.40	4	$80.34	256.7
65	60.6	28.84	1.02	58.26	8	$24.39	52.5
65	58.6	26.97	1.08	52.68	4	$34.59	77.5
65	56.6	25.16	1.14	47.47	4	$44.61	103.1
65	55.0	23.76	1.10	43.56	4	$52.55	124.1
70	65.6	33.80	1.10	73.91	8	$26.09	48.6
70	63.6	31.77	1.17	67.35	3	$49.91	71.7

[a] 1990 prices.

[b] Maximum internal pressure is a calculation based on a variety of safety factors. One of these factors is surface abrasion, which can significantly lower the potential pressure that a given tubing can withstand. The use of safety shields and fiber-tape wrapping of tubing is strongly recommended. For more information see Sec. 1.1.9.

Table 3-12

Duran Tubing (cont.)

O.D. (mm)	I. D. (mm)	Vol (ml)/ HT (cm)	± ml	Vol (ml)/ half-sphere	Pieces/ case	Cost ($)/ piece[a]	Max int pres (psi)[b]
70	61.6	29.80	1.23	61.19	3	$64.52	95.4
70	60.0	28.27	1.20	56.55	4	$56.71	114.7
70	56.0	24.63	1.22	45.98	4	$77.13	164.6
70	52.0	21.24	1.22	36.81	4	$96.22	216.6
75	70.6	39.15	1.18	92.13	8	$28.17	45.3
75	68.6	36.96	1.26	84.52	4	$40.45	66.7
75	66.6	34.84	1.33	77.34	4	$52.17	88.7
75	65.0	33.18	1.30	71.90	4	$61.25	106.6
80	75.0	44.18	1.44	110.45	4	$34.22	48.3
80	73.0	41.85	1.52	101.84	4	$47.07	68.5
80	70.0	38.48	1.68	89.80	4	$65.78	99.6
80	66.0	34.21	1.70	75.27	4	$89.22	142.5
80	62.0	30.19	1.70	62.39	4	$111.72	187.1
85	80.0	50.27	1.53	134.04	4	$36.30	45.4
85	78.0	47.78	1.62	124.24	4	$50.09	64.3
85	75.0	44.18	1.80	110.45	4	$69.94	93.4
90	85.0	56.75	1.62	160.78	4	$38.37	42.8
90	83.0	54.11	1.72	149.69	4	$52.93	60.6
90	80.0	50.27	1.92	134.04	4	$74.67	87.9
90	76.0	45.36	1.95	114.92	4	$76.18	125.6
90	72.0	40.72	1.97	97.72	4	$95.84	164.6
95	90.0	63.62	1.72	190.85	4	$40.45	40.5
95	88.0	60.82	1.82	178.41	4	$55.76	57.3
95	85.0	56.75	2.04	160.78	4	$78.83	83.1
100	95.0	70.88	1.81	224.46	4	$45.60	38.4
100	94.0	69.40	1.95	217.45	4	$53.64	46.3
100	93.0	67.93	1.93	210.58	3	$63.21	54.3
100	90.0	63.62	2.16	190.85	3	$89.19	78.7
100	86.0	58.09	2.20	166.52	3	$121.07	112.3
100	82.0	52.81	2.24	144.35	3	$152.41	146.9
105	99.0	76.98	2.05	254.02	3	$56.79	44.1
105	95.0	70.88	2.43	224.46	3	$93.21	74.8
105	91.0	65.04	2.48	197.28	3	$127.50	106.6
110	104.8	86.26	2.34	301.34	4	$51.83	36.3
110	104.0	84.95	2.32	294.49	3	$59.73	42.0
110	100.0	78.54	2.55	261.80	3	$97.50	71.3
110	96.0	72.38	2.61	231.62	3	$133.93	101.5
110	92.0	66.48	2.65	203.86	2	$169.15	132.6
115	109.0	93.31	2.43	339.04	4	$62.48	40.1
115	105.0	86.59	2.68	303.07	2	$102.85	68.0

[a] 1990 prices.

[b] Maximum internal pressure is a calculation based on a variety of safety factors. One of these factors is surface abrasion, which can significantly lower the potential pressure that a given tubing can withstand. The use of safety shields and fiber-tape wrapping of tubing is strongly recommended. For more information see Sec. 1.1.9.

Table 3-12

Duran Tubing (cont.)							
O.D. (mm)	I. D. (mm)	Vol (ml)/ HT (cm)	± ml	Vol (ml)/ half-sphere	Pieces/ case	Cost ($)/ piece[a]	Max int pres (psi)[b]
115	101.0	80.12	2.74	269.73	2	$141.02	96.8
120	114.0	102.07	2.54	387.87	4	$65.09	38.4
120	110.0	95.03	2.80	348.45	2	$107.27	65.1
120	106.0	88.25	2.88	311.81	2	$147.05	92.6
120	102.0	81.71	2.93	277.82	2	$186.03	120.8
125	115.0	103.87	2.93	398.16	2	$111.69	62.4
125	111.0	96.77	3.01	358.04	2	$154.28	88.7
125	107.0	89.92	3.08	320.72	2	$194.06	115.7
130	124.0	120.76	2.96	499.15	4	$70.91	35.4
130	120.0	113.10	3.06	452.39	2	$116.12	59.9
130	116.0	105.68	3.33	408.64	2	$160.31	85.1
130	112.0	98.52	3.22	367.81	2	$202.90	111.0
135	125.0	122.72	3.18	511.33	2	$121.34	57.6
135	121.0	114.99	3.47	463.79	2	$166.34	81.8
135	117.0	107.51	3.36	419.30	2	$210.93	106.6
140	134.0	141.03	3.41	629.92	4	$76.54	32.8
140	130.0	132.73	3.52	575.17	2	$125.76	55.5
140	126.0	124.69	3.82	523.70	2	$173.57	78.7
140	122.0	116.90	3.90	475.39	2	$219.77	102.6
145	135.0	143.14	3.65	644.12	2	$130.18	53.5
145	131.0	134.78	3.97	588.55	2	$180.00	75.9
145	127.0	126.68	4.05	536.27	2	$227.81	98.8
150	144.0	162.86	3.66	781.73	2	$93.98	30.6
150	140.0	153.94	3.78	718.38	2	$152.82	51.7
150	136.0	145.27	4.12	658.54	2	$211.22	73.3
150	132.0	136.85	4.21	602.13	2	$268.70	95.4
155	145.0	165.13	3.92	798.13	2	$158.76	49.9
155	141.0	156.15	4.26	733.88	2	$219.43	70.8
155	137.0	147.41	4.37	673.18	1	$277.36	92.1
160	150.0	176.71	4.29	883.57	2	$163.77	48.3
160	146.0	167.42	4.65	814.76	2	$226.27	68.5
160	142.0	158.37	4.75	749.61	1	$287.40	89.1
165	155.0	188.69	4.43	974.91	2	$168.79	46.8
165	151.0	179.08	4.81	901.36	2	$233.11	66.3
165	147.0	169.72	4.92	831.61	1	$296.53	86.3
170	160.0	201.06	4.83	1072.33	2	$173.81	45.4
170	156.0	191.13	5.22	993.90	2	$241.33	64.3
170	152.0	181.46	5.33	919.39	1	$306.56	83.6
180	170.0	226.98	5.13	1286.22	2	$92.60	42.8
180	166.0	216.42	5.55	1197.55	1	$255.47	60.6

[a] 1990 prices.

[b] Maximum internal pressure is a calculation based on a variety of safety factors. One of these factors is surface abrasion, which can significantly lower the potential pressure that a given tubing can withstand. The use of safety shields and fiber-tape wrapping of tubing is strongly recommended. For more information see Sec. 1.1.9.

Table 3-12

Duran Tubing (cont.)

O.D. (mm)	I. D. (mm)	Vol (ml)/ HT (cm)	± ml	Vol (ml)/ half-sphere	Pieces/ case	Cost ($)/ piece[a]	Max int pres (psi)[b]
180	162.0	206.12	5.94	1113.05	1	$324.81	78.7
190	180.0	254.47	6.01	1526.81	1	$195.26	40.5
190	176.0	243.28	6.16	1427.27	1	$270.98	57.3
190	172.0	232.35	6.57	1332.15	1	$343.97	74.4
200	190.0	283.53	6.95	1795.68	1	$263.67	38.4
200	186.0	271.72	6.80	1684.64	1	$366.80	54.3
200	182.0	260.16	7.25	1578.28	1	$466.40	70.5
215	201.0	317.31	7.67	2125.97	1	$394.92	50.4
215	197.0	304.81	7.83	2001.55	1	$502.73	65.4
225	211.0	349.67	8.04	2459.33	1	$413.66	48.1
225	207.0	336.54	8.89	2322.09	1	$527.34	62.4
240	222.0	387.08	9.89	2864.36	1	$563.66	58.4
250	240.0	452.39	9.33	3619.11	1	$434.24	30.6
250	236.0	437.44	9.93	3441.16	1	$602.43	43.2
250	232.0	422.73	10.51	3269.13	1	$767.56	55.9
270	260.0	530.93	10.31	4601.39	1	$469.40	28.3
270	256.0	514.72	10.97	4392.26	1	$651.36	39.9
270	252.0	498.76	11.61	4189.58	1	$831.78	51.7
300	290.0	660.52	13.34	6385.03	1	$522.92	25.4
300	286.0	642.42	14.08	6124.44	1	$726.28	35.8
300	282.0	624.58	14.79	5871.05	1	$926.57	46.3
315	301.0	711.58	15.05	7139.51	1	$762.98	34.1
315	297.0	692.79	15.80	6858.64	1	$975.51	44.1

[a] 1990 prices.

[b] Maximum internal pressure is a calculation based on a variety of safety factors. One of these factors is surface abrasion, which can significantly lower the potential pressure that a given tubing can withstand. The use of safety shields and fiber-tape wrapping of tubing is strongly recommended. For more information see Sec. 1.1.9.

Table 3-13

Duran Capillary Wall Tubing

O.D. (mm)	Bore (mm)	Vol (ml)/ length (cm)	± (ml)	Vol (ml)/ half-sphere	Pieces/ ctn	Cost ($)/ piece[a]	Max int pres (psi)[b] min wall[c]	Max int pres (psi)[b] max wall[c]
4	0.4	0.003	0.001	0.000	238	$0.92	721.6	743.1
4	0.6	0.005	0.001	0.001	244	$0.90	695.4	730.9
4	0.8	0.006	0.001	0.002	250	$0.88	661.8	713.0
5	0.4	0.003	0.001	0.000	65	$3.37	732.5	745.4
5	0.6	0.005	0.001	0.001	154	$1.42	716.1	737.2
5	0.8	0.006	0.001	0.002	156	$1.40	694.7	725.2
5	1.2	0.009	0.001	0.007	161	$1.36	639.1	690.2
6	0.4	0.003	0.001	0.000	106	$2.07	738.1	746.7
6	0.6	0.005	0.001	0.001	108	$2.03	726.9	740.9
6	0.8	0.006	0.002	0.002	110	$1.99	703.6	736.9
6	1.7	0.013	0.002	0.021	115	$1.91	595.2	671.9
6	2.2	0.017	0.002	0.045	122	$1.80	517.2	616.5
6	2.7	0.021	0.002	0.082	133	$1.65	432.8	551.4
7	0.8	0.006	0.002	0.002	79	$2.77	716.3	740.2
7	1.2	0.009	0.002	0.007	80	$2.74	685.4	723.1
7	1.7	0.013	0.002	0.021	83	$2.64	635.2	690.8
7	2.2	0.017	0.002	0.045	86	$2.55	574.8	647.9
7	2.7	0.021	0.002	0.082	91	$2.41	507.3	596.3
7	3.0	0.024	0.002	0.113	95	$2.31	464.5	562.1
8	0.8	0.006	0.002	0.002	60	$3.65	724.5	742.4
8	1.2	0.009	0.002	0.007	61	$3.59	700.8	729.0
8	1.7	0.013	0.002	0.021	63	$3.48	661.8	703.6
8	2.2	0.017	0.002	0.045	65	$3.37	614.0	669.5
8	2.7	0.021	0.002	0.082	67	$3.27	559.4	627.9
8	3.0	0.024	0.002	0.113	69	$3.18	524.1	600.0
9	0.8	0.006	0.002	0.002	47	$4.66	730.0	743.9
9	1.2	0.009	0.002	0.007	48	$4.57	711.3	733.2
9	1.7	0.013	0.002	0.021	49	$4.47	680.2	712.8
9	2.2	0.017	0.002	0.045	50	$4.38	641.6	685.0
9	2.7	0.021	0.002	0.082	52	$4.21	596.8	650.9
9	3.0	0.024	0.002	0.113	53	$4.13	567.6	627.8

[a] 1990 prices.

[b] Maximum internal pressure is a calculation based on a variety of safety factors. One of these factors is surface abrasion, which can significantly lower the potential pressure that a given tubing can withstand. The use of safety shields and fiber-tape wrapping of tubing is strongly recommended. For more information see Sec. 1.1.9.

[c] Because of the ranges in O.D. and Bore dimensions of capillary tubing, these two columns represent the extreme minimum and maximum bursting pressures based on the resultant variations of wall thickness.

References

[1] T.H. Black, "The Preparation and Reactions of Diazomethane," *Aldrichim. Acta*, 16 (1983), pp. 3-10.
[2] D.W. Koester and G.A. Sites, "Method for Loosening Frozen Joints," *Fusion*, 20 (1973), p. 45.
[3] E. Wheeler, *Scientific Glassblowing*, (New York: Interscience Publishers, Inc. 1958), p. 464.
[4] J. Butler, "Carbonated Beverages and Frozen Stopcocks,"*J. of Chem Ed.*, 64 (1987), p. 896.
[5] Pope Pacers, a publication of Pope Scientific, Menomonee Falls, WI 53052, (January 1968), p. 7.
[6] R. Thompson, "A Method for the Removal of Stuck Ground Glass Joints," *B.S.S.G. Journal*, 24 (1986), p. 96.
[7] C.H. Woods, "Removal of Broken, Evacuated Stopcock Plugs," *Chemists-Analyst*, 53 (1964), p. 53.

Chapter 4

Cleaning Glassware

4.1 The Clean Laboratory

4.1.1 Basic Cleaning Concepts

Cleaning glassware is the bane of chemistry; what gets dirty must get cleaned. However, just because something *looks* clean does not mean it *is* clean. If you do not know if glassware is clean, wash it. It is better to take the few extra minutes to be sure that something is clean than to spend hours on an experiment only to have contamination cause bizarre or inconsistent results. Attempting titrations with a base is futile if acid residues from a previous experiment were never properly cleaned from a burette. Worse, it's possible for unintended combinations of chemicals (and even water), to produce toxic fumes or explosions.

One common and simple test to determine the cleanliness of glassware is to examine how well water 'wets' the glass. Water will bead up on the walls of dirty glassware, but on clean glass walls, it 'sheets.' This test by itself should not, however, be used as the sole criterion for clean glass.

Despite the aforementioned precautions, glassware needs to be only as clean as is required for the work being done. Over-cleaning, or incorrect cleaning, wastes time, equipment, and money. You need to know your chemistry as well as the equipment being used. For instance, acids can be catalysts for certain organic reactions, therefore the use of an acid as a final rinse of your glassware will help such a reaction. However, using a base as a final rinse could prevent any reaction from taking place. In addition, the higher the precision of instrumentation, the cleaner the glassware that is required. The glassware for an instrument that has the detection ability of 0.1 ppm does not have to be as clean as an instrument with the detection ability of 0.001 ppm.

Cleaning glassware is invariably a multi-step process. If the contaminating material can be removed by soap and water, at least two more steps will be required: rinsing with water, followed by a distilled water rinse. If there is particulate material on the glassware, it needs to be brushed or wiped off before other cleaning processes begin. Similarly, organic solvents are used first to remove grease (i.e., stopcock grease) before water is used to remove salt deposits.

The basic tenet of cleaning is *like dissolves like*. Polar solutions dissolve polar materials and vice versa for non-polar solutions. Crystals cannot dissolve in oils, and grease cannot be cleaned by water. Lubricants based on chlorofluorocarbons require chlorofluorocarbon solvents for cleaning. Thus, for effective cleaning, you must know what you are cleaning to have an effective plan of (cleaning) attack.

Borosilicate glassware (such as Pyrex or Kimex) is more chemically resistant than soda-lime glassware. However, hydrofluoric acid, perchloric acid, and all bases can react with borosilicate glass and strip the surface glass off layer by layer. Given enough time, some organic tars and minerals in water can also react with the surface of laboratory glass.

Objectively, the reason we clean glassware is to reuse the glassware. Because some cleaning processes may dissolve glass, you need to be objective about which cleaning process you choose, and on what items to use that cleaning process. Glass dissolving is a cleaning process removing a layer of glass in a process called *stripping* and anything that was stuck to the glass is removed along with the glass. Some glass strippers are far more aggressive than others, and can do considerable damage to glassware in a short time. This damage may include any markings on the glass such as calibration lines. Such destruction may be just a nuisance in an item such as a beaker, but it can also make a volumetric flask absolutely worthless.

Another problem with glass strippers is that they are typically oxide-selective. Glass is composed of a combination of oxides. Because some of these oxides may be more susceptible to specific chemical attack than others, the result is an uneven surface. Such a pitted surface gets dirty easier, is more difficult to clean, and the greater surface area requires more time for removal of adsorbed gases (i.e., water) in vacuum systems.

It is obvious then that there is no single magic potion that can clean all glassware. Cleaning laboratory ware is an art and your knowledge of what solvents, or combinations of solvents to use for any given contaminant will improve with experience.

The best time to clean glassware is just after use. The longer dirty glassware sits, the harder it is to clean. Fortunately there are often waiting periods in most experiments when things that you are working on are cooling, heating up, or waiting to react. These waiting periods are excellent times to clean glassware. If an experiment has no waiting times, then try to have a large basin with a soapy water solution for used glassware handy (assuming that the contaminants will not react with soapy water). However, be sure the glassware is totally submerged, otherwise mineral deposits are likely to form on the glassware. Because mineral deposits cannot be removed with soap and water, more complex (and more dangerous) cleaning procedures will need to be used.

One common complication in laboratory cleaning is the proper rinsing of long and/or narrow pieces of equipment. This difficulty arises because once a narrow item is filled up with water (or whatever the rinsing and/or cleaning solution) no more liquid can enter the piece from its opening. As you can see in Fig. 4-1, any extra liquid in these containers just falls off the top. If the item has a sufficiently wide opening, there can be circulation within the piece. More specifically, this principle is all related to the force of the liquid, the viscosity of the liquid, the width of the stream, and the width of the container.

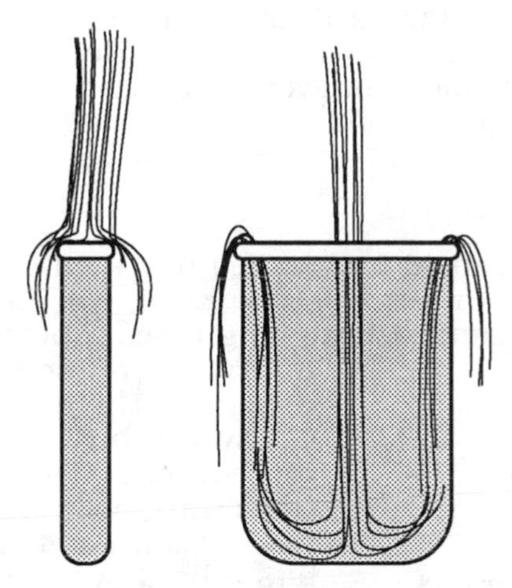

Fig. 4-1 Water flowing into narrow and wide containers.

Cuvetts and NMR tubes are two good examples of items that are difficult to rinse properly because of their narrow designs. If a jet of water is strong enough and narrow enough, it could enter, and rinse an NMR tube. However, such a stream is unlikely to be found in most laboratories. If a cuvett needs to be cleaned, a cuvett cleaner (see Fig. 4-2(a)) provides the required deep

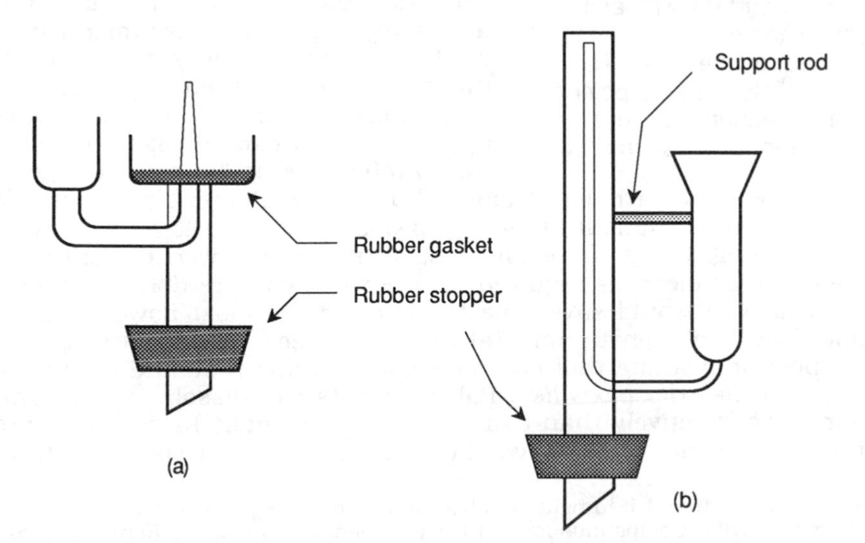

Fig. 4-2 Cuvett and acetone cuvett cleaner and NMR tube cleaner.

cleaning. The same goes for NMR tubes; an NMR tube cleaner (see Fig. 4-2(b)) provides deep rinsing that ensures no residue. Without the ability to squirt a rinsing liquid to the bottom of a tube, it is necessary to repeatedly fill up and pour (or shake) out the liquid many times to ensure the glassware is properly rinsed.

It is also possible to attach a flexible tube up to the distilled (or deionized) water source and add a pipette to the hose. By placing a pipette into test tubes, graduated cylinders, or even separatory funnels (which are upside down to provide draining), fast and efficient rinsing can be achieved.

4.1.2 Safety

Cleaning glassware should be handled one item at a time. Although you may be able to hold several items in one hand, it is better to wash one item at a time than to hurry and risk wasting time, equipment, and money (and injury to yourself) which is caused by breakage. There are many ways to clean glassware, but there is only one proper procedure, and that is to do it carefully. Glassware will invariably bounce one time less than the number of times it hits a hard surface. Star cracks (see Fig. 4-3), chipped edges, broken sections, or total destruction are all possible results of the "final landing".

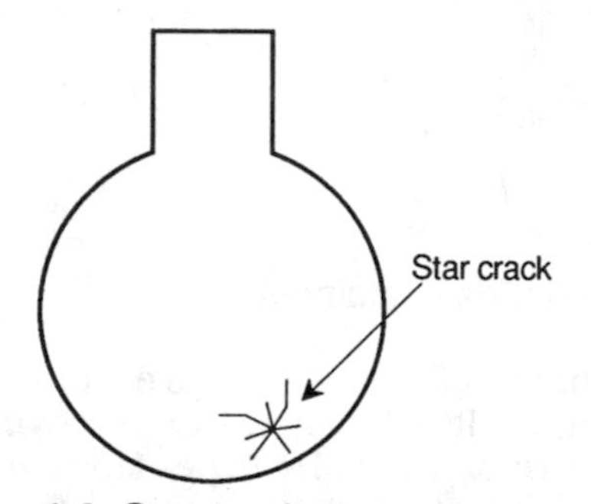

Fig. 4-3 Star cracks are caused by hitting one point of a laboratory flask against a hard surface with enough force to crack the glass, but not hard enough to destroy the flask. The crack lines radiate from the point of impact. They frequently occur in glassware left to roll loose in drawers.

Broken glassware, aside from being a financial loss, is dangerous. Labs often use beakers, funnels, graduated cylinders, and the like with chipped edges or broken ends. This practice is not safe, and many people have assumed that they could not be cut only to regret their naive optimism. Often, by simply fire-polishing the end of a chipped glass apparatus, you can salvage an item that would otherwise be dangerous to use.

Because there are various "strengths" of all the different cleaners, cleaning glassware can be thought of as a stepwise procedure. That is, as you try to clean something, you go from the least to the most powerful medium until the glassware is as clean as required for the work you are doing. Although it may seem that you would save time to only use the most powerful cleaner, that is not always safe, environmentally sound, or economically wise.

One aspect of cleaning that can save considerable time and energy is remembering that *like dissolves like*. Polar solvents can dissolve polar contamination far more effectively than can non-polar solvents.* In addition, proper selection of solvent material can avoid damaging the object you want to clean

* In general, as you go from left to right, and bottom to top in the periodic chart you go toward greater electronegativity. Compounds formed from elements with wildly differing electronegativities are more polar.

as well as preventing the introduction of possibly hazardous materials. For example, say you have a plastic container with a stick-on label which, after it is removed, leaves a sticky, gummy glue. You might try a hydrocarbon solvent to remove the sticky material, but that is likely to damage the plastic container. However, a light oil, such as WD-40® can dissolve and remove the adhesive.* The remaining oil film can easily be washed off with soap and water.

Regardless of the cleaning process, you should always wear a basic minimum of safety clothing. Lab coats do more than just protect your street clothes; if a dangerous chemical spills on a lab coat, the coat can be easily taken off at a moment's notice with no embarrassment. Street clothes often cannot be removed with equal ease and modesty, and can provide time for a toxic or dangerous chemical to soak through to your skin.

Eye protection is also critical. Although glasses provide some protection, when liquid splashes, it is seldom considerate enough to hit you straight on in the eye. Eye goggles fit over glasses and provide protection not only from the front, but from the top, sides, and bottom. If you wear contacts, it is strongly recommended to not wear them in laboratory environments. Although the use of goggles should prevent liquids from splashing your eye, it will not stop vapors. Contact lenses are made from an elastomer that is capable of adsorbing vapors, which will be in constant contact with the eye. Although the use of a fume hood will decrease the danger, it is best to avoid the problem altogether.

Gloves are good protection when using acids or bases for cleaning, and they should also be used with soap and water because they can offer, by removal, immediately dry, clean hands whenever they are needed. They also permit using much hotter water than can be used with bare hands. Typically, the hotter the cleaning solution, the faster and more effective the cleaning. Unfortunately, gloves cannot be used with hydrocarbon solvents because they are likely to dissolve the gloves. It is difficult to protect hands from the drying effects of hydrocarbon solvents. Although various hand creams are available to re-moisturize hands if necessary, it is best to try not to make contact with the solvents.

Never dispose of organic tars, Kimwipes®, paper with organic residue, or other solid material down a sink. Such materials should be placed in a waste container labeled for laboratory wastes. Never dispose of used cleaning solutions of any type down a sink without first checking with the safety officer. Different environmental laws exist in each state and sometimes within separate counties, so it is difficult to generalize waste disposal too broadly.

4.1.3 Soap and Water

Pre-preparation. Stopcock grease on a joint or stopcock must be removed before adequate soap and water cleaning can begin. It is best to remove excess stopcock grease physically with a Kimwipe®. Then, organic

* Likewise, one of the pitfalls of walking along a seashore beach is tar blobs sticking to the soles of feet. Again, the selection of a strong hydrocarbon solvent is not advised because such solvent are likely to dry out and/or damage human tissue, Some are carcinogenic. Surprisingly, common baby oil is a wonderful solvent for tar blobs and will not harm the skin.

greases (i.e., Apiezon) can easily be removed with halogenated hydrocarbons, such as chloroform or methylene chloride. Methylene chloride is a minor suspected carcinogen and is therefore safer to use than chloroform which is definitely a suspected carcinogen and known cause of liver damage. However, both of these are dangerous compounds and should be used in a fume hood. Acetone can also be used, but is not very effective on most stopcock greases. Heating glassware to about 400°C will also remove any organic greases. However, this heating may cause any remaining inorganic material to burn into the surface of the glass which may prove very difficult to remove. Be sure to remove any Teflon items before heating because burning Teflon fumes are extremely toxic.

Silicone grease can be mostly removed with pentane or methylene chloride, but this technique leaves a film residue. Such a residue will not affect most chemical reactions, but can wreak havoc for any future glassblowing work. It cannot be stressed enough how completely silicone grease must be removed before any glasswork is to be done. Silicone grease can be effectively removed with a base bath (see Sec. 4.1.6), but this process is glass-destructive and should not be used with any volumetric ware. These greases cannot be removed with heat because the silicone itself will not burn off and any remains on the glassware can only be removed with a three-minute soak in 5% to 10% HF.

Fluorocarbon greases (Krytox®) require a chlorinated fluorocarbon for removal. At the time of this writing, there are no environmentally safe forms of the chlorinated fluorocarbons available for cleaning. Heating (> 260°C) will remove the grease, but the fumes (lethal fluorine compounds) are highly toxic. For more information on cleaning these greases, see Sec. 3.3.8.

After the various greases are removed, it is then possible to continue with soap and water cleaning. However, whether organic solvents were used or not, glassware should be rinsed out first with a small amount of acetone and then water, before being placed in a soapy water solution.

Material. There is a variety of powdered and liquid washing compounds available on the commercial market that, when mixed with water, provide excellent cleaning. There are a few super-concentrated cleaners (that come as liquids) which are excellent for cleaning items by long-term soaking.

Preparation. Because there are as many different preparation techniques as there are commercial brands, it is best to read the instructions on the container of the cleaner you are using.

Use. Soap-based cleaning products are generally long-chain hydrocarbons that are negatively charged at one end. The hydrocarbon end of the soap molecule does not dissolve in (polar) water, but combines with other soap molecules to form 'sphere' like shapes called *micelles*. The outer surface of the 'sphere' has a charged end which allows it to freely associate within water. However, the hydrocarbon end is still able to dissolve other hydrocarbons. Thus, one end can grab dirt and the other end can grab water. Soaps and detergents do not make an oil soluble in water, rather with the assistance of agitation they aid its emulsification.

Soaps and detergents can also make water 'wetter' by lowering the surface tension of the water. In doing so, less energy is required to lift dirt off whatever it is on. Other agents within a cleaning solution may include materials that help emulsify oily matter, soften water, solubilize compounds, control pH, and perform other actions to assist cleaning action.

In general, these cleaning compounds work better in warm or hot water. Some scrubbing with a test tube brush, sponge, or the like, is usually needed, but some of the liquid cleaners only require soaking the glassware overnight. Do not use any rough-surfaced material (i.e., pumice, kitchen cleanser, rough plastic or metal scouring pads) that can scratch the glass. When glass is scratched, the rough surface provides an easier surface for contaminates to adhere to and is therefore much harder to clean (and maintain) in the future. This surface scratching is also why you should never use abrasive scouring cleansers on new porcelain surfaces in the home. If you do, you will destroy the surface and will have no recourse but to scour away for further cleaning. In addition to cleaning problems, abraded surfaces make glass much easier to break.

There is a nonabrasive scouring cleaner on the market called Bon Ami®. Most other cleansers contain silica as the abrasive agent. Bon Ami® contains feldspar and calcium carbonate, which are softer than glass and therefore cannot scratch glass surfaces.

After cleaning, copiously rinse glassware with tap water and follow with a deionized (or distilled) water rinse. Let the glassware stand upside down for storage or to dry. If your freshly-washed glassware needs to be immediately dried for an experiment, you may facilitate the drying by placing the glassware in a drying oven. Alternatively, you may pour about 10 ml of reagent-grade acetone or methanol into the glassware, swirl it around, and pour the remains into a container labeled "used acetone" or "used methanol," which can then be used for future cleaning.

Safety Considerations. You should always wear safety glasses. Because most cleaning compounds can be drying to the skin, it would be wise to wear rubber gloves. Also, wearing gloves allows the use of hotter water. Hot water can facilitate and improve the cleaning action of most cleaners.

Disposal. These soaps and detergents may all be washed down the sink. Some of the older types contain phosphates and, due to environmental concerns, should be phased out, or used as little as possible. Additionally some concentrated liquid cleaners can be reused many times before they need to be disposed, so check the instructions before you discard a cleaner.

4.1.4 Ultrasonic Cleaners

Ultrasonic cleaners can loosen dirt particles off the walls of solid material and are essential if there is any particulate matter in small crevasses or corners where it would be otherwise impossible to reach.

Ultrasonic cleaning works by emitting a series of very high-pitched sound waves that create a series of standing waves in solution. The waves alternately compress and decompress the cleaning solution. During the decompression stage, the liquid outgasses and millions of tiny bubbles are formed. During the compression stage, the bubbles implode. This implosion (called *cavitation*) is the destructive force that loosens dirt particles.

To use an ultrasonic cleaner, immerse the item to be cleaned in the cleaning solution within the device's tank. There are general purpose and industrial strength cleaning solutions as well as specific cleaning solutions for jewelry, oxides, and buffing compound removers. Turn the ultrasonic cleaner on for several minutes, and see if your item is clean. If not, repeat the process, and/or try an alternate cleaning method. After an item has been cleaned, rinse as you would with a soap and water wash.

Although uncommon, it is possible to damage glass items with ground sections in an ultrasonic cleaner. This damage can occur when items are left too long in the cleaner. Ground sections to be concerned about include ground joints, stopcocks, and even the lightly etched dots on flasks (used for writing on). The damage can range from simple cracks that radiate from the ground areas, to chipped-off pieces of glass.

One of the obvious restrictions with ultrasonic cleaners is that the object to be cleaned needs to be able to fit inside the cleaning tray. Commercial ultrasonic cleaners can be as small as 3½" diameter and 3" deep to as large as 19¾" x 11½" x 8".

4.1.5 Organic Solvents

There are three types of organic solvents that can be used for cleaning: non-polar, polar, and the halocarbons. They are all capable of removing adsorbed (soaked into the walls of a container) contaminants.

Nonpolar solvents, such as hexane can be used to dissolve nonpolar contaminants such as oils from glass.

Examples:

(various hydrocarbons)
Hexane
Ethane
Benzene

Polar solvents, such as the alcohols and ketones, are useful for the removal of polar contaminants, but also attach to adsorbed sites and thereby limit the amount of area available for undesirable materials. Thus, final rinses with polar hydrocarbons can be very beneficial.

Examples:

(various alcohols and ketones)
Ethanol
Methanol
Isopropanol
Acetone
Methyl ethyl ketone

Halocarbons, a class of polar solvents, are hydrocarbons with an attached halogen. There are commonly three types of halocarbon solvents: those based on chlorine, fluorine, and a combination of the two. They are all powerful degreasing materials and can be particularly effective in removing polar contaminants from glass. The chlorofluorocarbons are currently under review because they cause environmental damage to the ozone layer.[1]

Examples:

(various fluorocarbons, chlorocarbons, and chlorofluorocarbons)
Trichloroethylene
Perchloroethylene
Methylene chloride (dichloromethane)
Trichlorotrifluoroethane

By adding small quantities of a hydrocarbon polar solvent to a chlorofluorocarbon, the overall cleaning abilities of both may be substantially improved.

Pre-preparation. Ascertain whether the contamination is polar or nonpolar.

Material. Dichloromethane or acetone are good places to start because of their costs and safety. If there is limited or no success with either of these solvents, try other appropriate solvents.

Preparation. Be sure to remove any soaps or excess water that may be remaining in the glassware.

Use. Always use the lowest grade (Technical) of any solvent when cleaning glassware. Use 10 mls of a higher grade (Reagent) for a final rinse of the glassware. If the solvent has a high boiling point, select a suitable (polar or nonpolar) solvent with a low boiling point for the final rinse to facilitate evaporation.

Safety Consideration. Safety glasses should be worn. Because there is often a high percentage of synthetic fibers in clothing, it is a good idea to wear a lab coat. Many of the organic solvents used in the laboratory could damage your clothing. Organic solvents should not come in contact with skin because they are drying to the skin. Most organic solvents are irritating to some degree to most people. If an organic solvent is trapped against your skin, such as between a ring and finger or absorbed into clothing, the irritation will have a greater time to develop.

Keep in mind that a number of organic solvents are carcinogenic (i.e., carbon tetrachloride and benzene). Some may dissolve rubber gloves, so be careful of that possibility. *Always use organic solvents in a fume hood*. The vapors of many solvents can have minor to major health dangers. For instance, chloroform is a suspected carcinogen and known to cause liver damage. Chloroform used to be used as an anesthetic by putting patients to sleep. With improper ventilation, one could pass out from the vapors. Most organic solvents are flammable and should therefore be kept far from any open flame, sparks, or high heat sources.

Disposal. Acetone and alcohol can be saved for future cleaning, or recovered by distillation. None of the halogenated hydrocarbons should ever be poured down the sink. If poured on a rag to wipe a stopcock, you may be able to leave the rag in a fume hood to dry before disposing of the rag. However, such a rag may classify as a toxic waste and may require special handling and disposal. There are many laws that govern the disposal of organic solvents, so check with the safety coordinator where you work for specific information.

4.1.6 The Base Bath

The *base bath* seems like an ideal cleaning method. You carefully lower your item to be cleaned into the bath, let it soak for a period of time (one-half to several hours), take it out, rinse it, give it an acid rinse, rinse it in water once again, then give it a distilled water rinse. The base bath is the preferred method of cleaning when silicone stopcock grease is used, as it effectively removes that type of grease better than any other cleaning method.

However, there are disadvantages to the base bath. First, it has some safety hazards. The alcohol is a potential fire hazard and the bath's alkalinity is caustic to skin. The base bath is also a mild glass stripper. Therefore, glassware should not soak in a base bath for an extended period of time and the base bath should never be used for volumetric ware.

Pre-preparation. Wipe off any excess stopcock grease with a Kimwipe® tissue. Then, clean any glassware covered with organic stopcock grease with an organic solvent.* Next, use the standard general cleaning procedure with soap and water and/or a rinse with an organic solvent.

Material. Sodium hydroxide (NaOH) or potassium hydroxide (KOH) and ethyl alcohol.

Preparation. Mix one liter of 95% ethyl alcohol with a solution of 120 g of NaOH (or KOH) mixed in 120 ml of water.

Important note. *A base bath can be stored in a plastic (OK) or stainless steel (best) container.*** This solution is highly basic, and as such it will dissolve a glass container. The alcohol in a base bath is flammable, and it therefore presents a problem for plastic containers that would be likely to melt or burn.

Use. Separate all ground joints and stopcocks before placing an apparatus into a base bath, otherwise they may fuse and be inseparable. Place glassware (completely submerged) into the base bath and let it soak anywhere from a half-hour to several hours. This time period depends on the amount of cleaning required and the age of the base bath. After removing, rinse with water followed by a brief soak in a 2-3 molar solution of nitric acid. The acid soak will restore the hydroxyl groups on the glass surface and stop the base attack on the glass. The glassware should then be re-rinsed in water followed by a deionized water rinse. A final rinse of alcohol or acetone can be used to facilitate drying.

Base baths can be used over and over until they begin to show a decrease in effectiveness. Between use, a base bath should be covered to prevent evaporation and to prevent other glassware from inadvertently falling in.

Safety Considerations. Safety goggles should be worn. This liquid is highly caustic, so rubber gloves are a must. This liquid is also highly

* Organic stopcock greases are insoluble in a base bath and will prevent the base bath from cleaning any contamination beneath the grease.

** Because a base bath can dissolve aluminum, do not even try to determine whether a particular aluminum alloy is acceptable or not. The simplest and safest route is to use a stainless steel container.

flammable, so it is important to keep flames, sparks, high heat sources, or anything that could ignite the solution far away. Because a strong base can dissolve glassware, it is important not to let glassware sit too long in the base bath.

The long-term effects of glass remaining in a base bath are:

1) An etching (frosting) of the glass that could increase the ability of contaminates to "cling" to the glassware. Etching provides more surface area for dirt to cling to. Thus, it is more difficult to remove material from etched glassware.
2) The ability of high-vacuum stopcocks to maintain a vacuum may be lost.
3) The ceramic decals on glassware (including volumetric marks) may be removed.
4) Volumetric glassware is likely to exhibit increased volumes.

Disposal. As a base bath is used, it tends to lose volume both to evaporation and to the small loss incurred each time an object is removed carrying some base bath with it. The remaining base bath can be recharged by adding to it a new solution, so disposal is seldom necessary. If you wish to dispose of the remaining solution, as long as the base has been neutralized (check with litmus paper) and contains no heavy metals, it can be rinsed down the sink. However, check with local environmental and safety laws to see if any other concerns need to be addressed.

4.1.7 Acids and Oxidizers

Organic and carbonaceous materials can be easily removed with acids and oxidizers. In addition, other glass contamination, such as calcium and other alkali deposits, can be effectively removed with hydrochloric acid (HCl). On the other hand, because alkali deposits are removed from glass surfaces with this rinse, there will be a more porous surface on the glass than before the rinse. This greater porosity will mean greater water absorption and possibly greater cleaning cleaning problems in the future. This problem is more likely in porcelain containers and soft glass than borosilicate glass.

Pre-preparation. Wipe off any excess stopcock grease with a Kimwipe® tissue. Then, clean any glassware covered with organic stopcock grease with an organic solvent. Next there should be the standard general cleaning with soap and water and/or a rinse with an organic solvent.

Material. Hydrochloric acid (HCl), sulfuric acid (H_2SO_4), nitric acid (HNO_3), or chromic acid (Chromerge®). See the next section.

Preparation. Acids can be used as straight concentrated solutions or diluted to some degree with water. Some can be mixed (i.e., aqua regia: three parts hydrochloric and one part nitric acid). Another mixed solution is peroxysulfuric acid. It is made by mixing a few ml (i.e., 5 ml) of concentrated H_2SO_4 with an equal amount of 30% H_2O_2. Warming (by steam) can often increase the effectiveness of an acid or oxidizer.

Use. *All acids must be used in a fume hood.* Let the acid soak in the glassware for a short, or long, period of time (as necessary). Swirling and/or heating the acid or oxidizer by use of a steam bath (do not use a direct flame) can facilitate the action. Mineral deposits can often be removed by hydrochloric acid. Metal films can often be removed by nitric acid.

After cleaning with an acid or oxidizer, and after rinsing with water, it is good practice to use ammonia as a neutralizing rinse. This rinse should be done before a second rinse is made with water, and a final rinse with distilled (or deionized) water. The ammonia neutralizes the acid or oxidizer. However, if you are working with organic compounds, it is better not to do an ammonia rinse because many organic reactions are acid-catalyzed.

Safety Considerations. The use of a fume hood, eye protection, and protective gloves are a must. Lab coats are also recommended because these acids can destroy clothing and they can damage skin. If any acid gets in your eyes, wash them copiously with water and seek medical attention. If you wear contacts, after washing copiously with water, remove the contacts, rinse with more water, then seek medical attention. *Do not stop to remove contacts before rinsing with water.* The flushing water will probably remove your contacts for you. If you think you might have accidentally spilled any acid on any part of your body, wash copiously with soap and water just as a precaution. If you feel an itching or burning sensation after having worked with acids or oxidizers, wash the area copiously with soap and water. If the itching or burning sensation continues, seek medical attention.

Disposal. As long as the acid does not contain chromium, or is not hydrofluoric acid, it may be washed down the sink after proper neutralization.* Neutralization is done by first diluting the acid with water to less than 1 M followed by adding either solid sodium hydroxide or 5% sodium hydroxide while constantly mixing the solution until the solution is approximately neutral.[2] Any heavy metals that may have been dissolved by the acid must be removed before disposal. The best way to remove the metal is to participate the metal, neutralize the acid, and send the metal participate to a proper waste facility.

Any hazardous waste cannot be simply thrown away. Such materials must be stored by the user and picked up by companies licensed to remove such materials. For the sake of this planet, do not ignore environmental laws.

4.1.8 Chromic Acid

Because of the long history of chromic acid use, and because of the current environmental problems of heavy metals, chromic acid deserves separate comment.

Preparation. Mix 15-20g $K_2Cr_2O_7$ (potassium dichromate) and 50-100 ml water to dissolve. Then, slowly add 900 ml of concentrated H_2SO_4 (sulfuric acid). This solution gets very hot, so do not use a plastic container for mixing.

* Both hydrofluoric acid and chromic acid disposal are dealt with in sections 4.1.8 and 4.1.9 respectively.

Use. Chromic acid effectively removes organic contaminants and it has a long history of good service. Chromic acid is, by many standards, an excellent cleaner. By simply soaking your glassware in chromic acid, organic dirt and film deposits are removed. It is safe for glassware to be left in chromic acid for extended periods of time. It can be used over and over until it begins to turn green (it provides its own potency indicator). Chromic acid is safe for volumetric ware. It does not have the same problems of the *base bath* of dissolving the glass (and changing the volume) or of dissolving the ceramic lines on volumetric ware. There are, of course, safety precautions for its use because of its strong oxidation potential.

Chromic acid use is not recommended if you are doing experiments in conductivity because the chromic ion can coat the walls of the glassware.

Because chromic acid can be reused, it is often kept in large glass containers. It is important to keep a lid on the container that is (reasonably) airtight. This containment is not so much for evaporation concerns, but rather because sulfuric acid can absorb 30% of its weight in water, including water from the atmosphere. As it absorbs atmospheric water, its volume changes, and it can overflow the container.

Disposal. Disposal is best done by reducing the dichromate to the insoluble chromium hydroxide with a sodium thiosulfate solution. The process for neutralizing 100 ml of chromate solution is as follows:

1) Add sodium carbonate (about 180 g) slowly (while stirring) until the solution is neutral to litmus paper (color goes from orange to green).
2) Re-acidify this solution with 55 mL of 3 M sulfuric acid (color returns to green).
3) While continuously stirring, add 40 g of sodium thiosulfate ($Na_2S_2O_3 \cdot 5H_2O$). (The solution now goes to blue and cloudy.)
4) Neutralize the solution by adding 10 g of sodium carbonate. In a few minutes, a blue-grey flocculent precipitate forms.
5) Filter the solution through Celite or let stand for one week and decant the liquid. This liquid can be allowed to evaporate or be filtered through Celite. Regardless, the remaining liquid contains less then 0.5 ppm of chromium and can be washed down the sink (check local regulations).
6) The solid residue should be packaged, labeled, and sent to a proper disposal firm.[3]

Alternatives.[4] Fortunately there are a variety of alternatives to chromic acid so its use is really unnecessary. Some homemade alternatives include:

1) Combining H_2O_2 and H_2SO_4 to make peroxysulfuric acid. This solution does not have the convenience of a color change as it begins to lose its oxidizing potential. On the other hand, it does not have any heavy metals that are difficult to dispose of.
2) 3:1, a solution of three parts sulfuric acid to one part nitric acid.

Some commercial alternatives include:

1) ***Nochromix*** (Godax Laboratories, New York, N.Y. 10013), a powder, is mixed with concentrated H_2SO_4. The clear solution

turns orange as the oxidizer is used up and therefore provides the same 'indicator' advantage of chromic acid.

2) ***Eosulf*** (Micro, International Products Corporation) contains EDTA (an organosulfonate-based detergent, is nonacidic, has no toxic ions, and is biodegradable meaning there are less acids to store and work with and its significantly decreases hazardous chemical disposal.

3) **Phosphate based cleaning solutions** do very well as surfactants, and phosphate ions assist in solubilizing materials in polar solvents such as water. Unfortunately, these same phosphate ions also cling to the polar glass ions, cannot be easily removed, and interfere with some biochemical assays. Even if phosphate contamination did not interfere with your work, there are strong environmental concerns about using phosphate-based cleaners and how to safely dispose of them.

Although many labs may continue to use chromic acid, there are enough other choices* for cleaning organics off glassware for it not to be essential for use. If you use chromic acid, you may consider phasing out its use.

4.1.9 Hydrofluoric Acid

Hydrofluoric acid (HF) is an extremely dangerous but highly effective glass stripper. It is sometimes used as a cleaner because of its ability to remove layers of glass off an item's surface (called *stripping*). What remains underneath is very clean glass. Two of the best reasons to use HF is glassware that has burnt inorganic materials on its surface or glassware with contamination from radioisotopic work.

There are few occasions where HF is recommended over a safer glass stripper or a base bath (see Sec. 4.1.6). The biggest demand for HF as a cleaner is to remove burnt-on particulate matter from a glass surface. For example, if silicone grease was left on glassware that went through an oven (> 400°C), the nonvolatile remains of the grease are undoubtedly burnt into the surface. HF can remove the silicon grease remains, the decal identification markings, and some quantity of the glass. Thus, prolonged or repeated soaking of glassware in HF can easily remove enough of a glass item to make it useless. Volumetric ware should never be soaked in HF. It was found that cleaning a 100-ml flask with a 5% HF solution for five minutes increased the flask's capacity by 0.04 ml.[5]

Pre-preparation. All greases and oils must be removed before treating with HF. Baking the glassware in an oven of at least 400°C is very effective for removing any organics and other volatile materials. However, such baking can burn inorganics into the glass surface requiring more extensive soaking in HF which, in turn, could cause greater erosion of the glass.

Material. Hydrofluoric acid is available in plastic containers (in a concentration of 49%-52%). These containers should always be stored in a fume hood. Never store HF in glass containers.

* Each with their own range of safety concerns.

Preparation. HF should not be used at full (bottle) strength. A 5% to 10% solution is best. Once prepared, it should be stored in a polyethylene or other strong plastic container that can be closed. Because HF is a glass stripper, do not store the acid solution in a glass container. HF, at any concentration, should always be stored in a fume hood.

Use. It is important to wear gloves, safety glasses, and work in a fume hood!! If you do not have all of these items, do not use HF as a cleaning material! Pour the diluted HF from its plastic storage container into a plastic soaking tank (like a polypropylene beaker). Let the glassware soak in the HF for anywhere from 30 seconds to two minutes. Rinse for at least one minute (or five full flushes) with water. Then rinse with distilled or deionized water. The HF can be reused many times until a decrease in effectiveness is obvious.

Safety Considerations. Always wear rubber gloves with long arm covers when working with HF. It is hard to overemphasize how dangerous HF can be. One of the dangers in getting dilute HF on your skin is that it may initially be no more noticeable than water. If you get HNO_3 on your skin, you know it instantly by the burning sensation. That burning sensation is telling you something is terribly wrong. Because of the *seemingly* inconsequential aspects of HF, you might be inclined simply to rinse off any suspected HF. However, even with some washing, you may not be successful in removing all of the HF. **If any HF gets on your skin, wash with copious amounts of soap and water for ten minutes. If you get any HF under a fingernail, scrub thoroughly with a fingernail brush!!** The treatment for an HF burn under the fingernail invariably includes removal of the fingernail. Once in the skin, HF spreads throughout all the surrounding tissue. Hydrofluoric acid burns in appearance can be similar to fire burns, but the development is in slow motion, and can eat through to the bone. Remember, HF travels through body tissues doing damage to those tissues as it travels. Blindness is highly probable if HF gets in the eye. **Always wear protective goggles when using HF.**

The fumes from HF are also very dangerous and can cause burn blisters on the linings of the lungs. Although not pleasant, HF is not as sharply irritating to breathe as other strong acids (such as HNO_3), and you may not think to be repelled. **Always use HF in a fume hood.**

One safety trick is to add to your total HF solution about one or two percent nitric acid. This addition will not assist the cleaning properties to any significant amount, but it will add better warning characteristics to both the odor and any accidental skin contact. This practice is similar in concept to the addition of hydrogen sulfide to methane.

Disposal. Although HF is an acid, it cannot simply be neutralized and washed down the sink because the fluorine in the acid will attack any aluminum in the plumbing. Thus, HF, like chromic acid requires special handling. A 5% solution of HF can be safely treated by mixing with a 1:1 ratio of the following solution:[6]

2 gm aluminum nitrate ($Al(NO_3)_3 \cdot 2H_2O$)
0.8 gm $Ca(OH)_2$
1 liter of water

After mixing the HF and the above neutralizing solution in a 1:1 ratio, the resultant mixture can then be poured down the sink.

Extra Tip: For an extra-tough stain, use a mixture of 3 ml of storage bottle strength (49%-52%) hydrofluoric acid to 100 ml of concentrated nitric acid. This solution should be used for only a minute or two at most and must be done in a fume hood. Use the same safety considerations that you would use for standard hydrofluoric acid.[7]

4.1.10 Extra Cleaning Tips

Cleaning Fritted Filters. Because of the nature of fritted glass, its cleaning requires special mention. Fritted glass is crushed glass (Ace Glass uses crushed glass fibers) that is separated by size, compacted together, then heated until the pieces stick together. Because the structure is loosely connected, the frit maintains its porous nature which provides the frit its filtering abilities.

Because different materials get onto and in a frit, different cleaning processes are required to clean a frit. For this reason, storage is an important aspect of frit cleaning. If you know what got on a frit, you know how to clean it. Keeping a frit in a clean, dry, and dust-free area will simplify your cleaning needs. Before a fritted filter is first used, it should be rinsed with hot sulfuric acid and then with distilled water until the filtrate is neutral in pH. This procedure will remove any glass particles and dust that may be on the new frit.

After a frit has just been used, it is important to flush it immediately with distilled water in the opposite direction that it was used. Then proceed with a chemical cleaning. Once the fritted glass is cleaned, rinsed, and dried, store it in a clean, dry, and dust-free area. Some suggested contamination and appropriate cleaning solutions are as follows:[8]

Fats; Carbon Tetrachloride
Silver chloride; Ammonia
Albumen; Warm ammonia or hydrochloric acid
Nitric acid; Hot hydrochloric acid with potassium chlorate
Barium sulphate; Hot concentrated sulfuric acid
Organic materials; Potassium permanganate*
General cleaning; Potassium permanganate*

Because of the effects of basic solutions on glass, never let a basic solution (such as NaOH or KOH) remain in contact with a fritted glass filter. Rinse with hydrochloric acid and then with distilled water until the filtrate is neutral in pH.

Removing Tungsten from Glassware.[9] A dark film of tungsten is often deposited on the walls of components of glass vacuum systems having tungsten filaments. This deposit hinders observation within the glassware and can also cause an increase in wall temperature.

* Treat the filter with a 1% aqueous solution of potassium permanganate followed by a few drops of concentrated sulfuric acid. The heat of the reaction will increase the oxidizing power of the mixture.

Although a solution of HF, or a mixture of HF and HNO_3, can remove the tungsten stain, it also removes some of the glass, which can etch the glass and, if the HF solution is left too long, can make the glass dangerously thin for a vacuum system.

A tungsten coating can be removed quite successfully with a hypochlorite solution (Clorox® bleach, for example). A black opaque coating can be removed in a few minutes, and, if the solution is warmed, the process is even faster. The solution can be reused many times.

Removing Tar from Distillation Flasks.[10] As mentioned in the beginning of this section, the best time to clean glassware is immediately after it is used. The hardest glass pieces to clean are those that have been left for someone else to clean. Tar, however, is always hard to clean and is always left for last (which also makes it more difficult to clean). A simple solution for cleaning tar from the bottom of flasks is to invert the flask into a beaker of acetone and let the vapor of the solvent dissolve the tar (see Fig. 4-4). This technique can remove most of the tar deposit in a few hours. It can be hastened by heating the solvent in a steam bath (do not use an open flame).

The trick to this technique is getting the solvent into the neck of the upside-down flask. One method is to lightly heat the flask before inserting it into the solvent. Then, as the flask cools, the solvent will be drawn into the flask. Perhaps the simplest method is to place a flexible tube into the neck to allow air out of the flask as it is inverted into the beaker (be sure the tube is not soluble in the solvent).

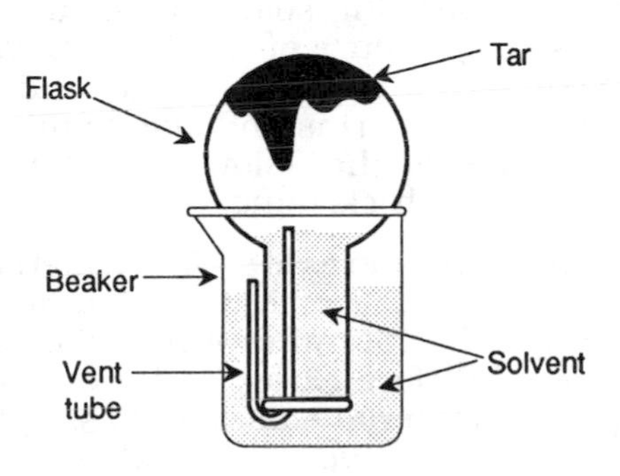

Fig. 4-4 Removing tar from a flask.

4.1.11 Additional Cleaning Problems and Solutions[11]

In the following cleaning procedures, all work should be done in a fume hood, hand and eye protection should be worn, and a lab apron is recommended. After cleaning, an item should be thoroughly rinsed in water followed by a rinse in distilled water. If an item needs to be dried, a final rinse in reagent-grade methanol, ethyl alcohol, or acetone can be used.

Iron Stains. Equal parts hydrochloric acid and water should be used.

Mercury Residue. Use hot nitric acid. The acid used should be collected and returned to your safety supervisor because it now contains a heavy metal and should not be rinsed down the sink under any circumstances.

Permanganate Stains. Use concentrated hydrochloric acid or a saturated solution of oxalic acid at room temperature.

Bacteriological Material. Soak the glassware in a weak Lysol® solution, or autoclave in steam. A number of sources recommend soaking the glassware in a 2% to 4% cresol solution. However, the EPA has recently identified cresol as a hazardous material. Because other options are readily available (see Sec. 4.1.1 through 4.1.8), there is no reason to use the material. If you have any cresol, contact a hazardous materials disposal firm in your area for removal.

Albuminous Materials. Soak in nitric acid.

Carbonaceous Materials. Soak in a solution of equal parts sulfuric and nitric acid. Be aware of the possible formation of dangerous compounds when organic material is present.

Magnesium Oxide Stains. Instead of HCl, try a 20% to 30% solution of sodium bisulfite ($NaHSO_3$). Because an acid is not being used, SO_2 is not liberated.[12]

4.1.12 Last Resort Cleaning Solutions[13]

The following solutions should only be used as last resorts for cleaning glassware because of the dangers each present. Each is meant to be used as a soaking solution for glassware and should be left in a fume hood at all times. After soaking, rinse off the solution with copious amounts of water. Before using any of the following materials, strongly consider whether the glassware is worth cleaning.

Alkaline Peroxide. This material is good for removing organics.

Place 100 g of NaOH (Sodium Hydroxide) into a 500-ml glass beaker. Slowly add 250 ml of water. This solution will get hot (120°C), bubble, and fizzle. Therefore gloves and eye protection are a must.

This material can etch glass if it is left to soak for an extended period of time, therefore you might wish to select an old, worn beaker in which to do the mixing. You must use a glass beaker because of the exothermic reaction that is created when the mixture is first mixed.

If this solution is in a container and allowed to get cold, it can begin to deposit out of solution. The solid can be removed by **slowly** adding water or a dilute acid.

Alkaline Permanganate. This material is good for removing siliceous deposits and acidic crud.

Place 500g to 1 kg of NaOH (Sodium Hydroxide) in a 5-liter beaker. Add 4 $^1/_2$ liters of water. Once dissolved, add 50 mg of $KMnO_4$ (Potassium Permanganate). This solution will be purple and, after leaving it overnight to allow it to become active, will turn green.

This mixture can be used up to one year. It should be covered to reduce evaporation. During that time some water will evaporate leaving a brown crust of MnO_4. This crust can be knocked back into the remaining solution. If too much water has evaporated, simply add water to original levels. If the solution leaves a brown deposit on glassware, the deposit can be removed with HCl (Hydrochloric acid). Besides damaging skin tissue, this solution can also stain skin, so be sure to use gloves.

Fuming Sulfuric Acid. This material is good for a wide variety of contaminants.

Use a small amount of heated sulfuric acid in a fume hood. Roll the item around so the acid comes in contact will all areas of the piece. Unless the item needs to soak (for example thirty minutes soaking time is recommended for the removal of silicone grease) immediately empty the glassware and rinse. Although more dangerous to use than a base bath, this technique can be used to remove silicone grease from volumetric ware without concerns about altering the volume of volumetric ware.

References

1 F. Sherwood Rowland, "Chlorofluorocarbons and the Depletion of Stratospheric Ozone," *American Scientist*, 77 (1989), pp. 36-45.

2 Margaret-Ann Armour, "Chemical Waste Management and Disposal," *J. of Chem. Education*, 65, (1988), pp. A64-8

3 *ibid.*

4 P.L. Manske, T.M. Stimpfel, and E.L. Gershey, "A Less Hazardous Chromic Acid Substitute for Cleaning Glassware," *J. of Chem. Ed.*, 67 (1990), pp. A280-A282.

5 L. Holland, *The Properties of Glass Surfaces* (New York: John Wiley & Sons, Inc., 1964), p. 295.

6 Dr. Cathy Cobb, personal communication.

7 Pope Pacers, a publication of Pope Scientific, Menomonee Falls, WI 53052, March 1966, p. 17.

8 *ibid*, p. 15.

9 Robert W. Burns, "To Remove Stains of Evaporated Tungsten from Inside of Glass Envelopes," *Rev. of Sci. Instrum.*, 44 (1973), p. 1409.

10 E. J. Eisenbraun, "Solvent Saving Procedure for Removing Tar from Distillation Flasks," *J. of Chem. Education*, 63 (1986), p. 553.

11 Kontes Catalogue, 1989, p. XIV.

12 R. E. Schaffrath, "A Suggestion for Cleaning Glassware Coated with MnO_2," *J. of Chem. Ed.*, 43 (1966), p. 578.

13 Page, P., "Cleaning Solutions for Glassware," *British Society of Scientific Glassblowers Journal*, 27, (1989), p. 122.

Chapter 5

Compressed Gases

5.1 Compressed Gas Tanks

5.1.1 Types of Gases

A gas is defined as any material that boils within the general ranges of STP (Standard Temperature [25°C] and Pressure [1 atmosphere]). Although there are many *compounds* that satisfy these conditions, only eleven *elements* satisfy these conditions (argon, chlorine, fluorine, helium, hydrogen, krypton, oxygen, neon, nitrogen, radon, and xenon).

There are two major groups of gases. The first group is known as the *non-liquefied gases* (also known as cryogenic gases). These gases do not liquefy at room temperatures nor at pressures from 25 to 2500 psig. They liquefy at very low temperatures (-273.16°C to ≈ -150°C). The second group of gases, known as the *liquified gases* can liquify at temperatures easily made in the laboratory (–90°C to ≈ -1°C) and at pressures from 25 to 2500 psig. These gases become solid at temperatures that the cryogenic gases become liquid. Carbon dioxide (dry ice), generally considered a liquified gas, could also be known as a *solidified gas* as it does not have a liquid state at STP.

As far as gas physics goes, there are only two types of gases. However, as far as gas shipping goes, there is a third classification known as the *dissolved gases*. This classification applies to the gas acetylene, which without special equipment, can explode at pressures above 15 psig. Because of this property efficient shipping of this gas becomes almost impossible. To avoid this problem, the gas is shipped dissolved in acetone and placed in cylinders that are filled with an inert, porous material. Under these special conditions, acetylene can be safely shipped at pressures of 250 psig.

Gases are stored and shipped in a compressed state in cylinders designed to withstand the required pressure. The cryogenic gases, such as oxygen and nitrogen, are shipped in both liquified and gaseous states, but the liquid state requires expensive equipment. Other gases, such as propane, are typically shipped and stored in their liquid states. The region above a liquified gas is an area called the *head space* which is the gas form of the liquified gas. The pressure of the gas in the head space depends on the vapor pressure which depends on the temperature of the liquid gas.

To control the rate of flow and releasing pressure, gases are released from compressed gas cylinders by regulators (see Sec. 5.2). In addition, a regulator can display the amount of gas remaining in a compressed gas tank, but not in a liquified gas tank.

The remaining cryogenic liquid in a tank can be estimated by observing any floats that may exist, but it can always be accurately determined by weighing the tank. First you must subtract the weight of the empty tank by the weight of the partially-filled tank (known as the *tare weight**), to determine the weight of any remaining liquid. Then you divide the difference by the weight of the liquid gas per liter to calculate the remaining volume.

Tank-weighing should be done with everything you plan to use with it (i.e., regulators, tubing, and so forth) attached, making sort of a '*ready-to-use tare weight.*' This ready-to-use weight means you do not have to strip the tank of all equipment to make remaining gas determinations. Otherwise, if it was originally weighed without a regulator but later weighed with one, you might assume that you have more gas than really exists.

5.1.2 The Dangers of Compressed Gas

Compressing a gas allows a lot of gas to exist in a small amount of area for transportation, storage, and use. It is physically easier to deal with a compressed gas tank that is 9 in in diameter and 51 in tall, than trying to store the approximately several hundred cubic feet** of gas contained within.

Regardless of the inherent danger of any given gas, once a gas is compressed its potential for danger takes on a completely different light. A completely safe gas compressed in a container at some 2000 lbs./in^2 could act like a bomb if improperly handled! Your safety, and the safety of your equipment, is therefore dependent on:

1) The quality of the construction of the compressed gas cylinders
2) The equipment selection to be used with the compressed gases
3) The proper use of the compressed gas cylinders and associated equipment.

The first two points of the above list can best be controlled by strict industry standards and the conscientious matching of equipment to the needs and demands of the user. The third requires education, and unfortunately the user often does not have the opportunity (or desire) to learn everything there is to know about how to use compressed gases safely. Therefore, there have been a variety of industry-standard "idiot proof" controls to minimize the possibility of mistakes. To make compressed gas cylinders consistently safe, reliable, and as idiot proof as possible, these strategies took the study, analysis, and deliberation of at least eighteen different private and governmental agencies (see Table 5-1).

The first level of safety for quality and control is the construction of compressed gas cylinders. The specifications for their construction in North America is defined by Department of Transportation (DOT) and Canadian

* All tanks that require weighing for volume content have their *tare weight* stamped on the side of the tank.

** A tank of this size will contain (for example) 282 cu ft of oxygen, 257 cu ft of nitrogen, or 290 cu ft of ultra-pure argon.

Transport Commission (CTC) regulations. Cylinders are made from carbon-steel or alloy-steel with seamless, brazed, or welded tubing that is formed by billeting (drawing flat sheets to a cylindrical shape) or by using punch-press dies. Ends are sealed by forging or spinning at great heat. Alternatively, closed ends are drilled out and a metal piece is plugged into the hole. In the U. S., one tank design can be used for a variety of gases (except for acetylene). In Europe, a tank can only be used for the gas it was designed for.

Generally, cylinders that are broad and squat in contour are for low-pressure service, such as the propane tanks used on automobiles or with campers. Those that are tall and thin are generally used for high-pressure containment, such as for oxygen, hydrogen, or nitrogen.

A series of letters and numbers are stamped on the shoulders of compressed gas tanks (see Fig. 5-1) to provide coded information for tank inspectors. These numbers tell under what codes the tank was made, the manufacturer, service pressure, serial number of the tank, when it was last inspected, and by whom. The manner in which the codes are laid out varies, and may be difficult for an untrained person to interpret. However, to the compressed gas industry, they are important for preventing mistakes from the misuse of the containers.

Table 5-1

Agencies involved in Standardization of Compressed Gas Tanks
American Gas Assoc.
American Petroleum Inst.
American Society for Testing and Materials
American Welding Society
Association of American Railroads
Canadian Transport Commission (CTC)
Compressed Gas Association (CGA)
Compressed Gas Manufacturer's Assoc. Inc.
Connections Standards Commission of the CGA (This was first known as the *Gas Cylinder Valve Thread Commission* of the CGA. Later it became the *Valve Standards Com.* of the CGA, and in 1971 it received its current name.)
Department of Transportation (DOT)
Interdepartmental Screw Thread Commission
International Standards Organization (ISO)
National Institute of Standards and Technology (This is the new name of the National Bureau of Standards)
National Fire Prevention Assoc.
Standards Associations (representing Great Briton, Canada, and the U.S.)
U.S. Army
U.S. Dept. of Commerce
U.S. Navy

5.1.3 CGA Fittings

To prevent gases from being attached to the wrong system, the Compressed Gas Association (CGA) implemented a variety of different fittings for attaching a regulator to a compressed gas tank. These fittings prevent a user from accidentally taking a regulator that is used (for example) for oxygen and attaching it to a hydrogen tank. Table 5-2 lists a number of gases and the appropriate CGA fittings or threaded connections that a regulator must have to be attached to a tank of such a gas (in this table, when a second CGA number follows a first entry, it is the CGA fitting for a lecture bottle). Alternatively, if you have a regulator and want to know what gases it can be used with—or what other gases use the same regulator—see Table 5-3.

If, after looking at Table 5-2, you wish to know what a fitting looks like, Fig. 5-2 illustrates the fifteen most common CGA fittings used on full-sized compressed gas tanks within the laboratory. Please note that *right-handed threads* close with CW rotation, and *left-handed threads* close with CCW rotation. Left-handed thread fittings can be easily identified by the notch on the closing nut as opposed to right-handed fittings, which are notch-less.

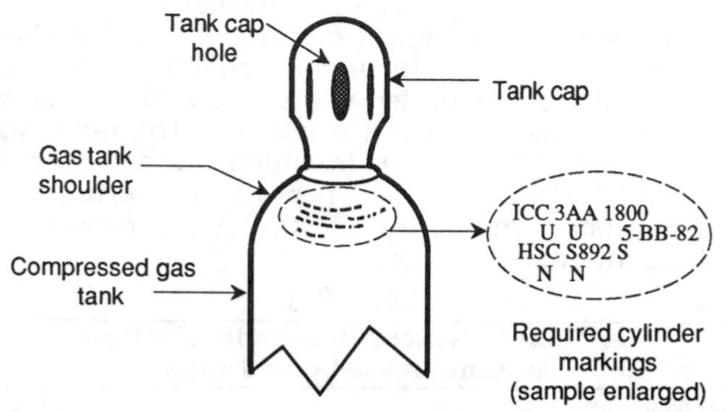

Fig. 5-1 An example of a compressed gas tank with cap.

Never spray or drip oil onto compressed tank cap threads to ease removal or to replace the tank cap. Minimally, the oil could contaminate the CGA fitting, but oil near compressed oxygen can explode! If you are unable to remove or replace a tank cap, obtain help from the manufacturer or the distributor of the compressed tank.

5.1.4 Safety Aspects of Compressed Gas Tanks

Sections of a compressed gas tank are designed to provide safety for potentially abusive conditions. The most common structural protection is the tank cap, which is placed over the valve, then screwed onto the threaded neck of the compressed gas tank (see Fig. 5-1). In addition, you can obtain separate foot-rings that help a cylinder stand up, and protective girdles for around the valve area.

Tank caps can occasionally be difficult to remove. The easiest way to remove one is to insert a large screwdriver a short way into the tank cap hole (see Fig. 5-1) being careful not to make contact with the main tank valve stem. Using the screwdriver as a lever, it is simple to remove a tank cap. Never use this technique to tighten a tank cap. Simply screwing on the tank cap by hand is sufficient—it is not necessary to forcefully tighten one down.

There are various color codes that are used with compressed gas tanks, but *only the color codes of the medical gas industry are consistent.* Note however, that the identifying colors from the U. S. and Canada are **not** consistent with each other (see Table 5-4).

Table 5-2

Common Laboratory Gases
Acetylene Connection 510
Air Connection 590 or 170
Allene Connection 510 or 110
Ammonia, anhydrous Connection 204/705 or 180
Ammonia, electronic Connection 660
Argon Connection 580 or 180
Arsine Connection 660
Boron trichloride Connection 660 or 180
Boron trifluoride Connection 330 or 180
Bromine pentafluoride Connection 670
Bromine trifluoride Connection 670
Bromotrifluorethlene 1/8" National Pipe Thread male outlet needle valve 112A
Bromotrifluoromethane Connection 320
1,3-Butadiene Connection 510 or 170
Butane Connection 510 or 170
1-Butane Connection 510
2-Butane Connection 510
Butenes Connection 510 or 170
Carbon dioxide Connection 320 or 170
Carbon monoxide Connection 350 or 170
Carbon tetrafluoride Connection 320
Carbonyl fluoride Connection 660
Carbonyl sulfide Connection 330 or 180
Chlorine Connection 660 or 110
Chlorine trifluoride Connection 670
Chlorodifluoromethane Connection 660
Chloropentafluorethane Connection 660
Chlorotrifluoroethylene Connection 660 or 170
Chlorotrifluoromethane Connection 320
Cyanogen Connection 660
Cyanogen chloride Connection 660
Cyclopropane Connection 510 or 170
Deuterium Connection 350 or 170
Diborane Connection 350
Dibromodifluoromethane Connection 660
1,2-Dibromotetrafluoroethane Shipped in cans, no CGA fitting needed
Dichlorodifluoromethane Connection 660
Dichlorofluoromethane Connection 660
Dichlorosilane Connection 678
1,2-Dichlorotetrafluoroethane Connection 660
1,1-Difluoro-1-chloroethane Connection 660
1,1-Difluoroethane Connection 660
1,1-Difluoroethyene Connection 320
Dimethylamine Connection 705 or 180
Dimethyl ether Connection 510 or 170
2,2-Dimethylpropane Connection 510
Ethane Connection 350 or 170
Ethylacetylene Connection 510
Ethyl chloride Connection 510 or 170
Ethylene Connection 350 or 170
Ethylene oxide Connection 510 or 180
Fluorine Connection 679

Table 5-2

Common Laboratory Gases (cont.)

Freon® 12 (Dichlorodifluoromethane)
Connection 660 or 170
Freon® 13 (Chlorotrifluoromethane)
Connection 320 or 170
Freon® 13B1 (Bromotrifluoromethane)
Connection 320 or 170
Freon® 14 (Tetrafluoromethane)
Connection 320 or 170
Freon® 22 (Chlorodifluoromethane)
Connection 660 or 170
Freon® 23 (Fluoroform)
Connection 320 or 170
Freon® 114 (2, 2 Dichlorotetrafluoroethane)
Connection 660 or 170
Freon® 116 (Hexafluoroethane)
Connection 320 or 170
Germane
Connection 350/660
Helium
Connection 580 or 170
Hexafluoroacetone
Connection 660
Hexafluoroethane
Connection 320
Hexafluoropropylene
Connection 660 or 170
Hydrogen
Connection 350 or 170
Hydrogen bromide
Connection 330 or 180
Hydrogen chloride
Connection 330 or 110
Hydrogen cyanide
Connection 660
Hydrogen fluoride
Connection 660 or 180
Hydrogen iodine
Connection 330 or 180
Hydrogen selenide
Connection 660
Hydrogen sulfide
Connection 330 or 110
Iodine pentafluoride
No CGA fitting needed
Isobutane
Connection 510 or 170
Isobutylene
Connection 510 or 170
Krypton
Connection 580
Methane
Connection 350 or 170
Methylacetylene
Connection 510
Methyl bromide
Connection 320 or 170
3-Methyl-1-butene
Connection 510
Methyl chloride
Connection 660 or 170
Methyl fluoride
Connection 350 or 170
Methyl mercaptan
Connection 330 or 110
Methyl vinyl ether
Connection 290
Monoethylamine
Connection 240/705
Monomethylamine
Connection 204/705 or 180
Neon
Connection 580 or 170
Nickel carbonyl
Connection 320
Nitric oxide
Connection 660
Nitrogen
Connection 580 or 170
Nitrogen dioxide
Connection 660
Nitrogen trifluoride
Connection 679
Nitrogen trioxide
Connection 660
Nitrosyl chloride
Connection 660
Nitrous oxide
Connection 326 or 170
Octofluorocyclobutane
Connection 660 or 170
Oxygen
Connection 540 or 170
Oxygen difluoride
Connection 679
Ozone
Connection 660
Perchloryl fluoride
Connection 670
Perfluorobutane
Connection 668
Perfluoro-2-butene
Connection 660
Perfluoropropane
Connection 660 or 170

Table 5-2

Common Laboratory Gases (cont.)	
Phosgene Connection 660	**Sulfuryl fluoride** Connection 660
Phosphine Connection 350/660	**Tetrafluoroethylene** Connection 350
Phosphorous pentafluoride Connection 330/660 or 180	**Tetrafluorohydrazine** Connection 679
Phosphorous trifluoride Connection 330	**Trichlorofluoromethane** Comes in drums with a 3/4" female outlet
Propane Connection 510 or 170	**1,1,2-Trichloro-1,2,2-tri-luoroethane** Comes in drums with a 3/4" female outlet
Propylene Connection 510 or 170	**Trimethylamine** Connection 204/705 or 180
Silane Connection 510	**Vinyl bromide** Connection 290 or 180
Silicon tetrafluoride Connection 330 or 180	**Vinyl chloride** Connection 290
Sulfur dioxide Connection 330	**Vinyl fluoride** Connection 320
Sulfur hexafluoride Connection 590	**Vinyl methyl ether** Connection 290 or 180
Sulfur tetrafluoride Connection 330 or 180	**Xenon** Connection 580 or 110

Table 5-3

Common Laboratory CGA Regulators	
CGA #	Chemical Gas Type
110	Allene
	Chlorine
	Hydrogen chloride
	Hydrogen sulfide
	Methyl mercaptan
	Xenon
170	Air
	1,3-Butadiene
	Butane
	Butenes
	Carbon dioxide
	Carbon monoxide
	Chlorotrifluoroethylene
	Cyclopropane
	Deuterium
	Dimethyl Ether
	Ethane
	Ethyl chloride
	Ethylene
	Freon® 12
	Freon® 13
	Freon® 13B1
	Freon® 14
	Freon® 22
	Freon® 23
	Freon® 114
	Freon® 116
	Helium
	Hexafluoropropylene
	Hydrogen
	Isobutane
	Isobutylene
	Methane
	Methyl bromide
	Methyl chloride
	Methyl fluoride
	Neon
	Nitrogen
	Nitrous oxide
	Octofluorocyclobutane
	Oxygen
	Perfluoropropane
	Propane
	Propylene
180	Ammonia, anhydrous
	Argon
	Boron trichloride
	Boron trifluoride
180 (cont)	Carbonyl sulfide
	Dimethylamine
	Ethylene oxide
	Hydrogen bromide
	Hydrogen fluoride
	Hydrogen iodine
	Monomethylamine
	Phosphorous pentafluoride
	Silicon tetrafluoride
	Sulfer Tetrafluoride
	Trimethylamine
	Vinyl bromide
	Vinyl methyl ether
240	Ammonia, anhydrous
	Monoethylamine
	Monomethylamine
	Trimethylamine
290	Methyl vinyl ether
	Vinyl bromide
	Vinyl chloride
	Vinyl methyl ether
320	Chlorotrifluoromethane
	Bromotrifluoromethane
	Carbon dioxide
	Carbon tetrafluoride
	1,1-Difluoroethyene
	Freon® 13 (Chlorotrifluoromethane)
	Freon® 13B1 (Bromotrifluoromethane)
	Freon® 14 (Tetrafluoromethane)
	Freon® 23 (Fluoroform)
	Freon® 116 (Hexafluoroethane)
	Hexafluoroethane
	Methyl bromide
	Nickel carbonyl
	Vinyl fluoride
326	Nitrous oxide
330	Boron trifluoride
	Carbonyl sulfide
	Hydrogen bromide
	Hydrogen chloride
	Hydrogen iodine
	Hydrogen sulfide
	Methyl mercaptan
	Phosphorous pentafluoride
	Phosphorous trifluoride
	Silicon tetrafluoride
	Sulfur dioxide
	Sulfur tetrafluoride

Table 5-3

Common Laboratory CGA Regulators (cont.)

CGA #	Chemical Gas Type
350	Carbon monoxide
	Deuterium
	Diborane
	Ethane
	Ethylene
	Germane
	Hydrogen
	Methane
	Methyl fluoride
	Phosphine
	Tetrafluoroethylene
510	Acetylene
	Allene
	1,3-Butadiene
	Butane
	1-Butane
	2-Butane
	Butenes
	Chlorotrifluoroethylen
	Cyclopropane
	Dimethyl ether
	2,2-Dimethylpropane
	Ethyl chloride
	Ethylacetylene
	Ethylene oxide
	Isobutane
	Isobutylene
	Methylacetylene
	3-Methyl-1-butene
	Propane
	Propylene
	Silane
540	Oxygen
580	Argon
	Helium
	Krypton
	Neon
	Nitrogen
	Xenon
590	Air
	Sulfur hexafluoride
660	Ammonia
	Arsine
	Boron trichloride
	Carbonyl fluoride
	Chlorine
660 (cont)	Chloropentafluorethane
	Chlorotrifluoroethylene
	Cyanogen
	Cyanogen chloride
	Dibromodifluoromethane
	Dichlorofluoromethane
	1,2-Dichlorotetrafluoroethane
	1,1-Difluoro-1-chloroethane
	1,1-Difluoroethan
	Freon® 12 (Dichlorodifluoromethane)
	Freon® 22 (Chlorodifluoromethane)
	Freon® 114 (2, 2 Dichlorotetrafluoroethane)
	Germane
	Hexafluoroacetone
	Hexafluoropropylene
	Hydrogen cyanide
	Hydrogen fluoride
	Hydrogen selenide
	Methyl chloride
	Nitric oxide
	Nitrogen dioxide
	Nitrogen trioxide
	Nitrosyl chloride
	Octofluorocyclobutane
	Ozone
	Perfluoro-2-butene
	Perfluoropropane
	Phosgene
	Phosphine
	Phosphorous pentafluoride
	Sulfuryl fluoride
668	Perfluorobutane
670	Bromine pentafluoride
	Bromine trifluoride
	Chlorine trifluoride
	Perchloryl fluoride
677	Argon
678	Dichlorosilane
679	Fluorine
	Nitrogen trifluoride
	Oxygen difluoride
	Tetrafluorohydrazine
705	Ammonia, anhydrous
	Monoethylamine
	Monomethylamine
	Trimethylamine

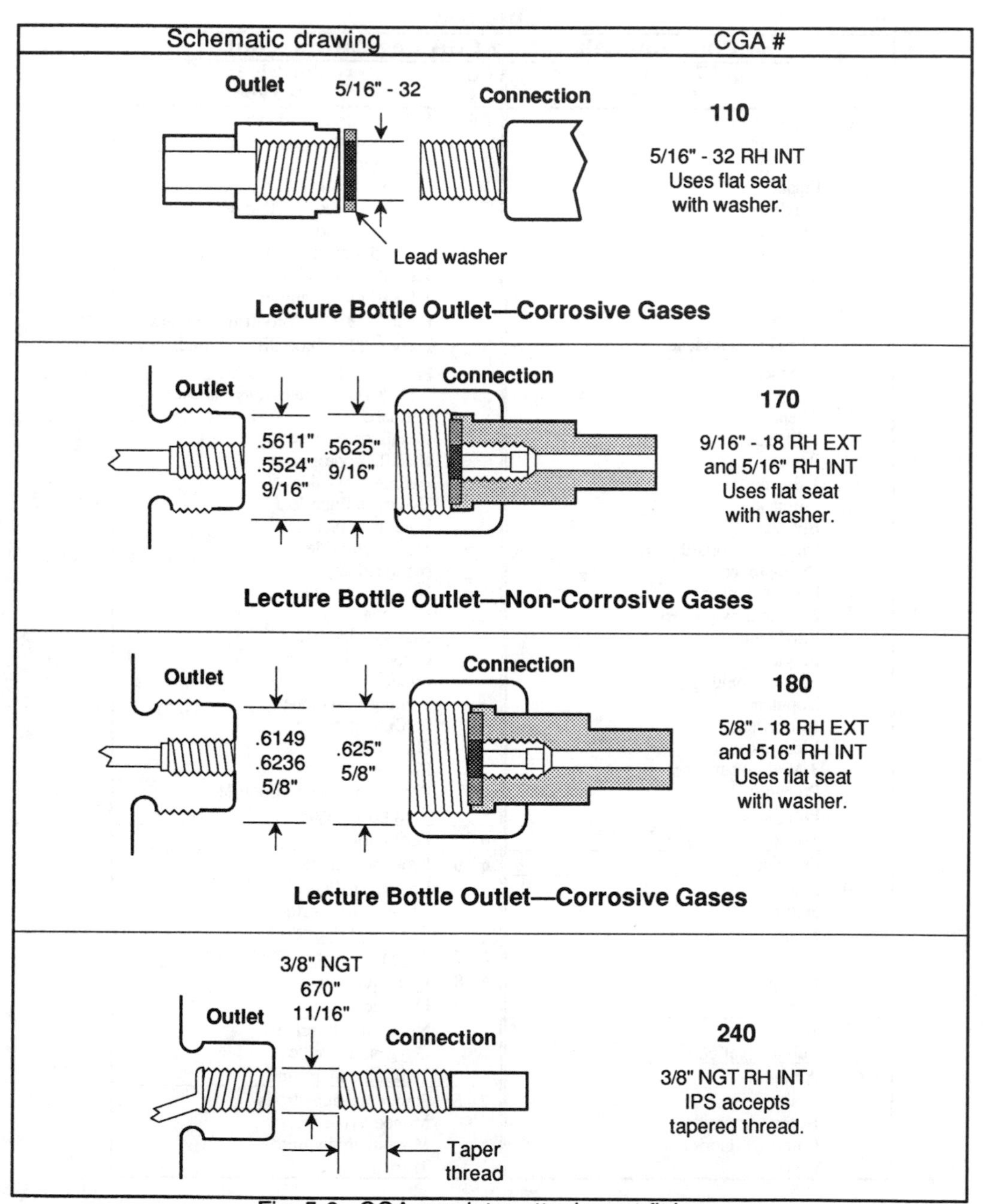

Fig. 5-2 CGA regulator attachment fittings.

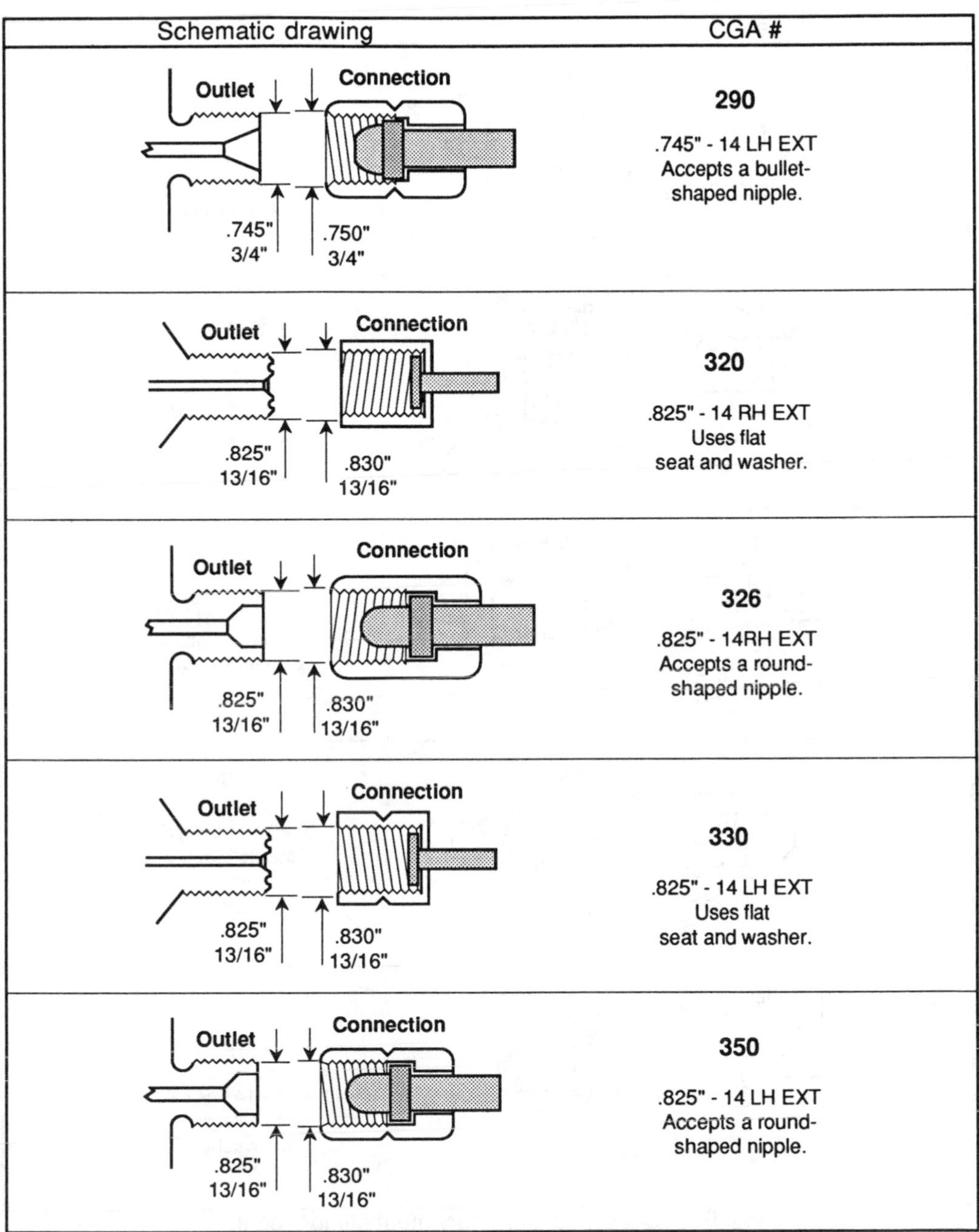

Fig. 5-2 CGA regulator attachment fittings (cont.).

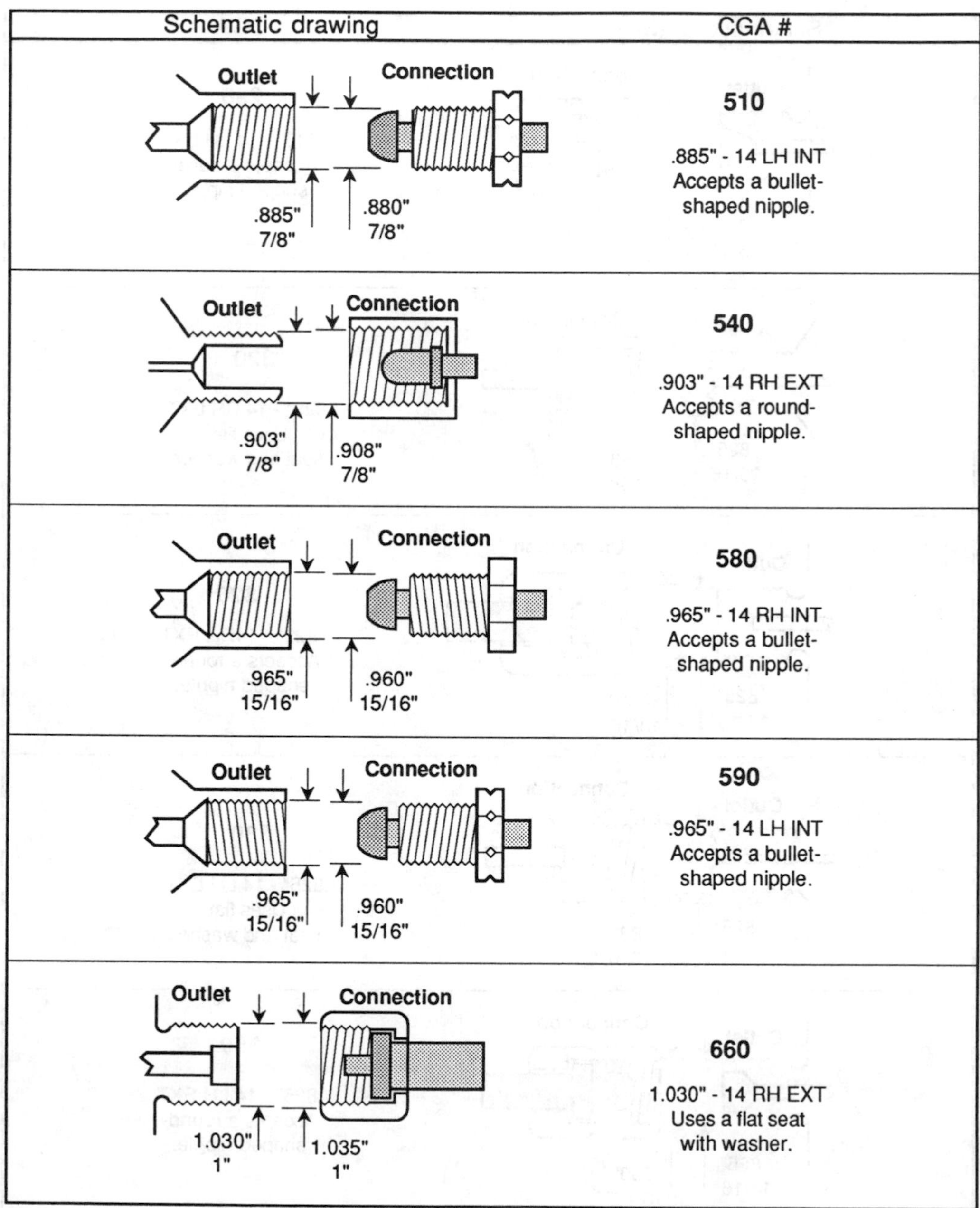

Schematic drawing	CGA #
Outlet; Connection; .885" 7/8"; .880" 7/8"	**510** .885" - 14 LH INT Accepts a bullet-shaped nipple.
Outlet; Connection; .903" 7/8"; .908" 7/8"	**540** .903" - 14 RH EXT Accepts a round-shaped nipple.
Outlet; Connection; .965" 15/16"; .960" 15/16"	**580** .965" - 14 RH INT Accepts a bullet-shaped nipple.
Outlet; Connection; .965" 15/16"; .960" 15/16"	**590** .965" - 14 LH INT Accepts a bullet-shaped nipple.
Outlet; Connection; 1.030" 1"; 1.035" 1"	**660** 1.030" - 14 RH EXT Uses a flat seat with washer.

Fig. 5-2 CGA regulator attachment fittings (cont.).

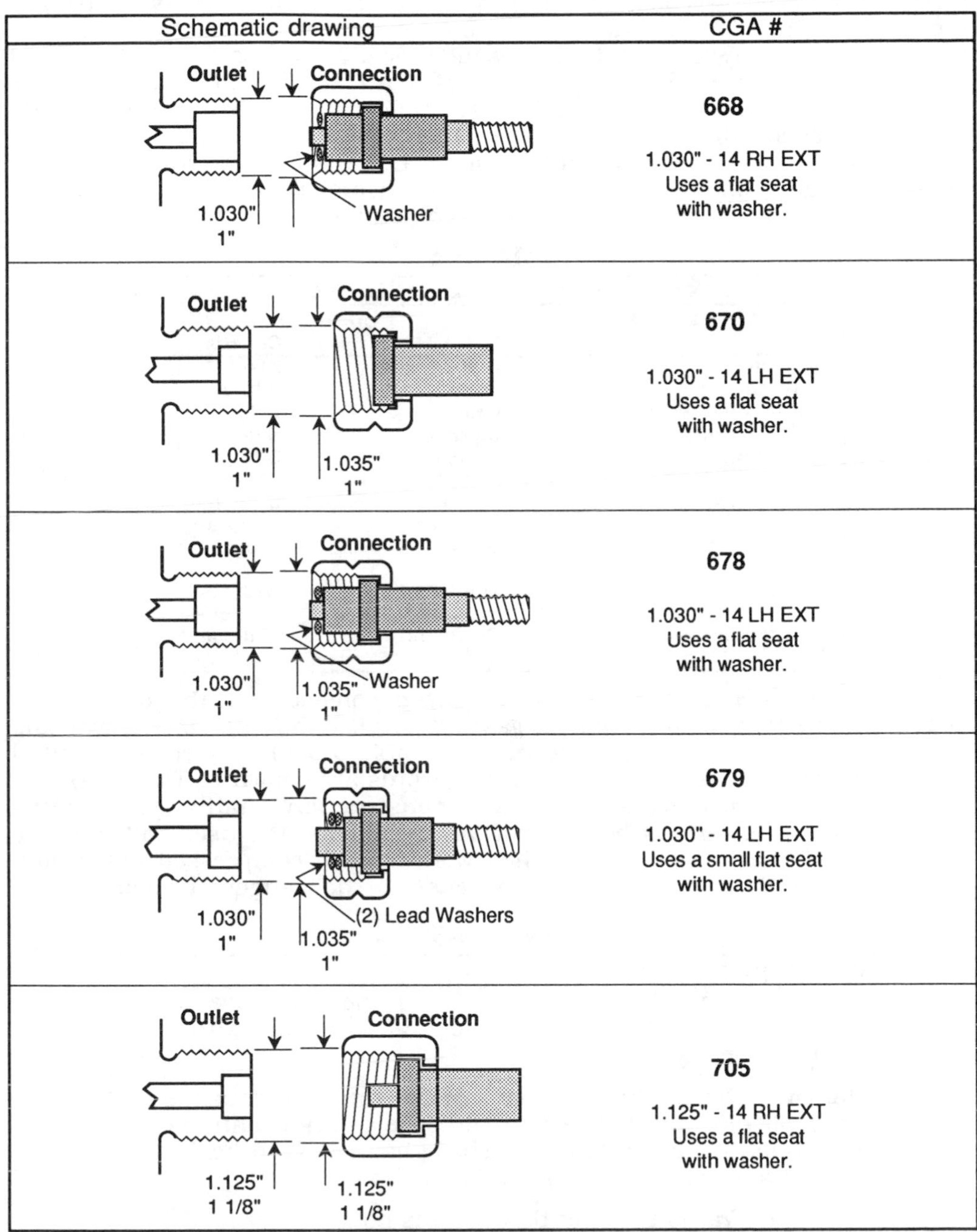

Schematic drawing	CGA #
Outlet, Connection, 1.030", 1", Washer	**668** 1.030" - 14 RH EXT Uses a flat seat with washer.
Outlet, Connection, 1.030", 1", 1.035", 1"	**670** 1.030" - 14 LH EXT Uses a flat seat with washer.
Outlet, Connection, 1.030", 1", 1.035", 1", Washer	**678** 1.030" - 14 LH EXT Uses a flat seat with washer.
Outlet, Connection, 1.030", 1", 1.035", 1", (2) Lead Washers	**679** 1.030" - 14 LH EXT Uses a small flat seat with washer.
Outlet, Connection, 1.125", 1 1/8", 1.125", 1 1/8"	**705** 1.125" - 14 RH EXT Uses a flat seat with washer.

Fig. 5-2 CGA regulator attachment fittings (cont.).

Compressed gas tanks of large industries (or of particular gas distributors) are often color-coded with their own specifications for easy ownership identification and/or gas type recognition. You should never trust the color markings of compressed gas tanks for identification unless the tank is used in medicine. The medical industries of both U. S. and Canada have established color codes for medicinal gas tanks, unfortunately they do not agree with each other (see Table 5-4). Otherwise, always depend on formal markings or labels for gas identification. If there is any question as to the contents of any compressed gas tank, do not use it: return the tank to your gas distributor for identification and/or replacement.

Table 5-4

Color Code of Medical Gas Cylinders

Type of Gas	Color	
	U.S.A.	Canada
Nitrogen	Black	Black
Oxygen	Green	White
Carbon dioxide	Gray	Gray
Nitrous oxide	Blue	Blue
Cyclopropane	Orange	Orange
Helium	Brown	Brown
Carbon dioxide-oxygen	Gray and Green	Gray and White
Helium-oxygen	Brown and Green	Brown and White
Air	Yellow	White and Black
Oxygen-nitrogen (other than air)	Green and Black	White and Black

Although the gas industry has not been consistent with color-coding (with the exception of the medical gas industry), is has agreed on warning signs with three specific *signal words* (see Table 5-5). The specific potential dangers associated with the various signal words are shown in Table 5-6

Regardless of compressed gas tank construction quality, the printed warnings, and the various "idiot proof" arrangements, the user still must be aware of some basic safety rules when working with compressed gases, and be able to apply them with common sense to the dangers of compressed

Table 5-5

DOT Signal Words
Danger: Warns the user that release of the gas in use to the atmosphere would pose an immediate hazard to health and/or property.
Warning: Warns the user that release of the gas in use to the atmosphere would not necessarily pose an immediate hazard, would nonetheless be extremely hazardous to health and/or property under certain conditions.
Caution: Warns the user that there is no danger from the gas in use beyond the dangers associated with using any gas that is under high pressure.

gases. There are two types of dangers when working with compressed gases: the danger presented by the specific gas you are using (flammable, toxic, oxidizing, and so forth) and the danger of having a metal cylinder under tremendous pressure standing a few feet from you. The next sections details potential dangers that can occur during general use and appropriate responses. In addition to the general potential dangers of compressed gas tanks, you should also be knowledgeable to the potential dangers of the gas you are using.

Table 5-6

Signal Words and Their Potential Hazards

Signal Word	Type of Hazard
Danger	Extremely hazardous gas and/or liquid Flammable Harmful if inhaled Poisonous Radioactive
Warning	Corrosive gas or liquid Extremely cold liquid and/or gas under pressure Extremely irritating gas Liquid causes burns Vigorously accelerates burning
Caution	High concentration in the atmosphere can cause immediate asphyxiation High pressure Liquefied gas under pressure

5.1.5 Safety Practices Using Compressed Gases

Always Work in a Well-ventilated Area. Perhaps the most subtle danger when working with compressed gases is asphyxiation. If one is working in an area with poor, or nonexistent, forced ventilation, and there is a leak of a non-harmful gas (non-flammable, non-poisonous), then there will be a resulting decrease of the percentage (within air) of available oxygen.* This decrease can lead to asphyxiation.

The most insidious thing about asphyxiation is that it comes on without warning. There can be feelings of sleepiness, fatigue, lassitude, loss of coordination, errors of judgment, confusion, and even euphoria preceding unconsciousness. Regardless, the victim will not be in a mental state to realize the danger and will be incapable of reacting properly, which involves getting out of the area to better air.

* Oxygen content in the atmosphere is normally about 21%. Because the partial pressure varies with the atmospheric pressure, the percentage of oxygen at higher altitudes will decrease relative to sea level. The flame of a candle will be extinguished when oxygen concentration is reduced to about 15 to 16%. A human will be rendered unconscious when oxygen concentration is reduced to about 12%. Prolonged exposure within a low oxygen percentage environment can lead to brain damage or even death.

IN CASE OF EMERGENCY: Anyone going into a room suspecting low oxygen concentrations should 1) arrange for a buddy system, and 2) bring in portable breathing equipment. Gas masks or filters will not help in an oxygen-poor environment. Obviously, if you see someone lying on the floor there is a desire to run and help them. Unfortunately that only places you in equal danger. Under no circumstances go into the room without calling the fire department, getting help, and attaching a rope onto yourself so your backup can drag you out.

Because most modern labs have adequate ventilation, asphyxiation is seldom a problem. However, when work is being done in a basement, in rooms that do not have an adequate ventilation system, or when a gas is heavier than air, every precaution should be taken such as the use of the buddy system, having portable air supplies available, and intercom systems with regular check-in times.

Always Secure Compressed Gas Tanks. A compressed gas tank should never be allowed to stand free. Instead, it should be supported by an approved tank support and the screws should be tightened by a wrench, not just by hand. If lying horizontal, the compressed gas tank should be prevented from rolling.

Just about everyone has heard stories of compressed gas tanks being knocked over, thus snapping off the main valve, turning the compressed gas tank into a rocket. The Compressed Gas Association claims that this orifice is too small for the tank to become a projectile. However, if the tank contains a flammable gas, the question of whether the tank will turn into a rocket is of secondary importance.

Even when a compressed gas tank is empty, it should be strapped down. The standard compressed tank used in most laboratories weighs about one hundred pounds and could be quite destructive if it were to fall over onto something or someone.

Keep Electric Lines Free from Compressed Gas Tanks. Compressed gas tanks are made of metal. Keeping electric lines away from compressed gas tanks helps to prevent spark development and potential shock to the user.

If you should come upon someone who is frozen by electric shock to an electrified compressed gas tank, do not handle the person with your bare hands. First, if possible, quickly ascertain the source of the electricity and if it is safe and easy to stop the current, do so. Otherwise, be sure you are not standing on, or in, water, and use something that is nonconducting to pull the victim from the electrified tank. Call for medical attention immediately.

If a tank regulator is connected to metal tubing that in turn is connected to any type of equipment, the tubing should be grounded (for example, to metal piping). This grounding can be done with bushing wire, or any heavy-gauge wire, and a screw clamp.

Keep Compressed Gas Tanks Away from High Heat. Compressed gas cylinders should not be subjected to atmospheric temperatures above 130°F. A compressed gas cylinder should never be subjected to a direct flame. If defrosting an iced or frozen tank, water above 130°F should not be used. If compressed gas tanks need to be left outside, they should be kept in a shaded environment.

Compressed Gas Tanks Should be Properly Moved. Compressed gas tanks should never be dragged, slid, or allowed to bang against one another. One should use a proper hand-truck or other suitable device for transporting the specific type of compressed gas tank. Compressed gas tanks should never be lifted by their caps, by magnets, or by ropes, chains, or slings unless the manufacturer has provided attachments (such as lugs) on the tanks.

Compressed Gas Tanks Should Be Properly Stored. When a variety of different gases are to be stored together, they should be separated by type. Combustible gases should never be stored directly next to sources of oxygen. They should also be separated as to whether they are full or empty. After a tank has been emptied,* it should be marked to make its status easily identifiable. One technique is to take a chalk, or crayon, and write "M T" on the tank signifying that the tank is "empty." Unfortunately, not all gas companies wipe this mark off and after the tank is returned, a full tank could therefore be displayed as "empty." A better technique is to loosely tape a piece of paper saying "empty" onto the tank. This paper will be removed by the gas company. Some compressed gas suppliers provide tags that can be hung on tanks which provide convenient identification of the gas supply (see Fig. 5-3). If you cannot obtain these tags from your supplier, they are very easy to make.

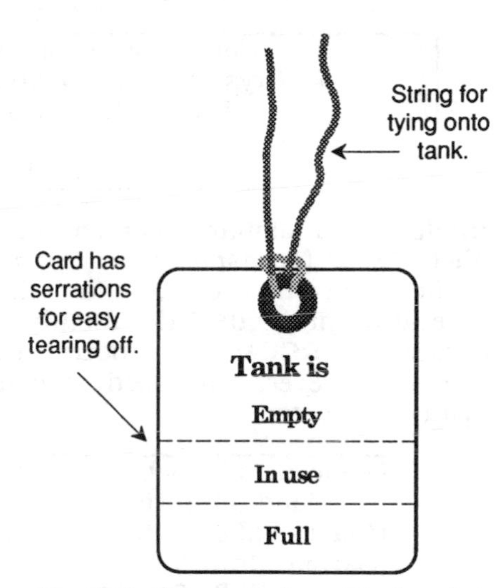

Fig. 5-3 Suggested card to be placed on compressed gas tanks.

When you empty a tank, other people's confusion and frustration can be prevented if *you* detach the regulator and identify the tank as empty before you call your gas people requesting a fresh tank. Likewise, if you need to remove the regulator on a partially-emptied compressed gas tank, replace the protective, and label the tank to indicate it is not a full tank, noting the remaining partial pressure.

Although it is not dangerous to store acetylene tanks on their side, such positions may allow solvent loss, which in turn can decrease the flame quality.

* A tank with less than 50 psig should be considered empty. It is not recommended to totally empty a tank because a completely emptied tank can develop condensation that can prematurely age the tank.

5.1.6 In Case of Emergency

In the U. S. there is an organization called **CHEMTREC** (Chemical Transportation Emergency Center) that has a 24-hour toll-free number to call for advice on chemical transportation emergencies. It is **1-800-424-9300**. For written or general questions, write:

Chemical Transportation Emergency Center 1825 Connecticut Ave. N.W., Washington D.C. 20009 or call 1(202) 887-1255

For questions specifically related to compressed gases, you may write or call:

Compressed Gas Association, Inc. Crystal Gateway Number 1, Suite 501 1235 Jefferson Davis Highway Arlington, VA 22202 1(703) 979-0900

Canada has a similar organization for chemical transportation emergencies called **TEAP** (Transportation Emergency Assistance Plan). It is operated as a public service by the Canadian Chemical Producer's Association through the cooperation and aid of the member companies who operate the Regional Control Centers (**RCCs**) throughout Canada. They are also available on a 24-hour basis. However, you need to call the center in the Canadian region where you are located.

Valleyfield, Quebec	514-373-8330
Maitland, Ontario	613-348-3616
Niagra Falls, Ontario	416-356-8310
Sarnia, Ontario	519-339-3711
Copper Cliff, Ontario	705-682-2881
Edmonton, Alberta	403-477-8339
Vancouver, British Columbia	604-929-3441

For further information about handling and/or shipping pressurized containers within Canada you can write to:

The Canadian Chemical Producers Association 350 Sparks Street, Suite 505 Ottawa KIR 758

5.1.7 Gas Compatibility with Various Materials

Gases can react with containers and tubing as well as with each other. The degree of reaction can range from premature aging to explosion. It is always best to prevent a possible reaction than to repair the damage of one. For example, Table 5-7 indicates that chlorine will not react with an O-ring made of Viton, but will with the others listed. Table 5-7 also indicates that acetylene will react with zinc and copper, but does not indicate that acetylene can also react with silver, mercury, or the salts of either of these compounds. Because it is impossible to list all gas and material combinations, you should check with your gas supplier for compatibilities of any gas and material before committing yourself to full-fledged experimentation, even though the information should be on the MSDS sheets.

Table 5-7‡

Compatibility of Various Gases and Materials*

Chemical Name and Formula		**Materials of Construction** Common Chemical Group**																	
		Metals							Plastics						Elastomers				
Name	**Formula**	a	b	c	d	e	f	g	h	i	j	k	l	m	n	o	p	q	r
Acetylene	C_2H_2	C1	S	S	I	U	U	S	S	S	S	S	I	I	S	C2	C2	C2	C2
Air	—	S	S	S	S	S	S	S	S	S	S	S	S	S	S	S	S	S	S
Allene	C_3H_4	S	S	S	S	I	U	S	S	S	S	S	I	I	S	S	S	S	I
Ammonia	NH_3	U	S	S	S	U	U	S	S	S	S	U	S	U	C3	U	S	S	U
Argon	Ar	S	S	S	S	S	S	S	S	S	S	S	S	S	S	S	S	S	S
Arsine	AsH_3	S	S	S	C5	I	S	S	S	S	S	S	S	I	S	S	S	S	U
Boron Trichloride	BCl_3	U	S	S	I	I	S	S	S	S	S	I	S	I	C3	I	I	I	I
Boron Trifluoride	BF_3	S	S	S	S	I	I	S	S	S	S	I	S	I	C3	I	I	I	I
1-3 Butadiene	C_4H_6	S	S	S	S	S	S	S	S	S	S	S	S	U	S	S	U	S	U
Butane	C_4H_{10}	S	S	S	S	S	S	S	S	S	S	S	S	U	S	S	S	S	S
1-Butene	C_4H_8	S	S	S	S	S	S	S	S	S	S	S	S	U	S	S	S	S	S
Cis-2-Butene	C_4H_8	S	S	S	S	S	S	S	S	S	S	S	S	U	S	S	S	S	S
Trans-2-Butene	C_4H_8	S	S	S	S	S	S	S	S	S	S	S	S	U	S	S	S	S	S
Carbon Dioxide	CO_2	S	S	S	S	S	S	S	S	S	S	S	S	S	S	S	S	S	S
Carbon Monoxide	CO	S	S	S	S	S	S	S	S	S	S	S	S	S	S	I	S	S	S
Carbonyl Sulfide	COS	S	S	S	S	I	S	S	S	S	S	S	S	I	I	S	I	I	I
Carboxide®	—	C4	S	S	I	I	U	I	S	S	I	I	U	U	C3	U	U	U	U
Chlorine	Cl_2	U	S	S	U	U	U	S	S	S	S	S	U	U	S	S	U	U	U
Deuterium	D_2	S	S	S	S	S	S	S	S	S	S	S	S	I	S	S	S	S	S
Diborane	B_2H_6	S	S	S	S	I	S	S	S	S	S	I	I	I	S	I	I	I	I
Dichlorosilane	H_2SiCl_2	I	S	S	I	I	I	S	S	S	S	S	I	I	S	I	I	I	I
Dimethyl Ether	$(CH_3)_2O$	S	S	S	S	S	S	S	S	S	S	S	S	U	S	S	S	S	I
Ethane	C_2H_6	S	S	S	S	S	S	S	S	S	S	S	S	I	S	S	S	S	S
Ethyl Acetylene	C_4H_6	I	S	S	S	I	U	S	S	S	I	S	I	I	S	S	I	S	I
Ethyl Chloride	C_2H_5Cl	S	S	S	U	I	S	S	S	S	S	S	U	U	S	S	S	S	U
Ethylene	C_2H_4	S	S	S	S	S	S	S	S	S	S	S	I	I	S	S	S	S	I
Ethylene Oxide	C_2H_4O	C4	S	S	C5	I	U	I	S	S	I	I	U	U	C3	U	U	U	U
Halocarbon 11	CCl_3F	S	S	S	C5	I	S	S	S	S	S	S	U	U	C3	S	S	U	U
Halocarbon 12	CCl_2F_2	S	S	S	C5	I	S	S	S	S	S	S	U	U	C3	S	S	S	S
Halocarbon 13	$CClF_3$	S	S	S	C5	I	S	S	S	S	S	S	U	U	C3	S	S	S	S
Halocarbon 13B1	CDF_3	S	S	S	C5	I	S	S	S	S	S	S	U	U	C3	S	S	S	S
Halocarbon 14	CF_4	S	S	S	C5	I	S	S	S	S	S	S	U	U	C3	S	S	S	S
Halocarbon 21	$CHCl_2F$	S	S	S	C5	I	S	S	S	S	S	S	U	U	C3	U	U	S	S
Halocarbon 22	$CHClF_2$	S	S	S	C5	I	S	S	S	S	S	S	U	U	C3	U	U	S	U
Halocarbon 23	CHF_3	S	S	S	C5	I	S	S	S	S	S	S	U	U	C3	I	I	I	S
Halocarbon 113	CCl_2FCClF_2	S	S	S	C5	U	S	S	S	S	S	S	U	U	C3	S	S	S	S
Halocarbon 114	$C_2Cl_2F_4$	S	S	S	C5	I	S	S	S	S	S	S	U	U	C3	S	S	S	S
Halocarbon 115	C_2ClF_5	S	S	S	C5	I	S	S	S	S	S	S	U	U	C3	S	S	S	S
Halocarbon 116	C_2F_6	S	S	S	C5	I	S	S	S	S	S	S	U	U	C3	I	I	I	S
Halocarbon 142B	$C_2H_3ClF_2$	S	S	S	C5	I	S	S	S	S	S	S	U	U	C3	U	S	S	S

Table 5-7

Compatibility of Various Gases and Materials* (cont.)

Chemical Name and Formula		Materials of Construction** Common Chemical Group																	
		Metals							Plastics						Elastomers				
Name	Formula	a	b	c	d	e	f	g	h	i	j	k	l	m	n	o	p	q	r
Halocarbon 152A	$C_2H_4F_2$	S	S	S	C5	I	S	S	S	S	S	S	U	U	C3	U	S	S	S
Halocarbon C-318	C_4F_8	S	S	S	C5	I	I	S	S	S	S	S	U	U	C3	S	S	S	S
Halocarbon 502	$CHClF_2$/ $CClF_2$-CF_3	I	S	S	C5	I	I	S	S	S	I	S	U	U	C3	S	S	S	S
Halocarbon 1132A	$C_2H_2F_2$	S	S	S	C5	I	S	S	I	S	S	S	U	U	C3	I	I	I	S
Helium	He	S	S	S	S	S	S	S	S	S	S	S	S	S	S	S	S	S	S
Hydrogen	H_2	S	S	S	S	S	S	S	S	S	S	S	S	S	S	S	S	S	S
Hydrogen Chloride	HCl	U	S	S	I	U	U	S	S	S	S	S	S	U	S	S	U	U	U
Hydrogen Sulfide	H_2S	U	S	S	S	I	I	S	S	S	S	S	S	S	S	U	S	S	S
Isobutane	C_4H_{10}	S	S	S	S	S	S	S	S	S	S	S	S	U	S	S	S	S	S
Isobutylene	C_4H_8	S	S	S	S	I	S	S	S	S	S	S	S	I	S	S	S	S	I
Isopentane	C_5H_{12}	S	S	S	S	S	S	S	S	S	S	S	S	U	S	S	S	S	S
Krypton	Kr	S	S	S	S	S	S	S	S	S	S	S	S	S	S	S	S	S	S
Methane	CH_4	S	S	S	S	S	S	S	S	S	S	S	S	I	S	S	S	S	S
Methyl Chloride	CH_3Cl	S	S	S	U	U	S	S	S	S	S	S	I	I	S	S	U	U	U
Methyl Mercaptan	CH_3SH	S	S	S	U	I	U	U	S	S	S	I	I	I	S	I	I	S	I
Natural Gas	—	S	S	S	S	S	S	S	S	S	S	S	S	I	S	S	S	S	S
Neon	Ne	S	S	S	S	S	S	S	S	S	S	S	S	S	S	S	S	S	S
Nitric Oxide	NO	U	S	S	S	I	S	S	S	S	S	I	S	I	S	I	I	S	I
Nitrogen	N_2	S	S	S	S	S	S	S	S	S	S	S	S	S	S	S	S	S	S
Nitrogen Dioxide	NO_2	I	S	S	S	I	I	S	S	S	S	I	U	I	S	U	U	U	U
Nitrous Oxide	N_2O	S	C6	C6	C5	S	S	S	S	C5	S	S	S	I	C3	S	S	S	S
Oxyfume® mixtures	—	C4	S	S	I	I	U	I	S	S	I	I	U	U	C3	U	U	U	U
Oxygen	O_2	S	C7	C7	U	S	S	S	S	C5	S	S	S	S	C3	S	U	U	S
Perfluoropropane	C_3F_8	S	S	S	S	I	S	S	S	S	S	I	I	I	I	I	S	S	I
Phosphine	PH_3	I	S	S	S	I	I	S	S	S	S	I	I	I	S	I	I	I	I
Phosphorous Pentafluoride	PF_5	I	S	S	I	I	I	S	S	S	S	I	I	I	I	I	I	I	I
Propane	C_3H_8	S	S	S	S	S	S	S	S	S	S	S	S	U	S	S	S	S	S
Propylene	C_3H_6	S	S	S	S	S	S	S	S	S	S	S	S	U	S	S	U	U	U
Propylene Oxide	C_3H_6O	I	S	S	I	I	I	I	S	S	S	I	U	S	C3	U	U	U	U
Refrigerant Gases	—	(see halocarbons)																	
Silane	SiH_4	S	S	S	S	I	S	S	S	S	S	S	S	I	S	S	S	S	S
Silicon Tetrachloride	$SiCl_4$	I	S	S	U	I	I	S	S	S	I	I	U	I	C3	I	I	I	I
Silicon Tetrafluoride	SiF_4	S	S	S	S	I	S	S	S	S	S	S	S	I	C3	S	S	S	S
Sulfur Dioxide	SO_2	U	S	S	U	U	U	S	S	S	S	S	S	U	S	S	U	U	S
Sulfur Hexafluoride	SF_6	S	S	S	S	I	S	S	S	S	S	S	S	I	C3	S	S	S	S
Trichlorosilane	$HSiCl_3$	I	S	S	U	I	I	S	S	S	I	I	U	I	C3	I	I	I	I
Vinyl Methyl Ether	C_3H_6O	S	S	S	S	I	U	S	S	S	S	I	I	U	C3	I	I	I	I
Xenon	Xe	S	S	S	S	S	S	S	S	S	S	S	S	S	S	S	S	S	S

‡ From the "Compatibility Chart" in the Specialty Gases Catalogue, Vol. 25, by LINDE, © 1991 by the Union Carbide Corporation. With permission.

* This chart has been prepared for use with dry (anhydrous) gases at normal operating temperatures of 70°F (21°C). Information may vary if different operating conditions exist. Systems and equipment used in oxidizer gas service (e.g., oxygen or nitrous oxide) must be cleared for oxidizer service.

S = Satisfactory for use with the intended gas (dry anhydrous) at a normal operating temperature of 70°F (21°C).

U = Unsatisfactory for use with the intended gas.

C1 through **C7** - Conditionally acceptable for use with the intended gases as follows:

- **C1** - Satisfactory with brass having a low (67-70% maximum) copper content. Brass with higher copper content is unacceptable.
- **C2** - Satisfactory with acetylene; however, cylinder acetylene is packaged dissolved in a solvent (generally acetone) which may be incompatible with these elastomers.
- **C3** - Compatibility varies depending on specific Kalrez compound used. Consult E.I. DuPont for information on specific applications.
- **C4** - Satisfactory with brass, except where acetylene or acetylides are present.
- **C5** - Generally unsatisfactory, except where specific use conditions have proven acceptable.
- **C6** - Satisfactory below 1000 psig.
- **C7** - Satisfactory below 1000 psig where gas velocities do not exceed 30 ft/sec.

I - Insufficient data available to determine compatibility with the intended gas.

**Materials of Construction are as follows:

	<u>**Metals**</u>			<u>**Plastics**</u> **(cont.)**	
a	=	Brass	**j**	=	Tefzel
b	=	303 Stainless Steel	**k**	=	Kynar
c	=	316 Stainless Steel	**l**	=	Polyvinylchloride
d	=	Aluminum	**m**	=	Polycarbonate
e	=	Zinc		<u>**Elastomers**</u>	
f	=	Copper	**n**	=	Kalrez
g	=	Monel	**o**	=	Viton
	<u>**Plastics**</u>		**p**	=	Buna-N
h	=	Kel-F	**q**	=	Neoprene
i	=	Teflon	**r**	=	Polyurethane

5.2 The Regulator

5.2.1 The Parts of the Regulator

A compressed gas tank, when filled, may contain up to 2,500 lbs/in^2 of pressure. These tremendous pressures cannot be throttled by the main tank valve to pressures that are usable in the laboratory. All the main tank valve can do is release or prevent the release of gas, not control the pressure with which a gas is released. A regulator is required to reduce the gas pressure in a controlled, stable, consistent, and reliable manner.

There can be one valve and 1 or 2 gauges on a standard high-pressure regulator. Fig. 5-4 displays all the possible gauges and valves which are described in the following subsections.

The Compressed Tank Valve or Main Valve (V - 1). This valve is not part of the regulator, but is part of the compressed gas tank itself. It rotates <u>CCW for open</u> and <u>CW for closed</u>. All this valve can do is let gas out (or in for filling). It cannot vary the pressure of the releasing gas. If the tank valve is to be used for a long period of time (> several hours), it is best to leave this valve completely opened. This position presses part of the valve against an internal seal to prevent leakage past the valve stem. When using the compressed gas tank for short periods of time, the amount of leakage is negligible, and several revolutions of this valve are sufficient. Never open the main valve of an acetylene tank more than one and one-half revolutions, and always open it slowly.

The Tank Pressure Gauge (P - 1). This gauge displays the pressure inside the tank. It represents the pressure coming into the regulator. This gauge is not necessarily on all systems, especially those that have a constant pressure, such as a *house air pressure System* (see below). This gauge is used to measure the pressure remaining in the pressure tank. Thus, if you notice you are low on gas, you can make arrangements for tank replacement before

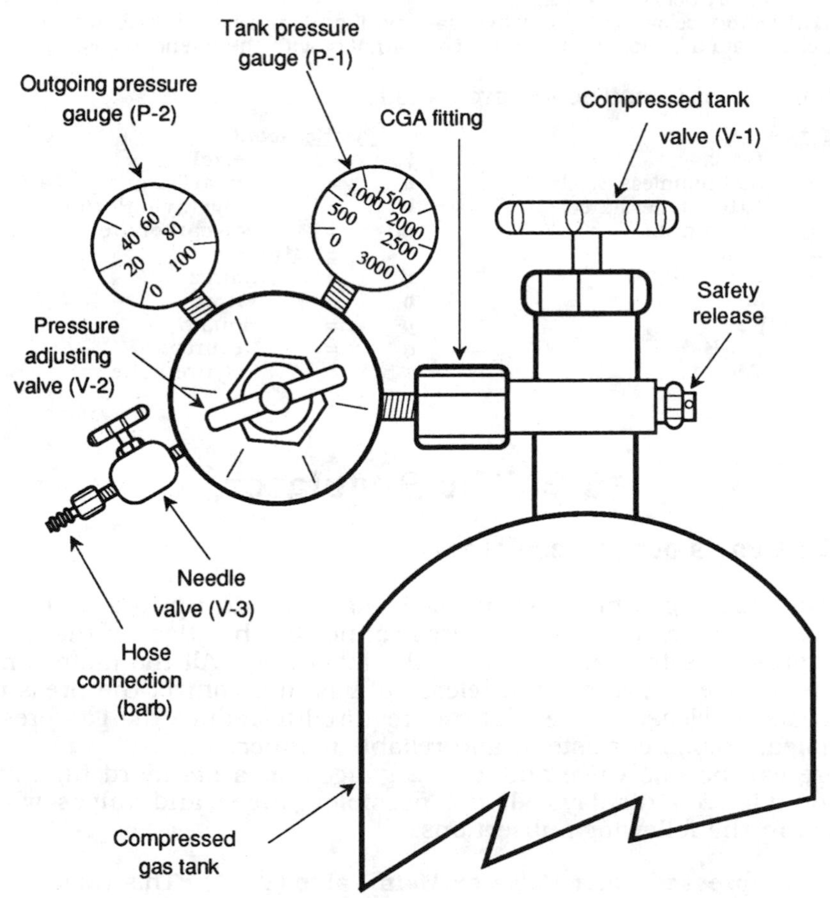

Fig. 5-4 Parts of the regulator.

an experiment has begun (and thereby avoid problems or an emergency situation). It is a good idea to observe the amount of gas you use per experiment or gas you use per day to better estimate your needs. This practice can prevent the down-time of waiting for a fresh tank caused by poor planning.

Gauges on liquefied gas tanks, such as propane, cannot indicate the remaining gas or the amount of gas used over a period of time. These tank pressure gauges only indicate the pressure of the gas (above the liquid) within the tank known as the head pressure.

The Pressure Adjusting Valve (V - 2). This valve adjusts the pressure that leaves the regulator. It rotates *CCW to decrease* pressure and *CW to increase* pressure. Before opening the tank valve (V - 1), rotate the valve CCW until it turns free. Open the tank valve, then rotate the pressure-adjusting valve CW to the desired working pressure. **Note**: The pressure from the regulator cannot be decreased unless gases are being released from the system. Therefore make sure that the needle valve (V - 3), and any other gas flow valves, are open to the atmosphere at least a small amount when attempting to decrease the regulator pressure.

The Outgoing Pressure Gauge (P - 2). This gauge displays the pressure that is set by the pressure adjusting valve (V - 2). If you are trying to decrease the pressure and do not see any change, be sure the system beyond the regulator is open to the atmosphere (see previous subsection).

The Needle (Flow Control) Valve (V - 3). This valve controls the volume of gas leaving the regulator. It rotates <u>CCW to open</u> and <u>CW to close</u>. If the regulator is connected to a glass system and/or rubber tubing, this valve should be closed when first opening up the pressure adjusting valve (V - 2) to prevent damage to the system from too great a pressure. The needle valve (V - 3) should be fully open if the system has a second needle valve further down the system, such as the valve of a glassblowing torch. The needle valve is not on all regulators, but can easily be added or removed.

The Hose Connection. This connection is called a barb by the welding and gas industry and is seldom on the regulator when purchased. Like the needle valve it can be easily added or removed. Copper or high-pressure tubing can be obtained and attached directly to the regulator with a fitting which provides a more secure system than that which can be obtained by forcing a plastic or rubber tube over a hose connection.

There are two types of regulators: single- and double-stage. Both work on the same principle of gas pressure pushing on one side of a diaphragm, and the pressure adjusting valve pressing a spring against the other side of the valve. The greater the spring pressure against the diaphragm, the greater the gas pressure available to leave the regulator. However, in a single-stage regulator, as the pressure within the tank goes down, you must back off the spring pressure (using the pressure adjusting valve) to maintain the same outgoing pressure. Otherwise the outgoing pressure slowly increases as the tank pressure decreases. The two-stage regulator avoids this problem by having two single-stage regulators linked end-to-end. The first maintains an outgoing pressure between 350–500 psig, which is the incoming pressure of the second. Because the pressure between the first and the second cannot vary until the tank pressure is below the outgoing pressure of the first stage, constant outgoing pressure from the second stage must be maintained.

5.2.2 House Air Pressure System

It is common for labs to have a high air pressure system built into benches, fume hoods, and other parts of the building. Sometimes the air pressure is too powerful for equipment to be attached. A simple regulator (see Fig. 5-5) can be attached directly on such a system using pipe fittings.

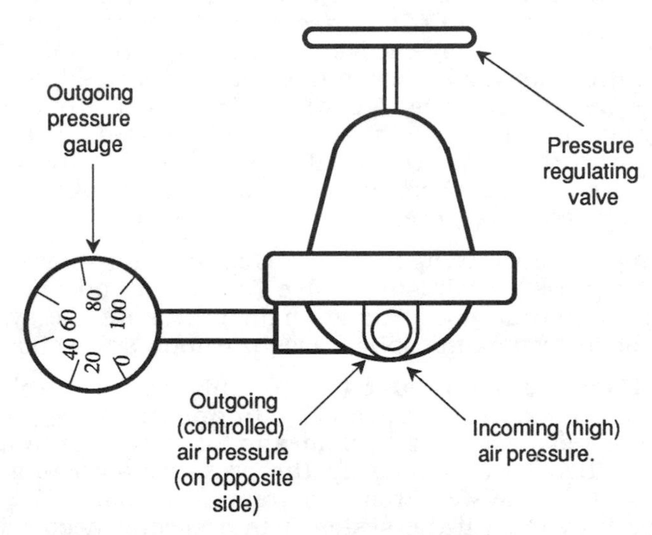

Fig. 5-5 Typical regulator used with a constant gas source such as house compressed air.

5.2.3 How to Install a Regulator on a Compressed Gas Tank

A few minutes of preparation will save you hours of clean-up time, replacement of equipment, and/or serious injury. Therefore, be sure to follow this general procedure:

1) Examine the edge of the regulator CGA fitting to determine whether the thread is right- or left-handed (see Fig. 5-6). Use an open end wrench, or large crescent wrench, to remove the regulator from the emptied tank.
2) Examine the CGA threads of the tank and of the regulator to be sure that they are clean of dirt and dust and free of grease. [If there is any grease on the threads of the tank, return the tank to the supplier. If there is any grease on the threads of the regulator, return it to an authorized repair dealer.]
3) Strap the tank securely into its proper position.

4) Stand to the side of the tank and quickly crack open and close the main tank valve. This action will force out any dust, dirt, or other particulate matter from the valve seating before you attach the regulator. This matter, if not removed, can get into the regulator and either shorten the regulator's life span, or get into the area where you are trying to deliver 'clean' gas. Do not stand in front of the valve port and be sure that the valve port is not aimed at anyone or anything. The force of air can easily blow papers or glassware a considerable distance.
5) Attach the regulator to the new tank tightly with a wrench as mentioned in step 1. Be sure to follow the direction of the threads. The fitting should be snug.

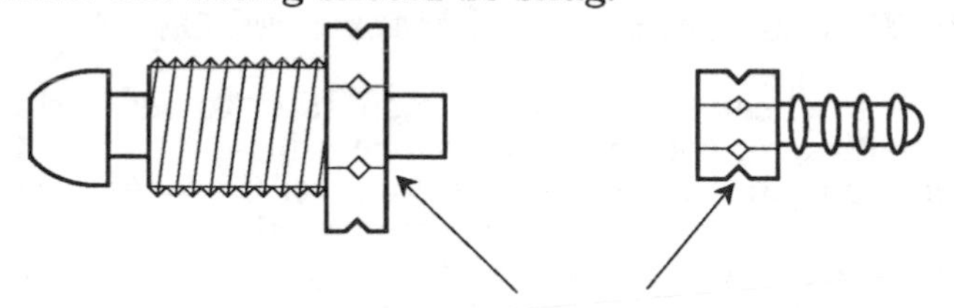

Left-handed nuts and fittings have a
notch cut into the corners to help
identify the direction of the thread.

Fig. 5-6 Samples of left-handed nuts and fittings.

5.2.4 How to Use Regulators Safely

Always open the main tank valve slowly. Intense pressure against the regulator's diaphragm can cause premature wear and aging. In addition, after the main tank valve has been turned off, the regulator should not remain under pressure for extended periods of time (i.e., several hours or beyond). The procedure of closing the tank valve (V - 1) and then bleeding the system to release any remaining gases will help extend the life of the regulator. This bleeding procedure is more critical with two-stage regulators which may trap a small quantity of gas within the first of their two stages. This gas can then burst out unpredictably and potentially dangerously.

Another practice for extending the life of a regulator is to unscrew (CCW) the pressure adjusting valve (V - 2) before opening the tank valve (V - 1), and before setting the regulator to the desired pressure. In addition, open the tank valve slowly to limit the amount of shock on the diaphragm. Continued shock on the diaphragm causes premature aging.

When opening the main tank valve to allow gas into the regulator, you should stand to the side (not in front or behind) of the gauges, preferably with the main tank valve between you and the regulator. If an old worn diaphragm bursts during the sudden pressure increase, glass and metal shards could spray out. This damage cannot happen if the pressure adjusting valve is unscrewed before you open the main valve because there is no pressure against the diaphragm.

Never use oil in or around a regulator. Oxygen regulators (CGA fitting 540) pose special dangers because oxygen decreases the ignition temperature of flammable materials. Oil around compressed oxygen can explode. Do not use plumbing 'pipe dope' on a CGA fitting. If a CGA fitting needs 'pipe dope,' there is something wrong with the fitting and it should not be used.

Do not use a regulator that has been dropped or shows any signs of physical abuse. If in doubt, have the regulator checked at the manufacturer's service center.

When storing a regulator, place it in a plastic bag to prevent dirt and dust from settling on, and in, the regulator. Dust and dirt can damage the diaphragm, or wedge within the sections, preventing complete closing of the diaphragm parts. Regulator cleaning and repair should be done by a service agency, and should not be done by inexperienced personnel.

5.2.5 How to Test for Leaks in a Compressed Gas System

The first leak to check is within the regulator: To see if there is gas leakage past the diaphragm, rotate the pressure adjusting valve CCW until it is fully open, then slowly open the main tank valve. If any gas leaves the regulator, the regulator is defective or broken and needs repair or replacement.

To see if you have a leak within the line of your pressure system, close all normal outlets. Then, after opening the main tank valve and rotating the pressure adjusting valve (V - 2) to the desired pressure, turn the main tank valve off. If you see a dropping off of pressure on the tank pressure gauge, you have a leak somewhere in your system.

To limit the region needing examination, rotate the pressure adjusting valve CCW one or two revolutions. If the reading on the tank pressure gauge drops, there is a leak in the compressed tank valve, the CGA fitting, or the tank pressure gauge. However, if the outgoing pressure gauge drops, there is a leaky outlet fitting, needle valve (if any), outgoing pressure gauge, hose, tubing leaving the regulator, or any other part of the system beyond the regulator. Finally, if the tank pressure gauge drops and the outgoing pressure gauge rises, there is a leak in the regulator itself. It is normal to see this last pressure rise when you are bleeding the pressure from a single-stage regulator just as the pressure in the regulator is almost extinguished.

If you know you have a leak but are not sure of the location, **do not** use a flame to aid your search.* If the gas is flammable, the dangers are obvious, but otherwise there is concern of burning parts and equipment. The best, and safest, technique is to spray, squirt, or drip a soapy water solution on the suspected area. Use either a diluted liquid dish soap** or a commercial solution such as Snoop®.*** The evidence of bubbles is a sure sign of a leak. However, be sure you witness bubble formation as opposed to bubbles just sitting there, which are likely to have formed during the application of the bubble solution.

5.2.6 How to Purchase a Regulator

Most regulators can provide uniform and accurate high-pressure output. Where regulators vary is in their output potential. A large gas demand re-

* Some people use the flickering of a flame as an indication of a leak.
** Liquid dish soap can corrode stainless steel, so rinse it off when completed.
*** SNOOP®, Product of the Nupro® Company, Willoughby, OH 44094.

quires a regulator with a large output potential. By selecting a regulator that only satisfies the minimum flow rate for your expected needs, you may be unable to achieve unexpected high-flow demands.

A wailing noise coming from your regulator is a symptom of using a large quantity of gas at a rate faster than the diaphragm within the regulator is capable of supplying. Unless you are able to use a smaller quantity of gas, the best solution is to purchase a larger regulator with a higher flow capacity (typically a two-stage regulator) and/or a better quality regulator.

If you absolutely need constant outgoing pressure, regardless of varying tank pressure or the varying rate of outgoing gases, select a two-stage regulator; otherwise, a single-stage is sufficient. Although the changes are subtle, single-stage regulators have less stable outgoing pressures as tank pressure drops or as you increase the rate of outgoing gases.

If you are using a corrosive gas, it is imperative to use a stainless steel regulator. Otherwise, a brass regulator is sufficient.

When you order a regulator, expect to receive only the regulator. Unless you specify that you also want a needle valve and/or a barb (hose connection), you probably will not get one! Be sure to ask the salesperson what you will be receiving with the regulator. This information becomes critical if you are ordering a regulator for a tank that has a left-handed CGA fitting such as a propane tank, because the fittings are also left-handed. Therefore, if you wish to add a hose connection onto the regulator after it has arrived at your lab, you might have difficulty finding a left-handed hose connection. You probably will have to go back to your mail-order supplier, or to a local welding supply house.

Chapter 6

High and Low Temperature

6.1 High Temperature

6.1.1 The Dynamics of Heat in the Lab

Heat is used in the laboratory for a variety of applications which include: speeding up chemical reactions, evaporating solvents, facilitating crystallization, softening or melting materials, and distillation by bringing chemicals to their vapor points.

To get heat from one point to another, heat (thermal energy) is transferred by three thermal processes: *conduction*, *convection*, or *radiation-adsorption*, or some combination of the three.

1) ***Conduction*** is the transfer of heat from one body to another by molecules (or atoms) being in direct contact with other molecules (or atoms). This process is how hot coffee heats a coffee cup.
2) ***Convection*** is the physical motion of material. An example of this process would be hot air rising and cold air settling. Rigid materials cannot have convection.
3) ***Radiation—Absorption*** is the result of heat energy being transformed into radiant energy. This energy process that gives us a sun tan.

There are five ways to heat materials in the lab: *open flame*, *steam*, *thermal radiation*, *passive electrical resistance* (such as hot air guns), and *direct electrical resistance* (such as hot plates). All of these heating methods (except thermal radiation) use conduction to heat the container holding the material to make the material hot.

6.1.2 General Safety Precautions

General safety precautions are important when dealing with hot materials. Because heat increases the activity level of chemicals, the chemicals become more dangerous when heated (nitric acid is dangerous, hot nitric acid is very dangerous). When you add the fact that heated chemicals can splatter or be ejected from a container, the dangers are compounded. In addition, the heat source and heated containers can also be dangerous.

Standard safety equipment is a must: eye or face protection, thermal gloves, lab coat, and closed-toe shoes should all be worn. Tongs, tweezers, or test tube holders should be used to transport heated containers.

There are other dangers present with the act of heating materials. They include:

1) *Toxic or dangerous gases*, produced from the material(s) being heated.
2) *Explosions*, caused by pressure buildup of trapped gases.
3) *Breakage and explosions*, caused by faulty or damaged apparatus or heating devices.
4) *Flash fires*, caused by flammable fumes ignited by a spark or flame.

Because of these potential dangers, heating operations should be done within a fume hood. The windows of the fume hood should have either tempered glass and/or plastic film coatings. The door of the fume hood should always be closed as low as possible, especially when in use. Place your hands underneath the door to work while protecting your face behind the window. However, do not use the fume hood door in place of eye protection—use both.

Unfortunately, a fume hood is not always available or practical. If you are unable to work in a fume hood but still require shielding, portable explosion shields are available and should be used.

Experiments or processes that use heat must constantly be monitored. If you need to leave the lab, even for a short time, be sure that someone else can monitor the operation in your absence. Problems such as chemicals boiling over, or the evaporation of materials from a container, can occur. In addition, water hoses that lead to condensers have a habit of slipping off or being turned off when no one is looking. To remedy this problem, the use of relays that turn off hear sources are recommended.

If you ever find a dry container on its heat source, it is likely to have developed thermal strain. In this condition, it is dangerous and likely to crack or break into pieces without any notice or warning. Unless you have the facilities to examine for strain, or to properly anneal the potentially strained glassware, throw it away (see Sec. 1.1.10).

6.1.3 Open Flames

Period movies, such as those starring Sherlock Holmes, often showed an oil lamp, always burning, always ready for heating a test tube. Contemporary movies now show a Bunsen, or Fisher, burner (see Fig. 6-1) in the lab, always burning, always ready for heating a test tube. If, as a child you had a student chemistry kit, it probably came with an oil lamp. The mere fact that you had it burning away on your kitchen table was proof to your younger siblings that you were doing chemistry.

While it is true that a burning flame provides good images and great ambiance, it is not a very good source of heat. You will not see open flames burning away in a laboratory for effect or ambiance. If an open flame is used for heating solutions in the laboratory, it is always shut off after use.

Fortunately, the use of oil lamps is now neither common nor necessary. Although there are modern substitutes for oil lamps, such as Bunsen and Fisher burners, they are seldom used and most heating is done by electric mantels and stirrer/hot plates. There are too many drawbacks associated with open flame heat sources for them to be considered the heat source of choice. The problems associated with these burners include:

1) Open flames may provide a too-intense and localized source of heat.
2) Open flames may provide an ignition source for flammable gases or other combustible materials.
3) Open flames may fill a poorly vented room with carbon dioxide.
4) The tubing connecting the gas outlet to the burner can leak emitting gas into the room.
5) Specific temperatures are hard to obtain and maintain, and it is simple to overheat materials.
6) It is simple for an open flame to heat things you do not want heated, causing injury or damage.
7) Lighting these burners in a fume hood can be very difficult. Although it is possible to ignite a burner with a striker (a flint and steel connected by a spring), it is not easy. An easier approach is with a disposable lighter. Probably the easiest technique for lighting a Bunsen burner in a fume hood is to have enough flexible tubing to take it out of the fume hood for lighting, and once lit, return it to its proper location.

Because of the problems associated with open flames, various states regulate their use in labs. Some states even ban the use of open flames in any

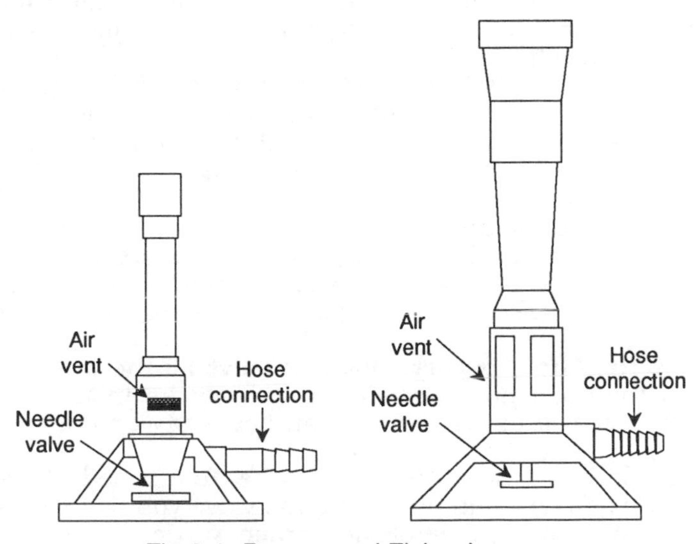

Fig 6-1 Bunsen and Fisher burners.

lab. Despite the problems associated with open flames, they are still used quite extensively in many labs because the advantages of open flames are:

1) Fuel is inexpensive.
2) Equipment is inexpensive and easy to set up.
3) They are easy to operate.

When you have a general chemistry lab with 20 to 30 (or more) students, providing expensive equipment to all can be a sobering jolt to safety considerations. Therefore, as long as open flames will continue to be used, the following are some safety insights for their proper use.

A burner is connected to its gas source with either amber latex or vinyl tubing. Amber latex tubing has one excellent property for use in connecting the burner to the gas supply—its excellent *memory* of its original shape. So, regardless of how long a hose is connected, the hose will always grip the hose connection it was originally slipped onto. Unfortunately, latex tubing ages over time and can easily burn. Both of these conditions, aging and burning, will eventually cause leaks in the tubing. If you use latex tubing, monitor it often for signs of age (cracking) or burns.

NEVER LOOK FOR GAS LEAKS WITH A MATCH. Your nose will probably offer the first indication that you have a leak. Then, you can either listen for the hissing of the leak, while occasionally twisting and moving the tubing to arrange the leak region, or coat the tubing with a soapy solution and look for bubbles.

Vinyl tubing has poor memory and with time begins to conform to new shapes. With time, vinyl tubing will lose its grip on hose connections and may leak or be pulled off easily. There are two solutions to this problem. The best solution is to use a hose clamp. This clamp will provide a secure attachment of the hose to the gas outlet and allow for (relatively) easy removal. Alternatively, you can soften the tubing by placing the end in boiling water for a minute or so. This method will allow you to shove the tubing onto the hose connection so far that the chances of accidental removal or leakage are considerably diminished.

Removing vinyl tubing is best done with a razor blade. Do not pull vinyl tubing off as the torque is likely to break the hose connection off.

Vinyl tubing does not age and is fire-resistant. However, a hot object may melt through the tubing causing a leak. Fortunately, because the tubing does not burn, it therefore is unlikely to ignite a combustible gas.

Needle valves on both burner types are threaded right-handed. These valves are in reverse angles during use, that is, you observe the valves from the top. This orientation may confuse some people. Just remember that in use, you close a needle valve by rotating it CW, and open the valve by rotating it CCW.

When first setting up a burner, close the valve all the way. Rotate the lab bench valve to the open position (checking for leaks in the tubing at this point is advised). Light a match, or get a striker ready and open the burner's needle valve.

Once the flame has ignited, it needs to be 'fine tuned.' By opening the valve you increase gas flow and by closing the valve you decrease gas flow. By rotating the air gates on the side of the burner, you can also increase or decrease the amount of air that is allowed to premix with the gas to intensify the flame. When you open an air gate, more air is available to combine with the gas, thus producing a bluer and hotter flame. As a gate is closed, less air is available to the gas, producing a whiter (yellower) and cooler flame.

Any flame that is predominantly white (or yellow) will smoke and deposit black (carbon) soot on any object that is placed over it. It is best to allow at least enough air in the gates to ensure a blue flame. A blue flame will not deposit soot. Additionally, a white (yellow) flame is not likely to have enough energy to accomplish any significant heating.

The main difference between the Fisher and Bunsen burner is size and heat dispersion. When trying to heat a test tube in a Fisher burner, you are likely to overheat the material. Likewise, when trying to heat a two liter flask with a Bunsen burner, you are likely to have a long wait. The exact demarcation line of which tool you should use for which type of heating job requires common sense and experience.

If you need to heat a test tube with a Bunsen burner, follow the following rules:

1) Be sure to hold the test tube with a proper test tube holder. Do not use fingers, rags, tweezers, or pliers.
2) Hold the test tube at an angle off of vertical.
3) Do not 'aim' the test tube at yourself or any other person. This positioning may be difficult in a crowded lab, but it is important. If the solution becomes superheated before boiling begins, the test tube can act like a cannon, shooting its hot contents up to twenty feet.
4) Heat along the bottom side gradually moving the test tube up and down the length of the solution (see Fig. 6-2).

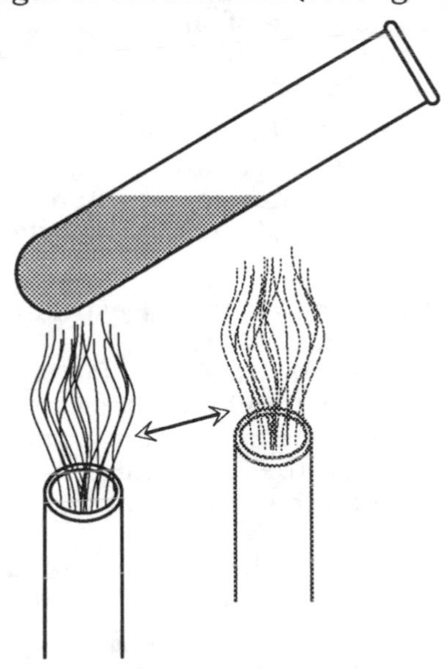

Fig. 6-2 Heating the whole bottom of a test tube.

It is preferable not to heat a beaker or Erlenmeyer flask with Fisher or Bunsen burners. The reason is discussed in detail in Sec. 1.1.7, which discusses the structural shapes of flasks. In essence, the problem is that the sharp angle of the side walls may not be able to disperse the thermal tension created by the expanding heated bottom. Round bottom flasks, on the other hand, have no side walls and therefore do not have this problem.

There are occasions where you will need to heat materials in a beaker because no other means of heating are available. In these occasions, do not let the flame come in direct contact with the glass. Rather, place the beaker on a wire screen or a wire screen with a ceramic square.* Even though the screen will prevent a direct concentration of heat at any one point on the beaker, use a soft blue flame, not an intense 'hissy' flame.

6.1.4 Steam

Steam is one of the safest and easiest techniques to provide even, uniform heat in the lab. Because steam can provide an even heat distributed uniformly across the entire bottom of a container, there are no fears of hot spots or solution superheating. In addition, the steam bath** is made up of simple components with no electrical wires or moving parts. Thus, there is nothing to wear out or to be damaged in a steam bath by accidental spills of heated materials.

Although steam can technically be any temperature greater than 100°C, under normal conditions, the user can consider the temperature to be about 100° C. Superheated steam (>250°C) is possible, but special insulation and equipment are required.

Regardless of *relative* safety, caution with the use and handling of steam is still required. Careful monitoring of the equipment for the duration of the experiment or process is still important. From a safety aspect, there are two things to keep in mind about steam;

1) Distilled water cannot be hotter than 100°C (at STP), but steam can be as hot as you can get it.
2) You cannot see steam. What people often call steam is in reality the condensed water of steam that has just cooled below 100°C.

These two facts point out one of the biggest dangers of steam: what you cannot see *can* hurt you. Caution is advised.

If you are fortunate enough to work in a facility that has self-generated steam, you can easily attach one end of a hose to the outlet and the other end to the steam bath or other steam delivery device. Be sure to use heavy-duty flexible tubing to withstand the high heat of the steam, and use as short a piece of tubing as possible so the steam will not condense before it gets to the steam pot.

Your delivery system is likely to be a *steam pot* (also called a *steam bath*) (see Fig 6-3). Steam pots may be made out of copper, brass, or cast alu-

* Some labs may still have wire screens with asbestos squares attached. These screens should be placed in bags and disposed of with the proper agencies. There are replacement screens available that use a ceramic material and are not hazardous.
** As opposed to the steam source.

minum. One of the benefits of using steam for heating is that the steam bath is designed to accommodate a wide variety of flask sizes; all that is necessary is to select a nested lid that is just smaller than the size of your flask. The lid should be selected before the steam valve is opened. Otherwise the container will become too hot to handle safely and further setup will become more difficult and dangerous. If you need to change a nested lid after it has been heated, be sure to use tweezers or tongs.

All steam pots have one tube allowing steam in, and another allowing condensed water out. Be sure that neither tube kinks or plugs up. A constricted steam tube may burst due to steam pressure. A constricted drainage tube will cause the pot to fill up with water which will halt the progress of your work. If the tube continuously kinks, replace it with a newer and/or heavier-wall tube.

If you do not have a built-in steam line, you can create your own steam. Boiling water over open flame, immersed electrically heated wires, or on a hot plate are all adequate techniques.

Very elaborate steam apparatus configurations have been written up by Lane, et. al.,[1] for superheated steam distillations. These superheated steam setups are fairly elaborate pieces of apparatus which combine steam generation and distillation in one piece. They are therefore not setups that can be pieced together in an afternoon. A more simple design of a steam generator has been proposed by Hagen and Barton.[2] It is a fairly inexpensive, yet robust design.

Whichever steam generator design you select, it must be continuously monitored. If a steam generator runs out of water, the unit may be damaged, and the experiment or process that uses the steam will stop.

The only water you should use in an electrically heated steam generator is deionized water to decrease (and/or limit) the corrosion of internal parts caused by water electrolysis. In addition, as salt concentration goes up, current conduction increases, which may lead to power failure.

Steam units in general, and electrical steam generators in particular, should always be placed away from easy access to limit the chances of accidental burns or unintended electrical contact.

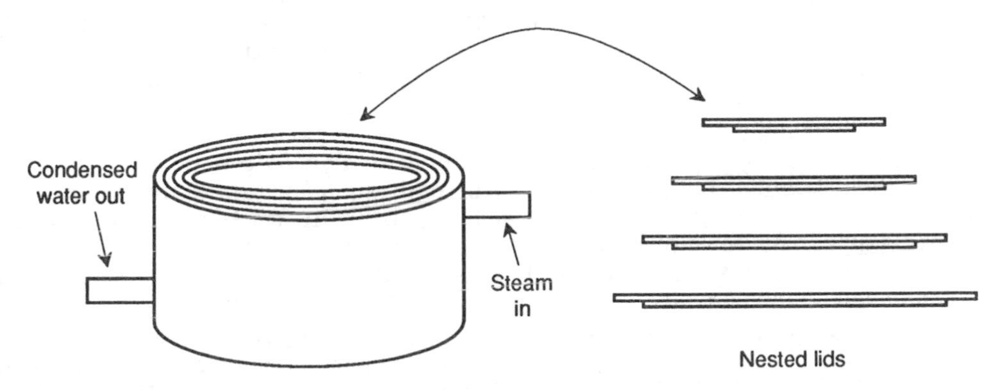

Fig. 6-3 The steam bath with nested lids.

6.1.5 Thermal Radiation

Thermal radiation is provided by heat lamps which emit large amounts of infra-red light. Besides keeping food warm at smorgasbord type-restaurants, heat lamps can be used to warm and/or dry samples.

Although heat lamps cannot provide great amounts of heat, they have the advantage of being able to work at a distance. Because heat lamps do not require contact or air to heat materials, they can be used to heat materials in a vacuum.

Do not stare at the light of a heat lamp as it can burn your retina. Similarly, do not leave your hands exposed to its light for more than a few seconds at a time as it will sunburn and damage your skin.

6.1.6 Hot Air Guns

Hot air guns are used to provide isolated, directed heat to individual targets. This method may be used to heat an individual sample on a vacuum line or to soften the grease on a stopcock.* Although one could use a standard hair dryer, it cannot provide the control, or endurance of an industrial-strength hot air gun.

There are two types of hot air guns: those that have a fan and those that do not. The obvious advantage of those with fans is that they are self-contained. The disadvantage is that the fan motor can create sparks, which could be hazardous depending on ambient gas conditions. The disadvantages of the motorless (no fan) design: you must have access to pressurized air, and the unit must have both a power cord and air hose for operation.

Motorized air guns allow extra air to enter the side of unit through an air gate to vary the amount of extra air that can enter the air gun. The less air that can enter the side, the hotter the air exhaust. A hot air gun can produce extensive heat which can easily peel paint or crack a window pane.

If you have pressurized air in your lab, you can use a motorless hot air gun. These units are smaller than standard hot air guns and are typically about the size of small home hair dryers. Motorless hot air guns only require heating filaments because they do not have fans. To use one, a flexible tube is attached to an air supply and the hot air gun. If your lab does not have plumbed-in compressed air, or the location of your outlet is too far away from where you want to use the hot air gun, another option is to use a compressed gas tank of air. Do not use compressed oxygen or any flammable gas for the air supply. You could use an inert gas, such as nitrogen for your air supply, but it would be costly.

Adjust the outgoing air pressure to between five to ten pounds of pressure. A needle valve on the regulator is required to regulate the amount of air passing the heating elements. Just like a built-in fan heater, the less air coming out of the regulator, the hotter the air out of the hot air gun, and vice versa. If the air supply runs out, you run the risk of burning the heating

* Any small-sized stopcock that requires heat to turn should be removed, cleaned and re-greased at the first opportunity. Large stopcocks may require a hot air gun for all rotation. Because heat ages stopcock grease faster than normal, the grease should be replaced two to three times more often than the other stopcock grease.

element. Always turn off the heating element before turning off the air supply. Otherwise, you run the risk of burning the heating element.

Probably the most common piece of worn or damaged equipment used in the laboratory is a hot air gun's electrical cord. These cords are frequently frayed due to their heavy use. Electrical cords should always be replaced as soon as there are any signs of fraying, and hot air guns deserve no exception.

6.1.7 Electrical Resistance Heating

One of the most versatile and robust heating techniques is the use of electrical resistance heaters. They are used in as varied heating approaches as hot plates, mantles, heat strips, immersion heaters, and even blankets.

Although some electrical heating units are self contained with their own electrical controllers (such as hot plates), many require separate controllers. If a heating unit has no apparent built-in mechanism with which to vary temperature, the unit will require a separate controller.*

Some electrical heating units that require controllers will have standard electrical plugs to connect into to controllers. Despite the fact that you can, **do not** plug these units directly into wall outlets. Your heat demands are likely to be far less than full electrical power, and you can only receive full power from a wall socket. Currently, more and more electrical heating units that require electrical controllers have special plugs that can only go into similarly-equipped controllers. Before purchasing a heating unit, verify with the supplier or manufacturer what type of fitting the unit is equipped with and whether it will be compatible with your existing equipment.

Some heating units such as mantles and hot plates can support the use of magnetic stirring bars which are magnetized rods covered in Teflon. Stirring bars can provide two functions. They can stir solutions to mix different materials together and assist the uniform and efficient heating of solutions. In addition, a stirring bar creates a vortex within a solution which provides a sharp point from which boiling can occur. Thus, using a stirring bar eliminates the need for boiling chips.

Magnetic stirring bars can be stored together, but they should be placed in a *clean, dry,* and *smooth* container. The clean and dry requirements are obvious, but smooth is equally important. A rough container (or dirt in the container) is likely to scratch the Teflon. Any scratches on the Teflon may trap dirt or chemicals which may affect future work. If you think that you will be unable to otherwise maintain a dirt-free container, consider keeping the tubes that the magnetic stirring bars are shipped in, and store the bars in the tubes.

Never rotate a spinning bar too fast because it is likely to cause splashing. Likewise, place the center of a flask over the center of the controller. Otherwise the spinning bar is likely to flip around like a wounded fish, causing a lot of splashing.

Electrical Heating Tapes. These units always require separate controllers. They come in various widths and lengths, and are always flexible.

* The controller is nothing more than an electrical rheostat that can vary the amount of electrical current the wires receive. This rheostat in turn also varies the amount of heat the wires can deliver. Some controllers have built-in (or external) thermocouples that enable the controller to maintain a specific temperature.

Depending on the insulating materials, they have upper heating temperature limits from 260°C to 760°C. Although the insulation may be resistant to chemical spills, the wires within may not. Sloppiness is therefore likely to be hazardous. Some of the insulations available should not be used against metal surfaces, so if you need to heat a stainless steel container, select suitable insulation.

Electric Heating Mantles. These mantles are like a pair of cupped hands and are designed to envelop individual round bottom flasks. Generally, each mantle size is designed to heat one size of flask, although there are a few mantle designs that can hold a variety of flask sizes and other glassware styles. Because the mantle securely supports the flask, it cannot tip or slide out of the heater. Like steam heating, the mantle provides heat equally along the entire lower surface of a flask. Some mantle designs provide insulation that completely covers an entire flask. This provides a uniform temperature throughout the flask preventing heat gradients within the solution and decreasing heat loss. Although the costs for separate mantles for each size flask can add up, mantles provide the best (and most uniform) heating for long periods of monitored, or unmonitored, usage.

Mantles, like electric hot plates, may have built-in controllers, some of which have thermostatic devices to set and control specific temperatures. In addition, some mantles have magnetic stirring mechanisms.

Similar to heating tapes, the insulation on mantles may be resistant to chemical spills but the wires within may not. Sloppiness is therefore potentially hazardous. If you have a mantle with a drainage hole on the bottom, be sure to provide some sort of collection basin underneath. Otherwise, chemicals may spill onto your lab bench.

Immersion Heaters. These devices are lowered directly into a solution for heating. Because the heaters come into direct contact with their solutions, the major consideration for selecting an immersion heater is the material used to cover the heating element. Because glass is so universally resistant to (most) chemical attack, glass is commonly used. However, glass is fragile and can easily be broken. In addition, glass should not be used in any alkaline, hydrofluoric acid, or hot phosphoric acid solutions as all three can dissolve glass.

The best types of immersion heater covers are made out of quartz glass. Because of quartz glass' tremendously low thermal coefficient of expansion, it is possible to remove a hot immersion heater from a solution and lay it on a cold bench top without fear of the unit cracking. Although this procedure is not safe either (someone could touch the heater and get burned), it is possible. Equally, it is possible for the immersion heater to have warmed up in the air, and later be immerse into a solution without fear of cracking. This is not a safe procedure (the steam or splattered liquid could burn the holding hand), but it is possible.

Some inexpensive immersion heaters may be made from borosilicate glass. Although they are not as capable of handling radical heat changes as quartz glass, they do quite well.

Surprisingly, as long as an immersion heater is not removed from the liquid being heated, and as long as the heated liquid is kept at moderate temperatures (20° to 30°C), even soft glass can be used as a cover for the heating element. For instance, immersion heaters for aquariums are often made from soft glass, and can survive because the water becomes a *heat sink* for the glass, preventing it from getting too hot and failing.

Immersion heaters with metal covers are more robust against physical abuse, and typically copper and stainless steel are used. Copper covers may be capable of some flexibility and can allow for adaptation to specific confined areas; however, they are more susceptible to chemical attack.

Hot Plates. These devices have a metal (cast aluminum, stainless steel, or some alloy), ceramic, or pyroceramic* top. Underneath the top is an electric resistance heater. Hot plates are used for heating flat bottom containers such as beakers and Erlenmeyer flasks. Because hot plate tops are nonporous, there are fewer concerns for spills affecting the heating elements as there are with heating mantles. Magnetic stirring devices are commonly included with hot plates.

Pyroceramic and ceramic tops are impervious to most chemical spills. The exceptions are typically any of the chemicals that can attack glass, such as hydrofluoric acid, alkali solutions, and hot phosphoric acid. Metal-topped hot plates can take more physical abuse than ceramic or pyroceramic tops. In addition, with metal, there are no concerns about scratches or cracks that, on a pyroceramic top, could lead to further deterioration. Metal-topped hot plates are however subject to corrosion from chemical spills.

Never place a cold beaker or Erlenmeyer on a heated hot plate, and likewise never take a hot beaker or Erlenmeyer and place it on a cold bench as either could cause the bottom of the container to break by thermal shock. One way to reduce the amount of thermal shock when transferring a container on or off the hot plate is to place it on something with less heat capacity, such as a wire screen.

Hot plates are also used to heat one type of material so that a second may be heated. These secondary heaters may include water baths, oil baths, sand traps, or aluminum plates. Water and oil baths are typically used to heat a drying flask on vacuum evaporators. Oil baths are more messy and require special clean up, but the oil (typically a silicone oil) will not evaporate during drying processes that can take up to several hours. Stronski[3] came up with a very simple and easy solution to limit water evaporation. Stronski recommended that the user float Styrofoam popcorn or chips up to three inches deep on water. This suggestion not only limited evaporation, but extended the maximum controlled temperature range.

The advent of various brands of microware laboratory equipment (originally developed by Mayo, Pike, and Butcher of Bowdoin College[4]), created a problem of how to heat these new smaller flasks. Two techniques have developed. The first involves placing the micro flasks into sand. The advantage of sand is that it can easily fill into all the external nooks and crannies of any container, and provide excellent, evenly distributed heat. The disadvantage of sand is that it is a poor heat conductor and can thus require a long time to be heated to a satisfactory temperature. In addition,

* Pyroceramic is a translucent material made by Corning that has both glass and ceramic properties.

once you remove a flask, it is not easy to reinsert a new flask into the 'filled in' hole.

An alternative approach is a solid aluminum block. This idea was developed jointly by Siegfried Lodwig of Centralla College and Steve Ware of Chemglass, Inc.[5] Aluminum conducts heat excellently and can be easily machined to accommodate different sized flasks. The success of aluminum heat blocks caused Lodwig to use them also for a variety of other heating techniques that had nothing to do with microware.

Because a variety of individual holes may be drilled into a single aluminum block, it is possible to easily rotate different sized flasks on a random basis with ease and have separate holes for thermometers as well. The thermometer allows you to easily monitor temperatures for various thermostat settings. These temperatures can easily be graphed to allow you to vary the temperature to predetermined settings.

Drying Ovens. These units are used to facilitate the drying and storage of glassware that needs to be dry and ready at a moment's notice. Glassware should never be wiped dry as the wiping process could easily contaminate the surfaces. There is no reason to dry glassware that will be getting wet immediately after washing. Similarly, there is no reason to place items in a drying oven that could otherwise drip dry. Glass typically has a certain amount of adsorbed water at all times. The only way to reduce this water is with heat and/or vacuum. The drying oven provides limited assistance for the former and will not only help to remove some of this water, but if the container is left in the oven, will help keep the water from returning as long as the glassware remains in the oven.

The last rinse before placing glassware into a drying oven should be with distilled or deionized water. Regular tap water may leave remains (seen as spots) that will contaminate the glassware. If possible, rotate the glassware so that liquid can drain during the drying process. Although it is acceptable to use technical grade acetone or methanol for a final rinse when not using a drying oven, never use either of them in conjunction with a drying oven. Flames or even an explosion can result from such a combination.

Drying ovens are typically set at 120°–130°C which is well below the recommended temperature of about 400°C needed to remove all adsorbed water. Contrary to popular misconception, it is completely safe to place any borosilicate volumetric ware in a drying oven. The temperature is way below what is needed to cause any distortion and resultant volume changes (see Sec. 2.3.5). Objects placed in drying ovens should remain three to four hours, so that the drying of the glass (the adsorbed water) can reach an equilibrium. Otherwise, if items look dry, they probably are dry.

Never place plastic items into a drying oven, especially items made of Teflon. Typically there is enough heat leaking through the top of a drying oven for plastic items to be placed on the top for drying. To prevent the plastic items from scratches or contamination, place them into an Erlenmeyer flask or beaker. When assembling Teflon stopcocks where the barrel was just removed from a drying oven, wait until the item is completely at room temperature before final assembly. A Teflon stopcock that is snug when warm, is likely to leak when cooled to room temperature.

6.1.8 Alternatives to Heat

Heat is typically used to bring a solution to its vapor point (boiling) so that it may be distilled, reduced, or purified. Although this procedure works in most cases, not all solutions can be safely heated. Some materials break down in heated conditions, and some do so violently. Fortunately, there are two alternative methods to obtain the vapor pressure of a material; one is to raise the heat, the other is to lower the pressure. By using a vacuum (see Chap. 7) to obtain the vapor pressure of a solvent, the solvent will boil off, leaving the material behind. This procedure can be done at room temperature, or slightly above room temperature.

The catch to the vacuum method is that you *must* have a controlled boil without which the material, and/or solvent are liable to be sprayed all over your vacuum system. Although a solvent can easily be pumped out of a vacuum system, it can cause serious problems if it remains in contact with stopcock grease, O-rings, or mechanical and/or diffusion pump oils. Any particulate material deposited within a vacuum line can only be removed from the vacuum line by disassembly and cleaning. With a glass vacuum system, such a cleaning may be difficult or impossible.

The standard approach to maintaining a controlled boil is to use boiling chips as one uses when boiling materials over a Bunsen burner. However, these chips have limitations as they may be very difficult (or impractical) to remove at a later time.

One of the best approaches, for maintaining a controlled boil, is to place a Teflon-coated stirring bar in a flask and then place the flask over a magnetic stirring device (as mentioned in Sec. 6.1.7). In a medium to high vacuum system (10^{-4} to 10^{-5} torr), this activity can be done in a static vacuum with a container sitting in liquid nitrogen near the material in question (see Fig. 6-4). The liquid nitrogen acts as a cryopump, trapping all the solvent boiling off, maintaining a constant pressure and therefore, a constant boil.

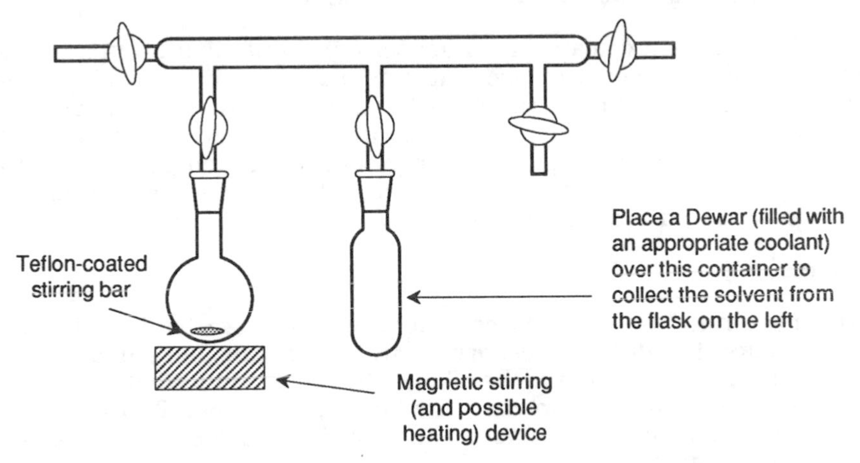
Teflon-coated stirring bar
Place a Dewar (filled with an appropriate coolant) over this container to collect the solvent from the flask on the left
Magnetic stirring (and possible heating) device

Fig. 6-4 A static vacuum can be used to remove solvents from samples.

6.2 Low Temperature

6.2.1 The Dynamics of Cold in the Lab

Cold can be used in the laboratory to prevent an experiment from getting too warm, slow the rate of a reaction, transfer materials in a vacuum system, allow for the separation of materials (with fractional condensation), or as a (cryogenic) vacuum pump.

The physical mechanisms of cold transfer are the same as heat transfer and use the same physical processes of conduction, convection, and radiation—absorption (for more information on these processes see Sec. 6.1.1).

Both hot and cold express different degrees of thermal energy and are directly related to each other. The decision of what is hot or cold is a subjective choice. Heat (energy) always transfers to cold (lack of energy). So, whichever object is relatively hotter will transfer some of its thermal energy to the object that is relatively colder. Thus, the water running through a condenser does not cool the condenser. Rather, the condenser heats the water which in turn leaves the condenser (by way of tubing) and is replaced by water, ready to be heated. As the heated water is removed, it takes that amount of heat energy away from the condenser. Although this distinction may seem inconsequential, it is fundamental to the understanding of what is taking place in the cooling process. Cooling could also be considered a 'removing heat' process.

To cool materials in the lab, you need to select cooling materials that are sufficiently cold and have the heat capacity to remove the necessary thermal energy. The following subsections describe techniques used to make things cold in the laboratory:

6.2.2 Room Temperature Tap Water (≈20°C)

Water is used as a coolant in condensers and diffusion pumps. For simple exothermic reactions, you can use water as a heat sink by placing glassware (containing a reaction) into a water bath. Alternatively, in an exothermic reaction, you can also use a cold-finger of running water to prevent overheating. However, all these examples can only cool to the temperature of tap water (usually room temperature). For every step below room temperature, more equipment and money are required.

6.2.3 Ice (0°C)

Many labs have freezers or ice machines to make ice.* If the ice is in cubes or blocks, it will be necessary to smash it into more usable, smaller, crushed pieces. The ice should be placed in some kind of cloth bag (or the like) during smashing, to prevent the ice pieces from flying around the room. Be sure to wear safety glasses any time you are smashing anything—and watch the fingers!

* If you constantly lose the scoop in the ice, tie one end of a string on the scoop and the other on the ice machine.

6.2.4 Ice With Salts (0°C to -96.3°C)

Various salts can cause freezing point depressions. These depressions are the results of the ions' colligative properties within solution. A significant freezing point depression is created not by any particular type of material, but rather by the number of particles you have in solution. The effects can be enhanced by achieving a supersaturation of material. For example, if you mix ice with a salt such as NaCl, you will end up with two particles within solution ($Na^+ + Cl^-$) (the temperature of ice water supersaturated with sodium chloride (23% by weight) is -20.67°C). If you place a salt such as $CaCl_2 \cdot H_2O$ into ice, you will end up with three particles within the solution ($Ca^+ + 2Cl^-$) (the temperature of ice water supersaturated with calcium chloride [30% by weight] is -41.0°C). Further, because you can get more methanol saturated into the ice water (68% by weight), you can achieve a greater freezing point suppression (-96.3°C). To a certain degree, you are limited by the solubility of a compound. Materials such as $Ca(NO_3)_2$ and H_2SO_4 have unlimited solubility. However, because of the corrosive nature of H_2SO_4 it is not a good choice for general use. Table 6-1 lists a wide variety of salts and compounds and their freezing point depressions by percentages of weight.

6.2.5 Dry Ice (Frozen Carbon Dioxide) (-78°C)

Dry ice comes in blocks that are wrapped in paper and are kept in specially insulated ice boxes. Because there is no liquid stage of CO_2 at STP, there is no fear of liquid leaking out of these paper containers (the triple point* of dry ice is at 5.2 atm and -57°C).

Never handle dry ice with your bare hands. At 1 atm, dry ice sublimes at -78°C. At these temperatures, severe tissue damage could result. You should always use heavy thermal gloves or tongs.

Dry ice, like water ice, can be smashed into smaller, more practical pieces. Like regular ice, it should be placed in a cloth bag for smashing to prevent ice chips from flying around the room.

6.2.6 Liquid Nitrogen (-195.8°C)

Liquid nitrogen is deceptively dangerous; it looks like water, but at its extremely cold temperature (-195.8°C), liquid nitrogen can do extensive tissue damage. A common lab demonstration is to place a flower into a Dewar of liquid nitrogen, remove the flower, and hit, or crush, the flower with a blunt object. The previously soft flower shatters as if made of fine glass. The same could easily be done with a finger.

Liquid nitrogen is shipped and stored in large insulated liquid/gas tanks (see Sec. 6.2.10). To dispense liquid nitrogen out of the large insulated storage tank, connect a metal tube to the "liquid" valve on the tank. Then place a receiving vessel over the end of the tube, and open the liquid port valve until the desired amount is obtained.

* The triple point of a compound is when the atmospheric pressure and temperature are compatible for the solid, liquid, and gas forms to exist in equilibrium at the same time.

Table 6-1‡

Freezing Point Depressions of Aqueous Solutions

Compound	% Soln. by Wt.	Temp °C
Acetic acid (CH_3COOH)	1.0	-0.32
	5.0	-1.58
	10.0	-3.23
	20.0	-6.81
	30.0	-10.84
	36.0	-13.38
Acetone (CH_3COH_3)	1.0	-0.32
	5.0	-1.63
	10.0	-3.29
Ammonium hydroxide (NH_4OH)	1.0	-1.14
	5.0	-6.08
	10.0	-13.55
	20.0	-36.42
	30.0	-84.06
Ammonium chloride (NH_4Cl)	1.0	-0.64
	5.0	-3.25
	10.0	-6.95
Amonium sulfate ($(NH_4)_2SO_4$)	1.0	-0.33
	5.0	-1.49
	10.0	-2.89
	16.0	-4.69
Barium chloride ($BaCl_2 \cdot H_2O$)	1.0	-0.22
	5.0	-1.18
	10.0	-2.58
	16.0	-4.69
Calcium chloride ($CaCl_2 \cdot H_2O$)	1.0	-0.44
	5.0	-2.35
	10.0	-5.86
	20.0	-18.3
	30.0	-41.0
Cesium chloride ($CsCl$)	1.0	-0.20
	5.0	-1.02
	10.0	-2.06
	20.0	-4.49
Ethanol (CH_3CH_2OH)	1.0	-0.40
	5.0	-2.09
	10.0	-4.47
	20.0	-10.92
	32.0	-22.44
	40.0	-29.26
	52.0	-39.20
	60.0	-44.93
	68.0	-49.52
Ethylene glycol (CH_2OHCH_2OH)	1.0	-0.15
	5.0	-1.58
	10.0	-3.37
	20.0	-7.93
	32.0	-16.23
	40.0	-23.84
	52.0	-38.81
	56.0	-44.83
Ferric chloride ($FeCl_3 \cdot 6H_2O$)	1.0	-0.39
	5.0	-2.00
	10.0	-4.85
	20.0	-16.14
	30.0	-40.35
Formic acid ($HCOOH$)	1.0	-0.42
	5.0	-2.10
	10.0	-4.27
	20.0	-9.11
	32.0	-15.28
	40.0	-20.18
	52.0	-29.69
	60.0	-38.26
	64.0	-43.02
D-Fructose (levulose) ($C_6H_{12}O_6$)	1.0	-0.10
	5.0	-5.44
	10.0	-1.16
	20.0	-2.64
D-Glucose (dextrose) ($C_6H_{12}O_6 \cdot 1H_2O$)	1.0	-0.11
	5.0	-0.55
	10.0	-1.17
	20.0	-2.70
	30.0	-4.80
Glycerol ($CH_2OHCHOHCH_2OH$)	1.0	-0.18
	5.0	-1.08
	10.0	-2.32
	20.0	-5.46
	36.0	-15.5
Hydrochloric acid (HCl)	1.0	-0.99
	5.0	-5.98
	10.0	-15.40
	12.0	-20.51
Lithium chloride ($LiCl$)	1.0	-0.84
	5.0	-4.86
	10.0	-12.61
	14.0	-21.04

Table 6-1

Freezing Point Depressions of Aqueous Solutions (cont.)

Compound	% Soln. by Wt.	Temp °C
Methanol (CH_3OH)	1.0	-0.28
	5.0	-3.02
	10.0	-6.60
	20.0	-15.02
	32.0	-28.15
	40.0	-38.6
	52.0	-58.1
	60.0	-74.5
	68.0	-96.3
Nitric acid (HNO_3)	1.0	-0.56
	5.0	-2.96
	10.0	-6.60
	19.0	-15.3
Phosphoric acid (H_3PO_4)	1.0	-0.24
	5.0	-1.16
	10.0	-2.45
	20.0	-6.23
	30.0	-13.23
	40.0	-23.58
Potassium bromide (KBr)	1.0	-0.29
	5.0	-1.48
	10.0	-3.07
	20.0	-6.88
	32.0	-12.98
Potassium carbonate ($K_2CO_3 \cdot 1\,\frac{1}{2}H_2O$)	1.0	-0.34
	5.0	-1.67
	10.0	-3.57
	20.0	-8.82
	32.0	-21.46
	40.0	-37.55
Potassium chloride (KCl)	1.0	-0.46
	5.0	-2.32
	10.0	-4.81
	13.0	-6.45
Potassium iodide (KI)	1.0	-0.22
	5.0	-1.08
	10.0	-2.26
	20.0	-5.09
	30.0	-8.86
	40.0	-13.97
Sodium chloride (NaCl)	1.0	-0.59
	5.0	-3.05
	10.0	-6.56
	20.0	-16.46
	23.0	-20.67
Sodium hydroxide (NaOH)	1.0	-0.86
	5.0	-4.57
	10.0	-10.47
	14.0	-16.76
Sodium nitrate ($NaNO_3$)	1.0	-0.40
	5.0	-1.94
	10.0	-3.84
	20.0	-7.81
	30.0	-11.28
Sulfuric acid (H_2SO_4)	1.0	-0.42
	5.0	-2.05
	10.0	-4.64
	20.0	-13.64
	32.0	-44.76

‡ From the tables of "Concentrative Properties of Aqueous Solutions: Conversion Tables" from the *CRC Handbook of Chemistry and Physics*, 52ND ed., (1971-1972,).Published by the Chemical Rubber Co. Cleveland, OH. With permission.

Liquid nitrogen should *only* be transported and held in double-walled, insulated containers. Transport containers should have narrow necks to avoid spillage.

Never leave liquid nitrogen in non-insulated containers (i.e., beakers). In a non-insulated container, liquid nitrogen will boil away very fast, and thereby require constant replacement (an economic loss). In addition, any water condensation will cause the container to freeze onto its resting surface making it difficult to move without breaking the beaker and increasing the potential danger. Finally, the most important reason to never place liquid nitrogen in a non-insulated container is that someone may inadvertently try to pick up the container and severely burn his or her hands.

6.2.7 Slush Baths (+13° to -160°C)

A slush bath can be described as a low-melting-point liquid (typically a hydrocarbon solvent) that is being kept in a partially frozen state by either liquid nitrogen or dry ice. The temperature will remain constant as long as you continue to add liquid nitrogen, or dry ice, to the bath to maintain its 'slushy' state. Table 6-3 is a comprehensive list of slush baths made of dry ice (CO_2) and liquid nitrogen (N_2). Duplicate temperatures indicate a choice of solvent or coolant.

To make a slush bath, pour the selected low-melting temperature liquid into a Dewar, then pour the coolant in while stirring briskly. A wooden dowel is wonderful for stirring because it will not scratch the Dewar's surface. There is no concern for contamination from the dowel because it is not likely to affect the performance of the slush bath.

During the mixing process, tremendous amounts of solvent fumes are likely to be given off along with condensed water from the air. Therefore, the original preparation of a slush bath should be done in a fume hood. Once the slush bath begins to reach equilibrium, the amount of vapors leaving the Dewar decreases and it is safe to remove the bath from the fume hood.

To make a *liquid nitrogen slush bath*, first pour the desired low-melting temperature liquid into a Dewar. Then slowly, while constantly stirring, add the liquid nitrogen until the desired consistency is achieved. Different low-melting liquids will have different viscosities. It is therefore desirable to know the potential viscosity limits of the slush bath you are using. This information is best retained by asking someone; otherwise, make a test slush bath before you set yourself up for an actual experiment or process. Knowing the viscosity of a certain slush bath is important, otherwise you may expect one slush bath to be as thick as another, and might add an excess of liquid nitrogen. If you add too much liquid nitrogen, it will become the predominant cooling medium and the resultant temperature will be cooler than listed in Table 6-3. However, as the liquid nitrogen boils off, the temperature will settle to the listed temperature.

To make a *dry ice slush bath*, be sure that your low-melting-temperature liquid has a freezing point above -78°C. Follow the same procedure as with the liquid nitrogen bath above, but use crushed dry ice. When crushing dry ice, use a hammer (not a pair of pliers or a wrench) and place the dry ice in a cloth bag so the pieces do not fly around the room. Always handle dry ice with thermal gloves or tongs, never handle dry ice with your bare hands or you might severely burn your hands and fingers. The primary advantage of using dry ice over liquid nitrogen is that it is less expensive, safer to work with, and more readily available. On the other hand, it is more difficult to work with. Similar to liquid nitrogen, if you add too much dry ice, it becomes the predominant cooling medium and the resultant temperature will be cooler than listed in Table 6-3. As the dry ice evaporates, the temperature will settle to the listed temperature.

By combining mixtures of organic solvents, it is possible to achieve temperatures that you may not otherwise be able to obtain because of lack of material or to avoid other materials. For example, with various combinations of *ortho* and *meta*- xylene and dry ice, you can achieve temperatures from -29°C to -72°C (see Table 6-2).

Regardless of which coolants or low-melting-point liquids are used to make a slush bath, no slush bath combination can be a perpetual (static) temperature system. Rather, a slush bath is a dynamic collection of materials that are either settling or boiling off. Thus, they require constant monitoring with (preferably) two liquid-in-glass thermometers* (or a thermocouple probe), consistent agitation, and occasional replenishing of coolant.

Table 6-2‡

Temperature Variations with Combined Organic Solvents		
% o-Xylene	**% m-Xylene**	**Temperature (≈°C)**
100	0	-29 (±2°C)
80	20	-32 (±2°C)
60	40	-44 (±2°C)
40	60	-55 (±2°C)
20	80	-68 (±2°C)
0	100	-72 (±2°C)

‡ Data interpolated from a graph from the article by A.M. Phipps and D.N. Hume, "General Purpose Low Temperature Dry-Ice Baths," *J. of Chemical Education*, 45 (1968), p. 664.

An alternative to the slush bath is the *coolant*, or *cooling bath*. These baths are handy when you may not have (or wish to use) the required low-melting-temperature liquid for a particular temperature. However, they require more work to maintain their specific temperatures. Liquid nitrogen is recommended for cooling baths because dry ice can be difficult to introduce to the bath in sufficiently small amounts. Like the slush bath, the cooling bath should be mixed in a fume hood.

The cooling bath differs from the slush bath because less coolant is used during initial mixing. Thus, the mixture never obtains the slushy state of the slush bath. The resultant bath temperature is warmer than the coolant, and by varying the amount of coolant you can vary the temperature. The trick is to select a low melting point liquid that is sufficiently low to provide a wide working temperature range. One good cooling bath liquid is methanol, which has a freezing temperature of -98°C, is reasonably safe to use,** and can easily be used for cooling baths. Another reasonably safe, low-temperature melting point liquid is petroleum ether (30°-60°) which has a freezing point of approximately -120°C, and can thereby provide a greater range of low temperatures.

The difficulty involved with cooling baths is that you control the temperature by varying the amount of coolant in the mix. This control requires constant attention and the slow, but constant addition of more coolant as the bath continues to warm during use. Despite the extra labor, many people prefer cooling baths because there is a greater choice of safe solvents. In addition, less solvents must be stored to obtain a wide range of temperatures.

* By using two liquid-in-glass thermometers, you can verify the quality of both thermometers by their agreement in temperature readings. If the temperatures do not agree, one of the thermometers may have a bubble in the stem or some other defect. Unfortunately, this trick does not let you know which is the defective one, but it provides a clue to the problem.

** Methanol is flammable and poisonous.

Table 6-3

Slush Bath Temperatures

Solvent*	CO_2 or N_2	°C	Solvent*	CO_2 or N_2	°C
p-Xylene	N_2	13	Isoamyl acetate	N_2	-79
p-Dioxane	N_2	12	Acrylonitrile	N_2	-82
Cyclohexane	N_2	6	Sulfur dioxide	CO_2	-82
Benzene	N_2	5	n-Hexyl chloride	N_2	-83
Formamide	N_2	2	Propylamine	N_2	-83
Aniline	N_2	-6	Ethyl acetate	N_2	-84
Diethylene glycol	N_2	-10	Ethyl methyl ketone	N_2	-86
Cycloheptane	N_2	-12	Acrolein	N_2	-88
Benzonitrile	N_2	-13	Amyl bromide	N_2	-88
Benzyl alcohol	N_2	-15	n-Butanol	N_2	-89
Ethylene glycol	CO_2	-15	s-Butanol	N_2	-89
Propargyl alcohol	N_2	-17	Isopropyl alcohol	N_2	-89
1,2 Dichlorobenzene	N_2	-18	Nitroethane	N_2	-90
Tetrachloroethane	N_2	-22	Heptane	N_2	-91
Carbon tetrachloride	N_2	-23	n-Propyl acetate	N_2	-92
Carbon tetrachloride	CO_2	-23	2-Nitropropane	N_2	-93
1,3 Dichlorobenzene	N_2	-25	Cyclopentane	N_2	-93
Nitromethane	N_2	-29	Ethyl benzene	N_2	-94
o-Xylene	N_2	-29	Hexane	N_2	-94
Bromobenzene	N_2	-30	Toluene	N_2	-95
Iodobenzene	N_2	-31	Cumene	N_2	-97
m-Toluidine	N_2	-32	Methanol	N_2	-98
Thiophene	N_2	-38	Methyl acetate	N_2	-98
3-Heptanone	CO_2	-38	Isobutyl acetate	N_2	-99
Acetonitrile	N_2	-41	Amyl chloride	N_2	-99
Pyridine	N_2	-42	Butyraldehyde	N_2	-99
Acetonitrile	CO_2	-42	Diethyl ether	CO_2	-100
Benzyl bromide	N_2	-43	Propyl iodide	N_2	-101
Cyclohexyl chloride	N_2	-44	Butyl iodide	N_2	-103
Chlorobenzene	N_2	-45	Cyclohexene	N_2	-104
Cyclohexanone	CO_2	-46	s-Butylamine	N_2	-105
m-Xylene	N_2	-47	Isooctane	N_2	-107
n-Butylamine	N_2	-50	1-Nitropropane	N_2	-108
Benzyl acetate	N_2	-52	Ethyl iodine	N_2	-109
Diethyl carbitol	CO_2	-52	Carbon disulfide	N_2	-110
n-Octane	N_2	-56	Propyl bromide	N_2	-110
Chloroform	N_2	-61	Butyl bromide	N_2	-112
Chloroform	CO_2	-63	Ethyl alcohol	N_2	-116
Methyl iodide	N_2	-66	Isoamyl alcohol	N_2	-117
Carbitol acetate	CO_2	-67	Ethyl bromide	N_2	-119
t-Butylamine	N_2	-68	Propyl chloride	N_2	-123
Ethanol	CO_2	-72	Butyl chloride	N_2	-123
m-Xylene	CO_2	-72	Acetaldehyde	N_2	-124
Trichlorethylene	N_2	-73	Methylcyclohexane	N_2	-126
Isopropyl acetate	N_2	-73	n-Propanol	N_2	-127
o-Cymene	N_2	-74	n-Pentane	N_2	-131
p-Cymene	N_2	-74	1,5-Hexadiene	N_2	-141
Butyl acetate	N_2	-77	iso-Pentane	N_2	-160
Acetone	CO_2	-77			

* All of these compounds are to some degree either poisonous, hazardous, toxic, and/or carcinogenic. If you do not know how to safely handle and use any of these materials, see the recommended reading in Appendix D for Chapter 6.

6.2.8 Safety With Slush Baths

Because the low-temperature melting point liquid in the slush bath is near its freezing temperature, there is little concern over toxic fumes as when working with some chemicals at room temperature. However, in the beginning, you must work in a fume hood because of the copious amounts of fumes released from the low-temperature melting point liquid. Once the slush bath is made, it is safe to remove it to the lab. However, it is best to leave the slush bath in the fume hood if at all possible. If the slush bath is used in the lab, move the slush bath to the fume hood immediately after your work is completed.

Never pour a slush bath down the sink. The low temperatures can destroy plumbing!!! Instead, let the coolant boil off in a Dewar in the fume hood. Later, the low-melting temperature liquid can be saved and reused.

Because slush baths achieve very low temperatures, protect your hands from the extreme cold. This protection presents some problems with the current temperature-protecting gloves available and problems inherent with slush baths. Thermal gloves made out of Kevlar,* a plastic, can be dissolved by some organic solvents. Thermal gloves made out of fiberglass are okay, but they are very slippery. In addition, broken glass fibers from fiberglass gloves can get under the skin and itch for many days. Asbestos gloves** are not allowed in many states. My personal choice among all these suggestions would be the Kevlar gloves.

Incidentally, when making slush baths out of organic solvents, do not use utensils (such as stirrers) that will dissolve in the organic solvent you are using. Although a thermometer may already be in the container and ready to stir with, do not use it as your stirrer as it is likely to break from the torsional forces of stirring the thick slush. Wooden dowels are excellent for mixing because they are strong and will not scratch the surface of a Dewar.

A potentially explosive situation can develop when an acetone slush bath is left sitting for an extended period. Over time, the acetone and dry ice separate and the acetone floats to the surface, whereas the dry ice settles to the bottom of the Dewar. The acetone soon warms up to near room temperature, but the dry ice remains near the slush bath temperature of -77°C. If any agitation causes the warmed acetone to cut into the dry ice slush on the bottom, a flume of boiled off CO_2 can erupt. This flume will carry the acetone layer that was on the surface in a large spray all around the area. If there is a flame or spark (from a motor) in the path of the acetone, this accident could have far greater consequences. This situation can be easily avoided by constantly mixing the solution. A safe alternative to the acetone slush bath is the ethanol slush bath. The ethanol slush bath is somewhat warmer (-72°C) but does not display the same potentially dangerous capabilities.

6.2.9 Containment of Cold Materials

There are two concerns for the storage of cold materials: longevity of the material and safety to the user. For example, if you place an ice cube on a lab

* Kevlar gloves are banana yellow with a surface like terry cloth.

** If your laboratory still has some asbestos gloves, you may need to check with your safety officer or the Department of Health in your area for proper disposal.

bench, it will melt. On the other hand, if you place an ice cube in an insulated container, it will also melt—but it will take longer. By providing insulation, you have added to the ice cube's longevity. If you hold an ice cube for an extended time, your hands will soon become so cold that eventually you will need to drop the ice cube. However, if you hold an insulated container containing an ice cube, there is no discomfort.

The reason for stating the obvious is to establish the purpose of specialized containers for containing cryogenic solutions. A properly-made container protects both the materials inside the container as well as the users outside. In addition, the container should be able to reasonably deal with the expected physical abuses that may be encountered within the lab. The final selection of a cryogenic container is based on its shape, design, construction material, use, and function. Although it is possible to use a Styrofoam cup to contain cryogenic materials such as liquid nitrogen, it is a poor choice for the demands of a laboratory. On the other hand, placing tap water in a Dewar may be a waste of money if you are only trying to cool an object down. As with most decisions in the lab, common sense must be used when making equipment selections. Ultimately, the selection of the quality, shape, and design of a coolant container may be based on six criteria:

1. ***Cost of coolant.*** If the coolant is inexpensive and readily available, you don't need a highly efficient container.
2. ***The coldness of the coolant.*** The greater the temperature difference of the coolant from the ambient room temperature, the better the quality of insulation required.
3. ***Use of the container.*** Will the container be stationary most of the time with little contact? Will the container be used indoors or out?
4. ***The handling abilities of the user.*** Is the user clumsy, or careful?
5. ***The operational use of the coolant.*** Will the coolant need to be left unattended for long periods of time?
6. ***The cost of the container.*** A 1-liter beaker costs ≈ \$5.00 at list prices. A 1-liter glass Dewar costs ≈ \$45.00. A 1-liter stainless steel Dewar costs ≈ \$95.00.

Dewars. Dewars are the best and most commonly used cryogenic containers in the laboratory. Their ability to maintain a temperature is exceptional. They are used in most labs where dry ice is found and in all labs where liquid nitrogen is found. Dewars are also found in many lunch boxes as Thermos bottles. Dewars are typically identifiable as a hollow-wall glass container with a mirror-like finish. That 'mirror' finish is a very accurate description because the silver coating on Dewars is the same as is used on mirrors.

As can be seen in Fig. 6-5, the Dewar is a double-wall glass container which is coated on the inside with a silver deposit. During manufacturing, after the Dewar is silvered, it is attached to a vacuum system and evacuated to about 10^{-6} torr before being 'tipped off' (see Sec. 8.2) to maintain the

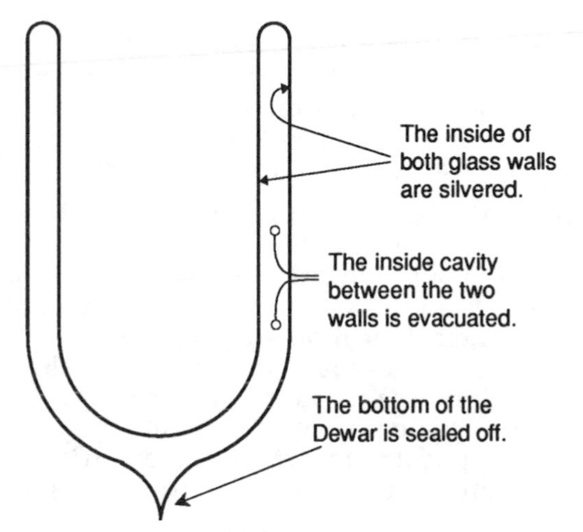

Fig. 6-5 A cross section of the Dewar.

vacuum on the inside. The Dewar achieves its temperature maintenance capabilities because of three different principles:

1) Glass is a poor conductor of heat, meaning that there will be very little temperature exchange from the coolant inside the Dewar with the rest of the container not in contact with the coolant.
2) The vacuum (within the double walls of the Dewar) cannot conduct heat. Because temperature cannot cross this 'vacuum barrier,' the cooling is further contained within the Dewar.
3) The silvered coating on the inside of the Dewar reflects radiation. The silvered coating prevents heat loss/exchange with the outside world.

For all of these features to be available in one package is an impressive feat. However, because glass can break under rugged conditions, it sometimes is preferable to use Dewars that are made out of stainless steel. Although stainless steel Dewars do not have all three of the heat exchange barriers of glass (stainless steel is not a good conductor of heat, but it conducts heat better than glass), they can stand up to far more physical abuse.

Glass Dewars should always be wrapped with fibered tape (not masking tape) to prevent glass from flying around in the chance the Dewar is accidentally broken. Some commercial Dewars have a plastic mesh. This mesh is acceptable, but wrapping with tape provides much better support to prevent flying glass. In addition, wrapping with white tape (such as any sport tape) provides one extra level of insulation for the Dewar than if the Dewar was wrapped with black tape (see Sec. 7.4.4). Dewars come in a variety of shapes and sizes as can be seen in Fig. 6-6.

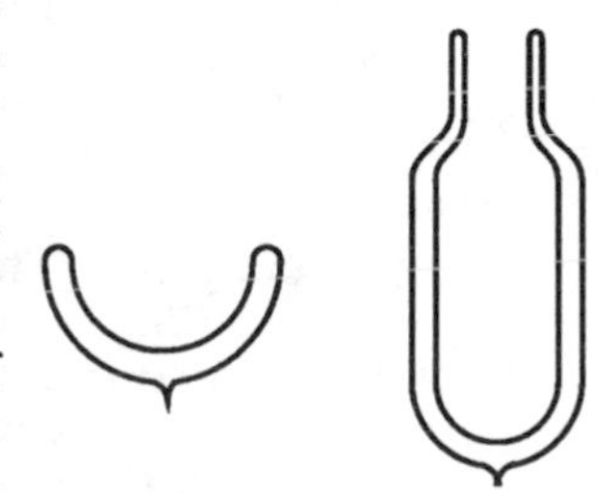

Fig. 6-6 Alternate Dewar shapes.

If frost appears on the outside of a Dewar, the vacuum within has deteriorated and the Dewar is no longer usable. This deterioration is typically caused by an imperfect tip-off at the base of the Dewar. It may take months or years for a Dewar to lose its vacuum, but once gone, the Dewar cannot effectively hold cryogenic fluids. It is possible however, for a glass shop to open the Dewar, clean, re-silver, re-evacuate, and re-tip off.

Foam Insulated Containers. Foam insulated containers are an inexpensive alternative to Dewars. They are double- (hollow-) walled containers that are filled with an insulating foam instead of a vacuum. These containers (which come in as many shapes and sizes as regular Dewars) are significantly less expensive but much less efficient than standard Dewars.

Foam insulated containers are adequate for some slush baths, but will require more effort to maintain the coolant. Foam insulated containers should never be used for long- or short-term storage of cryogenic liquids. They may be used for cryogenic liquid transport, but there will be significantly more loss of the liquid coolant (even in limited transport) than from a regular Dewar.

Beakers and Flasks. Beakers and flasks are the least effective containers for cryogenic materials because there is no insulation whatsoever. However, if the coolant is only water, ice, or a salt/ice mixture, not much insulation is required. There is little concern for rapid material loss with these coolant solutions because they are easy and inexpensive to replace. In addition it is (usually) safe to pick up these containers with your bare hands. If any ice forms on the sides of a container, it is simple to use gloves or tongs to pick up the beaker or flask and prevent possible skin damage.

6.2.10 Liquid (Cryogenic) Gas Tanks

Nitrogen, argon, and oxygen can be stored in liquid form in cryogenic gas tanks. As can be seen at the left of Fig. 6-7 they are large containers (about four and one-half feet tall and twenty inches in diameter). These tanks are in fact highly reinforced double-walled Dewars and can maintain the various cryogenic gases in liquid states (at room temperature) with a minimum of bleed-off.

Liquid nitrogen is used almost exclusively as a coolant in the lab. Liquid argon and oxygen are not used as coolants, but may be used in the lab, or in industry, when large quantities of these gases are required. The liquid form of the gas occupies much less space than the equivalent quantity of compressed gas. In addition, less time is lost in changing the equivalent number of tanks that would otherwise be required.

There are two major types of cryogenic tanks: one used primarily for liquid dispensing (see Figs. 6-7 and 6-8), and another used primarily for gas dispensing (see Fig. 6-7 and 6-11). Both are similar in size, both have rings of sheet metal around their tops to protect their valves from impact, and both have float devices on the top that indicate approximate liquid volume.

Both of the cryogenic tank designs have a tare (net) weight of ≈230-250 lbs., depending on design and manufacturer. When filled with a gas, the tank's weight is considerably greater: with nitrogen its weight is >300 lbs., with oxygen its weight is >400 lbs., and with argon its weight is >500 lbs.[6] Unless you have the proper training and equipment, never attempt to move a cryogenic gas tank by yourself. Should one of these tanks be tipped over and rupture, the potential damages and injuries could be extensive.

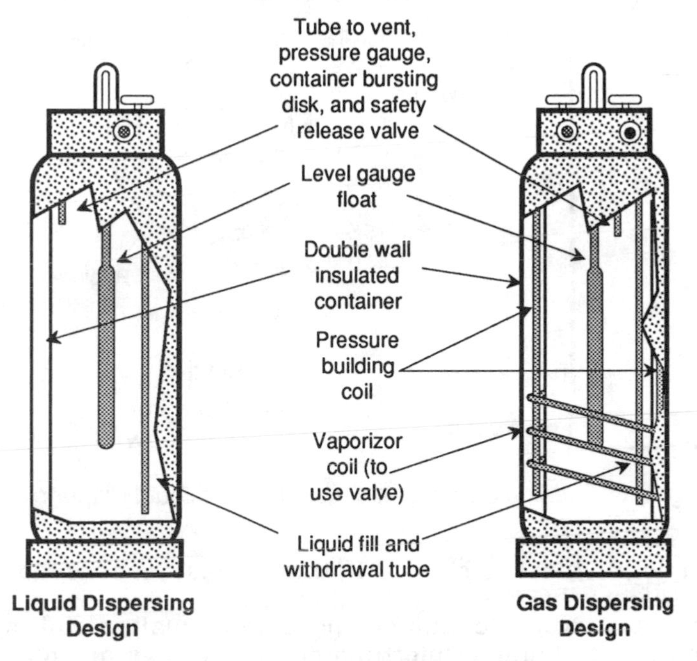

Fig. 6-7 The internal designs of liquid and liquid/gas dispersing cryogenic tanks.‡

‡ From the *Instruction Manual PLC-180A and PLC-180LP*, Figs. 3 and 8, by Cryogenic Services, Inc. Canton, GA 30114. With permission.

The Liquid Dispensing Tank. As opposed to high-pressure tanks filled with a highly compressed gas, cryogenic gas tanks hold the liquid form of a gas and are insulated to maintain the cryogenic temperatures necessary to maintain the gas in its liquid state.

Although the liquid dispensing tank can provide gas, it is best suited for dispensing liquids. The tank develops a head* pressure of 20 to 30 psig, which is sufficient to dispense the liquid gas (like a seltzer bottle) at a reasonable flow rate. If you remove the gas (as opposed to the liquid) at too fast a rate, the head pressure drops sufficiently that the amount of discharge equals the rate of gas creation, and the pressure drops to zero.

As seen in Fig. 6-8, the liquid dispensing tank has a gas port, liquid port, and a third port with a pressure gauge. Normal pressures are around 20 to 30 $^{lbs}/_{in^2}$. This pressure will change as liquid is dispensed or as the temperature conditions of the room change. Attached to the pressure gauge is a *pressure release valve* to prevent excess pressure build-up. Above a predetermined pressure, any excess pressure would pass out through the release

* The 'head' is the gas space within the container, above the liquid.

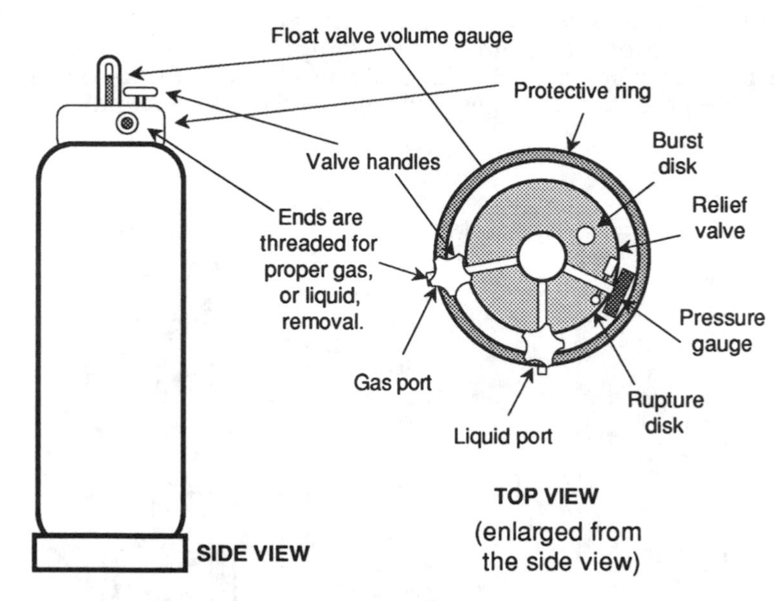

Fig. 6-8 The liquid (cryogenic) gas tank for liquid dispensing.

valve with a loud, hissing sound. If this release occurs do not become alarmed, the safety valve is simply doing its job.

Each port on a cryogenic tank should have a small, metal tag identifying each valve as gas or liquid. Unfortunately, these tags are often broken off leaving you with little idea of which port is which. Figure 6-8 shows one configuration that can be used for reference. However, not all manufacturers follow this pattern. If the tags are broken off, you have three ways of finding out which of the two ports will deliver gas or liquid, respectively;

1) Call the supplier of the tanks and ask it.
2) See if there are any tanks elsewhere in your building of that type and/or design. Because it is likely that any tanks from the same supplier will be of the same type, it is a safe guess that the valve setup will be similar.
3) Open and close the valve quickly several times to see what comes out. [Be sure that the area in front of the valve is free from materials that could be damaged by cryogenic temperatures. The first thing out of the liquid port will most likely be gas as it may take a moment or two for the liquid to arrive. Be sure you have adequate ventilation before doing this activity. Nitrogen and argon can displace air, leaving the user in an oxygen-poor environment. Do not use this technique with oxygen if there are any flames or sparks in the area.

The port in the middle (which may have a small metal clip hanging from it saying "Liquid") dispenses the liquid form of the gas within. To dispense the liquid into a Dewar, first attach a dispensing tube, which is available from your gas supplier. Place the opening of your Dewar over the end of the tube, and then slowly open the valve (see Fig. 6-9). Hissing and gurgling indicate that the liquid gas is pouring out the end of the tube. **Always point the open end of the Dewar away from your face and other people when filling so that any splattering cryogenic liquid will not fly into your face. You should wear safety glasses during this and any other operations with cryogenic fluids.**

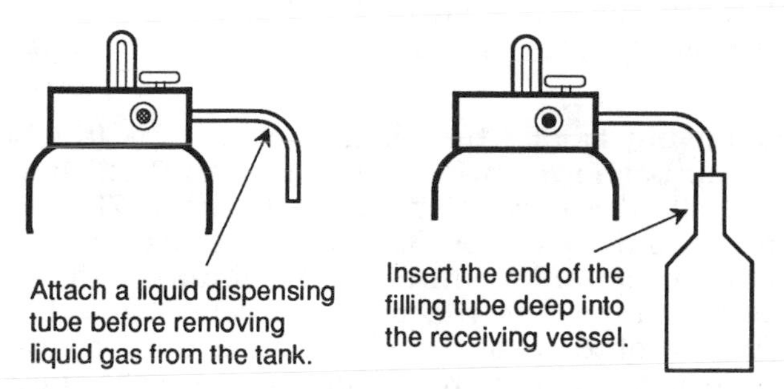

Attach a liquid dispensing tube before removing liquid gas from the tank.
Insert the end of the filling tube deep into the receiving vessel.

Fig. 6-9 Dispensing liquids from a liquid/gas tank.

Some labs may have a separate room where liquid nitrogen tanks are stored. [**Important: if you have such a room, be sure to leave the door open whenever transferring nitrogen into Dewars. The expansion of the nitrogen can easily leave a closed room oxygen poor!**] This tank setup might service one lab, a whole floor, or an entire building. To transport liquid nitrogen from a storage container to a lab, and provide short-term (one day) storage for the lab, there are various designs of transport Dewars (See Fig. 6-10). Transport Dewars are made of stainless steel, copper, plastic, or fiberglass. They are double-walled containers that are evacuated or foam-filled for insulation. The necks of transport Dewars should always be small compared to the size of their bodies. This design helps prevent splashing while transporting and facilitates pouring cryogenic fluids. When used for storage, a Styrofoam ball and pin should be loosely* placed in the neck to limit evaporation.

Fig. 6-10 Various designs of cryogenic liquid transport containers.

* If the cap is too snug, the expanding pressure (of the gas evaporating) may cause the cap to be ejected like a rocket or the container to explode.

It is essentially impossible to fill a short or shallow container from a liquid/gas tank because the liquid gas leaves the dispensing tube forcefully (there is limited ability to control the rate of flow), and once the liquid gas hits the bottom of the relatively warm container, it immediately (and forcefully) boils off and out of the container. In a taller container, the liquid gas does not eject itself from the container, but falls back to the bottom. Once back at the bottom, it can cool the container so that newly arriving liquid gas will not boil off. Eventually, incoming liquid gas will collect in a taller Dewar. A short or shallow Dewar can then be filled from the taller Dewar because the liquid can be slowly poured, limiting the amount of resultant ejecting fluids.

The Gas Dispensing Tank. The gas dispensing tank (the body is shown in Fig. 6-7 and the top is shown in Figures 6-11 and 6-12) is designed to provide gas continuously at a delivery pressure between 75 to 175 psig. It can provide a continuous supply of 250-350 cfh (cubic feet per hour) with bursts of up to 1000 cfh. These tanks can dispense liquid as well, but it is necessary to alter them by replacing the pressure relief valve from the supplied 235 psig valve to one of 22 psig. However, once the alteration is done, the tank cannot dispense gas at the greater pressure unless the tank is restored to its original condition.

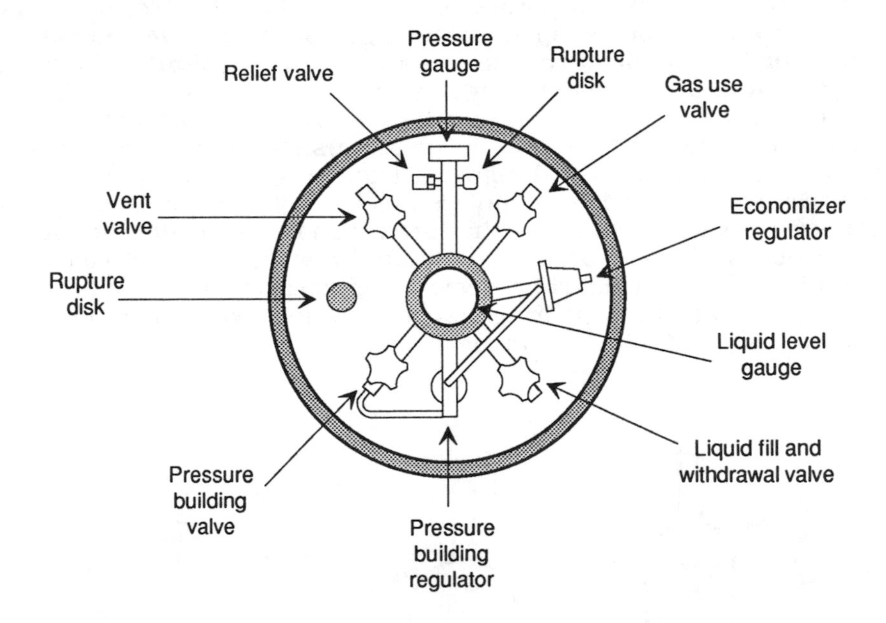

Fig. 6-11 Top view of pressure building liquid/gas tank.‡

‡ From *Instruction Manual for PLC-180A and PLC-180LP*, Fig. 2 from Cryogenic Services, Inc. Canton, GA 30114 . With permission.

The pressure building valve is connected to a series of heat transfer tubes within the casing that encircles the inner portions of the tank. By opening this valve, a much higher gas pressure can be achieved (while releasing the gas) than could otherwise be maintained. Oxygen and nitrogen tanks are set to deliver 125 psig, and argon tanks are set to deliver 75 psig. All gas dispensing tanks can be attached to an external vaporizer to provide gas at higher rates and/or pressures than may otherwise be available solely through the internal vaporizer.

An external regulator must be attached to a gas dispensing tank (just as you would to a high compression tank) to control the gas (tank) pressure to the desired (outflow) pressure. Liquid cryogenic tanks follow the same CGA (Compressed Gas Association) numbering standards (see Sec. 5.1.3) as compressed gases. For oxygen, use a CGA 540 fitting, and for nitrogen or argon, a CGA 580 is required.

To use a gas dispensing cryogenic tank, attach the correct regulator to the gas use outlet. Then, open the gas use valve and pressure building valve. Once the pressure is at least 125 psig, you may then adjust the regulator to the required delivery pressure and dispense the gas as needed. Do not handle any tubing from the tank as it will be very cold and can damage skin.

Cryogenic tanks have limitations that become immediately obvious if gas is withdrawn at a rate faster than the internal coils can maintain: frost develops on the outlet connections and/or regulator which decreases gas output. The problem can be resolved with an external heat exchanger. The external heat exchanger is first connected to the gas use valve. Then, the regulator is attached to the external heat exchanger.

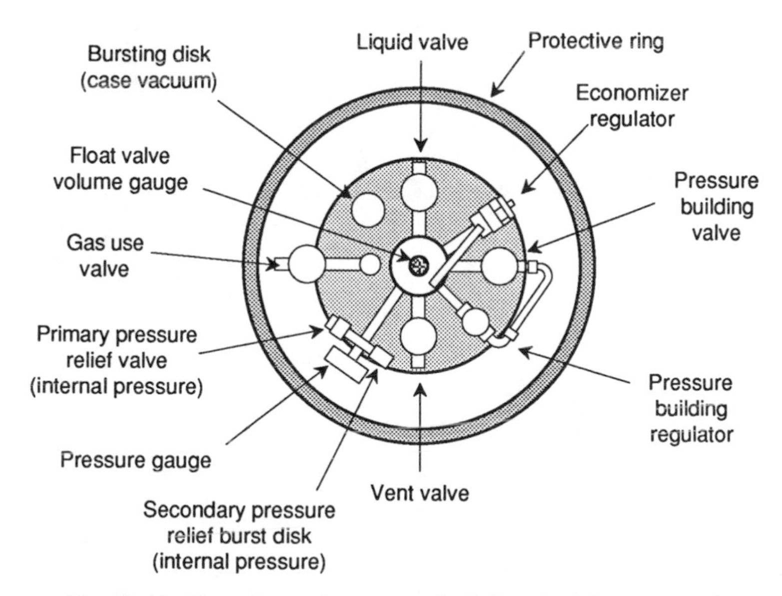

Fig. 6-12 Top view of pressure building liquid/gas tank.‡

‡ From *Instructions for XL-45 and 50*, Form TW-71B 1088, page 23, from Taylor-Wharton Cryogenics, Theodore, AL 36590. With permission.

Other options for gas dispensing cryogenic tanks are manifolds that can connect two to six cylinders together. These manifolds can provide flow rates of 250 cfh (cubic feet per hour), set up a reserve of gas for uninterrupted flow when changing cylinders, and (with an economizer circuit) can cut loss due to evaporation. For extra-high-capacity gas demands, there are external vaporizing manifolds which are combine of the external heat exchanger and manifold setup.[7]

References

1 R.K. Lane, P.D. Provence, M.W. Adkins, and E.J. Elsenbraun, "Laboratory Steam Distillation Using Electrically Generated Superheated Steam," *J. of Chem. Ed.*, 64 (1987), pp. 373-75.

2 J.P. Hagen and K.L. Barton, "An Inexpensive Laboratory Steam Generator," *J. of Chem. Ed.*, 67 (1990), p. 448.

3 R.E. Stronski, "Minimizing Evaporation from Constant Temperature Water Baths," *J. of Chem. Ed.*, 44 (1967), p. 767.

4 D.W. Mayo, R.M. Pike, and S.S. Butcher, *Microscale Organic Laboratory* (New York: Wiley, 1986).

5 S. N. Lodwig, "The Use of Solid Aluminum Heat Transfer Devices in Organic Chemistry Laboratory Instruction and Research," *J. of Chem. Ed.*, 66 (1989), pp. 77-84.

6 *Instruction manual for PLC-180A & PLC-180LP*, from Cryogenic Services Inc., Canton, GA 30114.

7 *Instruction manual for PGS-45 Portable Gas Supply System*, Form 13-109-D (February 1975), from Union Carbide Cryogenic Equip., Indianapolis, IN 46224.

Chapter 7

Vacuum Systems

7.1 How to Destroy a Vacuum System

Carpenters have a saying "measure twice, cut once." This saying implies that any extra time spent in preparation saves time and materials that may otherwise be wasted. Likewise, any time spent preparing an experiment, or equipment (including general maintenance), saves time and materials, and, in the long run, may also save a life.

This section lists fundamental potential pitfalls when working with a vacuum system. If you read no other section in this book, read this one. By following the rules and guidelines that you are directed to within this section, many hours, and perhaps weeks, of problems will be avoided. It is better to avoid the situation of "not having enough time to do it right, but plenty of time to repair the wrong." This list is not meant to be comprehensive, it cannot be. It is however, a collection of the more common disasters that occur on a laboratory vacuum system.

1) Blowing up a vacuum line by freezing air in a trap (see Sec. 7.4.3).
2) Sudden bursts of pressure in McLeod gauges that can cause mercury to spray throughout a line (problem) or break a line (big problem) (see Secs. 7.5.6-9).
3) Breaking stopcocks off vacuum lines (see Sec. 7.7.1, point 8).
4) Breaking off glass hose connections when pulling flexible tubing off (see Sec. 7.3.16).
5) Destroying the oil in a diffusion pump (see Sec. 7.3.1).
6) Destroying the oil in a mechanical pump (see Sec. 7.4.1).
7) Destroying a mechanical pump (see Sec. 7.3.4).
8) Wasting time re–evacuating vacuum lines (see Sec. 7.3.13).
9) Causing virtual leaks in cold traps (see Sec. 7.4.3).
10) Frothing the oil of a two-stage pump (see Sec. 7.3.3).
11) Breaking a cold trap off a vacuum line by not venting it to the atmosphere before removing the bottom (see Sec. 7.4.4).
12) Achieving a poor-quality vacuum when starting a vacuum system up for the first time (see Sec. 7.6.2).
13) Dissolving O-rings during vacuum leak detection (see Sec. 7.6.5).
14) Punching holes in glass with a Tesla coil (see Sec. 7.6.6).

15) Burning the filament of an ion gauge (see Sec. 7.5.20).
16) Imploding glass items on a vacuum system (see Sec. 7.7.1, point 2).
17) Placing items on a vacuum rack so that they fall into the rack (see Sec. 7.7.1, point 3).
18) Breaking vacuum system tubing when tightening two and/or three finger clamps on a vacuum rack (see Sec. 7.7.1, point 4).
19) Sucking mechanical pump oil into a vacuum line (see Sec. 7.3.4).

7.2 An Overview of Vacuum Science and Technology

7.2.1 Preface

In this book I have intentionally avoided equations whenever possible. The reason is simple: most people do not need, or use, equations when using equipment. This introductory section on vacuum technology contains equations. They are presented so the reader may better understand the relationship between the various forces in vacuum systems. None are derived, and none are used beyond presenting some basic points. If you are interested in the derivation of any formula, see the recommended books at the end of this chapter.

It is not necessary to know the material in this introductory section to run a vacuum system. It explains many terms and ideas that are used throughout the rest of this chapter, however. Because of all the the basic information contained within, I recommend that you read this section.

Vacuum systems are used in the lab for such purposes as: preventing unwanted reactions (with oxygen and/or other reactive gases); distilling or fractionally distilling compounds (a vacuum can lower the boiling point of a compound); or transferring materials from one part of a system to another (using cryogenic transfer). A vacuum system can also be used for more sophisticated processes such as thin film deposition, electron microscopy, and many other areas that could fill a book if listed.

The goal in creating a vacuum is to get rid of, or bind up, a significant amount of the gases and vapors (mostly air and water vapor) within a vacuum system. Regardless of the approach, the goal is a net reduction of pressure in the system.

A perfect vacuum is a lack of everything. This state we cannot achieve in the laboratory. Table 7-1 illustrates that by decreasing a vacuum to 10^{-3} torr (a good-quality vacuum in a standard lab) we can remove over 99.99% of the particles* that are present at room pressure. At 10^{-6} torr, a high vacuum, there is still quite a considerable number of particles remaining ($2.45 \times 10^{17}/m^3$). Despite the (seemingly) large quantity of particles remaining, this vacuum is still sufficient to successfully limit the chance of oxidation and/or unwanted reactions for many laboratory requirements. More than 95% of all

* For this discussion, water vapor has been ignored because its percent concentration varies on a day-to-day basis.

vacuum studies and technique work can be successfully achieved within the vacuum ranges of 10^{-2} to 10^{-6} torr. The other 5% (studies in the ultra-high vacuum range) consist of surface and material studies and space simulation, which require as little contamination as possible.

Aside from earthbound technological approaches to achieve a vacuum, the further away from the earth's surface you go, the less atmosphere there is and therefore, the greater the vacuum (relative to atmospheric pressure on the earth's surface) that can be achieved. In fact, on earth, someone standing on top of Mt. McKinley experiences vacuum greater than can be created with a standard vacuum cleaner at sea level.[1] Table 7-1 shows the approximate miles above earth to obtain various conditions of vacuum. As this table shows, outer space offers wonderful opportunities to produce vacuum conditions for experimental or industrial work. Outer space can provide an infinite vacuum system, an essentially contamination-free environment, and blip free (no pressure surge) conditions. The possibilities of ultra-vacuum research in space is discussed by Naumann in his paper on the SURF (Space Ultravacuum Research Facility) system.[2] Unfortunately, space is an opportunity that is as far from the expected uses of this book (no pun intended) as one can get. Back on earth we are still limited to using pumps, traps, oils, gauges, glass, metal-support clamps, elastomers, and other implements of laboratory vacuum systems.

Although vacuum systems are not as expensive as space travel, they are not inexpensive, and the greater the vacuum desired, the more expensive it will be to achieve. If your employer is concerned about the bottom line, it is important to consider your needs before you begin designing your system. Not only is there an increase in the cost of pumping equipment to cover the range from poor- to good-quality vacuums (mechanical to diffusion to turbomolecular to cryogenic pumps), but there are also the correspondingly increasing costs of support equipment (power supplies, thermocouple and ion gauges, and their controllers), peripheral equipment (mass spectrometry and/or He leak detectors), peripheral supplies (cooling water, Dewars, and liquid nitrogen), and support staff (technicians to run and/or maintain the equipment). These requirements can all add up.

Sometimes, you will have no alternative but to use comparatively expensive equipment. For example, a diffusion pump must be used in conjunction with a reasonably powerful (and therefore relatively more expensive) mechanical pump: if you were to use the diffusion pump with a small classroom demonstration mechanical pump, the diffusion pump would not work. On the other hand, do not assume that price is necessarily indicative of the best choice in equipment. You need to match components by their capabilities. That is, if your system is capable of only a low-vacuum, you are wasting money purchasing a gauge designed with high-vacuum capabilities (which can be significantly more expensive). The requirement of matching components is equally important whether you are purchasing vacuum equipment from scratch or adding to components you already own.

Because a perfect vacuum cannot be achieved, and the next best thing (outer space) is for all intents and purposes out of reach, there are definable limits to the ultimate pressure of any system you are planning. These limits are based on:

1) How much you can spend
2) How much maintenance you are willing/needing to do
3) What supplies and materials you have (or can have) available
4) Your vacuum experience and knowledge
5) What technical support you have available.

It is therefore important that you know what the demands and needs of your experimental work are. It is neither economical nor practical to have an elaborate vacuum system for simple vacuum needs.

7.2.2 How To Use a Vacuum System

The easiest way to explain a vacuum system is to explain each part of the vacuum system individually. However, the only way to use a vacuum system is to use all parts. Therefore there is a conflict of interest in coordinating a section that will explain the parts of vacuum systems in general and how to use *your* vacuum system in specific.

If you are starting to use a vacuum system for the first time, I recommend you skim this entire chapter. Afterward, study your vacuum system and see what components you recognize. Finally, re-read the sections of this chapter that are pertinent to your system. This method of study may seem like a lot of work before you turn on a switch or twist a stopcock, but the vacuum system you save may be your own.

Due to variations in equipment, controllers, and designs, what you see on your system will probably not be what you see in this book. You will have to accommodate and respond accordingly. In addition, re-read Sec. 7.1 and be sure you understand how accidents (disasters) can happen and how they can be avoided. That knowledge in itself will be the first major step to successful vacuum practice.

Table 7-1[3,4]

Pressure, Particles[a], and the Mean Free Path

Pressure (torr)	Quality of Vacuum	Miles Above Earth[b] (≈ miles)	# of Particles (per m^{-3})	Mean Free Path (m)
760	none	0	2.48×10^{25}	6.5×10^{-8}
0.75	medium	5	2.45×10^{22}	6.5×10^{-5}
7.5×10^{-3}	high	35	2.45×10^{20}	6.5×10^{-3}
7.5×10^{-6}	very high	50	2.45×10^{17}	6.64
7.5×10^{-8}	ultra high	90	2.45×10^{15}	664
7.5×10^{-10}	extreme ultra high	290	2.45×10^{13}	6.6×10^{4}

[a] This table is based on dry air because of the daily variations of water vapor.
[b] These approximate values are provided to allow the reader to appreciate the value, and difficulties, of space travel.

7.2.3 The History of Vacuum Equipment[5,6,7,8]

It is interesting, when looking back on history, to see how much has happened, and yet how little things have really changed. The following information is by no means comprehensive. At best, it is meant to provide the reader with an appreciation of how long ago some vacuum equipment was invented. For example, the McLeod gauge and Töepler pump were invented over a hundred years ago. Over the years, there have been many improvements made in such items as the ion gauge and diffusion pump, however, the principles on which they work are so fundamental that their importance and use have not been lost.

Aristotle, some 2000 years ago, was one of the first to make any serious comments about vacuum technology that have survived time. His statement, "*nature abhors a vacuum*" may not tell us very much about what is taking place in a vacuum, but it is as true now as it was then.

The first vacuum science work is credited to Evangelista Torricelli. Unable to do intended work with Galileo (who was busy with the church and the Inquisition at the time), Torricelli began his own experimentation. In the 1640's he demonstrated that when a pump was bringing water out of a well, it was operating by the force of the atmosphere pushing the water into the evacuated space of the pump rather than the force of the vacuum pulling the water up and out of the well.*

Torricelli proved that the force at work was the atmosphere pushing rather than a vacuum pulling with a simple experiment: if a void pulled material, a better void would pull better than a weaker void. He demonstrated that regardless of the quality of the pump being used, he could only draw water up a pipe some 33 feet. Pascal, a French mathematician and philosopher, later repeated Torricelli's work using glass tubes 46 ft long and wine. His work also demonstrated that the atmosphere could only support 33 ft of wine (whose specific gravity is similar to water).

Torricelli later made the first barometer by filling a 48 in sealed glass tube with mercury and inverting it. During this time, it was believed that the vacuum space created at the top of a barometer was a perfect vacuum. Obviously the concepts of vapor pressure and the various gas laws were a long way off.

The next major technological breakthrough came from Otto von Guericke, a German burgomaster, who invented a piston vacuum pump (which he called an *air pump*) that was capable of producing the best vacuum of the time. Guericke became sort of a vacuum showman, and is most known for the Magdeburg hemispheres. These structures were two hollow bronze hemispheres that were sealed by evacuating the space between the halves. Two teams of horses, each attached to a respective half, could not pull them apart. However, by opening the evacuated space to the atmosphere and allowing air into the sphere, the hemispheres fell apart on their own.**

* The difference between the two—a force pushing rather than a void—pulling is fundamental to understanding the movement of materials in a vacuum. Pressure is a force, and a vacuum is *not* an opposite (nor equal) force. A vacuum is simply a region with less pressure when compared to a region with greater pressure, and it is the molecules creating that pressure by pushing that creates a force rather than a lack of molecules creating a negative pressure by pulling.

** An interesting (and useful) way to duplicate this experiment is with an evacuated screw-top jar that you are unable to open. Take a knife, screwdriver, can opener (church key), or any prying device with which you can pry the lid away from the glass enough to momentarily distort the shape and allow air into the container. You will hear a momentary 'pssst' as air enters, as well as a click of the metal as its shape goes from concave to normal. It will now be very simple to open the jar.

Robert Boyle (a founder of the Royal Society of England) heard about Guericke's pump and experiments and decided to scientifically analyze them. Aside from making improvements upon the design of the air pump, he also developed the relationship between gas pressure, volume, and temperature. This relationship was later called Boyle's Law (see Eq. (7-2).

The 17th century saw the beginning studies of electric discharges within a vacuum by Nollet in Paris in the 1740's. These studies were later refined by Faraday in England during the late 1830's and by Crookes in the late 1870's.

Manometers were eventually used as measurement devices. By the 1770's, mercury was boiled to increase measurement accuracy. Thus, albeit crudely, both *outgassing* and *baking out* of a vacuum system were instigated.

In 1851, Newman developed a mechanical pump that achieved a vacuum of 30.06 in of mercury on a day that the barometer was reading 30.08 in. This pump was very impressive for the time. Vacuum technology was further enhanced by the invention of the Töepler pump in 1862, the Sprengel pump in 1865, and the McLeod gauge in 1874.

The 'Dewar', invented by Dewar in the early 1890's, utilized the combined effect's of evacuating and silvering the empty space of a double wall container. Later, having the ability to keep things cold, Dewar developed the basic concepts of cryopumps by superchilling charcoal for absorbing gases.

In the 1910's, Gaede invented the mercury diffusion pump that was later improved by Langmuir (from the General Electric Co.), and improved yet again by Crawford, and yet again by Buckley. In the 1920's, industry began substituting refined oil for mercury in diffusion pumps.

The new century saw an abundant development of vacuum gauges. Both types of thermal conductivity gauges were invented in 1906 (Pirani inventing the Pirani gauge and Voege inventing the thermocouple gauge). The hot cathode gauge was invented by Von Baeyer in 1909 and the cold cathode gauge was invented by Penning in 1937. The early 1950's saw improvements in both the hot cathode gauge (by Bayard and Alpert) and cold cathode gauge (by Beck and Brisbane). Redhead made improvements on both types of gauges a few years later.

Because 1945, the fields of vacuum physics, technology, and technique have become the backbone of modern industrial production. From energy development and refinement, to silicon chips in watches, games, and computers, all are dependent on 'striving for nothing.'

7.2.4 Pressure, Vacuum, and Force

Gas pressure can be loosely imagined as the pounding of atoms and molecules (a force) against a wall (an area). As molecular activity increases (for instance by heating), the atoms and molecules pound away with greater activity, resulting in an increase in pressure. In addition, if some gas molecules are removed, there is more room for the remaining ones to move around, and fewer are available to hit the wall. This activity results in a drop in pressure. We use atmospheric pressure (which varies itself) as a dividing line for describing what is a vacuum or pressure environment: that environment which is greater than atmospheric pressure is a pressure, that environment which is lower than atmospheric pressure is a vacuum. However, a positive pressure can only be defined by comparing it with something else that has less pressure, even if both are vacuums when compared to atmospheric pressure.

Units of pressure and vacuum should be identified as *force per unit of area*, and pressure units typically are mbar, psi, and kg/cm^2. Pressures below 1 torr ($\approx 10^{-3}$ atm) for many years were described in relationship to standard atmosphere* in various ways such as:

1) ***Millimeters of mercury*** (1/760 of a standard atmosphere).**

2) ***Torr*** (≈ 1/760 of a standard atmosphere).

3) ***Micron*** **(μ)** ***of mercury*** (1/1000 of a millimeter of mercury and/or 1/1000 of a torr).

4) ***Millitorr*** (also 1/1000 of a millimeter of mercury and/or 1/1000 of a torr).

To maintain the relationship of *force per unit of area* that is used in pressure and vacuum, the SI decided on the term *Pascal* (one Newton per square meter) as the vacuum unit. The numeric relationship between all of these designation can be seen in the Pressure Conversion Table, Table 7-2.

Table 7-2

To Convert	Pressure Conversion Table						
From \ To	Pascal (multiply by)	torr (multiply by)	atm (multiply by)	mbar (multiply by)	psi (multiply by)	kg/cm^2 (multiply by)	μ (multiply by)
Pascal	1	7.5×10^{-3}	9.87×10^{-4}	10^{-2}	1.45×10^{-4}	10.2×10^{-6}	7.5
torr	133	1	1.32×10^{-3}	1.333	1.93×10^{-2}	1.36×10^{-3}	1000
atm	1.01×10^5	760	1	1013	14.7	1.033227	7.6×10^5
mbar	100	0.75	9.87×10^{-4}	1	1.45×10^{-2}	1.02×10^{-2}	750.1
psi	6.89×10^3	51.71	6.8×10^{-2}	68.9	1	0.070307	5.17×10^4
kg/cm^2	9.81×10^4	735.6	0.968	981	14.2	1	7.35×10^5
μ	0.1333	1×10^{-3}	1.32×10^{-3}	1.33×10^{-3}	1.93×10^{-5}	1.36×10^{-6}	1

Vacuum and pressure measurements were all originally made compared to atmospheric pressure, or 'gauge' pressure. The term *psig* (pounds per square inch-gauge) refers to this comparison. Absolute pressure includes atmospheric pressure (14.7 psi) and is called *psia* (pounds per square inch-absolute). For example, your tire pressure is 35 psig or 49.7 psia. Generally, unless otherwise identified, the lone identification *psi* refers to gauge pressure.

7.2.5 Gases, Vapors, and the Gas Laws

There is no difference between a gas and a vapor except that a vapor is near its condensation temperature and/or pressure. Some gases (hydrogen,

* standard atmosphere is the average pressure that the atmosphere exerts at 0°C, sea level.

** Millimeters of mercury implies a length measurement. Because we are measuring a force, not a length, a different term was created called the torr (in honor of Torricelli). One torr is equal to 1 mm of mercury.

helium, nitrogen, oxygen, and other cryogenic gases) do not have a vapor state anywhere near STP and are sometimes called *permanent gases*. For more information on condensation as well as evaporation and equilibrium, see Sec. 7.2.6.

Of the three states of matter (solid, liquid, and gas), only gases have radically changing distances between molecules. When the distances between the molecules of a gas are different than what is found at STP (Standard Temperature and Pressure), we have either a positive or negative pressure (compared to atmospheric).

As far as a gas is concerned, it does not make a difference whether the size of the container, the temperature of the gas, or the amount of gas has changed. All these conditions separately or together can change the pressure within a container. Analysis of contained gases led to the following gas laws, most of which go by the name of the researcher who formally identified them:

1) All gases uniformly fill all spaces within a container. This space is called the volume (V). There cannot be an independent localized collection of gases exerting uneven pressures within a container.
2) All gases exert an equal pressure on all points of a container. Regardless of the shape of the container (on a static system), there can be no variation of gas pressure from any one point to any other point within that container.
3) All gases exhibit a direct relationship between the temperature and pressure for a given volume and given amount of gas. Assuming that nothing else changes, if the temperature increases, the pressure will increase a directly proportional amount.

Charles' Law

$$\frac{P}{T} = (\text{constant}) = \frac{P_1}{T_1} = \frac{P_2}{T_2} \qquad \text{Eq. (7.1)}$$

4) All gases exhibit an inverse relationship between the volume and the pressure of a gas for a given temperature and amount of gas. Assuming that nothing else changes, if the volume increases, the pressure will decrease a directly proportional amount.

Boyle's Law

$$PV = (\text{constant}) = P_1V_1 = P_2V_2 \qquad \text{Eq. (7.2)}$$

5) All gases of equal volumes, at the same temperature and pressure, contain the same number of molecules. The size of a gas atom (or molecule) has no relationship to pressure, temperature, or volume.

Avogadro's Law

$$\frac{V}{n} = (\text{constant}) = \frac{V_1}{n_1} = \frac{V_2}{n_2} \qquad \text{Eq. (7.3)}$$

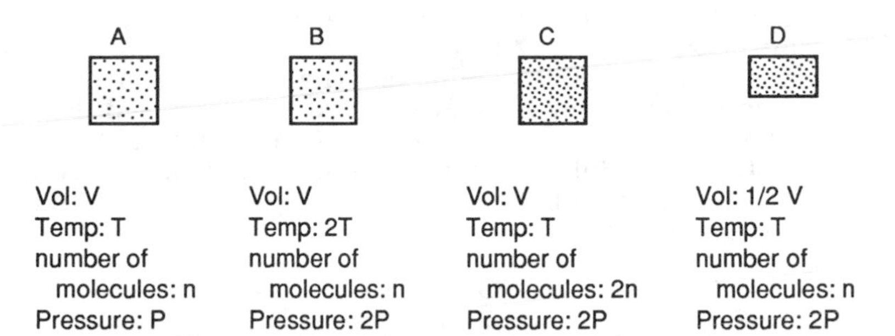

Fig. 7-1 Notice how you can double the pressure of a container by either doubling the temperature, the number of molecules, *or* halving the volume.

6) All gases exhibit a uniform relationship between the pressure and volume, temperature, and number of molecules present. This relationship is represented by the gas constant "R" (which is = 62.4 torr·l / moles·°K) when **P**ressure is in torr, **V**olume is in liters, **T**emperature is in K, and **n** is the number of moles.*

Ideal Gas Law

$$PV = nRT \qquad \text{Eq. (7.4)}$$

7) If more than one gas is in a container—and the gases are not interacting—each gas will exhibit its own characteristic pressure. The sum of these individual pressures equals the total pressure. The percentage of each partial pressure is equal to the percentage of that gas in the sample.

Dalton's Law of Partial Pressures

$$P_t = P_1 + P_2 + P_3 + \ldots \qquad \text{Eq. (7.5)}$$

Laws 3, 4, and 5 are shown pictorially in Fig. 7-1, where three different approaches to doubling the pressure are demonstrated. In Box A we have a given number of molecules at a given temperature and pressure in a container of one unit. In B, we have doubled the temperature, which causes the molecules to double their activity, which doubles the pressure. For Box C we have brought the temperature back to the original temperature, but have doubled the number of molecules within the container, which causes a doubling of the pressure. Finally, in box *D* we have the same original temperature, and the same number of molecules as in Box A, but we have decreased the size of the box by half, which doubles the pressure.

7.2.6 Vapor Pressure

The greatest vacuum that can be obtained within a system is solely dependent on the material with the greatest vapor pressure within the system.

* A mole is an Avogadro's number (6.023×10^{23}) of molecules. One mole is equal to one atomic mass, or the molar mass of a molecule or atom. For example, one mole of carbon (atomic mass 12) is equal to 12.01115 g.

Remember, a vapor is a gas that is near its condensation temperature and/or pressure. As can be seen in Fig 7-2, the evaporation, equilibrium, and condensation processes are dynamic conditions. That is, the molecules involved are not actually static after they have evaporated or condensed. For example, in evaporation, statistically more molecules are turning into a gas than those turning into a liquid (or solid); in condensation, statistically more molecules are turning into a liquid (or solid) than are turning into a gas; and, in equilibrium, the exchange is equal.

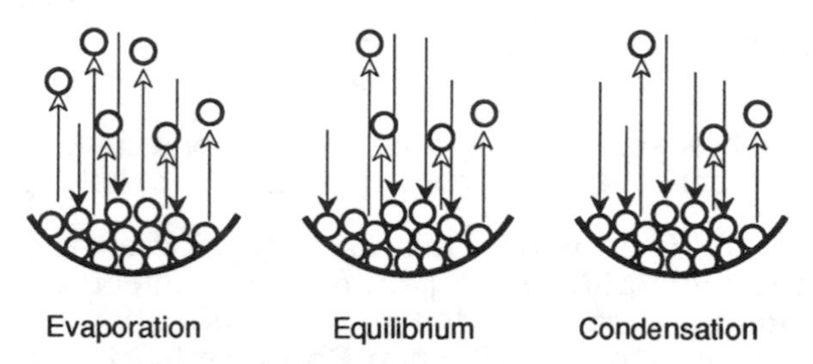

Fig. 7-2 Evaporation, equilibrium, and condensation are dynamic processes of vapor pressure.

The *vapor pressure* of a material is the maximum vacuum potential that can be achieved against its evaporation rate at a given temperature. As a material evaporates, its molecules are included among those in the air envelope around the material. Normally, to obtain a vacuum, you are trying to rid (or bind up) the gas molecules within a vacuum. As long as the molecules you are trying to get rid of are being replaced, you cannot improve the quality of the vacuum.

All compounds have different vapor pressures. The vapor pressure of any given compound is dependent on its temperature and the pressure of its environment. As temperature is lowered, the vapor pressure of any compound also lowers. Therefore, the easiest way to improve a vacuum is to lower the vapor pressure of the materials within the system by chilling with either water, dry ice, or liquid nitrogen. However, as pressure drops, vapor pressure increases, which is just the opposite of what you are trying to achieve. The only resolution to this conflict when trying to obtain as good a vacuum as possible is to use the best chilling mechanism available.

7.2.7 How to Make (and Maintain) a Vacuum

Aristotle's statement "nature abhors a vacuum" means that even if you are successful in creating a vacuum, your ability to maintain that vacuum can require an equal, if not greater, amount of work. When creating a vacuum, you must establish your needs, define (and understand) your conditions, consult with authorities before you make purchases, and understand (and accept) any compromises.

We often euphemistically refer to creating a vacuum as emptying a container of its contents. But, what does "empty" mean? We already stated that it is impossible to make a container void of contents, so what do we need to do to "empty" a container?

Table 7-3[9]

Composition of Dry Air[a]			
Gas	**Volume (%)**	**ppm**	**Partial Pressure (in torr)**
Nitrogen	78.08		593
Oxygen	20.95		153
Argon	0.934		7.1
Carbon Dioxide	0.031		0.24
Neon		18.2	1.4×10^{-2}
Helium		5.24	4.0×10^{-3}
Methane		2.0	1.5×10^{-3}
Krypton		1.14	8.7×10^{-4}
Hydrogen		0.5	3.8×10^{-4}
Nitrous Oxide		0.5	3.8×10^{-4}
Xenon		.01	7.6×10^{-5}

[a] The partial pressure of water vapor in air depends on temperature and relative humidity. For example, at 20°C (saturation vapor pressure of water =17.5 torr), and a relative humidity of 45%, the partial pressure of water vapor would be 0.45 times 17.5 torr or 8 torr.

There are three levels that should be considered when "emptying" a container. For example consider emptying a glass of water: First you take the glass and pour out the water. From a simple aspect, the glass is empty. However, there is still a film of water in the glass, so you dry the walls of the glass with a towel until they are dry to the touch. However, if you want the glass really really dry, you need to bake out the water that has adsorbed on the walls and is absorbed into the glass walls (water can soak into a glass matrix up to fifty molecules deep). Thus, when we talk about creating a vacuum, until you remove the *adsorbed* (surface concentration) and *absorbed* (material penetration) gases and vapors, you do not have an "empty" system.

So, say that you take a glass, pour out the contents, dry the walls, and bake out the glass so it is truly *empty*. The question is, will it now remain empty until water is poured back in? As far as vacuum science is concerned, no. As soon as the glass is exposed to the atmosphere at room temperature, the walls will re-saturate with water vapor and the glass will no longer be empty. To maintain a glass as 'empty,' it must be isolated from the atmosphere. Otherwise you must repeat the drying process.

Unfortunately, glass vacuum systems cannot be 'baked out' (as is done with metal vacuum systems) to remove adsorbed water. Baking is likely to damage stopcocks, rotary valves, or the glass walls of the system itself. Thus, glass vacuum systems are not practical if baking your system is required.

What typically happens with a glass vacuum system is that first a mechanical pump removes a great deal of the permanent gas particles. Then, greater vacuum is achieved with the combination of a diffusion pump (or similarly fast-pumping unit) and traps that bind up the various vapors within the system (for example oil, mercury, and water). The only way a system can achieve a vacuum lower than 10^{-6} to 10^{-7} torr is if the pump can remove water vapor faster than the water vapor can leave the walls. Most diffusion pumping systems cannot achieve this goal, but even if they could, there is a such substantial amount of water vapor within the glass that, unless the walls are baked, a better vacuum cannot be obtained.

Aside from adsorbed gases, there are six other sources of leaks in vacuum systems. All of these leak sources can make the job of maintaining a vacuum more difficult that obtaining a vacuum. An illustration of all the gas sources within a vacuum system can be seen in Fig. 7-3. Perhaps one of the most subtle sources of leaks into an ultra-high vacuum system is permeation of the glass by helium, present in the atmosphere at 5.24 ppm (see Table 7-3). Helium permeation* of glass can be useful for a standard leak (used with a He leak detector, see Sec. 7.6.9-11), but can be debilitating when trying to obtain extremely ultra-high vacuum with glass components.

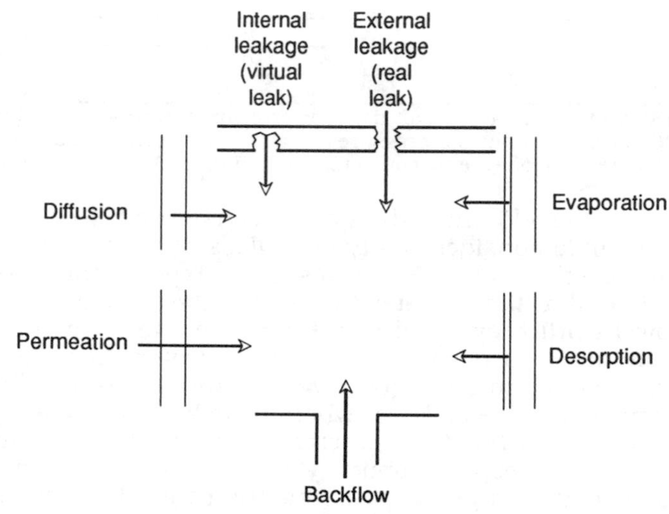

Fig. 7-3 Gas sources in a vacuum system.‡

‡ From *An Elementary Introduction to Vacuum Technique*, by G. Lewin, Fig. 19, © 1987 by the American Vacuum Society, American Institute of Physics, Inc. New York, N.Y. 10017. With permission.

Water adsorbs into the walls of a glass container, but that adsorption is the extent of its adhesion. Some adsorbed molecules react chemically with some types of containers in a process called chemical adsorption (chemisorption). For example, carbon monoxide chemisorbs with palladium, but not with gold). The bonds resulting from chemisorption can hold molecules to the surface with far greater force than would exist with only physical attraction. It is also possible for a molecule (that normally would not chemisorb with the container wall) to break up when hitting the wall's surface. At that point the molecule's constituent parts chemisorb with the container walls. When an adsorbed gas reacts with the materials of a container, it is called *reconstruction* (for example, the reconstruction of iron with oxygen is rust).

* Permeation (and absorption) are both conditions that only apply when the penetrating molecule is much smaller than the molecules of the wall material. Any size molecule can be involved in adsorption.

7.2.8 Gas Flow

There are three basic types of gas flow: *turbulent*, *viscous*, and *molecular*. The type of flow passing through a given system is dependent on both the Mean Free Path (MFP) of the molecule(s) and the diameter of the container (tube) through which they are flowing. A useful formula when talking about MFP is the Knudsen number (*Kn*), defined in Eq (7.6).

$$Kn = \frac{L}{d} \qquad \text{Eq. (7.6)}$$

where L = MFP
and d = Diameter of the tube in question

When a system is first brought into vacuum conditions from atmospheric pressure, the flow is turbulent (see Fig. 7-4). At this time, the *Mean Free Path* (MFP) is approximately 9×10^{-7} cm, which is considerably smaller than any tube the gas is likely to be in. All interactions are gas-gas which means there is a greater likelihood that a molecule of gas will hit another molecule of gas than it will hit a wall (see Fig. 7-5).

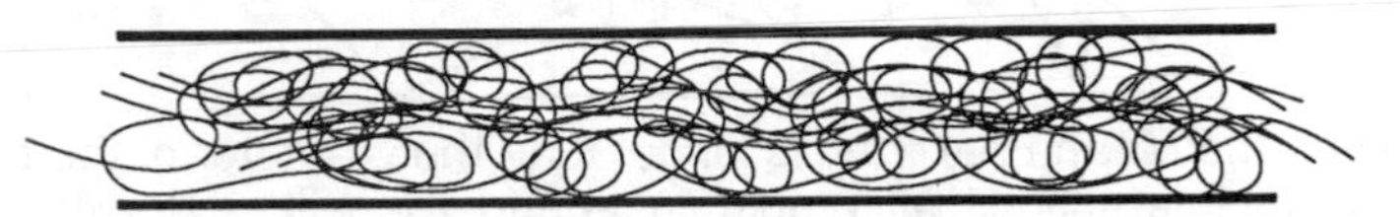

Fig. 7-4 Turbulent flow is primarily a gas-gas interaction.

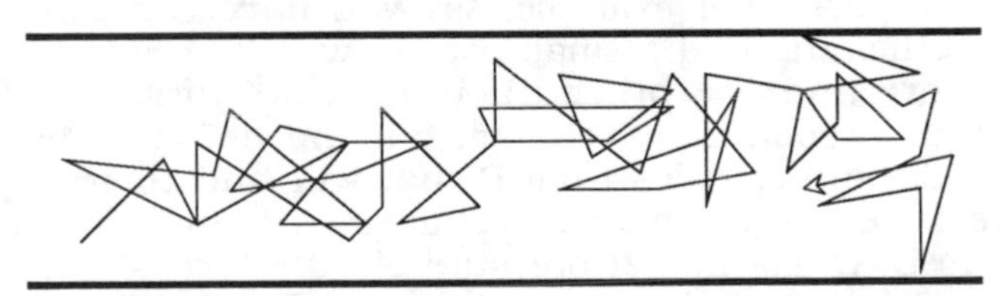

Fig. 7-5 When the Mean Free Path is fairly short, a gas molecule to more likely hit another gas molecule than the walls of the container. This situation is known as a gas-gas interaction.

The transition to viscous flow is complex and is dependent on flow velocity, mass density, and the viscosity of the gas. Viscous flow is similar to the flow of liquids through pipes: The flow is fastest through the center of the tube while the sides show a slow flow and there is zero flow rate at the walls (see Fig. 7-6). The gas interactions in viscous flow are gas-gas and gas-wall, or in other words, a molecule is equally likely to hit a wall than another molecule.

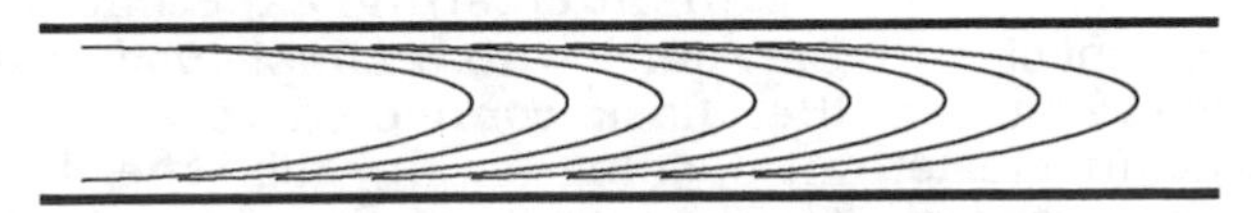

Fig. 7-6 Viscous gas flow.

When *Kn* of the system is <0.01, the flow is probably viscous. The transition between viscous and molecular flow is fairly straightforward: When $1 > Kn > 0.01$, the flow is in transition to molecular flow; when $1 < Kn$, there is a molecular flow.

One of the more interesting characteristics of gas-wall interactions is that not only are the gas reflections not specular (mirror-like), but the molecules can bounce back into the direction that they bounced from (see Fig. 7-7). This bounce-back is partly because, at the molecular level, wall surfaces are not smooth but very irregular. In addition, there is likely to be a time delay from the time a molecule hits a wall to the time it leaves the same wall. At the molecular level, when a molecule hits a wall surface, instead of reflection (like a billiard ball), the process is more likely to be adsorption (or early condensation). When the molecule leaves a wall surface, the process is desorption (or evaporation), thus explaining why there is the random movement and time delay for molecular reflection.

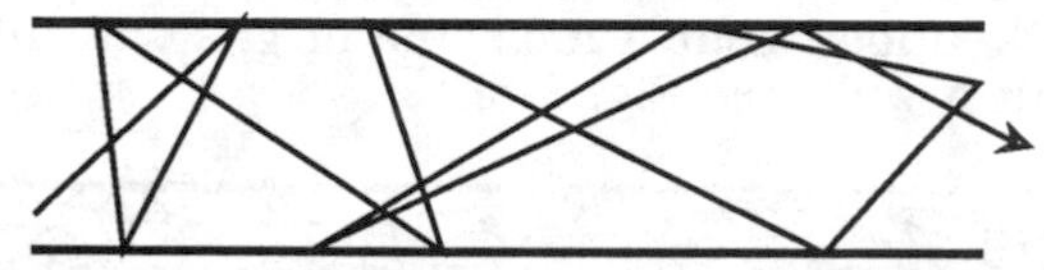

Fig. 7.-7 When the Mean Free Path is longer than the diameter of the container of the gas, gas-wall interactions will predominate and you will have molecular flow.

The gross movement of molecules in a high vacuum state is a statistical summation of the parts. For example, say you have two containers that are opened to each other through a small passageway (see Fig. 7-8). One of them (A) is at 10^{-3} torr and the other (B) is at 10^{-5} torr. The movement of molecules in both is completely random, but the *net* movement will be from A to B. There will be molecules from B that will find their way into A, even though the pressure is greater in A. But again, the net molecular movement will be from A to B. Once the net pressure of 10^{-4} torr is achieved, there will still be movement between the two containers. Eventually, according to the first principle of gases presented earlier, there will be the same number of molecules in B as in A. There will always be a greater number of A molecules than B molecules, but the number of molecules on either side of the system will eventually be the same.

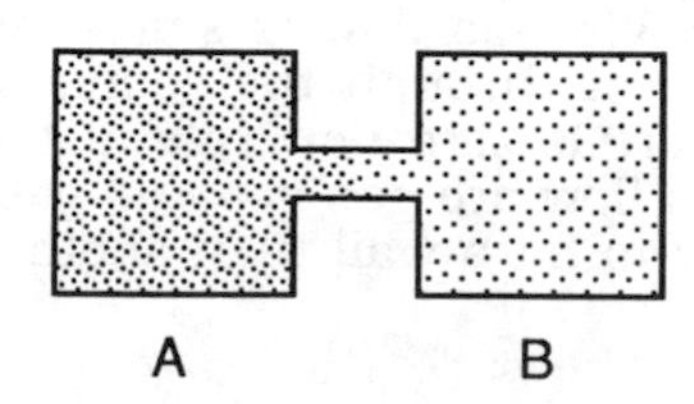

Fig. 7-8 The movement of molecules from one vacuum container to another (of greater vacuum) is statistically random.

The time necessary for the molecules to travel from A to B (or B to A) depends on the abilities of molecules to pass through a tube. This passage is dependent on three things: the pressure difference between A and B, the diameter of the connecting tube, and the length of the connecting tube.

An interesting experiment was done by Barbour[10] with a simple vacuum system of a 5-liter flask connected to a thermocouple with an opening for dry air. The flask was adapted to receive one of five different tubes (of different diameters and lengths) and in turn was connected to a pumping system (see Fig. 7-9).

Barbour evacuated the 5-liter flask and then filled the system with dry air. He then re-evacuated the flask and calculated the time it took to go from 10^{-2} to 10^{-3} torr. His results are shown in Table 7-4. From his description, it is unknown whether he dried and/or pre-evacuated the tubes to decrease the effects of water vapor, which could have slowed the pumping speed somewhat. This slowing effect would have become more pronounced as the surface area increased dramatically such as with Tube E. Regardless, the data effectively show the effects of tube diameter versus length on pumping speed. Note that Tube E never achieved 10^{-3} torr after three hours of trying to stabilize to the lower pressure.

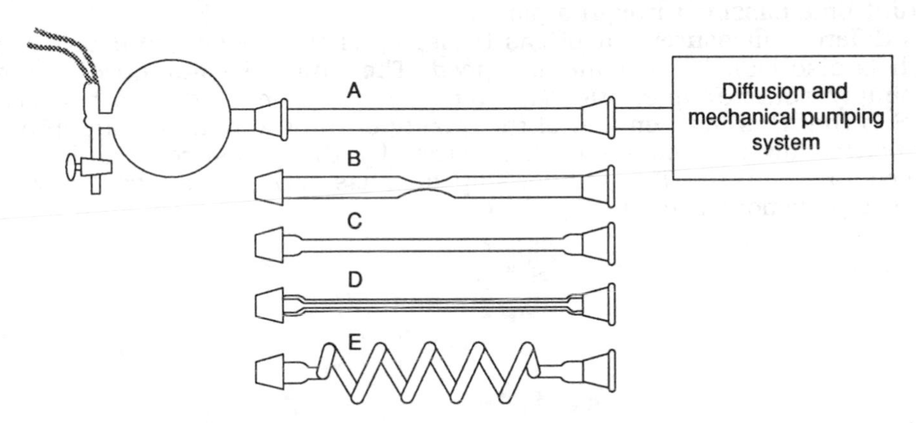

Fig. 7-9 Transport of gases through tubes of different lengths and diameters.‡

‡ From *Glassblowing for Laboratory Technicians*, 2nd ed., Fig. 10.2, by R. Barbour, © 1978 by Pergamon Press Ltd. With permission.

Table 7-4‡

The Relation of Tube Length and Diameter to Pumping Time (see Fig. 7-9)

Tube	Length	Bore	Time
Flask only, no tube			14.1 sec
A	50 cm	20 mm	21.5 sec
B	50 cm	20 mm w/ small section of 7.3 mm	30.8 sec
C	50 cm	10 mm	64 sec
D	50 cm	3 mm	1800 sec
E	150 cm	3 mm	10,800 sec

‡ From *Glassblowing for Laboratory Technicians*, 2nd ed., Table 6, by R. Barbour, © 1978 by Pergamon Press Ltd. With permission.

Whether turbulent, viscous, or molecular flow, the fastest transport through tubes is when the diameter is large and/or the length is small.

Fortunately, at a certain point, larger diameters or shorter tubes make no significant difference in performance.

7.2.9 Throughput and Pumping Speed

There will always be a net transport of gas from one end of a system to another *if* there is a pressure difference between the two parts. The quantity of gas (at a specific temperature and pressure) that passes a given plane in a given amount of time is called the *throughput* or *mass flow rate* (*Q*). When throughput is equal to zero, you have a steady-state condition. Throughput is measured in pressure-volume per unit of time, such as torr-liters per sec or 1 Pa-$M^3/_s$ = 1 W (watt). Interestingly enough, because force is required to move the gas, throughput can also be considered as the amount of energy per unit time passing through a plane.

A different measurement of gas transport is the *volumetric flow rate* (*S*), which is also called the *pumping speed*. The units of volumetric flow rate are simply *volume/time*, such as liters per sec. The difference between Q and S is that S is independent of the quality of the vacuum. Throughput and volumetric flow are related to each other by the pressure P and are expressed in Eq. (7.7). The relationship of mass flow, volumetric flow, and pressure is demonstrated in Fig. 7-10.

$$S = \frac{Q}{P} \qquad \text{Eq. (7.7)}$$

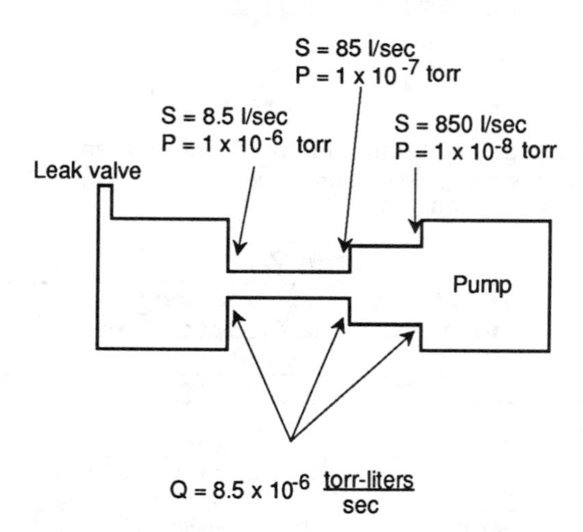

Fig. 7-10 A system in a steady-state condition showing the relationship between Q, S, and P.‡

‡ From "*An Introduction to the Fundamentals of Vacuum Technology*," Fig. 7, by H.G. Tompkins, © 1984 by the American Vacuum Society, American Institute of Physics, Inc., New York, N. Y., 10017. With permission.

In a typical vacuum system, tubes are connected to tubes, which are connected to traps, which are connected to other tubes, and so forth. The transport of gas through a system of these tubes is called *conductance* (C). Every tube within the system offers a different conductance depending on its length and diameter.

Conductance is the throughput (Q) divided by the difference between the pressure going into the tube (P_1) and the pressure going out of the tube (P_2) (see Eq. (7.8)). Although conductance and volumetric flow rates have the same units, conductance is used to describe the ability, or efficiency, of a gas flowing through a section (one or more) of tubing, while volumetric flow rate deals with the amount of gas that can go past a single plane of the system.

$$C = \left(\frac{Q}{P_1 - P_2}\right) \qquad \text{Eq. (7.8)}$$

Vacuum systems are generally not made with one tube and one easy way to calculate conductance. Rather, vacuum systems are constructed with many tubes of different sizes. The total conductance of an entire system is calculated by adding all the reciprocals of the conductance of the various parts (see Fig. 7-11) and is shown in Eq. (7.9).

$$\frac{1}{C_t} = \frac{1}{C_1} + \frac{1}{C_2} + \frac{1}{C_3} + \frac{1}{C_4} \qquad \text{Eq. (7.9)}$$

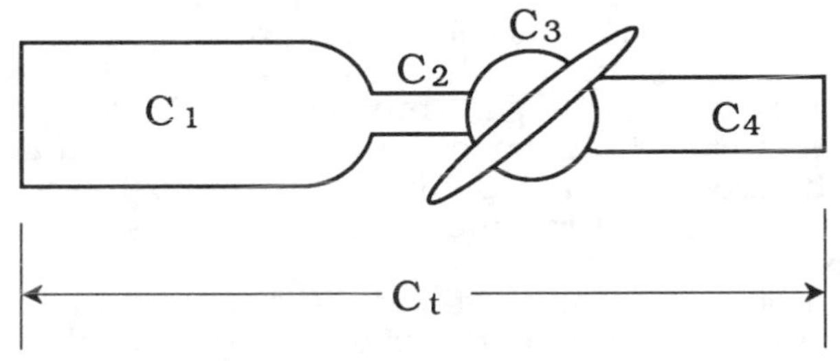

Fig. 7-11 The total conductance is the addition of the reciprocal conductance of all the parts (see Eq. [7.9]).

7.3 Pumps

7.3.1 The Purpose of Pumps

A pump's function (when creating a vacuum) is to remove gases and condensable vapors from the chamber where you want the vacuum. There are two ways to remove gases and vapors. The one that typically comes to mind is gas transfer, which physically removes gases and vapors from one area and release them into another. The other approach for removing air is to trap the offending gases and vapors. Once trapped, the vapors are still in the region where you are working but are bound up in such a fashion so as not to impede the work in progress. The two approaches of gas removal, pumps, and corresponding pumping ranges, are shown in Table 7-5.

Table 7-5

Pump Families and Characteristics			
Gas Transfer	**Gas Capture**	**Pumping Range**[a]	**Relative Speed**
Aspirator		760-10 mm Hg	slow
Rotary (single-stage)		$760\text{-}10^{-3}$ torr	slow
Rotary (double-stage)		$760\text{-}10^{-4}$ torr	slow
	Sorption	$760\text{-}10^{-3}$ torr	slow
Roots Blower		$10^{1}\text{-}10^{-5}$ torr	moderate
Turbomolecular		$10^{-2}\text{-}10^{-10}$ torr	high
Diffusion (oil)		$10^{-3}\text{-}10^{-11}$ torr	high
	Getter	$10^{-3}\text{-}10^{-10}$ torr	high (dpnt on gas)
	Sputter-Ion	$10^{-5}\text{-}10^{-10}$ torr	high (dpnt on gas)
	Cryosorption	$10^{-4}\text{-}10^{-13}$ torr	high

[a] These pumping ranges are only general levels, not limits or potentials. There are many factors that can affect pumping range and speed, including the nature of your work and your system's configuration and design.

The most common level of vacuum required within a laboratory is not a very significant vacuum. In fact, a high ($\approx 10^{-3}$ torr) to very high ($\approx 10^{-6}$ torr) vacuum will satisfy some 95% of most laboratory vacuum needs. The levels of ultra-high vacuum (whose demands are usually reserved to surface studies, space simulation, and particle accelerators) are seldom found in many research laboratories. Thus, in the chemical research laboratory, the pumps that are usually found are the gas transfer type. More specifically, one is likely to see aspirators, diffusion pumps, and rotary vane and/or piston pumps. Although turbomolecular pumps (turbopumps) and cryopumps may also be found in the laboratory, their high initial cost, and higher level of technical support, excludes them from being the first pump of choice. The other pumps mentioned in Table 7-5 are usually used in specific industrial or highly specialized research. With all due respect to these fine pumping alternatives, discussions within this book are limited to aspirators, rotary (vane and piston) pumps, and diffusion pumps. For further information on the other pumps mentioned, refer to Appendix D.

The concept of "you get what you pay for" can be altered for vacuum pumps to "the less you want, the more you pay." With that thought in mind, you do not want to pay for vacuum you do not need. Note however, that the stated limit of a pump is usually its ideal limit, which is dependent on the type of gas being pumped and whether the vacuum gauge is connected at the pump inlet. These conditions are unlikely to be like *your* system and therefore, you should obtain a pump that is rated better than your *needs*. Again, the onus is on you to recognize what you need. Keeping up with the Jones' is probably the best approach to take. If someone has a system like the one you want, ask the person if they are satisfied with their components such as the pump, or in retrospect, should the pump have been bigger or smaller?

Finally, in regard to pump power: even though you can get a small two-cylinder car to pull a load of bricks, it is more efficient to get a truck. Being penny-wise in selecting a pump may be being pound foolish in the loss of wasted time and equipment in the laboratory.

7.3.2 The Aspirator

The *aspirator* (also called a *water-jet* pump) is one of the simplest and most economic means of obtaining a vacuum in the laboratory. It has no moving parts of its own (see Fig. 7-12), uses no electricity, and relies only on moving water to function.

The aspirator is not a very fast pump (1-2 gallons/minute), nor does it create a powerful vacuum ($\approx$10 mm Hg) (with proper liquid nitrogen trapping, aspirators can achieve vacuums as great as 10^{-2} torr).[11] It is ideally suited for emptying large containers of liquids down a sink or for supplying the vacuum necessary for a filter flask.

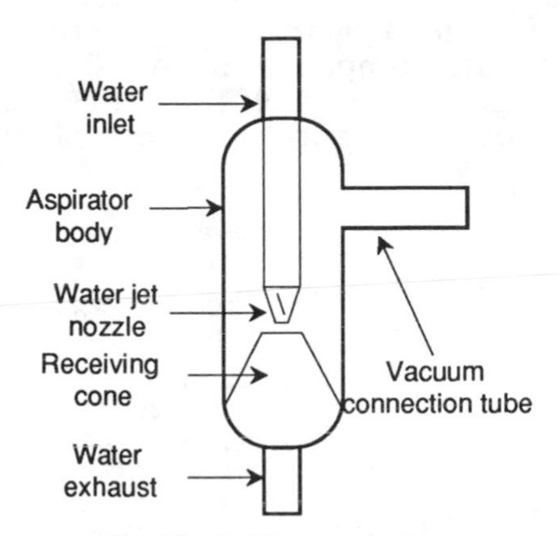

Fig. 7-12 The aspirator.

Aspirator operation is accomplished simply by attaching the device you wish to evacuate with suitable flexible vacuum tubing to the tube on the side of the aspirator. Then, turn on the water and immediately a vacuum will exist through the vacuum connection tube. The aspirator can operate with water pressure ranging from 10 to 60 lbs, depending on design and construction. Generally, the greater the water pressure, the greater the vacuum created by the aspirator. At a certain point (depending on aspirator design) the vacuum potential will level off and no greater vacuum can develop despite increasing the force of the water. Whatever the aspirator design, the vapor pressure of water will always be the upper limit of the aspirator unless some form of trapping is added to the system.

If an aspirator has the appropriate attachment, it can be attached directly to a sink faucet so water exhaust can go down the drain. To prevent splashing that occurs when the water hits the sink's bottom, place a one-half to one-liter beaker under the aspirator filled with water. When the aspirator's water hits the pre-filled beaker, the force will be absorbed and no splashing will occur. Otherwise, if the aspirator is going to be used at some location with no adequate drainage, a tube should be connected to the water exhaust so that the water can be channeled to a proper receptacle.

The aspirator's pumping mechanism is quite sophisticated, and it all begins by water streaming past the water jet nozzle. By decreasing the internal diameter at the point where the water leaves the nozzle, there is an increase in water speed passing this point. Because the MFP of the gas and vapor molecules within the aspirator is much less than the pump dimensions, aerodynamic shear causes air movement in the desired direction. This air movement will occur regardless of whether there is direct contact of the air with the water or not. Additionally, because the boundary between the rushing water is quite turbulent, air is physically trapped and removed from the aspirator body. Finally, the last mechanism is based on the fact that the faster air moves, the less its density. As moving (and therefore less dense) air is removed, non-moving (and therefore more dense) air is drawn in.

7.3.3 Types and Features of Mechanical Pumps

There are many mechanical pump designs, the most common of which is the rotary pump. It is named for its use of rotating internal parts that collect, compress, and expel gas from a system. There are two different types of rotary pumps: vane (both single-and double-vane, see Fig. 7-13) and piston (see Fig. 7-14).

The vane pump, by definition, uses a spring-induced vane to separate the atmospheric gases from an evacuated system. Regardless of whether the vanes sweep in motion (as in the double-vane pump) or remain still and let an eccentric drum roll past (as in the single-vane pump), the vanes permit motion within the pump while maintaining a pressure difference of up to

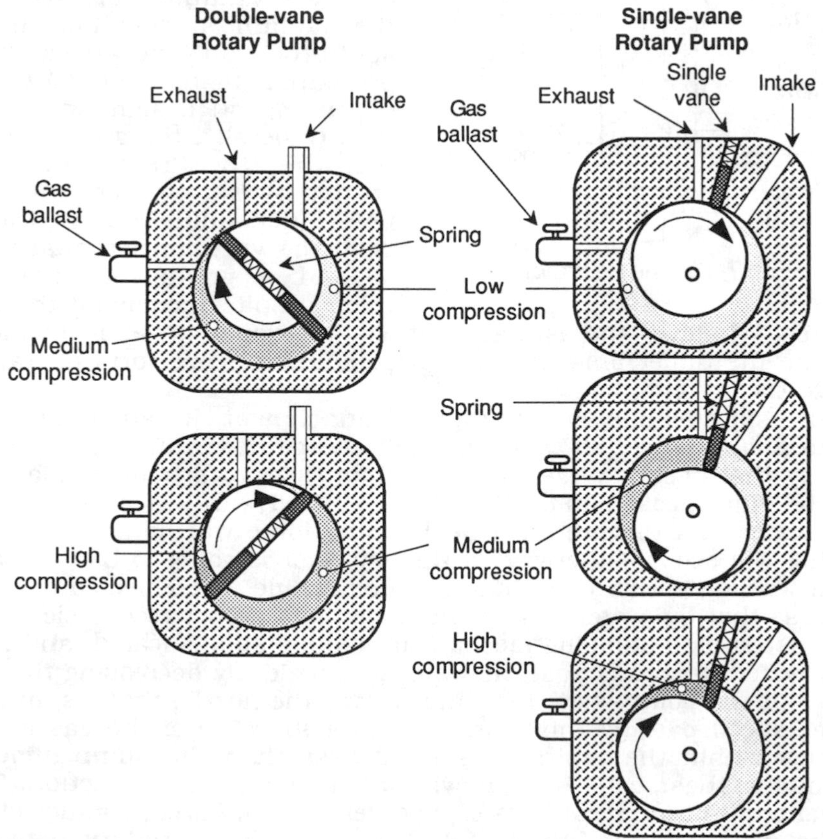

Fig. 7-13 This series of drawings displays how gas is drawn into single-and double-rotary vane mechanical pumps, compressed, and expelled into the atmosphere.

10^{-3} torr. One of the obvious differences between the double- and single-vane pumps is that the single-vane pump's main rotor spins eccentrically. This eccentric spinning creates a vibration, and if the pump is placed or mounted too close to sensitive instruments, this vibration can affect their operation. The double-vane rotary pump rotates evenly and quietly with little vibration. Pump vibration is important if the pump is mounted in a location close to sensitive instruments that could be affected by vibration.

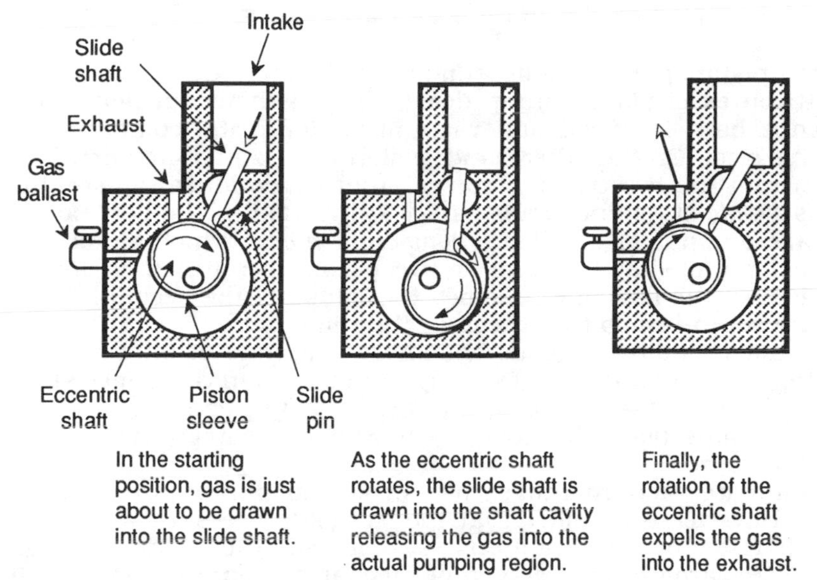
Intake
Slide shaft
Exhaust
Gas ballast
Eccentric shaft
Piston sleeve
Slide pin
In the starting position, gas is just about to be drawn into the slide shaft.
As the eccentric shaft rotates, the slide shaft is drawn into the shaft cavity releasing the gas into the actual pumping region.
Finally, the rotation of the eccentric shaft expells the gas into the exhaust.

Fig. 7-14 This series of drawings displays how gas is drawn into a piston mechanical pump, compressed, and expelled into the atmosphere.

The piston pump, like the single-vane pump, also has an eccentrically rotating drum. However, rather than a vane pressing against the drum, there is a free rotating sleeve around the drum that rolls around the pump's inner cavity as the drum rotates. By rolling around the pump's inner cavity, there is no sliding of moving parts as the vane pumps. Connected to the rotating drum is a sliding shaft with a hole in it that allows air into the pump only when the drum is in the correct position, and prevents air coming in at all other times. Properly-made piston rotary pumps are counterbalanced so as to have very minimal amounts of vibration.

There are inherent weaknesses in the design of vane pumps that are overcome with piston pumps. For example, on each sweep of a vane, oil is brought from the atmospheric side to the vacuum side, and needs to be re-outgassed on each sweep. A piston pump's roller does not sweep oil around. Rather, it rolls over the oil without pushing it aside, thus significantly reducing the outgassing problem. In addition, the slide shaft (within the piston pump) maintains the internal sections of the pump as two separate sections. This separation further prevents the oils from mixing. Within the pump, one section is maintained at the vacuum of the chamber being evacuated, and the other at the pressure for expelling gas into the atmosphere. Because vanes are constantly scraping the internal walls of the pump, extra heat is generated. This heat produces extra wear on the pump, its parts, and the oil. On the other hand, the extra heat helps to outgas the oil to some degree.

Finally, because the piston does not need to make actual contact with the internal sections of the pump (only contact with the oil is required) there is less wear than with vane pumps. Because the piston pump parts do not make physical contact, they can tolerate larger sized particulate matter (that may accidentally enter the pump), and are also more forgiving of internal scratches.

Rotary piston pump studies made by Sadler[12] demonstrated that the above attributes of piston pump design helped them out-perform vane design pumps both in speed and vacuum achieved (he compared pumps of comparable size). On the other hand, piston pumps had greater vibration.

Aside from the two types of rotary pumps mentioned, you will often see the terms *single*-stage and *double*-stage mentioned in the context of pumps. A two-stage pump is also called a *compound pump*, which simply refers to a mechanical pump that has one or two pumps connected in line together. The exhaust of the first one becomes the intake of the second. The rotors of the two are set in-line so that they use the same shaft.

A two-stage pump can usually provide a decade of improvement over a single-stage pump and is generally quieter than a single-stage pump (although the use of a gas ballast (see Sec. 7.3-5) can increase the noise). Two-stage pumps typically have a (somewhat) greater purchase price and maintenance cost.

The ultimate vacuum ranges of mechanical pumps are around 10^{-3} to 10^{-4} torr. However, remember that the "ultimate pressure" of a given pump is dependent on pump conditions, system conditions, and the gas being pumped.[13] Regardless of the type of mechanical pump design, neither the pumping mechanism nor the vapor pressure of the oil is the limiting factor in obtaining low pressure. Rather, the natural limitation of mechanical pumps is due to the solubility of air (and any other gases that are being pumped) in the pump oil (for more information on the relationship of pump oils to mechanical pumps, see Sec. 7.3.7).

There is not much that can be done about the fact that air in the oil raises the vapor pressure, and therefore raises the potential pressure obtainable with single-stage pumps. However, two-stage pumps provide a solution: by keeping the chamber above the oil in the first stage (on the inlet side of the pump) in a vacuum state, there is less air to saturate the oil. Thus, a two-stage pump can go beyond the physical limits of a single-stage pump. In addition, two-stage pumps fractionate their oil as they work. The oil on the inlet (vacuum) side maintains its purity while the oil on the atmospheric side tends to collect the contaminants and products of oil degradation. This design helps maintain the capability of the pump between oil changes.

The design of two-stage pumps makes them poor choices for pumping air at high pressures (i.e., atmospheric) for an extended amount of time. The excess air pressure tends to cause the oil in both sections of the pump to froth and foam, which in turn may cause the bearings and/or the seals to overheat. Thus, two-stage pumps make poor roughing pumps. It is for this reason that some systems require two mechanical pumps. One pump is a single-stage *roughing* pump to get a large system (or a system with large quantities of moisture) down to a pressure that a diffusion pump would work. This system is then joined by a two-stage pump that can deliver the proper fore pressure for the diffusion pump. The second pump is called the *fore* pump. If the system is small, and can be pumped down to about 1 torr within several minutes, then the roughing pump and the fore pump can be the same pump. However, if your work tends to cycle between atmospheric and vacuum on a repeated and continual basis, the addition of a roughing pump is recommended.

Pumps can also be divided into direct-drive or belt-driven models. Direct-drive models run faster and quieter. They are smaller and lighter. Belt-driven models are more easily serviced, obtain a lower ultimate pressure, and have less backstreaming. In addition, because they run cooler, they often have longer service lives (providing they have received proper maintenance).

For all pump types or designs, there are three different types of motor housings, each providing different levels of safety. Therefore, the environment that pump will be used in should also be considered when selecting a pump:

1) ***Open Motors*** provide an inexpensive housing that is to be used in a clean dry environment (no protection necessary).

2) ***Totally Enclosed Fan Cooled Motors*** are used where moisture or dirt is prevalent (protects the motor).

3) ***Explosion-Proof Motors*** are required to meet the National Electrical Code Standards for use in hazardous conditions. These are used where gas, dust, or vapors may be in hazardous concentrations and a spark from the motor could cause a fire or explosion.

The final pump characteristic is pump size. To obtain a pump that is too large and/or too great a pumping power for a given need can be a waste of money. However, if a pump is not large or powerful enough, there is a further waste of extra money spent when the proper size pump must be purchased. In general, it is better to err too large than too small. A simple formula can be used to aid in the selection of a pump size:

$$S = \frac{V \times F}{t}$$ Eq. (7.10)[14]

Where
S = pumping speed (liters per minute)
V = chamber volume (liters)
t = time of evacuation desired (minutes)
F = pump down factor
= (2.3 x log ($^{760}/_{\text{desired}}$ chamber pressure in torr)), and
= (see Table 7-6).

Table 7-6‡

Desired Chamber Pressure in torr						
100	10	1	0.1	0.01	0.001	0.0001
F = 2.03	4.33	6.62	8.93	11.2	13.5	15.8

‡ From "*Vacuum Products*" by Welch Vacuum, p. 125, © 1988 by Welch Vacuum Technology, Inc. With permission.

For example, if you have a vacuum line of about 20 liters that you wish to bring to 10^{-2} torr in 10 minutes, then:

$$S = \frac{20 \times 11.2}{10} = 22.4 \text{ liters/minute}$$

This determination is assuming that you have a perfectly-designed vacuum system with no narrow passageways, no extra chambers, no bends, no contamination, and no traps, baffles, or stopcocks. Not much of a vacuum system, but it sure would be fast. Because you are working in the real world, the real pumping time will be slower. To compensate for this slower time, it can be helpful to multiply the above result by 2 or 3 to ensure that the pump will be adequate to handle your system's needs. However, as a tool for approximating the minimum size needed, the above formula is usually sufficient. The final decision of what pump to purchase, as in all selections, will depend on what you are willing to settle for and/or what you can afford.

In many laboratories, the demands on a standard mechanical pump are less critical than in industrial use. A lab is generally more interested in obtaining a low pressure than the speed and/or volume pumped. Typically, the primary consideration in the laboratory is that the mechanical pump achieves the vacuum required for diffusion pump operation.

Most manufacturers suggest the size of fore pump to be used with their diffusion pump. However, although the manufacturer will usually recommend a mechanical pump of their own design, you probably already own a pump that is sufficient, and it may be necessary to mix and match machines.

To match a fore pump with a diffusion pump use Eq. (7.11):

$$S_f = \frac{S_d P_d}{P_f} \qquad \text{Eq. (7.11)}$$

where
- S_f = fore pump speed
- P_f = fore pump operating pressure
- S_d = diffusion pump speed
- P_d = diffusion pump operating pressure.

Whatever you get for the fore pump speed from Eq. (7.11), you should use a pump with at least double that figure to accommodate pressure fluctuations and make up for line impedance within the vacuum system.

In summation, the following suggestions should be considered when selecting a mechanical pump:

1) For table-top demonstrations, and/or small experiments, a direct drive pump is excellent. Its portability, low vibration, and generally quiet operation is important.

2) For general experimentation, large manifold systems, and industrial operations, belt-driven mechanical pumps should be considered. They can easily be placed in pans (to catch any oil leaks) on the floor, and they can be left on for days and/or weeks of continuous operation.
3) Double-vane pumps have the least amount of vibration.
4) Piston pumps are not made with the same close tolerances as vane pumps so they are more forgiving for particulate matter developing within the pump oil. They also exhibit greater performance than other pumps of comparable size.

The size of the pump however, is still up to the needs and demands of the user.

7.3.4 Connection, Use, Maintenance, and Safety

At first glance, mechanical pump use seems so straightforward that little needs to be said—and for the most part that's true. However, failure to observe some simple rules can result (at least) in inefficient running of the pump and (at worst) pump destruction.

Connection. Always place your pumps in some type of tray so that any leaks will be contained and will not create slick spots. This placement is especially important if the pump is positioned on the floor. By placing the pump on the floor (and in a tray), any vibration effects from the pump on instruments sitting on lab benches will be limited.

The most efficient attachment of a pump to a vacuum system is with a tube as short in length and as large in diameter as practical. Unfortunately, it is not uncommon to see a five-foot coil of vacuum hose between the mechanical pump to the vacuum system. This setup is not only a waste of tubing, but it also causes a significant decrease in pumping efficiency.

Pump accessibility is equally important. The sight gauge (see Fig. 7-13), pump belt, exhaust line, electrical connections, and oil drain should be easily accessible. If there is any chance that there could be particulate matter coming from the system, filter traps* should be placed at the pump intake.

All exhaust from a mechanical pump should be vented to a fume hood regardless of the room's ventilation quality or the type of pumped gases. Each time you bring new samples into vacuum conditions, your system is pumping at atmospheric pressure.** Because pump oils have low vapor pressures, and pump oils themselves are considered non-toxic, there is little concern for breathing pump oil mist. However, there may be dangers from trapped vapors within the pump oils. Regardless, there is little reason to breathe the pump oil mist if it can be avoided. Check with the manufacturer or distributor of your pump for an oil mist filter for your pump. If you use a condensate trap, be sure you position your exhaust line so that material does not drain back into the pump (see Fig. 7-13).

* These traps do not significantly affect the flow rate, but without them, destructive wearing of the pump mechanism could result. Contact your pump supplier for filters to fit your specific pump.

** This pumping tends to froth the oil and the addition of an oil mist filter is imperative.

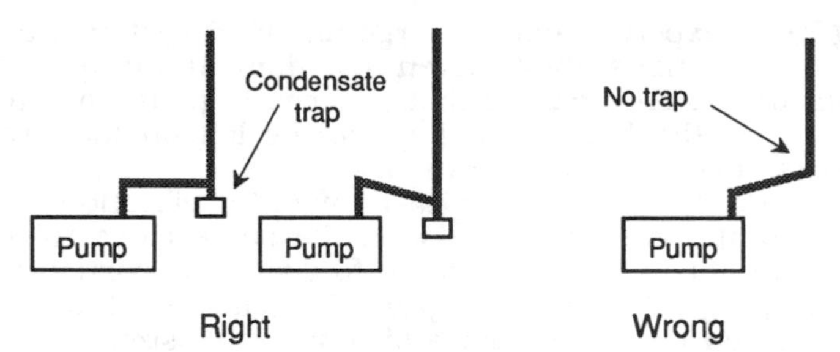

Fig. 7-15 The proper orientation of a mechanical pump's exhaust condensate trap.‡

‡ From Vacuum Vol. 31, N.S. Harris, "Practical aspects of constructing, operating and maintaining rotary vane and diffusion-pumped systems," 1981, Pergamon Press, Plc. With permission.

Use. When you first start a mechanical pump, a gurgling noise will sound. This noise is to be expected when a mechanical pump is pumping against atmospheric pressure and should subside in a few seconds to minutes depending on the size of the system. If the sound does not change, several causes may be present: there may be a stopcock (or valve) on the system open to the atmosphere, part of the system may have been broken off, or the lower portion of a cold trap may have slipped off the trap into a Dewar. Gurgling can also indicate that the pump oil is low. If this case applies, the pump should be stopped, vented, and then filled to the proper level. Finally, a somewhat different gurgling sound can be caused by an open gas ballast, although this situation may is sometimes be confused with a major air leak. A gurgling noise from a mechanical pump cannot be caused by a minor (or small) leak.

Before shutting off a mechanical pump, the pump should be vented to atmospheric pressure. Although some mechanical pumps have a check valve designed to prevent reverse flow, it is not a good idea to depend on such a valve. If a vacuum is held on the inlet side of a pump with the pump off, the pump oil may be drawn up and into the vacuum line. Therefore, it is best to treat all pumps as if they did not have such a valve. When turning off a mechanical pump, you need to vent (to the atmosphere) the section of the vacuum line connected to the pump. Your pump should be separated from the rest of the line when venting so that the rest of the line is not exposed to the atmosphere. If the vacuum system is exposed to the atmosphere, it will take longer to re-pump down because moisture from the air will have absorbed on the walls of the system.

Maintenance. Probably the most common problem with mechanical pumps in labs is inattention to the pump oil. More pumps have died because of LSETCOI (*Let Someone Else Take Care Of It*) than any heinous intentional act of destruction. The reason is simple: changing pump oil is not fun or glamorous. The proper maintenance of a mechanical pump however, is as necessary as cleaning glassware, keeping a proper lab notebook, and backing up a computer hard disk.

If you are supplied with a used mechanical pump, and do not know (or trust) its history, check the oil, both for level and quality. On the side of any mechanical pump is a sight gauge (see Fig. 7-16). The oil level should be between the 'high' and 'low' sight lines. If it is too high, some oil should be drained from the pump. If it is too low, some oil should be added.

When examining pump oil quality, it can be beneficial to compare the used oil with unused oil. To do this comparison, keep a small sample of pump oil in a closed jar for inspection. The oil in the pump should be discarded if the oil appears cloudy with a whitish, emulsified appearance, is discolored, or especially if it has tar or particulate flecks.

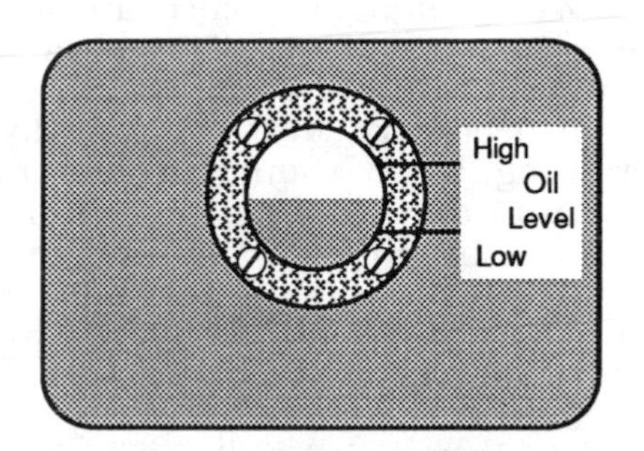

Fig. 7-16 Check the gauge window on mechanical pumps to see that the oil level is adequate.

If you have the facilities to check the viscosity of pump oil, the oil should be about 300 SSU at 100°F (SAE 20 W), ± 100 SSU.[15] If the oil's viscosity is beyond that range, it should be changed.

If you cannot check the viscosity of the pump oil, you should be able to cross-check the oil by the current pressure obtained by the pump vs. its fresh (new) oil pressure. If there is a 100-mtorr, or greater, increase in vacuum using fresh oil as opposed to the current vacuum, change the oil. You can maintain a log book (see Fig. 7-17) of your lab's pump history to have a record of pressure changes over time.

There can be no rule for how often pump oil should be changed. It cannot be done on an "every 2 to 3 thousand mile" basis, or even after 20 to 30 hours of operation. It may be necessary to change the oil as often as every day or as seldom as once every several years. It all depends on how often you use your system, what types of chemicals your system is being exposed to, how effective your trapping systems are, and how effectively you are using your gas ballast (if any) (see Sec. 7.3.5). If your system requires an oil change every six months, it does not mean that you only have to check it every six months. Monthly, or even weekly oil quality checks are advised for any system. A log book, kept in the lab, with sections that note important data (as shown in Fig. 7-17) will help in lab maintenance. [This log book will also be of value when looking for leaks.]

As ironic as it may seem, a pump that is run continuously may require fewer oil changes than a pump that is constantly cycling between atmospheric and vacuum pressures. This condition may be true even if the second pump is run the same number of hours because the cycling pump will be exposed to a greater amount of condensable vapors, which are more likely to severely affect the pump's oil.

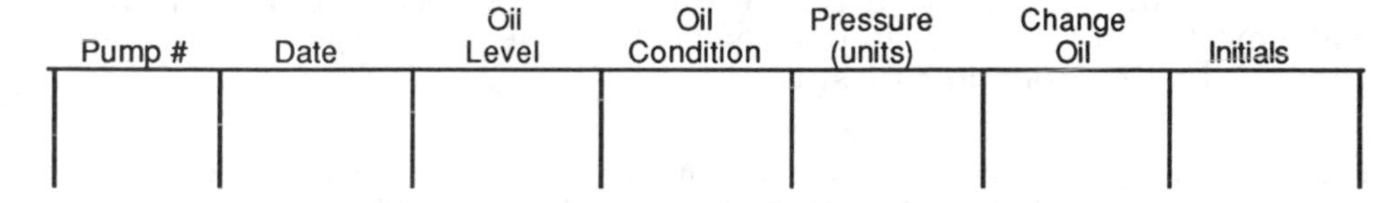

Pump #	Date	Oil Level	Oil Condition	Pressure (units)	Change Oil	Initials

Fig. 7-17 A suggested log book layout for routine mechanical pump maintenance in the laboratory.

When changing a mechanical pump's oil, first run the pump for a short time to warm the oil. Warm oil will drain more efficiently from the pump. Open the side valve and let the oil pour directly into a container (use a funnel if necessary). If it is physically possible, tipping the pump may speed oil draining, but it should not be required. The pump may retain small pockets of pump oil within small sections of the pump. These pockets can be emptied out by *partially** closing the exhaust port, and turning the pump by hand. If you start up the pump with a fast click-on and click-off to remove this last bit of oil (as opposed to rotating the pump by hand), the oil may spurt out with surprising force and create a mess.

Return the old oil to where ever used oil is to be returned. Although mechanical pump oil by itself is not toxic, during use it becomes a repository for condensable vapors that may have passed through it. While changing the oil, wear rubber gloves and safety glasses. **Never pour used mechanical pump oil down the sink or throw containers of it in the trash. Because the oil could be carrying toxic materials, it should not be sent to a trash dump or landfill. If there are any questions, check with your pump oil supplier, your safety coordinator, and/or your city's Department of Waste Management.**

Pour pump oil into the exhaust port. The amount of oil required is dependent on the individual pump and should be stated in the pump manual. If you cannot find the manual and have to 'wing it,' pour slowly and allow ample time for the oil to show up in the observation port.

If there are any spills when emptying or refilling a pump, clean them up immediately. Pump oils are not toxic, but such spills become slick and slippery, and are considered accidents just waiting for someone to walk by.

Emptying the old oil from a pump and refilling the pump with new oil is generally sufficient when changing a pump's oil. However, if pump oil is particularly filthy, empty the old pump oil and refill the pump with a flushing fluid.** Let the pump run for ten minutes with the flushing fluid against a load, then drain and refill with the correct oil. A flushing fluid can only be used with hydrocarbon pump oils. Alternately, fill the pump halfway and let it run for a minute, but no longer (never partially fill a pump, and let the pump run for any extended period of time). A final alternative to filthy pump oil is to fill the pump up with pump oil and let it run. By alternately plugging and unplugging the inlet for some five minutes, the oil will agitate and aid the flushing of any undesirable materials and/or old oils within the pump. If you use this technique, refill and empty the pump one extra time before refilling the pump for use.

Any oil seen around the drive shaft is an indication that a drive shaft seal may need replacement. Unusual noise and/or heat may indicate bearings that may need to be changed or a failure of the internal lubrication system. Typically, the first section of a pump that will show wear from abrasive particulate matter that may have entered will be the discharge valves and will be indicated with drops in pressure.

* Completely closing the exhaust port can cause the oil to come out in forceful spurts.

** A flushing fluid is a special hydrocarbon liquid designed specifically for "flushing out" mechanical pumps. It can be obtained from most mechanical pump, and mechanical pump oil manufacturers and suppliers. Do not use a hydrocarbon solvent, such as acetone because you will destroy your pump.

If you have to change the belt, be sure to align the pulleys both parallel and on the same plane (see Fig. 7-18). After the belt is on, check its tightness by firmly placing your finger on the center of the belt. The deflection should be between ½ in to ¾ in (see Fig. 7-19). If the belt is too loose it may slip on the pulleys which causes friction and heat, leading to premature failure. On the other hand, if the belt is too tight it may break, even within a week of constant use.

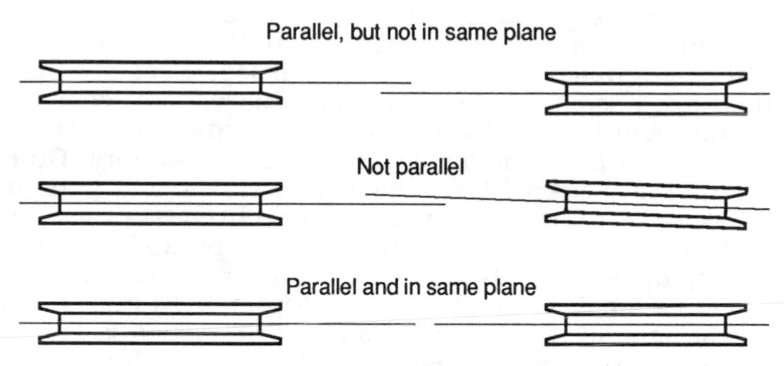

Fig. 7-18 Alignment of mechanical pump belt pulleys.

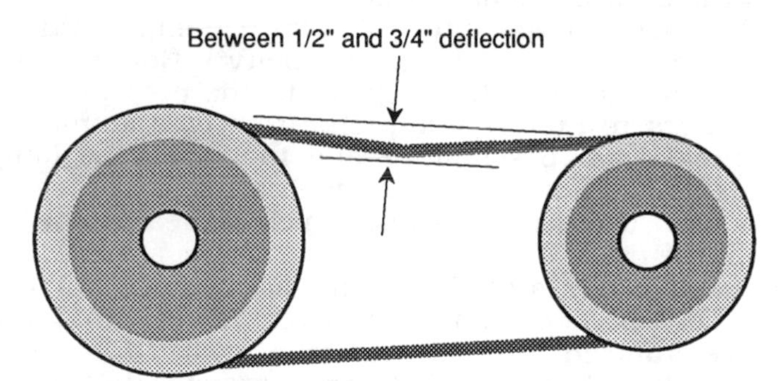

Fig. 7-19 Acceptable deflection of the mechanical pump belt is between ½ in and ¾ in.

7.3.5 Condensable Vapors

A working pump is constantly trapping condensable vapors within the pump oil. This trapping is caused both by the pump's churning action which physically mixes the oil and gases, and the high pressures within the pump during the compression stage of the pump cycle. During the compression stage, the gases and vapors must be brought to a pressure greater than atmospheric before they can be expelled into the atmosphere. Condensable vapors in a vacuum are in a gaseous state. However, as they are compressed within the pump, some of them may condense out. Minimally, the condensable vapors (such as water) will decrease the vapor pressure of the pump oil. On the other hand, other condensed liquids (such as hydrocarbons) can mix, emulsify, and/or break down, the pump oil. They can also directly destroy a pump by chemical attack, or indirectly, by poor pump performance, cause extra wear and tear on the pump parts.

There is no *one* good way to prevent condensable vapors from affecting a mechanical pump. There are however, two directions that one can take in dealing with the problem: one is to prevent them from getting to the pump, the other is to prevent them from affecting the pump once they are present. Neither is the best approach, and usually it takes combinations of the two to deal effectively with the problem. An alternative approach is to constantly change the pump oil. This solution however is not cost-, or time-, effective.

Traps, either of chilled or chemical design (see Sec. 7.4 on traps and foreline traps), are used to prevent condensable vapors from getting to a mechanical pump and its oil. Because of relative cost and ease of use, cold traps are probably the most commonly used in the laboratory. Other types of traps that are used are coaxial traps, molecular sieves, and particulate traps (see Section 7.4 for more information on the effectiveness of various traps).

To prevent condensable vapors that reach a pump from affecting the mechanical pump oil, a gas ballast (also called a vented exhaust) is used. The gas ballast allows a small bit of atmosphere (up to 10%) into the pump during the compression stage so that the gas from the system is only part of the gas in the pump at the time of greatest compression. Thus, at the time of compression, the total percentage of condensable vapor within the pump is much less than there would be otherwise.

Ballasting decreases the potential vacuum a pump could normally produce (about one decade of performance capability*). However, it dramatically improves its performance over the long run in the presence of condensable vapors. Plus, it helps to protect pump oils from contamination, which decreases pump breakdown possibilities and increases the longevity of the pump oils. Incidentally, running a pump with a ballast causes the pump to run a bit hotter than it otherwise would which decreases the potential gas-carrying ability of the oil.

For maximum efficiency, a gas ballast should be open when first pumping a vacuum system or when operation of a system is likely to develop condensable vapors. Once sufficient vacuum has been achieved and the majority of condensable vapors have been removed from a system, the gas ballast can be closed, although it does not hurt to leave a ballast open all the time.

There is a tendency, when work is over, to turn everything off. Normally this approach would be the proper one to take. However, it is better to leave a mechanical pump on, pumping against a dry nitrogen purge, after regular use. The purge should be held at about 300μ for as long as a day to help 'flush out' any condensate from the pump oil. This extended pumping will not help pump oil already broken down, but it will help decrease any more pump oil from further destruction by expelling any remaining contaminating material. It also helps to remove water vapor from the pump oil. When letting your pump run for such an extended period of time, always let the pump pull against a load. In other words, never let a pump run with the inlet

* If you can obtain 10^{-4} torr without a ballast, you may only be able to obtain 10^{-3} torr with a ballast. If you are not obtaining expected performance and you have a ballast, you may want to check if the ballast was inadvertently left open.

open to atmospheric pressure because the pump oil will froth and lose its protective characteristics—which can ruin the pump.

Although gas ballasts can be found on both single- and double-vane pumps, as well as piston design pumps, they are more likely to be found on two-stage pumps than single-stage pumps. It is interesting to note that a vane pump tends to run noisier with the gas ballast open, whereas a piston pump tend to run quieter with the gas ballast open.

There is an alternate approach to preventing condensable vapors from entering mechanical pump oils, that is, maintaining the pump at relatively high temperatures. The high heat prevents the vapors from condensing within the pump despite the high pressure. However, due to the inherent dangers of this type of pump, as well as the problem that pump oils begin to deteriorate after being maintained at high temperatures, this approach is seldom used.

7.3.6 Traps for Pumps

All pumps should be protected from materials within the system, the system should be protected from all pump oils, and a diffusion pump (if any) should be protected from mechanical pump oils in the foreline.* These protections can be achieved with traps. More information on traps is presented in Sec. 7.4, but the following information is important for general operation.

Any water or hydrocarbon solvents should be removed from traps before beginning vacuum operations. Any materials in a trap when first beginning operation will be drawn into the pumps. Therefore, having no material in the traps to begin with removes this possibility.

A buildup of water or hydrocarbon solvents during operation can decrease, or cut off, throughput of gas through a trap. If this situation occurs, close off the trap from the rest of the vacuum line and pump. After this procedure, vent the trap to the atmosphere, let the trap come to room temperature, and empty the trap. In some cases it may be necessary to remove the trap before it has thawed and place it within a fume hood to defrost. It is always advisable to have some extra lower sections of cold traps to exchange with one that is being cleaned to limit your downtime. If you expect to remove frozen trap bottoms often on a glass system, it may facilitate operations if O-ring joints are installed on the system rather than standard taper joints because O-ring joints are easy to separate, even if a trap is cold.

Cold traps must be used if mercury is used in your system (such as manometers, diffusion pumps, bubblers, or McLeod gauges) and if your mechanical pump has cast aluminum parts. Mercury will amalgamate with aluminum and destroy a pump. Even if your mechanical pump does not have aluminum parts, the mercury may form a reservoir in the bottom of the mechanical pump, which may cause a noticeable decrease in pumping speed and effectiveness. Aside from a cold trap between the McLeod gauge and the system, place a film of low vapor pressure oil in the McLeod gauge storage bulb. This oil will limit the amount of mercury vapor entering the system that makes its way to the mechanical pump.[16] In addition, an oil layer should be placed on the mercury surface in bubblers and other mercury-filled components.

* The foreline is the section of the vacuum system between the high-vacuum pump (i.e., diffusion pump) and the fore pump (i.e., mechanical pump).

When first starting up a vacuum system, let the pumps evacuate the traps for a few minutes before setting the traps into liquid nitrogen. Otherwise you are likely to trap oxygen in the traps and create a potentially dangerous situation when the pumps are turned off (see Sec. 7.4.3).

7.3.7 Mechanical Pump Oils

In this section on mechanical pumps, there has been constant attention placed on the protection and/or maintenance of pump oils because of the incredible demands placed on these oils. Ideally, all mechanical pump oils should:

1) Be thermally stable
2) Be chemically inert
3) Exhibit a low vapor pressure over a wide temperature range
4) Lubricate
5) Maintain the same lubricity and viscosity over a wide temperature range
6) Maintain a vacuum seal between sections
7) Cool heated sections of the pump
8) Trap vapors and particulate matter from the vacuum system
9) Be compatible (and not interfere) with the environment, and protective from the environment
10) Be non-toxic.

In addition, mechanical pump oils can be specialized for use in specific environments such as those with high-oxygen contents. Some are blended for use in specific types of pumps such as direct-drive pumps, belt-driven pumps, and rotary-piston pumps. As with most things, no single product fits the bill for all circumstances. Thus, there are many varieties, grades, and types of mechanical pump oils.

Various manufacturing processes produce oils with different characteristics. Each of these characteristics varies the oil's tolerance of what it can withstand and still provide safety and protection to the pump. Among the properties that can be altered and enhanced are:

1) **Molecular weight**: It is better to have a highly-refined fluid composed of a narrow range than a wide range of molecular weight.

2) **Vapor Pressure**: Although the vapor pressure of an oil itself is seldom a limiting factor of what the vapor pressure in use will be, it still provides a benchmark level to compare various oils.*

3) **Pour point**: This property is critical if the pump will be used in cold environments. An oil with too high a pour point could

* Regardless of the pump oil used, the vapor pressure of a pump oil (off the shelf) is seldom the limiting factor in the potential vacuum a mechanical pump can achieve. For example, when diffusion pump oils are used in mechanical pumps, they tend to exhibit a decreased ability to pull a vacuum due to the partial pressure of the dissolved gases within the pump oil. Thus, due to the heat of the pump and contamination in the oil, the vapor pressure of the oil can be greater than its stated pressure by a factor of 10 to 100.

prevent a pump from starting and require the pump to be artificially heated to begin operation. On the other hand, an oil with too low a pour point may burn off if used in too hot an environment.

4) **Viscosity**: Pumps with tight tolerances, such as vane pumps, require a low viscosity oil whereas pumps with low tolerances, such as piston pumps, require a higher viscosity oil.

5) **Fire point**: Any pumping of pure oxygen (or a high percentage oxygen mixture) can cause most oils to explode or degrade rapidly. Therefore, pumping these materials requires a very high (or nonexistent) fire point.

With all the varieties of mechanical pump oils available, there is one further complication: pump manufacturers are selective about the oil that goes into their pumps. Most pump manufacturer want to sell you their brand of pump oil(s). This tactic is not so much greedy than it is an attempt to ensure (what they feel is) a high-quality oil goes into (what they believe is) a high-quality product. It is the best way they have to prevent you from destroying your pump. If there are any questions about the selection of pump oils, check with your pump's manufacturer. The manufacturer usually would rather help you for free than have one of his or her pumps look bad—even if it is your fault! However, be warned: if you place a non-approved oil into a pump, a manufacturer can void any warranty associated with the pump! If you have been advised to use a pump oil on a new pump that is different than what the operating instructions recommend, be sure to find out (in writing) what effects this oil may have on the warranty.

Despite the fact that all oils are made for use in mechanical pumps, it is not always possible to simply pour one type out and pour a different one in. Some mechanical pump oils are specifically designed for specific pump designs, and different pump oil solutions (i.e., hydrocarbon and phosphate ester) should never be mixed. Further complications can arise if you are changing one pump's service to a different type of service. Depending on the intended service, it may be necessary for a pump to be returned to the manufacturer and rebuilt. For example, if you need to pump pure oxygen, a mechanical pump must be altered at a factory. This alteration will include changing the lubricating oils in the bearings and seals, which if not done, could cause an explosion. In addition, because some oils are more viscous than others, the pump may need alteration to increase or decrease clearances between parts.

Just like auto engine oil, mechanical pump oil breaks down and needs to be replaced. And, similar to auto engine oil, mechanical pump oil should be replaced over a period of time, even if it has not broken down. Although the quality of an oil may not have diminished, replacing old oil also removes particulate matter and collected chemicals. Particulate matter, if soft, can turn pump oil into a gel. Because the viscosity of this gel will be greater than the original oil, the pump will be operating at a higher temperature and is more likely to fail. Hard particulate matter can scratch the insides of a pump causing leaks and decreasing performance potential. Chemicals typically get trapped in a pump by entering as vapors that were not stopped by a trap.

Vapors in the vacuum state will condense out during the compression stage within a pump. Once condensed out, a condensed chemical's vapor pressure can be sufficiently high (i.e., higher than atmospheric pressure) and may prevent the opening of the pump's exhaust valve, which lowers the pump's pumping ability. Condensed gases can also react with pump oil to form gum deposits or simply degrade the oil. In addition, although a *pump oil* may be impervious to acids, the *pump containing the oil* is not likely to be. Thus, if your work produces acid fumes, periodically test the oil's pH.

7.3.8 The Various Mechanical Pump Oils

There are five primary types of mechanical pump oil.* The advantages and disadvantages of these various materials are explained below and in Table 7-7.

The Hydrocarbon Oils. These oils are the most common and least expensive. They are composed of paraffinic, naphthenic, and aromatic hydrocarbons and provide excellent sealing qualities. When these oils are properly distilled, they can exhibit low vapor pressures (10^{-4} to 10^{-6} torr). However, they oxidize quickly when heated, have a tendency to form sludge and tars, and have a tendency to foam if pumping against atmospheric pressure (as opposed to a vacuum).

By fractionating or vacuum- (and double vacuum-) distilling the oils,** finer grades have been developed that exhibit widely varying properties. Furthermore, oils from different geographic locations have different (both good and bad) properties. Blends of these oils can create new oils of wildly varying compositions and qualities.

Many of the negative properties of hydrocarbon oils can be overcome by the use of additives which "inhibit oxidation, reduce foaming, disperse contamination, reduce wear, depress the pour point, and increase the viscosity index."[17] On the other hand, these additives tend to increase the vapor pressure of the oil. Some manufacturers add a coloring agent to the oil that neither helps nor hinders the oil, but such oil may be more appealing.

Fortunately, most standard mineral oils work extremely well for a wide range of work. Although they are more likely to break down, especially in harsh environments, their relatively inexpensive costs make them easy to replace. Because some conditions will cause a hydrocarbon oil to change into a tar, there are flushing fluids available that help to remove tar deposits when changing the oil. In addition, flushing fluids help remove particulates and improve the removal of vapors remaining after draining the original oil. Flushing fluids can only be used on hydrocarbon oils.

The Phosphate Ester Oils. These oils react slowly with water and therefore require more frequent changes than hydrocarbon oils. These oils are considered fire-resistant, not fire-proof. Before the advent of the fluorocarbon oils, they were used for oxygen rich environments.

* Despite their high success in diffusion pumps, the silicone oils do not have the lubricity characteristics that are required for use within mechanical pumps. Some mechanical pump oils with a silicon base have been formulated, but they have not exhibited any improvements over other oils currently in use.

** Vacuum-distilled grades are often called mineral oils.

Table 7-7

Mechanical Pump Oil Types		
Pump Oil Type	**Good Points**	**Negative Points**
Chlorofluorocarbon (CFC) (ex.: HaloVac®)	Extremely non reactive and will only ignite under extreme circumstances. Less expensive than perfluoropolyethers.	Should not be used in pumps with aluminum parts. Heating over 280°C will produce HF gas. Currently, can only be removed with chlorofluorocarbon solvents, most of which are being phased out. Viscosity varies with temperature. Has a rather high vapor pressure.
Chlorophenyl-methylopolysiloxane (a.k.a. Chloro-siloxane)	Particularly useful in low temperature operations because of low viscosity.	Unless required for low temperature operations, value is negligible.
Hydrocarbons (ex.: too many to list)	These pump fluids are the cheapest and provide good, all-round protection for the pump. They can be specialized for specific pumps such as direct-drive, belt-driven, piston, vane, and for specific uses as well.	Should not be used in systems that will be pumping pure oxygen. In harsh environments, should be changed very often.
Perfluoropolyether (PFPE) (ex.: Fomblin® and Krytox®)	Non-flammable. Can be used for the pumping of pure oxygen. Maintains stability in strong acids, bases, and halogens. Initially expensive, but can be regenerated to reduce overall costs. Viscosities and pour points are similar to hydrocarbon oils.	Heating over 280°C will produce HF gas. Currently, can only be removed with chlorofluorocarbon solvents, which at the time of this writing are being phased out because of environmental concerns. While not attacked by most gases, some are absorbed causing very acidic solutions. This quality makes neutralization filtration a must. The oil should be periodically replaced to remove the toxic materials built up within it.
Phosphate Ester (ex.: Fyrquel®	Has a higher flash point than hydrocarbon oils and is considered fire-resistant, not fire-proof. Its use was more common until the advent of fluorocarbon fluids.	Reacts slowly with water from water vapor when in use causing a decrease in potential vacuum. Thus it is necessary to change the pump oil more often than would be required with hydrocarbon oils.

The Chlorofluorocarbon Oils (CFCs).* These oils are non-reactive, do not form tars, do not break down, and are non-flammable—therefore they can be used in pumping pure oxygen. On the other hand, they have greater viscosity changes with temperature than most other oils. In cold temperatures the viscosity can be great enough to prevent a pump from starting. In addition, if these oils are subjected to temperatures greater than 280°C, highly toxic fluorine compounds are produced. As an extra safety precaution, there should be no smoking anywhere near pumps using CFCs. It is easy for a contaminated finger to pass a CFC onto a cigarette, which in turn will be burnt and inhaled!

* The acronym "CFC" stands for all chlorofluorocarbons, including those that are specifically harmful to the environment and those that are quite benign.

The chlorofluorocarbon oils should not be used with old pumps or Sargent-Welch, direct-drive pumps, both of which may have aluminum components (if in doubt, check with the manufacturer). A momentary seizure of an aluminum part can cause a highly localized temperature increase, which when in contact with a chlorofluorocarbon oil, can cause an explosion.[18]

Incidentally, thermocouples can be affected by CFCs and display a reading three to five times greater than the real pressure. The gauge itself is not affected because the difference is due to the greater thermal conductivity of a CFCs to air. The user need only to divide the thermocouple reading by a factor of three to five to obtain actual pressure.

The Perfluorinated Polyether Oils (PFPE). These oils are the most expensive oils available. They are extremely nonreactive, do not form tars, do not break down, are non-flammable, and can therefore be used in pumping pure oxygen.

If these oils are subjected to temperatures greater than 280°C, highly toxic gases can be produced. Therefore, there should be no smoking anywhere near pumps using PFPEs. It is easy for a contaminated finger to pass a PFPE onto a cigarette, which in turn will be burnt and inhaled!

The excellent stability of these oils creates a new problem. Among the duties of pump oils is trapping vapors and particulate material from a system. Acids and particulate matter remaining within a pump oil can destroy a pump, even though the oil may demonstrate that it has not yet broken down. Fortunately, PFPE oils can be purified (reclaimed) which dramatically decreases the high initial costs of these fluids. Therefore, proper changing schedules, using fresh oil, do not necessarily mean a tremendous increase in cost beyond your original investment.

7.3.9 Storing Mechanical Pumps

If a mechanical pump is going to be unused for any length of time, the pump oil should be changed before storage. If the pump will be disconnected from a vacuum system, the inlet and exhaust tubes should be plugged with neoprene stoppers.* Rubber stoppers are more likely to disintegrate, and a cork may crumble allowing particulate materials into the pump which could damage it once it is restarted.

7.3.10 The Limitations of Mechanical Pumps and the Demands of High-vacuum Pumps

Mechanical pumps can effectively remove ≈99.99% of the air from a vacuum system. The last 0.01% is a combination of outgassing (mostly water vapor) and leaking. Considerable water vapor clings tenaciously to the walls of a vacuum system and is removed slowly. Hablanian[19] describes a specially designed vacuum system that is specially baked and trapped and is capable of achieving inlet pressures of $\approx 10^{-8}$ torr with a mechanical pump. A vacuum of this quality cannot be achieved with a standard laboratory vacuum setup.

A small mechanical pump can only have speeds of 2-5 $^{l}/_{sec}$. However, a comparable diffusion pump can achieve speeds of 100 to 300 $^{l}/_{sec}$. It is easy

* A natural rubber stopper will deteriorate from contact with the pump oil.

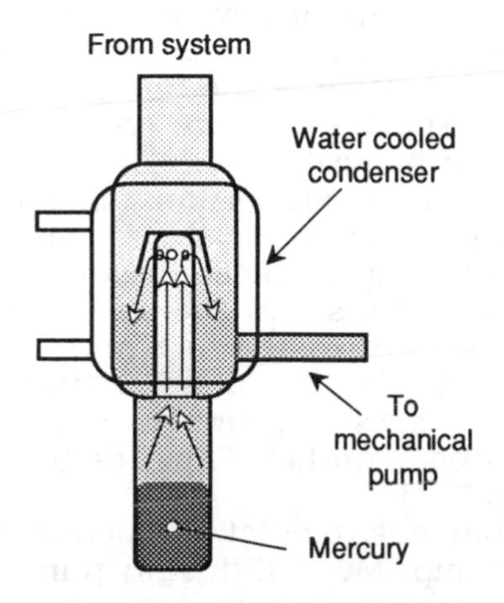

In a mercury diffusion pump, the mercury is heated to the point of vaporization. This vapor travels up into the condenser area where it is ejected at supersonic speeds from little holes. The vapor knocks any wandering gas molecules down toward the mechanical pump outlet which can then expel them from the system. The vapor later condenses and collects in the heating pot for reuse.

Fig. 7-20 A mercury diffusion pump.

to see why the assistance of an auxiliary pump is required to effectively obtain vacuums beyond 10^{-4} torr, such as those in the *high* or even *ultra-high* vacuum range. Although it is possible to obtain larger and faster mechanical pumps, the cost/efficiency ratio starts to go down and it becomes less expensive to use a mechanical/diffusion pump combination.

Cryogenic and turbo pumps are known for their cleanliness and effectiveness. However, these pumping systems are very expensive and require significant training for proper operation. The most common auxiliary pump used in the laboratory is the diffusion (or vapor) pump.

7.3.11 Diffusion Pumps

A diffusion pump connected to a vacuum system by itself can accomplish nothing. It must be used in tandem with another pumping mechanism that serves two functions: it must achieve a vacuum sufficient to achieve the vapor pressure of the heated diffusion pump liquid (about 10^{-2} to 10^{-3} torr), and it must remove the air trapped by the diffusion pump. The upper pressure range for a diffusion pump to operate is limited by the vapor pressure of the heated oil (or mercury) at the jets. Unless the second pump can achieve this pressure, the diffusion pump will not operate.

The basic principle of a diffusion pump can be shown in a simple single-stage mercury diffusion pump (see Fig. 7-20). The beauty of the diffusion pump is its lack of moving parts. On the system side of the pump (at about 10^{-2} to 10^{-3} torr, or better), gas molecules wander around, limited by their mean free path and collisions with other molecules. The lowest section of this diffusion pump is an electric heater that brings the diffusion pump liquid up to its vapor pressure temperature.* The vapors of the diffusion pump liquid are vented up a central chimney where, at the top, they are expelled out of vapor jets at supersonic speeds (up to 1000 $^{ft}/_{sec}$). Below these jets is

* Relative to the pressure within the system.

a constant rain of the pumping fluid (mercury or low vapor-pressure oil) on the gases within the vacuum system. Using momentum transfer,* gas molecules are physically knocked to the bottom of the pump where they are trapped by the vapor jets from above. Finally, they are collected in a sufficient quantity to be drawn out by the mechanical pump.

Pumping speed within a diffusion pump is proportional to the area being pumped. When diffusion pump vapors are traveling their fastest, they can make the pump pump faster. Unfortunately at these greater speeds, they have the least compression ratio, which means that the collected gases may not have sufficient pressure to be drawn out by the fore pump. By shaping the insides of the diffusion pump accordingly, (i.e., a larger tube around the first stage and less around subsequent stages) engineers can manipulate the vapors of the pump to get the best of both worlds. This design can be seen in the three-stage pump shown in Fig. 7-21.

The diffusion pump in Fig. 7-20 has only one jet. By definition, this design would be called a single-stage pump. Metal diffusion pumps can have from one to six stages, whereas glass diffusion pumps can have one to three stages. Pumps with several stages allow each stage to be aided by the work of the preceding stage. The first stage (at the inlet of the pump) can offer a high pumping speed. This high speed however comes at the cost of a low compression ratio. Because the gas flow is constant, the next stage does not need the pumping speed that the first did and can therefore provide a higher compression ratio. This process continues so that the lowest stage can provide the highest compression ratio needed for the fore pump. This setup is possible because the last stage has the lowest pumping-speed requirement.

Diffusion pumps can be made out of glass or metal. Metal diffusion pumps are more durable, can be made to more exacting standards, and therefore can provide a more reliable vacuum straight out of the box. Metal diffusion pumps can be (reasonably) easily removed from a vacuum line and be completely dismantled for cleaning. They are designed to be attached to metal vacuum systems. To attach a metal diffusion pump to a glass system, a glass to metal flange is required.

If ordering such a flange, be careful that you order a glass that is compatible with the glass on your system. Most glass systems are made out of Corning - Pyrex [7740], Kimble - Kimex [KG-33], or Schott - Duran [8330] glass. A glass-to-metal flange using any of these glass types is compatible to any other of the set. Some glass or metal systems have glass-to-metal components. The metal may be stainless steel machined to receive any of the three above-mentioned glasses, or Kovar®, which is an iron, nickel, and cobalt alloy. The glass sealed to this alloy is either Corning 7052, Kimble K-650, or Schott 8250 glass. These three types of glass are compatible with each other, but not with the former three. There must be a third glass (forming a graded seal) between the glass fused to the Kovar, and common laboratory borosilicate glass. Therefore, be sure you know what type of glass you need before ordering a glass-to-metal flange.

* Momentum transfer is possible because momentum = (mass) x (velocity), and the velocity of the diffusion pump fluid when it leaves the pump jets is typically at supersonic speeds. Thus, it is easy for the diffusion pump liquid to knock around any size molecule.

Glass diffusion pumps provide (relatively) easy attachment to a glass vacuum system, are mostly free from attack by corrosive substances, provide easy observation of the materials inside the pump, and can be cleaned, with some difficulty.

Fig. 7-21 displays three different designs of glass diffusion pumps (that use low-vapor-pressure oil). These three designs were selected to demonstrate the addition of stages. They are also among the most commonly-seen diffusion pump designs. Although the three pumps shown are all air-cooled, it is possible to obtain most small diffusion pumps as air- or water-cooled. The advantage of an air-cooled diffusion pump is that there is no water hose connection to slip off the pump and no overheating occurs if the building water is shut off without warning. However, in a warm, non-ventilated room, air-cooled diffusion pumps can lose their efficiency. Water-cooled pumps work efficiently regardless of room conditions.

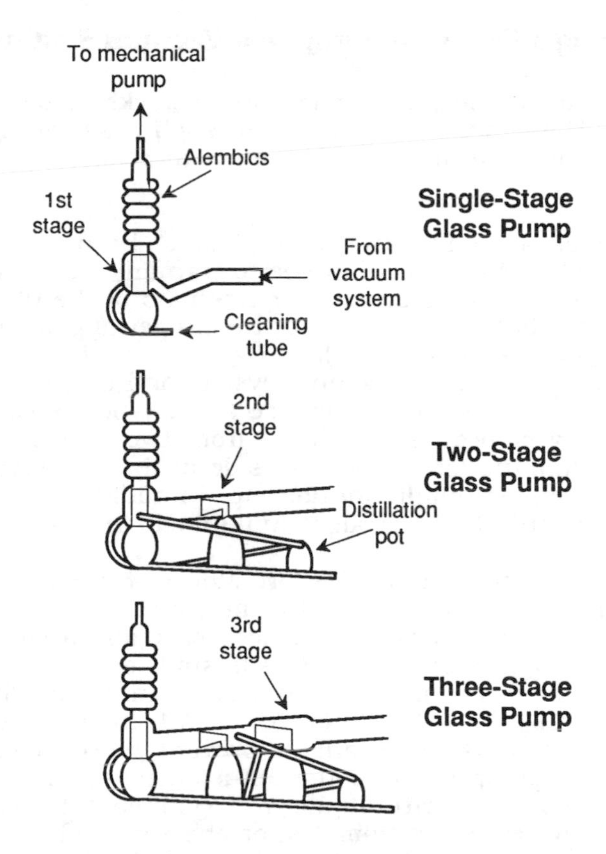

Fig. 7-21 Various forms of glass diffusion pumps. The potential speeds and ranges of glass diffusion pumps are shown in Table 7-8.‡

‡ From *Journal of Chemical Education*, 42 (1965) A445-59. With permission.

Table 7-8‡

Glass Diffusion Pump Capabilities		
Pumping Stages	**Speed (ℓ/sec)**	**Range (torr)**
1	8	$2 \times 10^{-3} \rightarrow 5 \times 10^{-6}$
2	26	$2 \times 10^{-3} \rightarrow 8 \times 10^{-7}$
3	29	$10^{-3} \rightarrow 8 \times 10^{-8}$

‡ From *Journal of Chemical Education*, 42 (1965) A445-59. With permission.

The data in Table 7-8 show that the addition of stages dramatically increases the speed of a pump while only moderately affecting the net vacuum potential.[20] These figures however are taken from a vacuum gauge placed at the inlet end of the pump, the location which provides the *best potential vacuum* that can be expected from any system.

7.3.12 Attaching a Diffusion Pump to a Vacuum System

As previously mentioned, a diffusion pump works in tandem with a mechanical pump. The simplified alignment of such a system is vacuum line, diffusion pump, mechanical pump. However, in reality, the setup is a bit more complex.

1) A trap (i.e., a liquid nitrogen trap) must be placed at the inlet (system) side of a vacuum line to trap any vapors coming from the line as well as any lighter fractions of the diffusion pump fluid (see Sec. 7.4 on traps for the proper alignment and use of traps on a vacuum system).
2) The tubing between the vacuum system and diffusion pump must have as few restrictions (in size and shape) as possible. Glass or rotary stopcocks with bores from twelve to fifteen millimeters minimum (on the system side of the diffusion pump) will help to ensure minimum hold-up. In addition, tubing in these areas should be as straight (unkinked, with as few bends) as possible.
3) Because the pressure on the fore side of the diffusion pump is considerably higher than the pressure on the system side, large bore stopcocks are not necessary and a stopcock with a bore of eight to ten millimeters is sufficient.
4) Oils used in mechanical pumps have significantly higher vapor pressures than the oils used within diffusion pumps. Therefore, it is important to prevent backstreaming of mechanical pump oils into the diffusion pump. This barrier can either be done with liquid nitrogen cold traps, molecular sieves, water-cooled thimbles, or chevron baffles.
5) A vacuum system needs to achieve a sufficiently low pressure before the diffusion pump can begin operation. A mechanical pump that works in tandem with a diffusion pump is called a fore pump. Fore pumps must be capable of achieving $\approx 10^{-2} \rightarrow$

10^{-4} torr and have sufficient speed to effectively work with a diffusion pump.

Most mechanical pumps with sufficient speed and ability are two-stage pumps. However, it is not good to run a two-stage mechanical pump for an extended period of time without pulling on a load. Therefore, a fairly large system that would require running a mechanical pump for over five to ten minutes before a $\approx 10^{-1}$ torr vacuum was achieved should include a smaller roughing pump. Once the vacuum has been 'roughed' down, the fore pump can be started to bring the vacuum line to a low enough pressure to start the diffusion pump.

Fortunately, the vacuum system in most laboratories is small enough that the fore pump can double as the roughing pump with no problems. However, cycling your vacuum line back and forth from atmospheric pressure can put a significant strain on the fore pump, so a roughing pump should be used. That way every time you need to pull from atmospheric to vacuum conditions, all that is necessary is to close off the line to the diffusion pump and open the line to the roughing pump. Then, once the pressure is within the range that the diffusion pump can operate, close the valve to the roughing pump and open the line to the diffusion pump.

A glass vacuum system requires a glassblower to assemble and fuse the various sections together (if you have not had any experience working with glass, do not start with a vacuum system). This technique provides the best approach to assembly because it provides a solid, leak-tight seal between all parts. The main problem with this permanent form of connection is if you ever need to dismantle the items for any reason, you again need the services of a glassblower. In addition, you should be forewarned that glass can be worked only so many times by a glassblower after which it begins to devitrify easily and lose its strength and chemical durability. The attachment and disassembly of any glass apparatus should be only used in case of emergency, not as regular lab practice.

The second best method of attachment is with O-ring seals. These seals can provide an easy method of attachment and disassembly. However, O-rings can leak if the alignment between the two members is not perfect. In addition, some O-ring elastomers may decompose if left in contact with diffusion pump fluids for an extended time.

The third form of attachment is with some type of ground joint. It would be wonderful if this attachment could be made with a full-length standard taper joint (see Sec. 3.1.1), but alignment of several joints that allow for easy assembly and disassembly of several parts is seldom possible. Ball-and-socket joints would seem like an alternative, but they are not really acceptable for high-vacuum work, and should not be considered. However, the use of ball-and-socket joints is not impossible. Rather, because there is such a strong possibility for leaks, their reliability is always in doubt, and the use of these joints for vacuum line assembly should never be your first choice.

7.3.13 How to Use a Diffusion Pump

In the stylized drawing of Fig. 7-22, you see the layout of a vacuum line, traps, roughing pump, fore pump, diffusion pump, and related stopcocks (because this drawing is stylized, it's not intended to look exactly like your vacuum system). First find the parts on your line that are represented by the parts in Fig. 7-22. If not every item shown is on your line, determine which parts your line and this drawing have in common, and proceed accordingly. The following general guidelines should be followed when using a vacuum system with a diffusion pump (see Table 7-9 for diagnostic checks and corrections of diffusion pump problems):

1) The system must be in a vacuum state before the diffusion pump can work.
2) The diffusion pump should be bypassed in the early stages of the system's pump-down, the cooling water for the diffusion pump **must** be turned on early, and the diffusion pump **must** be in a vacuum before the diffusion pump heaters are turned on. If your air-cooled diffusion pump has a fan, turn it on now. Some diffusion pumps are wired so that turning on the heater also turns on the cooling fans. If you have this setup, verify that the fan is working when the heater is turned on.

   If you are starting a glass diffusion pump for the first time and need to adjust the heat (from a voltage regulator), first place the heat at a very low (10%-20% of full) setting and let it stabilize (this procedure will also help to outgas the oil). Then, over the course of time, increase the setting some 10 units at a time letting the temperature stabilize at each setting. When you see the 'wet marks' of vapor condensing on the horns inside the pump, the temperature is set. If the oil is furiously pumping away, the voltage regulator is on too high and needs to be reduced. Once the temperature is established, mark your voltage regulator for future reference. If the pump is water-cooled, connect a relay to shut off power to the pump in the event of a water pressure loss.

   Metal diffusion pumps require no adjustment, they are factory-set to provide the right temperature. The important thing is to make sure they are filled with the proper amount of oil. After attaching to a vacuum system, simply plug the outlet of the metal diffusion pump into an appropriate (110-120 V or 220-240 V) outlet. If the pump is water-cooled, connect a relay to shut off power to the pump in the event of a water pressure loss.
3) Let the mechanical pumps evacuate the traps before they are placed in liquid nitrogen to prevent freezing oxygen within the traps.
4) When shutting down the system, isolate the diffusion pump and the system from the mechanical pump(s). Vent the mechanical pump(s) after shutting them off.

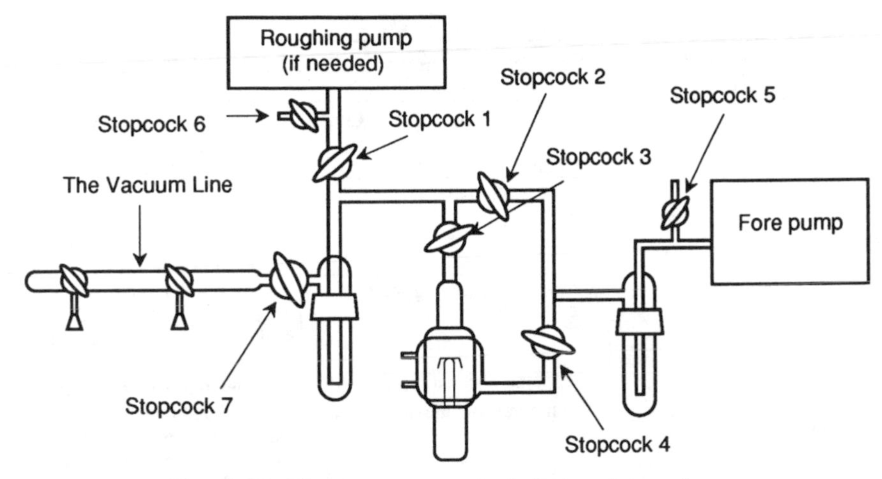

Fig. 7-22 The arrangement of stopcocks and traps for a vacuum line with a diffusion pump.

Aside from the generalized description above, the following procedures should be used for operating a large vacuum system with a diffusion pump. If you are beginning a pump-down operation, proceed as follows:

1) Check to make sure all stopcocks are closed.
2) Turn on the roughing and fore pumps.
3) Open Stopcock 1, 4, and 7.
4) After pressure in the vacuum line is 10^{-1} torr or better, close Stopcock 1 and open Stopcock 2.
5) Place Dewars with liquid nitrogen under cold traps.
6) Be sure cooling water is running in (or that the fan is blowing) the diffusion pump and turn the diffusion pump's heater on.
7) After pressure in the vacuum line is 10^{-2} to 10^{-3} torr or better, close Stopcock 2 and open Stopcock 3.

Notice that once the mechanical pumping was transferred over to the fore-pump, the roughing pump was not turned off. Leave the roughing pump run, because there may be ancillary parts of a vacuum system (such as the drawing of mercury back into a McLeod gauge) that are operated from the vacuum of the roughing pump. In addition, the roughing pump typically receives the bulk of the water vapor and other condensable vapors. Letting the pump run (against a load) helps to remove the condensed vapors from the pump oil. If the pump has a ballast, leave it open. Otherwise, be sure that nothing else on the vacuum line requires the use of the roughing pump before it is turned off.

If the line you are working with is small and does not require a roughing pump, proceed as follows:

1) Check to make sure that all stopcocks are closed.
2) Turn on the fore pump.
3) Open Stopcock 2, 4, and 7.

Table 7-9‡

Diffusion Pump Fault Diagnosis and Correction[21]

Fault	Probable Cause	Correction
Poor ultimate pressure	Leak in system, virtual or real	Locate, rectify or repair
	System dirty	Clean system
	Contaminated pump fluid	Verify and replace
	Low heat input	Check heater voltage, check for continuity of elements, burned out elements, poor electrical or thermal contact.
	Insufficient cooling water	Check water valve, water pressure, check tubing for kinks and/or obstructions and back pressure.
	High pressure between the diffusion pump and the mechanical pump.	Check for leaks between mechanical pump and diffusion pump, check mechanical pump performance, breakdown, and/or pump fluid.
Low speed (extra long pumping time)	Low heat input	Check heater
	Low fluid level	Check and top off fluid, if necessary
	Malfunctioning pump assembly, improperly located jets, damaged jet system	If metal system, check and rectify; if glass system; check for particulate material. If necessary, empty the fluid, clean the pump, and replace with new fluid.
Inlet pressure surges	Heater might be too high	Verify and reduce heat
	Trap has trapped material too close to top of liquid nitrogen level	When starting system, do not fill dewars to top until the system has settled.
	Fluid outgassing	Allow fluid time to outgas
	Leak in system ahead of pump inlet	Verify and repair
High chamber oil contamination	Virtual leak in system	Verify and correct
	High pressure between the diffusion pump and the mechanical pump.	Check for leak in foreline, poor mechanical pump performance, breakdown of pump fluid
	Prolonged operation above 10^{-4} torr without adequate trapping.	Review procedures for starting vacuum system and use of traps
	Incorrect system operation and air-release procedures	Review procedures for starting vacuum system

‡ From *Vacuum* Vol. 31, N.S. Harris, "Practical aspects of constructing, operating and maintaining rotary vane and diffusion-pumped systems," 1981, Pergamon Press, plc. With permission.

4) Place Dewars with liquid nitrogen under cold traps.
5) Be sure the cooling water is running in the diffusion pump and turn the diffusion pump's heater on.
6) After pressure in the vacuum line is 10^{-3} torr or better, close Stopcock 2 and open Stopcock 3.

To turn *off* a vacuum line with a diffusion pump, do the following:

1) Close Stopcocks 2, 3, 4, and 7. Although there are none displayed in Fig. 7-22, any stopcocks that go to vacuum gauges or other peripheral equipment should also be closed.
2) Turn off the heater to the diffusion pump, do not turn off the cooling water or fan until the pump has been off for at least an hour, or is cool to the touch.

If the line you are working with has a roughing pump, the shut down procedure is as follows:

1) After shutting off the roughing and fore pumps, open Stopcocks 5 and 6 to vent the pumps (this action prevents the mechanical pump oil from accidentally being sucked into the vacuum line).
2) If there is material in the cold trap near the vacuum line that needs to be removed, open Stopcock 1 to vent the trap. Then remove the trap so that it may defrost in a fume hood. It is unlikely that the cold trap near the fore pump will need to be removed for cleaning.

If the line you are working with does not have a roughing pump, the shut-down procedure is as follows:

1) After shutting off the fore pump, open stopcock # 5 to vent the pump (this prevents the mechanical pump oil from accidentally being sucked into the vacuum line).
2) If there is material in the cold trap near the vacuum line that needs to be removed, open Stopcock 2 to vent the trap. Then remove the trap so that it may defrost in a fume hood. It is unlikely that the cold trap near the fore pump will need to be removed for cleaning.

Until you need the vacuum that can be achieved with a diffusion pump, the water content on the walls of a vacuum system is mostly irrelevant. However, once you need to achieve 10^{-5} torr and beyond, the presence of water adsorbed into the vacuum system's surface will slow down, and even prevent, achieving greater vacuums. Because of this idiosyncrasy, when first trying to bring a vacuum system down to 10^{-6} torr, you may need to pump on the system from a half to an entire day before you have removed enough water to obtain 10^{-5} torr. Even with auxiliary pumps, water vapor is difficult to remove. The most effective way to remove it is to heat the entire vacuum system and drive the water off the system walls. This approach cannot effectively be done to glass vacuum systems because of damage to stopcocks, valves, and glass connections, although localized heating (with a heat gun) can be done on glass with limited success.

Once you have successfully removed the bulk of water from the walls of the vacuum system, do not allow it to return. One easy and effective demonstration of the effect of water on vacuum is to pump a vacuum system down to some established level after it has been vented with atmosphere. Then, using dry nitrogen or argon, vent the system back to atmospheric pressure. Now, re-pump the system back to the same vacuum as before. It should take about one-tenth the time.[22] This example demonstrates why the ability to bake out a vacuum system improves the pumping speed by speeding up the removal (outgassing) of water vapor from the system's walls. It also demonstrates that once a vacuum system has been successfully pumped down, you do not want to re-expose it to the atmosphere. If you need to expose sections of your vacuum system to the atmosphere (for example traps or mechanical pumps), section off these parts with valves and stopcocks so that the rest of the system can remain in a dry vacuum state.

IMPORTANT: The diffusion pump works with heat, (sometimes) water, and electricity. Be sure: 1) Your water connections to the condenser on the pump and their plumbing attachments are secure and the hoses show no signs of cracking, folding, or tearing; 2) the electrical connections are located in a position where they cannot be accidentally touched; and 3) there is nothing combustible (that could be set off with a spark) in the area where you are working.

7.3.14 Diffusion Pump Limitations

Because of design, setup, and/or operation, there are four factors that can affect the ultimate pressure of a diffusion pump:[23]

1) Back diffusion of the pumped gas against the vapor stream
2) Saturation vapor pressure of the pump fluid or of decomposition products
3) Evolution of gas from the pump components
4) Dissolved gases in the pump fluid being released when heated.

Back diffusion is when the pressures at the outlet and inlet have established a constant ratio, and are analogous to the *compression ratio* found in mechanical pumps. Back diffusion is the lowest for light gases (hydrogen, helium) and increases markedly for heavier gases (nitrogen). Although it is possible to design pumps that can take advantage of these factors, there is nothing that the user can do about them, and the effects are at most minimal in only the ultra-high vacuum range.

Saturation vapor pressure can be caused by either back-streaming or back migration. Although diffusion pump oils have low vapor pressures ($10^{-5} \rightarrow 10^{-8}$ torr), in use they break down* into multiple vapor pressures. The resulting higher vapor pressure oils may backstream into the system, while the rest of the oil remains in the pump. Some of the low vapor pressure fractions may turn into a sludge. In a poorly-designed or operated system, a considerable amount of higher vapor pressure oil may leave the diffusion pump and drift into the system or be drawn out by the mechanical pump. This loss of the oil from the pump can be a significant problem. The effects of this loss might not seem all that important, but if the loss is great enough to begin exposing heating elements, other more significant damage can occur. By including baffling and/or traps, this loss is greatly minimized. Unfortunately, the more effective the baffling, the more the pump slows down. Some baffling can decrease pumping speed by up to 50%[24] (which is why some of the stated pump speeds are irrelevant).

* Break-down of diffusion pump oils can be accelerated by exposing hot mineral oil to the atmosphere.

Back migration is caused by improper cooling around the orifice of the pump. With this problem, pump oil that had condensed begins to re-evaporate. This problem is more significant with ultra-high vacuum systems.

Evolution of gas (outgassed) from pump parts, or attachments during use will decrease pumping potential. Any outgassing beyond general water vapor or high vapor pressure oil components is generally of greater concern to ultra-high vacuum systems. However, you need to be conscious of the types of materials you are placing on high-vacuum systems. For example, some stopcock greases are adequate for student quality labs, but will fail tremendously in high-vacuum systems. In addition, connections in high vacuum systems need to be solid and permanent. Flexible elastomer tubing and epoxies cannot sufficiently seal against leaks, nor can their high outgassing rates be contained by typical pumping systems.

Factors 1 and 3 are of greater concern to systems that are trying to obtain ultra-high vacuum conditions. Factors 2 and 4 can affect attempts of high and ultra-high vacuum demands. The gases dissolved within pump oil from contamination during condensation and later released when pump oil is re-heated is the easiest of these four problems to prevent and the hardest to eliminate once established. To prevent contaminates from entering a pump, establish (preferably) liquid nitrogen traps between the system and the pumps.

7.3.15 Diffusion Pump Oils

The six major categories of diffusion pump fluids are explained in Table 7-10 (based on information from Laurenson[25]). Diffusion pump oils, like mechanical pump oils, need to be protected to maintain their properties. Fortunately for diffusion pump oil, by the time it is typically exposed to materials within the vacuum system, the bulk of any contaminating materials are already removed. However, sudden exposure to the atmosphere can destroy hot hydrocarbon pump oils (or even cause them to flash or explode) and damage many others. Silicone oils, on the other hand, can easily survive contact with oxygen while hot. In addition, contact with undesirable vapors can speed the disintegration process of other pump oils. When operating in peak condition, a perfected diffusion pump oil should:

1) Be thermally stable
2) Be chemically inert
3) Exhibit a low vapor pressure over a wide temperature range
4) Trap vapors and particulate matter from the vacuum system
5) Be compatible (and not interfere) with, and protective of, the environment
6) Be non-toxic.

This list differs from mechanical pump oil requirements primarily in the exclusion of lubricity demands. Because there are no moving parts in a diffusion pump, there are no lubricating requirements whatsoever, (within some metal diffusion pumps there may be some rust prevention requirements under certain environments).

Table 7-10

Diffusion Pump Fluid Types		
Diffusion Pump Fluid (Vapor Pressure in torr)[a]	**Good Characteristics**	**Poor Characteristics**
Mercury ($\approx 10^{-3}$)	Does not break down at high temperatures or in oxidizing environments.	Cannot be used in systems of electrical processing (i.e., silicon chip processing). If not properly trapped and vented, mercury can be a significant health hazard.
Mineral Oils (5×10^{-5} - 10^{-8})	Very economical, can safely be used in mass spectrometers.	If exposed to the atmosphere while hot, can oxidize quickly. Can easily be broken down by heat. Tend to form conducting polymers.
Silicone Fluids (10^{-6} - 10^{-9})	Are thermally stable and are resistant to oxidation and reasonably resistant to chemical attack. Are also reasonably priced.	Form insulating polymeric layer when irradiated by electrons so they cannot be used where physical electronic equipment is being pumped, such as in helium leak detectors.
Polyphenyl ether (Santovac 5) ($< 10^{-9}$)	Thermally stable, good oxidation resistance. Forms conducting polymers under energetic particle bombardment. Good for mass spec. and ultra-high vacuum.	Not very chemically resistant. Relatively high cost.
Perfluoro polyethers (Fomblin) (Krytox®) (3×10^{-8})	If exposed to too much heat, decomposes to a gas rather than breaking down. Is resistant to oxidation and chemically resistant with few exceptions. No polymers are formed under energetic particle bombardment. Can be regenerated for reuse.	Provides somewhat lower pumping speeds than other oils. High initial cost. Above 300 - 350°C, breaks down into aggressive and toxic compounds. To effectively remove it, chlorofluorocarbons must be used.
Miscellaneous fluids (esters) (ethers) (sebacates) (phthalates) (napthalenes) (10^{-6} to 5×10^{-9})	Generally good thermal and oxidation resistance, resistant to most chemicals (depending on fluid used). Usually conducting polymers formed under energetic particle bombardment. Generally low in costs.	(Watch out for exceptions to the "Good" list.)

[a] Depending on the temperature, grade ,and type used.

Of the six diffusion pump oil types, none has all of the above properties. In general, lighter oils pump faster than heavier oils, but heavier oils can achieve lower ultimate pressure. When your work demands varying pump oil requirements, it sometimes is easier to have separate vacuum systems than to simply change pump oils because the pump oils are seldom compatible, and mixing may impair potential peak performance.

To change a diffusion pump fluid requires the complete removal and cleanup of all previous fluid before adding a new, different oil. However, before adding the oil, be sure the type of oil (or mercury) that you wish to use will work in the pump that you have. Some oils require specific tolerances or pump designs for optimum performance. One example (in the extreme) is the use of mercury in an oil pump or vise versa. Mercury *may* work in some types of pumps but the performance will not be very good. Oil, on the other hand, will not work in any mercury diffusion pump design.

The decision to use mercury- or oil-based diffusion pump fluid may be academic to those living in areas where mercury use is banned, but for others, such decisions are beyond idle curiosity. Mercury does not break down on contact with air when hot (although it may oxidize somewhat), it will not react with most compounds,* and gases do not dissolve within mercury to the same degree as they do with oils. On the other hand, mercury is a health hazard and once mercury has spilled on the floor, (it is said) the only way to *truly* remove mercury is to burn the building down.

Silicon oils do not break down on contact with air while hot, but on the other hand, they can polymerize and develop an insulating film on electronics. Thus, their use is not acceptable for instruments such as mass spectrometers (including He leak detectors). Some specific properties of a spectrum of diffusion pump fluids are shown in Table 7-10.

7.3.16 Diffusion Pump Maintenance

Because there are no moving parts within a diffusion pump, there are no parts to wear out. However, as the need for higher vacuums increases, so is the need for a greater regard to cleanliness. This need is important for the pumps, pump oils, and the entire system. Not only does the vacuum potential go down as dirt piles up, but so does pumping speed.

The most common problem with hydrocarbon diffusion pump oil is its fractionation into multi-vapor pressure components. As pump oil breaks down, it develops both lower and higher vapor-pressure characteristics. Oils with high vapor pressures can potentially drift into the system, although they are more likely to be effectively removed from the system by being trapped in the alembics of the central vertical tube, in the cold trap between the system and the diffusion pump, or in the cold trap between the diffusion pump and the mechanical pump. If not trapped, they are free to travel into the vacuum line itself or into the mechanical pump. Diffusion pump oils that collect in a mechanical pump are not likely to have any significant performance effects (as opposed to the degrading effects of mechanical pump oil collected in diffusion pumps).

In two- and three-stage glass pumps, a special distillation pot is added onto the end to separate fractionated oil. The three (or four) pots on two- and three-stage glass diffusion pumps are all wired in series. The first, and largest, stage of the pump has the longest resistance heating wire and therefore the greatest amount of heat. The last (and smallest) distillation pot has the shortest wire and therefore provides the least amount of heat. As in the single-stage pump, low boiling oils (those with a high vapor pressure) are trapped in the alembics above the first stage. Each successive stage allows oils with higher boiling points to pass through connecting tubes toward the distillation pot. The lower heat of the distillation pot is not hot enough to significantly heat the oil, leaving heavy tarry oils remaining.

* Mercury amalgamates with a few metals such as gold and aluminum, so attention must be paid to metal part selection for use on vacuum systems.

Before cleaning any diffusion pump, it is important to remove and/or unplug any electrical leads. Water should be turned off and removed if necessary. It is always best to cut off rubber or plastic tubing with a razor blade and replace it, rather than to try and pull the tubing off a hose connection.

To clean a metal diffusion pump, it must be removed from the rest of the system. Pour the used oil (or mercury) into a proper receptacle. Do not throw the mercury away because it is a toxic waste (a heavy metal) Fortunately, mercury may be reclaimed and reused. As far as diffusion pump oils, check with the health and safety and/or environmental officer in your institution and/or the waste disposal management of your city. Be sure to mention any hazardous materials that may have been absorbed by the pump oil during its operation to the proper authorities.

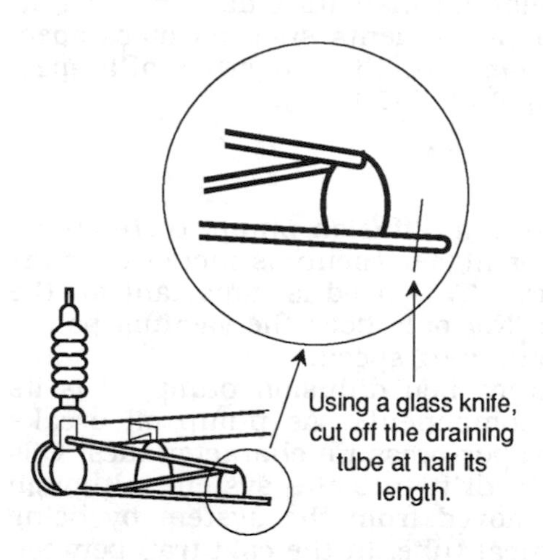

Using a glass knife, cut off the draining tube at half its length.

Fig. 7-23 How to drain the oil from a glass diffusion pump.

Because it is not recommended to repeatedly remove a glass diffusion pump that has been fused onto a vacuum system, an alternative approach is available: On the bottom of all glass diffusion pumps should be a lone glass tube that seems to do nothing. Its purpose is to drain the oil from the diffusion pump without requiring pump separation from the system. First unplug the diffusion pump from its electrical source, then, with a glass knife, scratch and snap off the drainage tube at ≈ **half its length**. Assuming that the pump was properly installed with a slight pitch toward the cleaning tube, all the oil should drain from the opened tube (see Fig. 7-23). To remove any oil trapped in the alembics, a flexible tube attached to a filter flask and a house vacuum can suck out the majority of the oil. The rest of the oil can be flushed out with solvent or base bath if the oil is silicon-based (see Sec. 4.1.6).

If you are removing a hydrocarbon oil, after the oil has sufficiently drained, use a cotton swab to clean any oil filling the drain tube. If you wish, pour a solvent through the pump to flush out the rest of the oil. You may want to temporarily plug the open end of the drainage tube with a cork to let the oil soak in the solvent. Because the sharp ends of the (recently cut) tube will prevent the cork from obtaining a good seating, the broken end of the drainage tube should be fire-polished (see Sec. 8.2.3). Fire-polishing is likely to burn contamination into the glass, so leave room to remove this burnt end before extending the length of the tube. Now, pour solvents or a base bath into the pump to soak the internal parts for about an hour or so.

If your pump used silicone oil and has left crusty remains that will not drain, a base bath is recommended (see Sec. 4.1.6). Place a cork in the end of the drainage tube (fire-polish as before), pour in a base bath solution, and let the pump sit for an hour or two. Because a base bath is highly flammable, be sure to unplug all electrical components of the pump before you begin this type of cleaning process.

After the base bath has been drained, let the pump soak for a few minutes with an acid rinse (to stop any alkaline reactions on the glass surface). After three or four water rinses, follow with a distilled water rinse and finally some methanol to speed the drying process. Do not blow air through the pump to speed the drying process as most compressed air is full of oils and other particulates (although dry nitrogen is acceptable). Alternatively, you can place the house vacuum hose to the pump and draw the ambient air through the pump. Remember, any acids or bases must be neutralized before disposal.

Hydrocarbon oils are generally easier to remove than silicon oils, but many of the materials of these oils are considered by the EPA to be toxic, such as chloroform or methylene chloride. For seriously contaminated oils, and/or oils that are very tarry, something that is more heavy-duty such as decahydronapthalene or trichloroethelene may be needed, but the latter is also considered to be a toxic waste. Each of these solutions should be followed by acetone and ethanol rinses.[26] Another diffusion pump oil, polyphenylether, can be dissolved in either trichloroethelene or 1,1,1-trichloroethane.[27] Again, both of these are considered by the EPA to be toxic materials.

Be aware that the old oil from a pump (and any solvent used to clean out the old oil) contains toxic materials from the vacuum system. For example, if the system had a McLeod gauge, it is likely that the old oil is contaminated with mercury. The amount of contamination concentration determines how the oil or solvent can be disposed of. Unfortunately, because of the possibility that specific EPA-established concentration levels will change before you read this book no disposal procedures are provided. Therefore, contact the EPA, or local regulatory agencies, to verify the various toxicity levels and the proper disposal procedures for materials of those levels.

If you are using a fluorinated oil such as fluoropolyether, do not fire polish the end of the drainage tube and do not seal any glass onto the area until you have thoroughly cleaned the area. Any remaining grease that is heated higher than 280°C will turn into toxic fumes and could be lethal. You must use a fluorinated solvent such as trichlorotrifluoroethane or perfluorooctane to clean the system. The biggest problem when using fluorinated materials (either as stopcock or joint grease or as pump oil) is the environmental hazard of the solvents needed for cleaning. During the time this book was written, substitutes and alternatives for these solvents were being planned and created, but none had received final approval. By the time you read this book, it may not only be important for the environment to check on the availability of these substitutes, it may be illegal not to. The best recommendation that can be made now is for you to call the manufacturer of the solvent and see if a substitute is available.

Several recommended procedures for working with solvents begins with using as little solvent as possible. In addition, many solvents can be reused. For example, a solvent used for a final rinse can later be used for an earlier rinse. After a solvent has been reused, it can be distilled, or roto-evaporated, so you end up with a smaller amount of contaminated liquid. Because disposal costs are based on volume, any decreased amount of waste can provide significant cost savings.

After the cleaning has been completed, it may be necessary to remove the end of the drainage tube to remove any burnt deposit. Then, add an extension of the draining tube and close it off at about 1 to 1½ in (see Fig. 7-24). Now the pump can be refilled. The reason for cutting the drain tube in half (as mentioned) is to provide distance from the diffusion pump onto which you can fuse an extension. If an extension were to be sealed directly onto the pump, extensive annealing of the pump would be required.

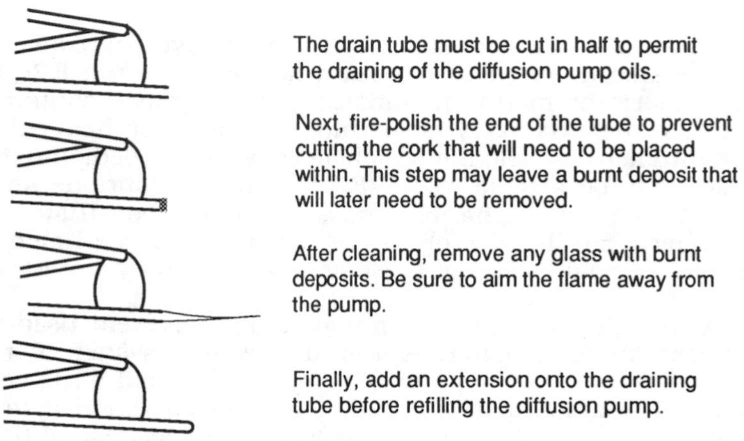

The drain tube must be cut in half to permit the draining of the diffusion pump oils.
Next, fire-polish the end of the tube to prevent cutting the cork that will need to be placed within. This step may leave a burnt deposit that will later need to be removed.
After cleaning, remove any glass with burnt deposits. Be sure to aim the flame away from the pump.
Finally, add an extension onto the draining tube before refilling the diffusion pump.

Fig. 7-24 The procedure for closing the drain tube on a glass diffusion pump.

The cleaning of a mercury diffusion pump* is somewhat simpler because mercury does not break down as most pump oils do. However, mercury gets dirty, and a dirty mercury pump still needs to be cleaned. After you have drained the mercury out of a *glass* diffusion pump, refill the pump with ≈ a 6 molar nitric acid solution and let it sit until the mercury has been removed. **Do not pour this liquid down the sink! Check with local waste management and/or your health and safety officer**. The pump should be flushed with distilled water and then rinsed with methanol for drying.

To remove any mercury deposits from a metal vacuum pump, it is important to check with the manufacturer because some cleaning techniques may destroy the pump.

Alembics and distillation pots are not necessary on mercury diffusion pumps because mercury does not fractionate like oils. Although mercury and oil diffusion pumps use the same principle to function, they differ markedly in design. Because there is no fractionating ability in mercury pumps, oils cannot be used in mercury pumps. However, mercury can be used in an oil-designed metal pump on a limited basis—albeit with a noticeable loss of performance. Never use mercury in a diffusion pump with exposed heating elements because the mercury will short out the pump.

*Save any mercury taken from McLeod gauges, manometers, and diffusion pumps as it can be sold to your mercury supplier for purification, and reuse. One company that provides this service is Bethlehem Apparatus Co., Inc., 890 Front St., Hellertown, PA 18055.

7.3.17 Toepler Pumps

It is fairly easy to move condensable gases from one section to another within a vacuum system by placing liquid nitrogen (within a Dewar) around a trap or sample tube. The condensable gas will travel to the trap, condense out, and/or freeze. Non-condensable gases (helium and hydrogen for example) cannot be moved in this manner and require a Toepler pump.

Toepler pumps are used for the collection and transfer of non-condensable gases. They are not capable of creating great vacuums, but they can be very effective pumps regardless. A 500 cc-size Toepler pump can remove 99.9% of a two-liter gas bulb in 10 minutes.

A Toepler pump is a piston pump with mercury as the piston. The mercury piston is powered by both the vacuum from a vacuum system and the atmosphere. The actual operation of a Toepler pump involves evacuation followed by re-admission of air out of, and into, the Toepler pump. There have been a wide variety of techniques to automate this process because manual operation can be quite tedious.

The most common automation mechanism is to use the electrical conduction abilities of mercury to trigger switches that operate the pump. An example of such a Toepler can be seen in Fig 7-25. In operation, the gas to be pumped enters the *inlet* at the top of the *piston chamber*. The pumping action begins when air is allowed into the lower chamber by the stopcock at *C*. The air forces the mercury up into the *piston chamber* and into the *Inlet* and *exhaust* tubes. Finally, the mercury pushes up the *float valves* preventing the mercury from going out the *inlet*. Once the mercury makes contact with

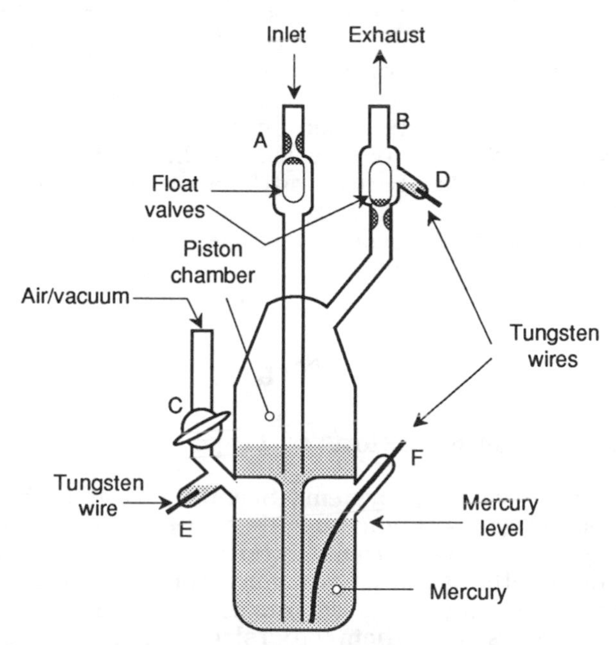

Fig 7-25 General layout of a typical electrically operated Toepler pump.

the *Tungsten wire* at *D*, the circuit with the *Tungsten wire* at *F* is complete. Then, the stopcock at *C* is rotated to evacuate the lower portion of the pump. The mercury finally is drawn back into the lower portion of the pump. The *float valve* at *exhaust Tube* (at *B)* prevents gas being pumped from re-entering the Toepler pump. As the mercury fills the lower chamber, it fills the small chamber (with *Tungsten wire* at *E*) which completes the circuit with the *Tungsten wire* at *F*. This completion activates the stopcock at *C* to re-admit air into the pump and the process repeats until it is manually turned off.

The major problem with this type of setup is when the mercury makes (and un-makes) contact with the *Tungsten wire* at *D*. A spark may form that reacts with the gas being pumped through the *exhaust*. Mechanisms to by-pass this problem have included either placing the air/vacuum cycling on a timer, thereby by-passing the need of electrical contacts, or placing a photo-sensitive relay on the side of the *exhaust tube*, which is activated when the opaque mercury fills the transparent glass tube.

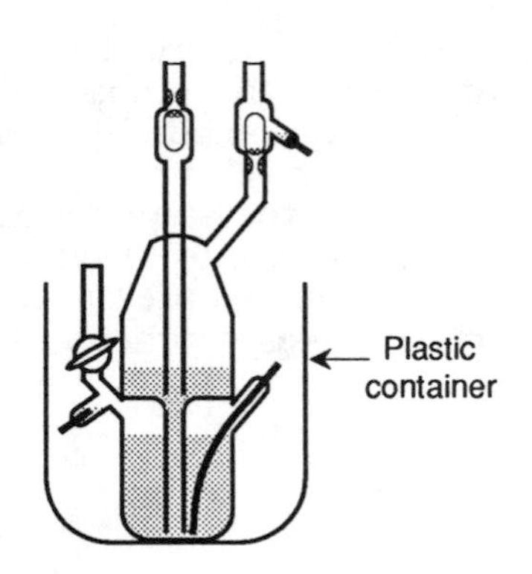

Fig 7-26 By placing a Toepler pump (or any mercury-containing device) in a plastic container, spills are avoided and it's protected from an accidental bump.

As with most items on a vacuum line that contain mercury, place the Toepler pump in a secondary container that is firmly attached to another surface to both contains any mercury that may spill from an accident as well as protect the pump from accidental bumps (see Fig 7-26). Plastic containers (such as plastic milk cartons) are particularly good because the mercury will not affect plastic: it might, however amalgamate with the metals in a metal can. In addition it is easier to get mercury out of a plastic container (with smooth walls) than a metal one (with a narrow rim). The plastic tub can be glued onto the table with some epoxy. The epoxy will stick better if you roughen up the bottom surface of the plastic container with sandpaper. Because not all epoxies stick to plastic, test the epoxy before assuming that it will hold.

7.4 Traps

7.4.1 The Purpose and Functions of Traps

Traps are used on vacuum systems because of their abilities to remove (or bind) condensable gases. Remember that to obtain a vacuum, you must remove everything from a given area. However, like the environmental adage "you cannot throw anything away," you may not want everything that is removed from a vacuum system passed into the pumping system, or into the air you breath. Solvents (and their vapors) can impair or severely damage mechanical pumps by dissolving seals or by breaking down mechanical pump oils. Condensed vapors in pump oil can impair a pump's performance. Volatile compounds that pass through a pump can expose technicians to an unacceptable health risk. Thus, the use of traps in vacuum systems can protect equipment **and** the people who are using the equipment.

Traps are supposed to prevent exhausts from being released into the working environment. However, they should not be relied on over the possible negligence of the user(s), or other possible accidents causing failure. Therefore ***all exhausts from mechanical pumps should be vented to fume hoods*****.**

Depending on their design and placement, traps can have three completely different uses and functions. The first use that typically comes to mind is containing contamination by preventing liquids, or condensable gases, from getting to where you do not want them to be. Traps are also used in high-vacuum systems to bind up condensable vapors (which helps improve the potential vacuum). Finally, traps can be used with successively cooler coolants (see Sec. 6.2 for different coolant materials) to lower the vapor pressures of materials (which can be used to separate compounds by their condensation or freezing points).

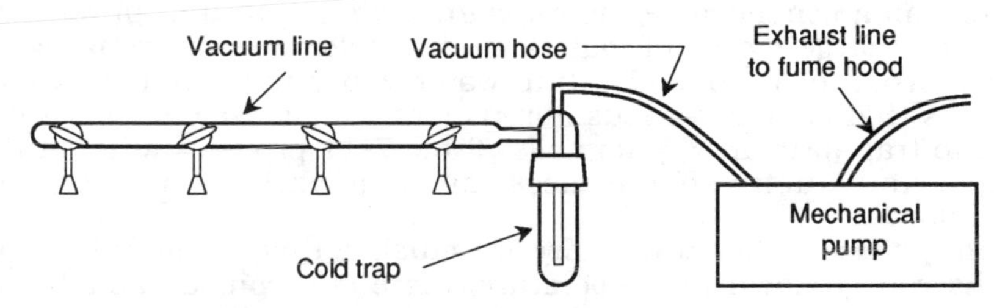

Fig. 7-27 A cold trap is worth the trouble on even a simple vacuum system.

The simplest way to protect a pump, pump oil, and pump seals; reduce maintenance time; and protect the worker from the effects of condensable gases in a vacuum system, is to place a trap between the vacuum line and pump (see Fig. 7-27). For example, although water vapor is not likely to damage a mechanical pump or worker, it can still impair pump performance by hydrating the pump oil. This problem can be resolved (after the fact) by keeping the pump running for an extended period of time (overnight) against a small dry air (or dry nitrogen) leak. However, this problem can easily (and should) be prevented by trap use at the outset.

Not only can a trap prevent condensable vapors from damaging materials on their way out of a vacuum system, as the quality of a vacuum line improves, they also prevent pump oil vapors from getting into the system. Remember that as a vacuum increases to the level of molecular flow, there is nothing to stop particles from drifting 'upstream.' If your system has diffusion and mechanical pumps, you should include a trap between the two pumps in addition to the cold trap between the system and the diffusion pump. Mechanical pump oils, which have a higher vapor pressure than diffusion pump fluids, will decrease the effectiveness of the diffusion pump. This problem is easily solved by placing a trap between the two (see

Fig. -28). The use of properly designed and placed cold traps can allow diffusion-pumped vacuum systems to achieve vacuums in the region of 10^{-9} torr[28] and greater!*

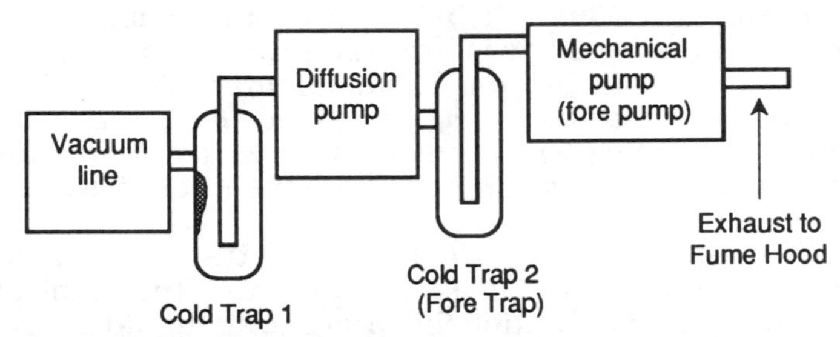

Fig. 7-28 Cold Trap 1 protects the diffusion pump, mechanical pump, and the user from materials that could otherwise drift in from the vacuum line. Cold Trap 2 protects the diffusion pump from any oils that may drift from the mechanical pump.

7.4.2 Types of Traps

There are four basic trap types: *cold traps* that rely on condensation or freezing to trap a condensable vapors; *particulate traps* that physically block the passage of large pieces of materials; *molecular sieves* primarily used to trap hydrocarbon oils—they also trap water vapor to a limited extent; and, *coaxial traps* that contain various fibrous materials loosely strung within the trap base to trap [primarily] pump oils. Table 7-11 provides a list of the various traps and their effectiveness against various vacuum system contamination.

Cold traps use either water, dry ice slush baths, or liquid nitrogen as coolant. As you go down in temperature, there is a considerable increase in price but an increase in potential performance is realized as well. A liquid nitrogen cold trap provides the most comprehensive protection available in any single trap design. However, although liquid nitrogen cold traps are easy to maintain, they are not inexpensive to operate, require constant maintenance, and may require constant observation (of the coolant level). Cold traps are capable of serving a duel function: As they rid the system of condensable vapors, they also help serve a pumping function by containing vapors that otherwise would contribute to an entire system's pressure.

Molecular sieve traps have low maintenance and low long-term costs. Their main disadvantage is that they require regeneration of the sieve by baking at 250 to 300°C every 300 hours of use or less. However, the bakeout cannot be done with the traps attached to the vacuum line. Otherwise, already-trapped hydrocarbons are dumped right back into the system. They can be baked out on the system if a dry nitrogen purge (of some 300μ pressure) is passed through the system while the baking is occurring. Because this type of trap requires a long downtime during the bake-out period, this design may not be acceptable to all users. Some molecular sieves allow for easy removal of charging material, others do not. Roepke and Pung[29] devised a simple homemade molecular sieve with a removable charge unit.

* This vacuum region can only be achieved on special (usually metal) systems that can be baked.

Table 7-11‡

		Trap Selection											
		Coaxial Traps					Molecular Sieve		Cold Traps			Particulate Traps	
		Stainless Steel Element	Copper Element	Activated Alumina Element	Activated Charcoal Element	Magnetic Element	Synthetic Zeolite Charge	Activated Alumina Charge	Liquid Nitrogen	Dry-ice Slurry	Water Cooled	Polyester Element	Drop-out
TA *	Prevent Oil Backstreaming	☆☆	☆☆☆	☆☆☆	N/R	☆☆	☆☆☆	☆☆☆	☆☆☆☆	☆☆☆	N/R	N/R	N/R
TA *	Trap Water Vapor	N/R	N/R	☆☆	N/R	N/R	N/R	N/R	☆☆☆☆	☆☆☆	N/R	N/R	N/R
TA *	Trap Organics (1)	N/R	N/R	N/R	☆☆	N/R	N/R	N/R	☆☆☆	☆☆☆	☆☆	N/R	N/R
TA *	Trap Acidic Vapors	N/R	N/R	KK	N/R	N/R	N/R	N/R	☆☆☆	☆	N/R	N/R	N/R
TA *	Trap Iron-containing Vapors	N/R	N/R	N/R	N/R	☆☆☆☆	N/R	N/R	N/R	N/R	N/R	>0.01 mm	>0.02 mm
TA *	Trap Particulates	(2)	(2)	N/R	N/R	(2)	N/R	N/R	N/R	N/R	N/R	>0.01 mm	>0.02 mm
TM **	Time Interval to Replace Elements	Six Months	Six Months	Vapor Load?	Vapor Load?	Load?	Vapor Load?	Vapor Load?	–	–	–	–	–
TM **	Regeneration in situ	No	No	No	No	No	Yes	Yes	–	–	–	–	–
TM **	Need Only Clean Trap	No	No	No	No	No		–	Yes	Yes	Yes	Yes	Yes
TM **	Refrigerant Hold Time (hrs)	–		–	–	–	–	–	2 to 4	>24	Cont.	–	–
TC ***	Trapping Mech. Operating Temperature	Room Temp.	Room Temp.	Room Temp.	Room Temp.	Room Temp.	Room Temp.	Room Temp.	-198°C	-79°C	1 to 20°C	Room Temp.	Room Temp.
TC ***	Limiting Pressure	10^{-4}	10^{-4}	10^{-4}	10^{-4}	10^{-4}	$<10^{-4}$	$<10^{-3}$	10^{-5}	10^{-3}	10^{-4}	10^{-3}	10^{-3}
TC ***	Can Lower Pump Base Pressure	–	–	Yes	–	–	Yes	Yes	Yes	–	–	–	–

‡ From "*Vacuum Products*" by Welch Vacuum, p. 107, © 1988 by Welch Vacuum Technology, Inc. With permission.

☆ - Fair ☆☆ - Good ☆☆☆ - Very Good ☆☆☆☆ - Excellent N/R - Not Recommended

* **T**rap **A**pplications (Ratings are qualitative guidelines, if you have questions, call the manufacturer of your pump)

** **T**rap **M**aintenance

*** **T**rap **C**haracteristics

(1) Examples: paraffins, solvents, organic acids, alcohols.

(2) Coaxials can trap large particulates, but are not designed for this purpose.

Particulate traps are simply physical barriers, or screens, that catch particulate matter the screens in clothes dryers catch lint. Piston rotary pumps have looser tolerances than vane pumps and therefore are more forgiving to particulate matter. However, you should not depend on this quality in an effort to ignore particulate traps. If you use liquid nitrogen traps, be forewarned that they do not stop particulate material which can do phenomenal damage to mechanical pumps of any kind.

Coaxial traps utilize an adsorbing material drawn out to a fine fiber to provide as much surface area as possible. The various different materials available provide varying capabilities. For example, copper provides the best absorption capabilities, but is not very durable and needs to be replaced often. Stainless steel and bronze will survive better in tougher environments, and activated alumina must be used if organics or pump oils need to be trapped. Like molecular sieves, these traps work at room temperature.

7.4.3 Proper Use of Cold Traps

Many people overlook the fact that liquid nitrogen cold traps have a proper and improper orientation. The trap orientation is not critical if the vacuum system operation does not generate copious amounts of condensable vapors. If the liquid nitrogen trap is attached the wrong way, there can be a build-up of frozen material in the center tube resulting in a reduction of the trap's throughput. This reduction can eventually cut off all flow of gases through the trap (see Fig. 7-29). If the orientation is properly established, the build-up of frozen vapors collects on the outer wall leaving the center tube free for gas transport.

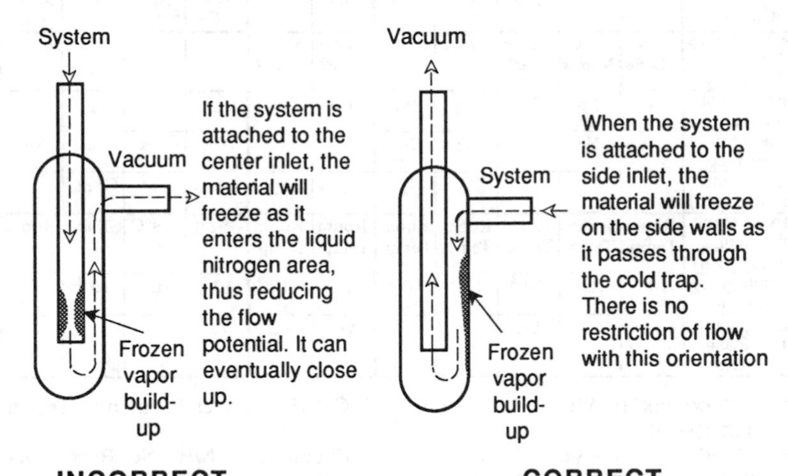

Fig. 7-29 If a cold trap is improperly oriented, the system may clog up.

Improper selection of coolant for a cold trap may artificially limit the potential vacuum of your system. For instance, the vapor pressure of water (which is often the primary condensable vapor in all vacuum systems) is quite high without any cold trapping, moderate at dry-ice temperatures, but negligible at liquid nitrogen temperatures (see Table 7-12). If your vacuum needs are satisfied within a vacuum of 5×10^{-4} torr, you can safely use dry ice (and save money because dry ice is less expensive than liquid nitrogen). Another temperature option for a coolant is the slush bath (for more information on coolants see Sec. 6.2).

Table 7-12‡

Vapor Pressure of Various Substances as a Function of Temperature

Temperature		Vapor Pressure, torr			
°C	K	Water	NH_3	CO_2	Hg
100	373.1	760	.....	.....	2.7×10^{-1}
50	423.1	93	.....	.....	1.3×10^{-2}
0	273.1	4.6	3,220	.....	2×10^{-4}
–40	233.1	0.1	540	.....	1×10^{-6}
–78.5*	194.6	5×10^{-4}	42	760	3×10^{-9}
–120	153.1	10^{-7}	0.2	10	10^{-13}
–150	123.1	10^{-14}	6×10^{-4}	6×10^{-2}	
–195.8**	77.3	$\approx 10^{-24}$	10^{-11}	10^{-8}	

‡ From *Vacuum Science and Engineering* by C.M. Van Atta, © 1965, McGraw-Hill, Inc., p. 331. With permission.

* Sublimation temperature of dry ice at 760 torr (one atmosphere).

** Boiling point of liquid nitrogen at 760 torr (one atmosphere).

Starting a vacuum system should include the following trap procedures for safe and efficient vacuum system operation:

1) Prevent Air (Oxygen) from Freezing within the Cold Trap. One of the more common laboratory accidents can occur when air (oxygen) is frozen in a cold trap. This accident is caused by placing a cold trap within liquid nitrogen while there is still sufficient air within the trap to be frozen. Later, once the liquid nitrogen is removed (either intentionally or is boiled off) the frozen air vaporizes. The excess pressure created by this frozen air can result in an explosion of the entire line—or stopcock plugs being blown across the room. Most labs have "horror stories" of these potentially dangerous, and always costly, occurrences.

To prevent oxygen from freezing in cold traps, be sure that there is no air in the trap when pouring liquid nitrogen around the cold trap. If you suspect that you may have air frozen in the cold trap, be sure the frozen air has room to expand. There are three ways to prevent explosions from frozen air:

1) Leave that section of the vacuum system in a continuous state of vacuum, never bringing it up to room pressure.
2) Begin pumping on the system with a mechanical pump for a few minutes to make sure that the system is below $\approx 10^{-1}$ torr before placing liquid nitrogen around the cold trap.
3) Leave a stopcock open [to the atmosphere] at the end of an experiment to vent the system after the work is done.

Suggestion 1 should always be practiced whenever possible. The re-entry of atmosphere into a vacuum system reintroduces copious amounts of moisture onto the vacuum system's walls and reintroduces gases back into the liquids within your system (oil and/or mercury). The next time the system is used, the walls will need to be re-dried and the liquids re-outgassed. Thus, a greater amount of time than would otherwise be necessary will be required the next time you wish to obtain a vacuum.. This extra time will also place more wear and tear on your pumps and expose them to extra condensable vapors.

Suggestion 2 should always be done as standard practice. Assuming your traps are emptied of liquids and condensable vapors, there should be nothing available to damage the pumps. Therefore, the practice of running the pumps for a few minutes (before placing liquid nitrogen around the cold traps) helps ensure that no air* within the system is available to be frozen.

Suggestion 3 has limited practicality. It can only be effective after work on the vacuum line is complete, and you wish to shut down the system. This plan will not help vacuum lines left unattended and/or whose liquid nitrogen has boiled off. Regardless, if frozen oxygen boils off faster than can be released by the vent, the results would be the same as if no vent were available. Thus, the potential for disaster is still present.

In practice, use Suggestions 1 and 2. It is better to insure that no air freezes within the system than to develop techniques to deal with problems that should be avoided.

2) Limit the Amount of Moisture Near the Top of your Cold Trap: A common error when first starting up a vacuum system is pouring liquid nitrogen too high into a Dewar. People often overfill Dewars in the early start-up process because the liquid nitrogen boils off very quickly and there is a desire to overfill. This overfilling can cause either a limit in the potential vacuum of a vacuum system, and/or create unaccountable pressure blips.

The problem originates because the liquid nitrogen high in the Dewar freezes moisture in the upper regions of the cold trap (see Fig. 7-30). Later, as the liquid nitrogen boils off, the moisture at the top of the trap evaporates (because nothing is keeping it frozen) and the pressure in the system rises, thus becoming a virtual leak (see Sec. 7.6 on Leak Detection), which is very difficult to find because no one thinks to look for a leak within a cold trap.

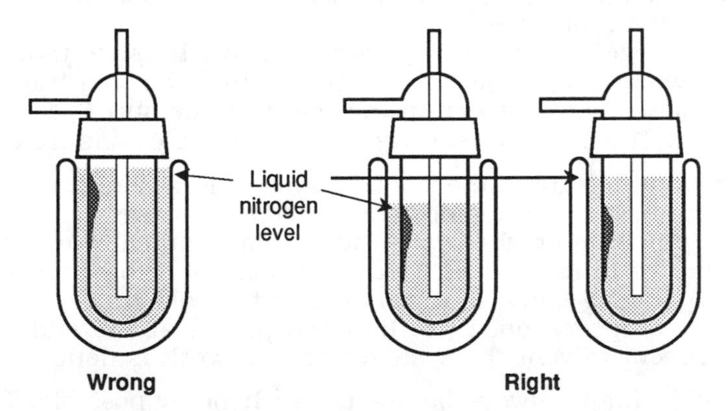

Fig. 7-30 By filling the liquid nitrogen to the top of the Dewar when first cooling the cold trap, the user risks limiting the maximum pressure the system can achieve. However, by keeping the level low in the beginning, the user can keep any frozen material locked up.

* By "no air," I'm not trying to imply there is a perfect vacuum, rather that the amount of air remaining is well within levels that could cause problems.

Fortunately, this problem is easily prevented by maintaining a low liquid nitrogen level in the Dewar when starting your system. Later, by maintaining a high liquid nitrogen level, any moisture that is trapped will stay frozen.

This *virtually leaking cold trap problem* is more likely to occur when a vacuum system is first being started, or when a system has been down for an extended time because more moisture is in the system. The problem is more difficult to control when you are constantly cycling from atmospheric pressure to a vacuum state because of the moisture constantly being brought into the system and the greater strain on the liquid nitrogen. The problem will always exist if your work creates copious amounts of condensable vapors stopped within the cold trap.

7.4.4 Maintenance of Cold Traps

There are two types of cold trap maintenance: 1) maintaining the liquid nitrogen (or the coolant of choice) at a proper level while the trap is in use, and 2) maintaining clean traps to insure the best throughput.

Maintenance of Liquid Nitrogen. A variety of automatic filling devices have been designed to maintain the liquid nitrogen level within a Dewar. Many of these devices are available commercially, and many others are sufficiently simple for lab construction. If you use an automatic filling device, try to limit the length of the filling lines. If the filling lines are overly long, the warmer gas (which precedes the oncoming liquid nitrogen) may warm the Dewar somewhat before the liquid nitrogen arrives and may cause small pressure blips. In lieu of mechanical or electrical automatic filling devices, periodic inspection and manual refilling is always available.

Perhaps more important is limiting the amount of liquid nitrogen lost by boiling off. First, use a good quality Dewar that is larger in radius than the cold trap by about 2 cm on all sides. All Dewars should be wrapped with tape* to prevent flying glass in case of implosion. If you ever notice frost on the outside of a Dewar, it is likely that the protective vacuum within the Dewar has filled up with atmosphere. This occurrence is not uncommon and is caused (typically) by a poor quality tip-off of the Dewar when it was evacuated.

Additional liquid nitrogen protection can be obtained by placing some insulation on top of a Dewar, which otherwise is exposed to the atmosphere. This insulation can be as simple (yet effective) as cardboard placed over the top of the Dewar, leaving a cut-out for the cold trap. Styrofoam, a more efficient insulation material, can also be cut out and placed on the top of the Dewar (see Fig. 7-31(a)). Alternative approaches include cutting Styrofoam into a cork shape (see Fig. 7-31(b)), or sprinkling crushed-up Styrofoam on top of the liquid nitrogen (see Fig. 7-31(c)). Each of these approaches has some positives and negatives: Approach (a) can be knocked off easily, and provides the least amount of insulation. However, it is easy to make and allows for easy examination of the liquid nitrogen level. Approach (b) provides

* In a study by Brown[30], it was found that white protective wrapping tape made a small, but noticeable, difference in how long liquid nitrogen would remain in a Dewar as opposed to no tape or black electrical tape.

excellent insulation abilities, but is not as easy to make and is more difficult to put into place. It is also more difficult to examine the liquid nitrogen level. Approach (c) is the easiest to make (just crush up some Styrofoam), but can be messy and locating the level of liquid nitrogen can be confusing or difficult. Regardless of which approach you use, there must be a route for built-up gases to escape, otherwise the plug could be blown out with some force.

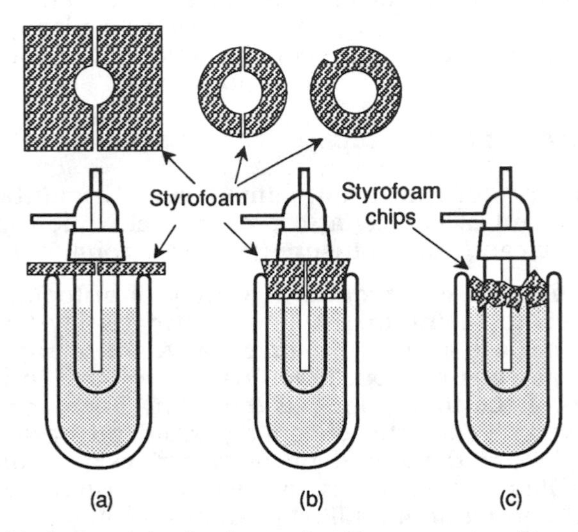

Fig. 7-31 Various methods of covering Dewars using Styrofoam. Be sure that whichever approach is used, you allow a route for gases to escape.

Maintain a Clean Trap. During vacuum operation, a trap may become so filled that vacuum operation becomes impaired and emptying the trap is necessary. To empty a trap, it must be removed from the system, and to remove the trap, it must be vented to the atmosphere. If the trap is not vented, separating the lower removable part of the trap from the upper section could be like separating the Magdeburg hemispheres (see the historical review in Sec. 7.2.3). If brawn is greater than brain, you could damage your system. Unless there is a reason to vent the entire system, do not. When a system is exposed to the atmosphere, moisture can re-saturate the walls of the system and gases can be re-sorbed into the vacuum liquids (diffusion pump oils and/or mercury). Extra (wasted) time will be required to return the system to its original vacuum condition. In addition, if your diffusion pump uses hydrocarbon oils, air will destroy the diffusion pump oils (if they are hot) and they will then have to be replaced. All this wasted time and money can simply be avoided when first constructing vacuum system by including a venting stopcock on the trap side of the system to vent the trap to atmospheric pressure while leaving the rest of the line in a vacuum (see Fig. 7-32).

Incidentally, if your work creates a constant build-up of material in the cold traps, have an excess supply of trap bottoms available. This preparation allows you to remove a filled trap bottom and transfer it to a fume hood while immediately replacing another on your vacuum line, thus significantly cutting down the amount of "down-time" on your system.

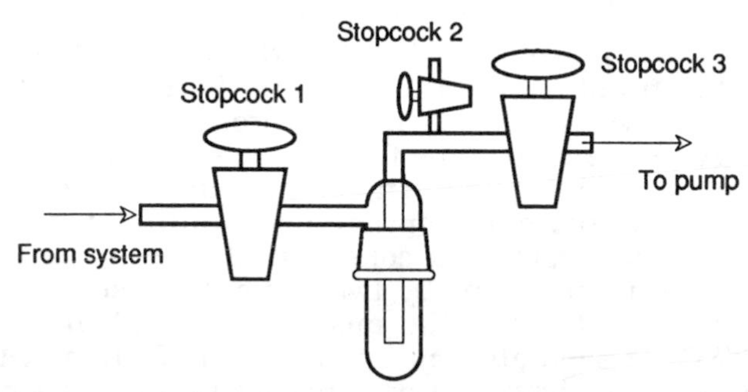

Fig. 7-32 Using valves or stopcocks to separate the various parts of your system allows you to open sections of your system while maintaining a vacuum in the rest of the system.

Following Fig. 7-32, if you need to temporarily remove the base of a trap while a system is in use:

1) Close Stopcocks 1 and 3.
2) Open Stopcock 2.
3) Remove base of cold trap.
4) Replace base of cold trap.
5) Close Stopcock 2.
6) Open Stopcock 3.
7) When pump quiets down, replace Dewar under cold trap.
8) Open Stopcock 1.

Removing the base of a cold trap while it has been in liquid nitrogen can be difficult, especially if the trap uses standard taper joints. The cold temperatures can make (even fresh) stopcock grease sluggish and firm. The easiest way to separate the base from a trap is to let the trap come to room temperature. The joint may also be easier to separate if you aim a hot air gun around the entire circumference of the joint (do not aim the hot air gun at one spot and expect it to heat the other side). If you expect to remove a frozen trap often, consider using O-ring joints which are easy to separate regardless of temperature.

When shutting down a vacuum system, close off your system from the trap section. That way, as trapped compounds warm up and go into a vapor state, they will not be able to drift into the rest of the vacuum line. You should also vent your pump to the atmosphere. Most pumps do not have good check valves near their oil reservoirs. If they are shut off with a vacuum on the vacuum side, the mechanical pump oil can be drawn up into the system. So, to shut down a vacuum system (see Fig. 7-32), it is recommended that you:

1) Close Stopcock 1.
2) Remove coolant (i.e., liquid nitrogen).

3) Turn off Pump.
4) Open Stopcock 2.

7.4.5 Separation Traps

All of the traps mentioned so far protect the vacuum line, pumps, pump liquids, and/or the people using the system. There can be other traps on vacuum systems whose function is not for protection, but rather they are tools for chemistry. Separation traps fall into this category and can separate a mixed compound into different fractions by using individual freezing temperatures in coolants of these appropriate temperatures.*

Separation traps are typically a collection of interlinked U-shaped traps attached off the main vacuum line by two stopcocks (see Fig. 7-33). This arrangement allows separations of the mixed compound into as many traps your system has. Once separation is complete, any faction of the separation may be removed from the system at any time and in any order. The contents within a trap may even be sent back to the main holding trap for further separation. The following will provide a generalized procedure for utilizing such a separation process:

1) Attach your sample to the system at Stopcock 14.
2) Open Stopcocks 14, 12, 13, 1, 3, 5, and 7 to evacuate the separation line and all the 'U' traps.
3) After the extension line on the vacuum system has been evacuated, close all stopcocks except 14 and 12.
4) Place liquid nitrogen (in a Dewar) around the holding trap.
5) Stopcocks 15 and 1 can be opened, and using the cold from the liquid nitrogen as a sorption pump, transfer the mixed compound into the holding trap (you may want to lightly heat the original compound to facilitate the transfer)
6) Close Stopcock 12 and 1.
7) Place Dewars, with the appropriate temperature slush baths, under the other traps on your system. The Dewar closest to the holding trap should have the warmest of the cold temperatures, and the Dewar farthest away should have the coldest bath.
8) Remove the Dewar from the holding trap, empty it, and replace the Dewar (this procedure allows a slow warming of your mixed compound). Open Stopcock 9 (or 2 and 3) and allow the compound with the highest vapor pressure to freeze out.
9) After a sufficient amount of time has elapsed, open the stopcocks to the successive traps in similar succession.
10) Once the material in the holding trap has successfully passed into the other traps, close all stopcocks. Then, transfer the purified materials, one by one (from the lowest vapor pressure to the highest), back into containers such as those that originally held the mixed compound.

* See Sec. 6.2.8 for how to make varying temperature slush baths.

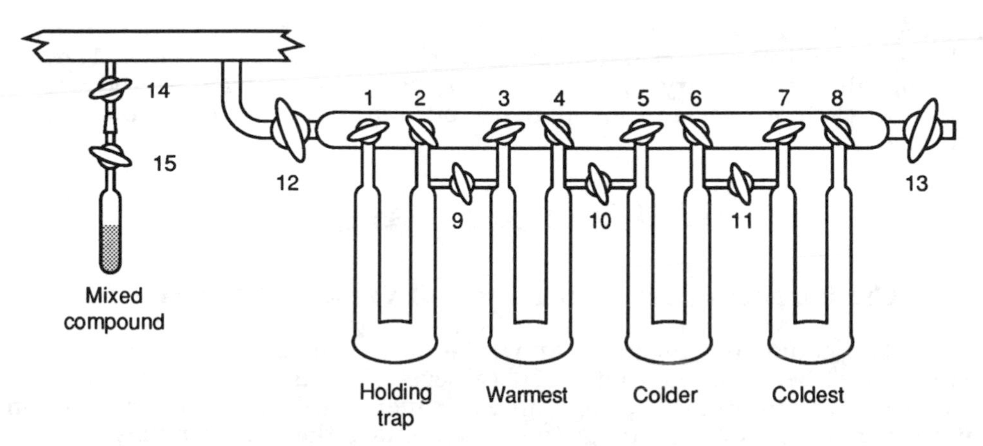
14
15
Mixed compound
1
2
3
4
5
6
7
8
12
9
10
11
13
Holding trap
Warmest
Colder
Coldest

Fig. 7-33 Separation traps connected to a vacuum line.

7.4.6 Liquid Traps

The most efficient way to prevent the liquid from an oil or mercury pressure gauge or bubbler from spilling into the rest of your vacuum system is to place a liquid trap between the liquid container and the vacuum system. The liquid trap design is fairly straightforward (see Fig. 7-34), and the installation of one is strongly recommended.

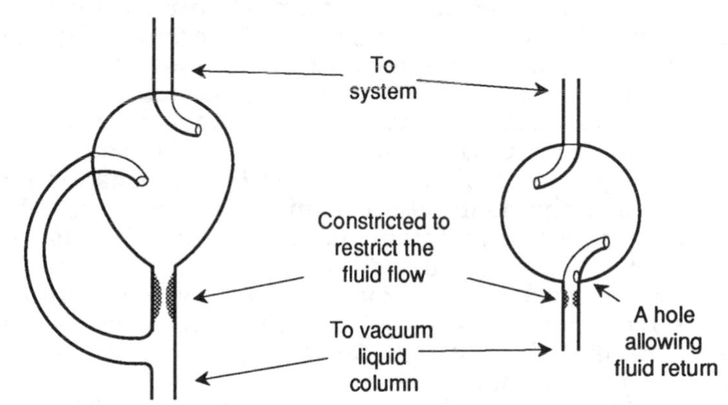
To system
Constricted to restrict the fluid flow
To vacuum liquid column
A hole allowing fluid return
In this trap, fluids travel the easier route of the side arm (rather than the constricted center tube) and splash into the pear-shaped flask. From there they pour back into the container from where they originally came.
In this design, fluids are splashed against the inside wall where they can flow back into the tube from the hole on the side of the lower tube.

Fig. 7-34 Two liquid trap designs.

The beauty of liquid traps is that once in place they require no further oversight, care, or maintenance. Once you have seen the damages caused by a manometer or a McLeod gauge that has 'burped' you understand the value of liquid traps. However, the value of liquid traps can be over-emphasized, and they should not be used as panaceas for clumsy vacuum work. They will not stop all the mercury (or other fluids) that are being battered around within a system, so do not depend on liquid traps to make up for carelessness.

7.5 Vacuum Gauges

7.5.1 The Purposes (and Limitations) of Vacuum Gauges

The two common reasons to know the quantitative level of vacuum in a system are either to know that a general level of vacuum has been achieved, or to see what quantitative effects an experiment has had upon the vacuum. The former requires a type of 'go/no-go' reliability that may or may not have accuracy requirements. The latter is important with experiments such as gas creation or changes in vapor pressure. For proper analysis of experimental data, highly accurate readings are required. Regrettably, the greater the vacuum desired, the less reliable, or consistent, a vacuum reading is likely to be.

To obtain accurate and consistent low-, high-, and ultra-high vacuum readings require years of experience and training beyond the scope of this book. This book explains the basic operation of common gauges, and (hopefully) will give you the background to obtain the years of experience. Accuracy, however, is a somewhat loose term when used for vacuum gauges because actual (real) pressure is typically experimentally determined.[31] Accuracy can depend on factors as diverse as system use, experimental technique, the gauge angle of attachment, external magnetic fields near the gauge, and ambient gases within the vacuum system. Accuracy for gauges is formally defined by the ISO (The International Standards Organization) as "the closeness of agreement between the result of a measurement and the true value of the measurement."[32] To maintain the best possible accuracy, re-calibration is required to maintain the closest possible agreement to the manufactured stated accuracy. In addition, re-calibration may be required each time a new gas species is introduced to a system. Unfortunately, all vacuum gauges tend to show a loss of accuracy over time, and this process continues over the life of the gauge. Re-calibration cannot restore this lost accuracy, nor can re-calibration prevent error.

There are two basic ways for a vacuum gauge to 'read' a vacuum: *direct* and *indirect*. For example, say that on one side of a 'wall' you have a known pressure, and on the other side of the 'wall' you have an unknown pressure. If you can know that a certain amount of deflection implies a specific level of vacuum, and you can measure the current wall deflection, you can then determine the pressure *directly*. This process is used with mechanical or liquid types of vacuum gauges. On the other hand, if you know that a given gas will display certain physical characteristics due to external stimuli at various pressures, and you have the equipment to record and interpret those characteristics, you can infer the pressure from these *indirect* measurements. This indirect method is how thermocouple and ion gauges operate.

Each method of vacuum reading has its advantages and disadvantages. No single method of vacuum reading is entirely easy and/or comprehensive, and no vacuum gauge can read a full range of pressures or be wholly accurate on all types of gases. Thus, the vacuum gauge(s), controllers, and other peripheral equipment you decide to use on your system are dependent on what vacuum range you need to read, what level of accuracy you need, and cost.

Once you have decided on which vacuum gauge to use, you then need to consider where, and how, it should be attached. The vacuum gauge should preferably be attached near the area where the vacuum work will be performed. Pressure is determined by the number of molecules per cc, and at low pressures, a vacuum gauge located far away from where you are working may give a misleading value. Despite Avogadro's Law, at different pressures there may be a random distribution of molecules within the vacuum. These differences can cause gauges to provide inaccurate pressure readings. There are three reasons for this problematic condition:

1) In molecular flow, there is a time factor before pressure can equalize within a vacuum system. Small diameter (and/or long length) tubing can compound this problem.
2) If the entry tube to the gauge is too narrow (and/or long) the statistical movement of molecules to enter the gauge may be inaccurately low. On the other hand if the angle of the tube to the gauge is in direct line with a molecular stream, the count of molecules could be inaccurately high.
3) Ion gauges have pumping capability providing lower pressures in the gauge region than areas located some distance away (see Sec. 7.5.19).

Finally, be aware that if you read 3.0×10^{-3} from your vacuum gauge, that figure is not likely to be what your 'real' vacuum is. Likewise, if you are trying to duplicate an experiment that is pressure-specific, and despite repeated attempts the experiment fails, your vacuum reading may be at fault, not you. Vacuum gauges, by their nature and design, are always inaccurate. In standard laboratory conditions, vacuum measurements that are ± 10% reliable are very difficult, and those that are ± 1% are essentially impossible.[33] The reasons for this imprecision can include:

1) The gauge may be inherently inaccurate.
2) The gauge may be calibrated for a different gas than what is in your system.
3) The elements of the gauge may be contaminated.
4) The gauge may not be receiving a sufficient amount of time for proper equilibration.
5) The misalignment of parts within a gauge can alter the accuracy of readings.
6) Different components, or types of filaments, within the same gauge can alter the accuracy of readings.

Always have a valve (or stopcock) between your vacuum system and vacuum gauge (and any other component of your vacuum system) to allows unneeded sections to be shut off when not in use and limit contact with potentially corrosive materials. The less contamination a gauge is exposed to, the longer the accuracy and reliability of the gauge will be maintained.

There are five major families of vacuum gauge design. The gauge families are the *mechanical* (see Sec. 7.5.2), *liquid* (see Sec. 7.5.4), *thermionic* (see Sec. 7.5.14), *ion* (see Sec. 7.5.19), and *other* (in this book I have only made passing reference to the *momentum transfer gauge* in Sec. 7.5.24). Each family has its own strengths and weaknesses, and many vacuum systems often will have two or more gauge families represented on a single vacuum line. This mixture is partly to allow uniform pressure readings from atmospheric to high vacuum (a single gauge cannot do both) or to provide periodic cross-reference and calibration.

7.5.2 The Mechanical Gauge Family

The easiest way to see a quantitative vacuum measurement of a system is to look at a dial and read it. This direct reading can be done with mechanical gauges. The surface of mechanical gauges that come in contact with gases within a vacuum system can be made out of metal, or glass (borosilicate or quartz), both of which can be fairly impervious to chemical attack.*

Mechanical gauges rely on diaphragms, bellows, Bourdon-tubes, and capsules that are squeezed, pushed, pulled, twisted, and turned by the various forces within a vacuum system. Mechanical gauges mechanically transfer this distortion to dials, mirrors, or pointers for reading. Many mechanical gauge designs are quite accurate.** For recording purposes, or to translate extremely minor amounts of distortion into readable figures, a displacement transducer can be connected to the moving part of a gauge.

The advantages of mechanical gauges are:

1) Ease of use and the ability to read a positive *and* negative pressure
2) The ability to provide a constant (as opposed to intermittent) reading
3) The ability to provide continuous recording during use (if properly equipped)
4) If the proper construction material is selected, there can be limited, or no, reaction between the gauge and materials within the vacuum system.

The disadvantages of mechanical gauges are:

1) Their inability to read below approximately 10^{-2} torr (although very accurate specially-designed mechanical gauges can read to approximately 10^{-4} torr)
2) They are sensitive to temperature changes, and may be sensitive to pressure changes***

* If there is any doubt whether the gases in your system may attack the materials of the vacuum gauge, ask the component's manufacturer.

** For an unidentified type of quartz Bourdon manometer, an accuracy of about ± 0.01% over 20% to 100% of its range, with regular re-calibrations, is reported.[34]

*** Most gauges are sealed to the outside world and thus should be unaffected by *atmospheric* changes.

3) They are subject to the effects of hysteresis, or rather, the tendency of materials to 'remember' distortions and not return to their original shapes (mechanical gauges with glass distortion membranes will not have symptoms of hysteresis). Hysteresis of vacuum gauges can be diagnosed by the following symptoms:
   a) The sensitivity of the diaphragm may change with time.
   b) After prolonged use (especially with radical temperature and/or pressure changes) the diaphragm will show signs of aging.
   c) The 'zero point' often drifts.

Mechanical gauges can be easily attached onto metal vacuum systems, however, due to the construction materials of mechanical gauges, it is often impossible to make a direct seal onto a glass vacuum system. If necessary, a Swagelok® or Cajon® Ultra-Torr® may be used for making a glass-to-metal seal (see Sec. 3.1.5). The glass-to-metal seal may then be fused onto the vacuum system.

7.5.3 Cleaning a Mechanical Gauge[35]

There are two primary maintenance problems with mechanical gauges: dirt and/or grease. Both are results of improper trapping. If you can see dirt or oil on the dial, it is very dirty and will require cleaning. If your work has (or can create) particulate matter, install line filters to stop materials before they can get into areas of the system that may be damaged (such as mechanical pumps), or are hard to clean (such as gauges). If it is possible to open up the gauge, gently blow the particulate matter away with a dry nitrogen spray. Do not use 'plumbed-in' compressed air because these airs always carry moisture and oil vapors.

Oil on the dial is indicative of oil migration from the pumping system. Improper and/or inadequate traps or user error are to blame for this problem. Clean with an appropriate solvent for the type of grease found. Rinse and dry the gauge before re-installing back onto the line. Methanol is a good final rinse as it dries quickly and cleanly.

7.5.4 The Liquid Gauge Family

The liquid gauge family is identified simply as vacuum gauges that have some liquid (usually mercury or a low vapor pressure diffusion pump oil) directly in contact with the vacuum. The amount of liquid movement is directly proportional to the force exerted on it, and the (measured) amount of movement is read as the vacuum. Because mercury has traditionally been used for vacuum measurement, the term "mm of mercury" is commonly used even with non-liquid gauges. However, the SI system of measurement is changing the vacuum unit to the Pascal. The environmental and health concerns of mercury are limiting, and will eventually end its use.

Low vapor pressure oils can be substituted for some operations, but calibrations for density must be made so that their measurements can be interpreted as mm of mercury, torr, or Pa. Unfortunately, vacuum measurements can take a considerable amount of time when using oil because it takes a long time for a film of oil to settle from the walls of a manometer.

Despite the limitations and problems that mercury oil present, perhaps the biggest problem with the liquid gauge family is that its members can be very difficult to keep clean because the liquid is in direct contact with materials in the vacuum system. As mercury becomes dirty or contaminated, it tends to stick to the walls of glass tubing, thus decreasing its accuracy. As oil becomes dirty, its density can change and may provide inaccurate measurements. Realistically though, any vacuum system whose upper range is 10^{-3} torr is not likely to be significantly affected by limited contamination.

On the other hand, a liquid trap is essential between **any** liquid vacuum gauge and the manifold itself. The simplest accident can cause hours of needless delay as the vacuum line is cleaned out. Normally a simple splash trap is sufficient (see Sec. 7.4.6). Vacuum systems that go below 10^{-5} torr should have cold traps placed between their liquid vacuum gauges and vacuum systems. This placement will keep vapors from contaminating either side of the system.

When a vacuum system is first started and brought from atmospheric pressure to a vacuum state, parts within the vacuum system need to be outgassed. When liquids are outgassed, they boil. This process needs to be done slowly, otherwise the liquids will 'bump' (boil) violently, possibly causing the mercury (or oil) to splash across the system. This splashing could waste *some* time minimally by trying to get the liquid back into the gauge, or more time by requiring the system to be cleaned up, or maximally, an extended period of time could be wasted by breaking the system. Therefore, *slowly* open the stopcock to your liquid manometer for outgassing (Sec. 7.5.8 provides specific information on how to let the liquid outgas in a McLeod gauge). It is not necessary to open the stopcock all the way to full open. If the stopcocks to the liquid vacuum gauge are kept closed to atmospheric pressures, the outgassing process needn't be repeated. However, if the gauge is brought to atmospheric pressure for an extended time (about a day) gases can re-enter the gauge and the full outgassing process must be repeated.

7.5.5 The Manometer

Second only to the mechanical gauge as the easiest device to measure and read a vacuum (and decidedly easiest in construction) is the liquid manometer (See Fig. 7-35). A well-made mercury manometer, kept very clean, can measure vacuums of up to 10^{-3} torr. This sensitivity can be increased by up to 15 times if a liquid with less density, such as diffusion pump oil, is used. However, diffusion pump oil is far more difficult to keep clean and can require either a very tall (and thereby impractical) column, or a manometer of very limited range. In addition, because of the strong surface tension between diffusion pump oil and glass, long waiting periods between readings are required as the oil settles into place.

The mechanics of a U-tube manometer are simple: "The difference between the levels of two interconnected columns of liquid is directly proportional to the difference between the pressures exerted upon them, assuming

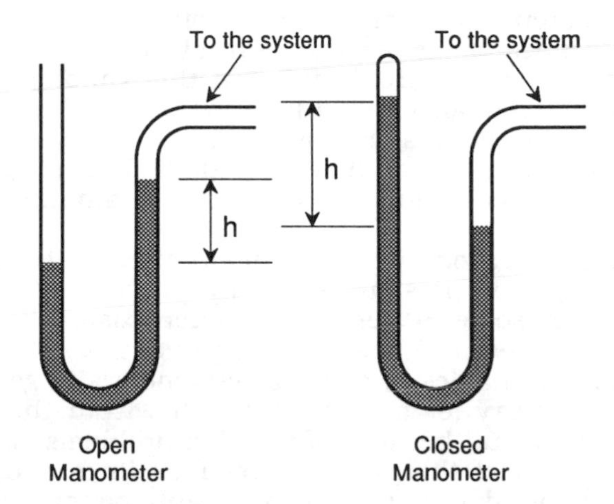

Fig. 7-35 Open manometers are used to measure pressures relative to atmospheric whereas closed manometers are used for pressures far smaller than atmospheric.‡

‡ From *Vacuum Science and Engineering*, Fig. 3-1, by C. Van Atta, (© 1965 by McGraw-Hill, Inc., New York, N.Y.). With permission.

equal capillary effects on both tubes."[36] In practice, one end of a manometer is attached to an unknown pressure and the other to a reference pressure that is known. For all practical purposes, the known value needs only to be a much smaller order of magnitude than the unknown value.

By maintaining a manometer at a constant temperature, the liquid within will maintain a consistent density. From 0° to 30°C the density of mercury changes about 0.5 %. Therefore, a consistent reading of accurate vacuums can only come from a consistent room temperature. Manometer reading techniques (see Fig. 7-36) are consistent and simple regardless of the design and type of manometer. All that is necessary is some basic subtraction, a good metric ruler, and a steady eye. The manometer is read by observing and measuring the difference of the mercury column heights within the manometer.

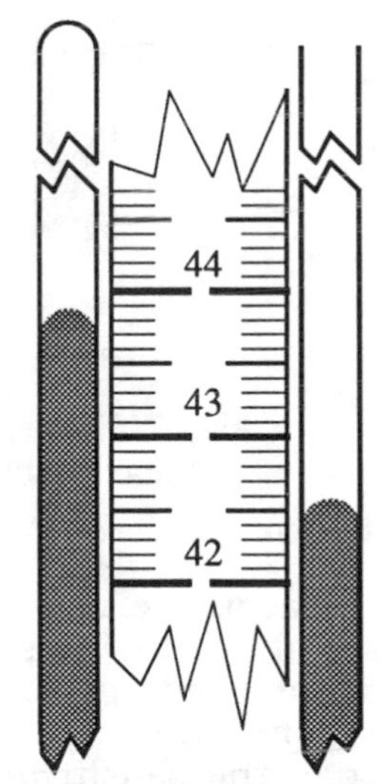

Fig. 7-36 In this example, take the measurement from the closed side (439 mm) of the manometer, and subtract from it the measurement from the side of the manometer connected to the vacuum system (426 mm) to obtain the vacuum reading (13 mm Hg). Remember, avoid parallax problems, and with mercury, read the top of the meniscus.

The choice of mercury for manometers is often a matter of convenience, or rather the acceptance of the least amount of inconvenience. Mercury has a rather high vapor pressure (10^{-3} torr), but this vapor pressure is also at the upper ranges of what can be read by a manometer. Low vapor pressure oils (as used in diffusion pumps) can be used, but these oils wet the walls of a manometer and can take a long time to settle before reading can be made. Mercury is fairly nonreactive and retains a limited amounts of condensable vapors.

One of the biggest reasons why mercury is used is because it is a very heavy liquid. Atmospheric pressure can only push mercury about 76 cm, while it pushes water some 30 feet. Manometers using lighter density liquids can be so tall they become beyond inconvenient.

Fortunately, the mechanics of reading manometers is generally irrelevant to the amount of mercury (or any other liquid) within the device. Reading manometers can be difficult because of parallax problems. This difficulty can be complicated by the fact that manometers lack lines encircling them like burettes. Carroll[37] found that by placing graph paper behind a mercury manometer, it was possible to see problems of parallax by the line's reflection off the glass and mercury (see Fig. 7-37).

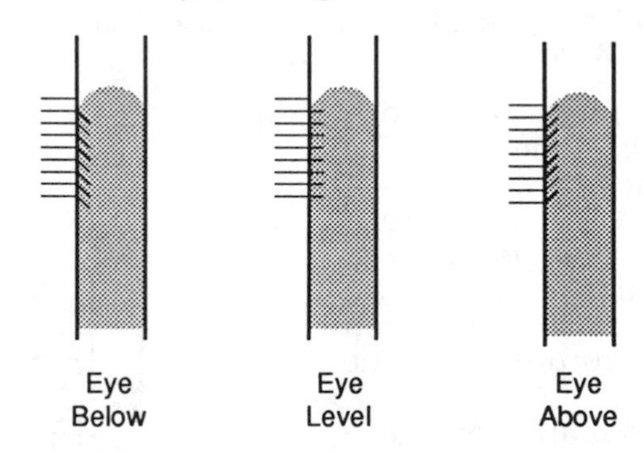

Fig. 7-37 Using graph paper lines to help avoid parallax problems while reading manometers.‡

‡ From the *Journal of Chemical Education*, 44 (1967), p. 763 . With permission.

For *very* accurate manometer readings, the NIST (National Institute of Standards and Testing) has a compiled list of eight possible errors that can develop while reading manometers.[38] These errors introduce extremely small variations and therefore are not included here. If you need to read a manometer to the sensitivity required for these variations to show up, it is better to use a different gauge type with higher sensitivity.

At first glance, it seems that one ought to be able to attach a vacuum line 77 or 78 cm above the lowest part of a gauge because at STP it is not possible for a vacuum to pull mercury any higher than 76 cm. While it is true that a vacuum cannot pull mercury higher than 76 cm, momentum can. If the mercury is being outgassed, or there is an unexpected pressure surge, momentum can carry the mercury farther than your worst nightmares. Therefore, always include a liquid trap on all liquid systems (see Sec. 7.4.6).

7.5.6 The McLeod Gauge

Rather than using mercury as a piston that is pushed about by the forces within a vacuum system, the McLeod gauge traps a known volume of gas of unknown pressure and compares it to a known volume of gas at a known pressure using Boyle's law:

$$P_o V_o = P_f V_f \quad \text{Eq. (7.12)}$$

Where P_o = original pressure
V_o = original volume
P_f = final pressure
V_f = final volume.

The advantages of the McLeod gauge are:

1) The wide range of pressures it can read
2) Its great accuracy (McLeod gauges are used to calibrate electronic gauges)
3) Readings are unaffected by the type of gas species within the system (although condensable vapors can affect readings).

The disadvantages of the McLeod gauge are:

1) It can only read the pressure at single points in time, not continuously as can mechanical or electronic gauges.
2) The McLeod gauge uses mercury which can be a nuisance because the gauge is difficult to clean. In addition, backstreamed mercury can sometimes affect your work.
3) The McLeod gauge has no ability to compensate for condensable vapors.

McLeod gauges are always made of glass. Figure 7-38 shows two typical McLeod gauge designs. There are many different configurations and designs of the standard McLeod gauge, but all have the same general configuration:

1) A lower bulb for mercury storage
2) An upper bulb whose volume has been accurately determined
3) One capillary tube extending from the vacuum bulb with its other end fused shut (in Fig. 7-38 it is labelled Capillary B).
4) A second capillary tube (in Fig. 7-38 it is labeled Capillary A), both ends of which are attached to a larger tube.
5) A larger tube which connects the McLeod to the vacuum system.

Ideally, the McLeod gauge should only be open to the system when a vacuum reading is being made to limit the amount of mercury vapors backstreamed into the vacuum system as well as the amounts of materials that drift into (and contaminate) the McLeod gauge from the system. The mercury in a McLeod gauge must be kept clean because dirty mercury tends to stick on the walls of the capillary tubing and leave trails that prevent accurate readings. It is much easier to keep the mercury in a McLeod gauge clean than it is to clean the capillaries of a McLeod gauge.

A liquid trap can be placed between the McLeod gauge and the rest of the system to prevent mercury from accidentally spraying throughout your system. If you do not want condensable vapors affecting the McLeod gauge

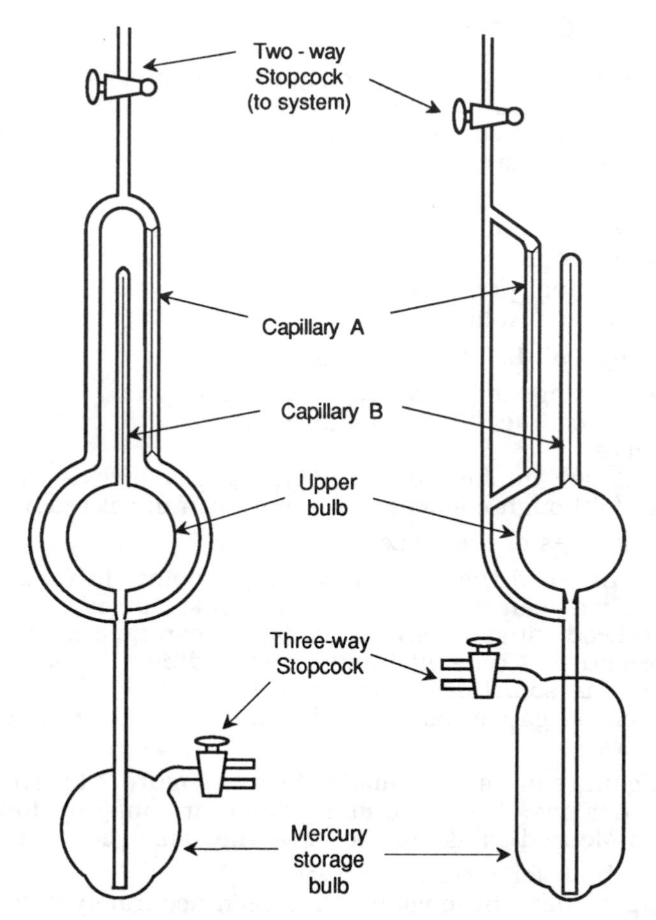

Fig. 7-38 Two common configurations of the McLeod gauge.

readings, or mercury vapors to enter your system, a cold trap can be placed between the liquid trap (shown in Fig 7-39) and the main vacuum line.

Unfortunately, placing traps between the McLeod gauge and vacuum system increases the time needed to make a reading because the physical restrictions decrease the throughput and increase the time for gas equilibration. The efficiency lost by slower McLeod gauge readings (due to traps) must be balanced by other needs such as limiting the amount of mercury that gets into your system and/or keeping condensable gases out of the McLeod gauge.

An interesting problem occurs when a liquid nitrogen cold trap is placed between the system and McLeod gauge. The cold trap becomes a cryopump and draws mercury vapor from the gauge. This effect, called "mercury streaming," effectively limits the number of molecules that can enter the gauge (like salmon trying to swim upstream) and inaccurately low readings may result. Only readings in vacuums less than 10^{-4} torr are significantly affected, and atoms of higher molecular weights produce larger errors (i.e., Xenon at 30 % inaccurate readings).[39]

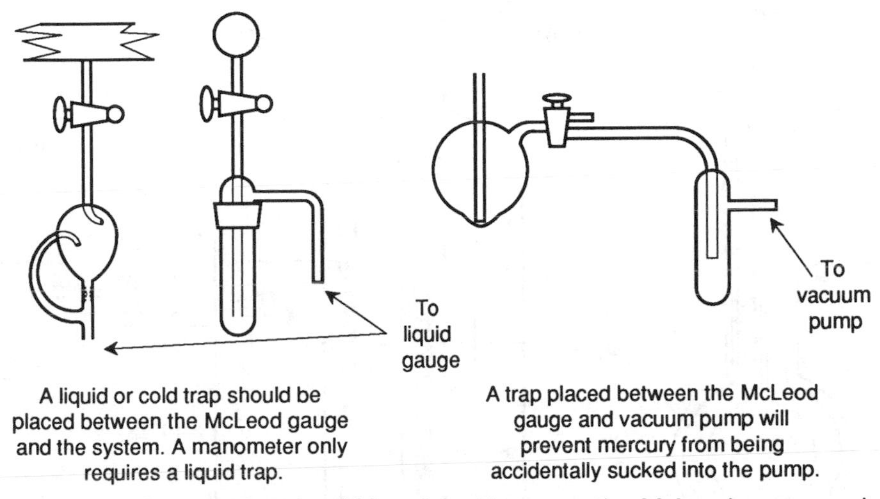

A liquid or cold trap should be placed between the McLeod gauge and the system. A manometer only requires a liquid trap.

A trap placed between the McLeod gauge and vacuum pump will prevent mercury from being accidentally sucked into the pump.

Fig. 7-39 Some traps should be placed between the McLeod gauge and the rest of the system to prevent mercury from spilling into your system.

7.5.7 How to Read a McLeod Gauge

There are two different methods of reading a McLeod gauge: one uses a *linear scale*, the other a *square scale*. Because the square scale is primarily used, I will limit my explanation to its use. Refer to Fig. 7-40.

Always leave the stopcock leading to the McLeod gauge closed during experimental work to limit contamination entering the gauge and mercury from entering the system. When a pressure reading is desired, open the two-way stopcock that separates the McLeod gauge from the vacuum system. After the reading is completed, return the two-way stopcock to its closed position. Once the two-way stopcock is opened, slowly and carefully open the three-way stopcock (Position 1) to let mercury rise into the upper sections of the McLeod gauge by allowing atmosphere into the lower portion of the gauge.

As mercury rises within the McLeod gauge, it is drawn up into the two capillaries. At this point, it is important to keep your eye on the capillary that is open on both ends (Capillary A). When the mercury gets closer and closer to the (inside) top of the second capillary (Capillary B), place your finger on the end of the open tube of the three-way stopcock to prevent air from entering the gauge. Your finger can provide precise control of the incoming air. Once the mercury reaches the height h_1, air is prevented from entering and a reading can be made. This method may take a bit of practice, but it soon will become relatively easy.*

* If you overshoot the mark, simply turn the three-way stopcock to the #2 position for a moment pulling the mercury level down. It is not necessary to bring the mercury all the way back to the storage bulb, simply bring it below the h_1 level.

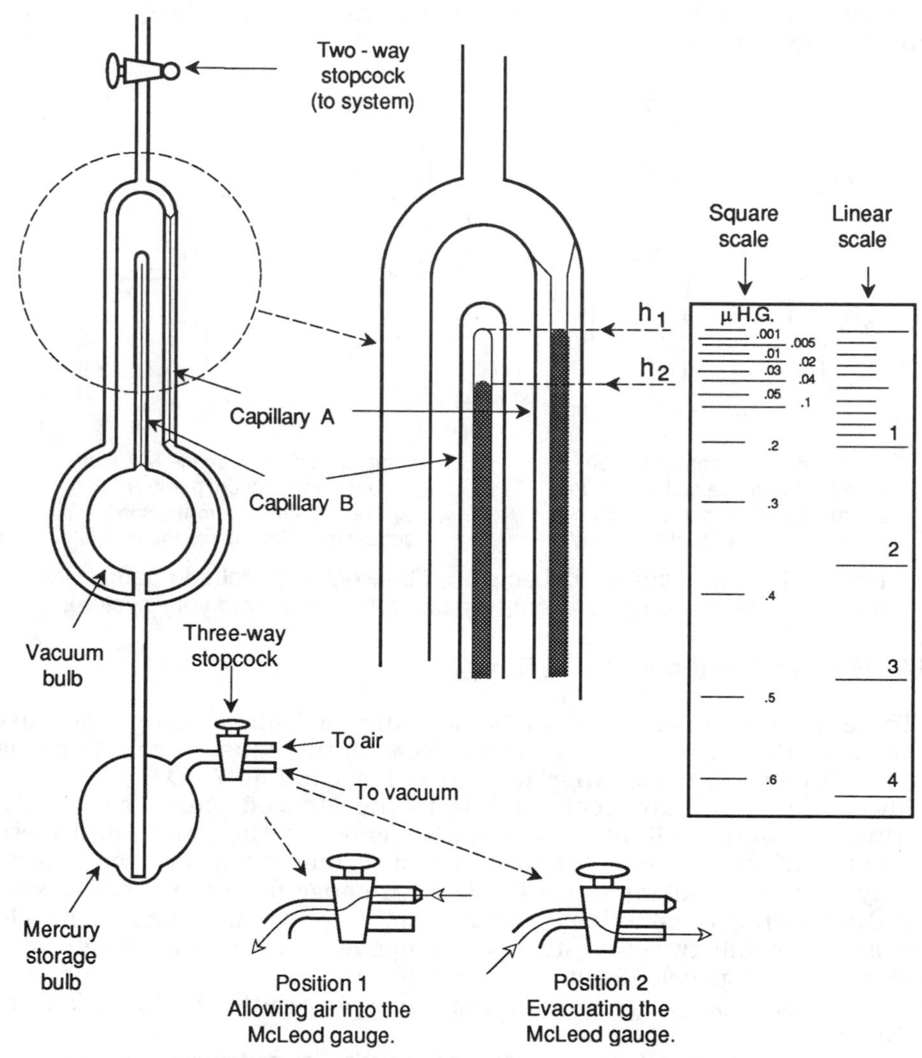

Fig 7-40 How to use and read the McLeod gauge.

Once the mercury in Capillary A is the same height as the inside top of Capillary B, reading the pressure is done by simply drawing a visual line from h_2 (the top of the mercury in Capillary B) over to the reading scale and reading your vacuum. In Fig. 7-40, the pressure is 0.04 μ Hg, 4 X 10^{-5} mm Hg, or 4 X 10^{-5} torr.

After reading the pressure, rotate the plug of the three-way stopcock 180° to the #2 position to draw the mercury back into the storage bulb. Once the mercury is back in the storage bulb, turn the three-way stopcock 90° to a closed position. Once a vacuum reading has been made, turn the two-way stopcock connecting the McLeod gauge and the vacuum system 90° to a closed position.

Despite being used for the calibration of other gauges, McLeod gauges are not perfect. There can be errors from physical limitations such as capillary depression phenomena, sorption and desorption from the gauge walls, and gas condensation within the gauge. Operator errors also exist such as difficulties of sighting the mercury heights properly. At 10^{-4} torr, the accuracy of a McLeod gauge is ±3%. Assuming no operator error, at 7.5×10^{-7} torr (the upper limits of a McLeod gauge), inaccuracy can be ±15%. Specially-adapted McLeod gauges can have their accuracy improved up to ±0.9% and ±3.5%, respectively.[40]

7.5.8 Bringing a McLeod Gauge to Vacuum Conditions

When first starting up a vacuum system from atmospheric conditions, do not expose a McLeod gauge to a sudden vacuum. Opening up the wrong stopcock at the wrong time can cause violent outgassing of the mercury. The mercury may be inadvertently sprayed throughout your system (despite traps) with enough force to break a line.

To bring a McLeod gauge to vacuum conditions, and before turning the vacuum pumps on, be sure that the two-way and three-way stopcocks are rotated to their off positions. Then, turn on the vacuum pump and

1) Slowly open the two-way stopcock a small amount allowing the mercury to slowly rise from the storage bulb, up to the bottom of the second bulb, then close the stopcock.
2) Slowly open the three-way stopcock to evacuate the McLeod gauge, drawing the mercury back into the storage bulb.
3) Repeat steps 1 and 2 several more times until the mercury enters the upper bulb very slowly and evenly. The number of repetitions is directly related to the volume of your system. Thus, the larger your system, the more repetitions of steps 1 and 2 will be required.

7.5.9 Returning a McLeod Gauge to Atmospheric Conditions

If you must return a McLeod gauge to atmospheric conditions, such as would be necessary for periodic re-greasing of the stopcocks, the following procedures should be observed:

1) Slowly open the two-way stopcock (to system) allowing mercury into the McLeod gauge.
2) Close the 2-way stopcock.
3) Open your system to the atmosphere.
4) Partially open the three-way stopcock to the atmosphere (Position 1).
5) Slowly open the two-way stopcock letting the mercury fall back into the storage bulb.

7.5.10 The Tipping McLeod Gauge

Fig. 7-41 shows a *tipping* McLeod gauge. All of the same components found in a standard McLeod gauge are there (except the three-way stopcock), but in a significantly different configuration. It is much easier to use this gauge than a standard McLeod gauge, but it is less accurate, and cannot read as low a vacuum. Depending on the design, its lowest reading is from 0.050 to 0.001 mm Hg.

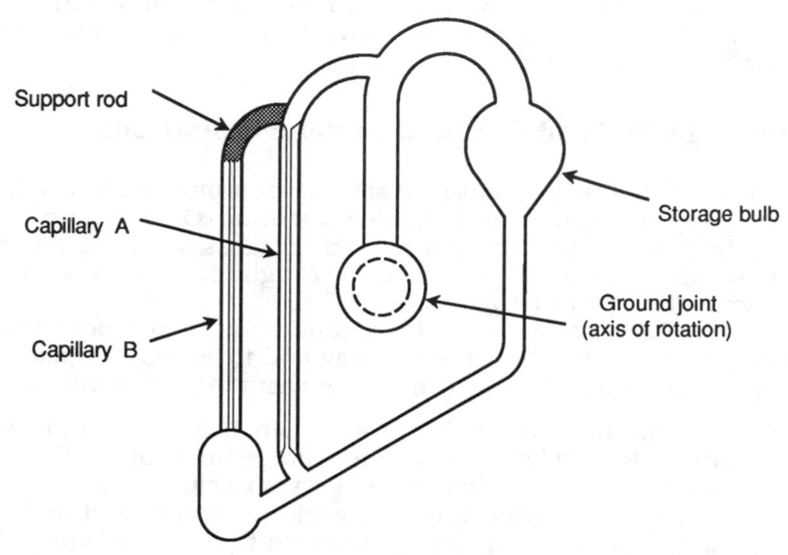

Fig. 7-41 The tipping McLeod gauge.

The tipping McLeod gauge can be mounted directly onto a vacuum system or can sit on a metal stand and connect to a vacuum system via a flexible vacuum hose. Regardless of the mounting technique used, there should be a two-way stopcock between the gauge and vacuum system that allows the user to shut off the McLeod gauge when it is not in use. This configuration limits the amount of mercury that backstreams into the vacuum system and the amount of material from the vacuum system that backstreams into the McLeod gauge.

To operate the tipping McLeod gauge (see Fig. 7-42):

1) Open the two-way stopcock separating the gauge and vacuum system (not shown in Fig. 7-40).
2) Rotate by hand the gauge from the rest position to the vacuum measurement position.
3) Read Capillary B just as in Fig. 7-40.
4) After the measurement is read, release the gauge and it will rotate back to the rest position by itself.
5) Close the two-way stopcock off.

Do not let a tipping McLeod gauge flop around too freely. Although the stopcock grease will cause some amount of drag, twisting the gauge too quickly from one position to another is likely to cause the mercury to spill into the closed capillary where it can be difficult to remove.

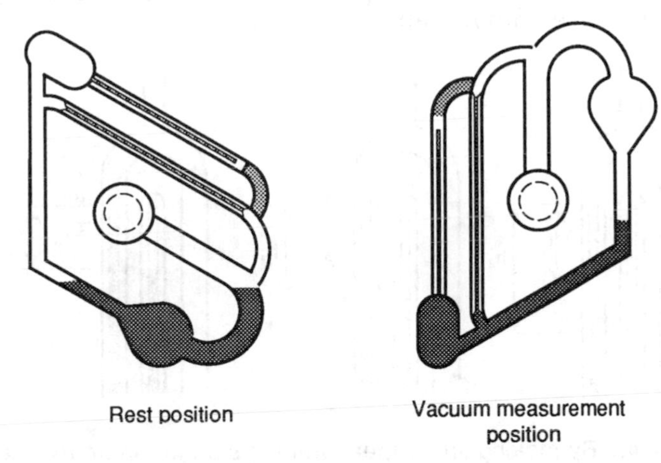

Fig. 7-42 How to use a tipping McLeod gauge.

7.5.11 Condensable Vapors and the McLeod Gauge

Condensable vapors are gases near their condensation temperatures or pressures. In the low pressures of a vacuum system they are gases. Examples of fluids that form condensable vapors at STP are water, alcohol, and mercury.

The McLeod gauge works on the ideal gas law principle that the pressure of a gas increases proportionally as its volume decreases. However, condensable vapors do not obey the ideal gas law and therefore, under standard use, the McLeod gauge cannot accurately measure them. Obviously, condensable gases with low vapor pressures at STP do not affect the readings of McLeod gauges significantly enough to alter the validity of the readings.

When a two-way stopcock is opened, all the gases (condensable and noncondensable) within the vacuum system can pass into the McLeod gauge, the vacuum bulb and Capillary B. When the three-way stopcock is rotated, the gases that were in vacuum are brought to atmospheric pressures. Any condensable gases that 'condense' at higher pressures will occupy less space. This 'less space' will show up as a better-quality vacuum than really exists within your system. The McLeod gauge will indicate a lower reading equal to the vapor pressures of the condensable gases at their ambient temperatures.

To determine whether your readings from a McLeod gauge are being affected by condensable vapors, take measurements of h_1 and h_2 for three to four levels (see Fig. 7-43). If the product of $(h_2 - h_1)h_2$ is not the same but $(h_2 - h_1) = P$, (a constant) seems to hold, then you have at least one condensable vapor present.* With accurate measurements, Wear[41] found it possible to identify the condensable vapors, estimate their concentrations, and determine the vacuum system pressure and partial pressure due to non-condensable gases. To avoid condensable vapor problems in McLeod gauges, limit the amount of condensable vapors that enter the McLeod gauge by using chilled or chemical vapor traps.

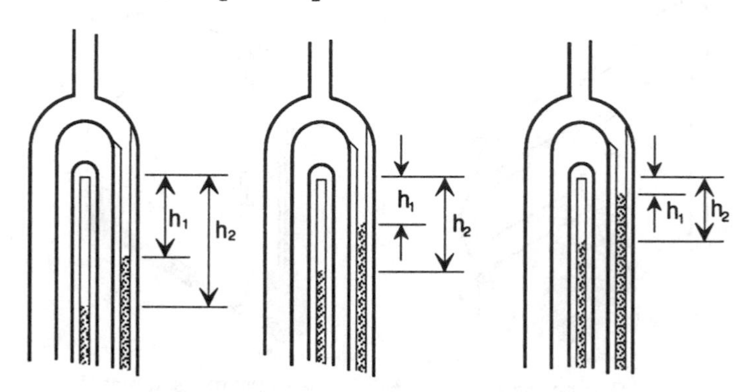

Fig. 7-43 By making linear measurements as the mercury rises within a McLeod gauge, it is possible to determine whether there are condensable vapors within the McLeod gauge.

7.5.12 Mercury Contamination from McLeod Gauges

There are two mechanisms for contamination from McLeod gauges: contamination which is backstreamed into the system from the McLeod gauge, and contamination which comes from the McLeod gauge storage bulb during the evacuation process.

Research by Carstens, Hord, and Martin[42] displayed a relatively high level of mercury being pumped out of a McLeod gauge during the evacuation process. Typical readings, in mg/m^3, were

Residual level	0.2
Evacuating reservoir	0.4

However, once a layer of low vapor pressure oil was added to cover the mercury in the storage bulb, the readings went down by a decade:

Residual level (with oil)	0.02
Evacuating reservoir (with oil)	0.05

The research authors pointed out that because the levels were erratic and low, the actual health dangers were also rather low.

One should also remember that the oil within the mechanical pump also becomes a trap for emitted mercury vapors. However, the mechanical pump

* To make these measurements, a special vacuum system is required. A complete derivation of the formulas and procedures can be found in the book *The Design of High Vacuum Systems and the Application of Kinney High Vacuum Pumps* by C.M. Van Atta, © 1955 by Kinney Manufacturing Div., New York Air Brake Company.

becomes only a temporary trap for the vapors. Once in the pump, the mercury collects in little pools and is slowly emitted into the atmosphere. The fact that pump performance is impaired by mercury pools has been confirmed by the author's personal observation.

Stopping mercury migration into a system can be achieved by placing a cold trap between the gauge and system. Stopping mercury contamination from a McLeod gauge's storage bulb can be achieved with a thin layer of low vapor pressure oil over the surface of the mercury. This oil will not affect the gauge's performance in any way.

7.5.13 Cleaning a McLeod Gauge

Invariably, the mercury within a McLeod gauge will get dirty. The telltale evidence is when the mercury does not cleanly run down the glass tubing and/or you see a film on the surface of the mercury. Traps are the best way to avoid this problem.

A McLeod gauge must be removed from a system when it is being cleaned. This removal probably will involve the talents of a glassblower. The mercury should be carefully poured out of the gauge and sent to a mercury distiller for cleaning. The grease should be removed by an appropriate solvent. Silicon grease should never be used on a McLeod gauge as it requires far too much maintenance and replacement.*

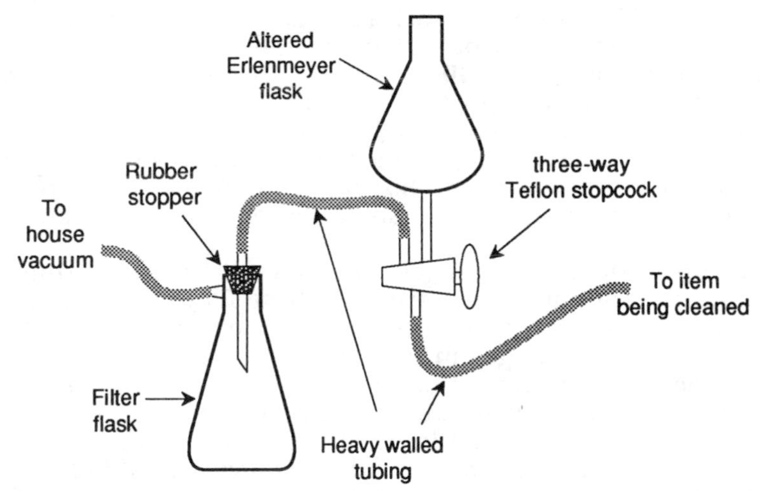

Fig 7-44 Vacuum apparatus for cleaning hard-to-clean apparatus.

Dissolve any mercury remaining in the gauge with nitric acid. Heating the nitric acid will facilitate the cleaning.** **Use gloves and work in a fume hood!** After draining, and thorough rinsing with tap water, there should be a

* Silicone stopcock grease requires cleaning and replacement every month or two whether the stopcock is used or not.

** Nitric acid containing mercury is a toxic waste. It must be saved and disposed of under proper conditions and with licensed firms.

final rinse with distilled water. If you choose to extend the process one more step with a methanol rinse to facilitate drying, be sure that all the nitric acid has been removed because the acid is not compatible with organic materials.

It can be very difficult to get liquid into various parts of a McLeod gauge because of the closed capillary tubing. One solution to this problem is to use a vacuum cleaning setup as shown in Fig 7-44. A Teflon stopcock is used because it requires no grease.

When the stopcock shown in Fig. 7-44 is turned to one position, a vacuum will be created in the item to be cleaned. Rotating the stopcock 180° will allow the cleaning or rinsing solution to be drawn into the piece. Then by rotating the stopcock one more time while holding the item in a draining position, the liquid can be removed into a filter flask.

Do not break apart a McLeod gauge for easier cleaning. The calibration of the various parts is extremely accurate. Some McLeod gauges have sections that are intended to be dismantled and have ground glass joints between sections. Some tipping McLeod gauges have plastic end caps (on the closed tubes) to facilitate cleaning.

7.5.14 Thermocouple and Pirani Gauges

Thermocouple and Pirani gauges both use the physical characteristic of heat (within a vacuum) to infer the amount of vacuum within a system. If an object is hot, the only way it can cool down is to transfer its heat to the surrounding area by conduction or radiation. The standard way this transfer happens is that the object conducts its heat through the air (and anything else it is touching), and radiates its heat with IR radiation to other surfaces. In a vacuum, there is less air through which a hot object can conduct its heat, and therefore it can only lose heat through IR radiation. Both gauges have fast response times, and are excellent for determining pressures between 1 and 10^{-3} torr. They also can be used in leak detection (see Sec. 7.6.8).

Thermocouple and Pirani gauges both have filaments within them that are exposed to the vacuum of the system. These filaments are always under small, constant electric loads and, because of their resistance, they get hot. At higher pressures, the air/gas in the system conducts all the heat from the wire. As the vacuum increases, less heat can be lost through (air/gas) conduction, and more heat is then maintained by the wire. Once the vacuum is high enough, heat is lost primarily by conductance from the wires holding the filament and IR radiation, thus creating the lower limits of the gauge.

Both gauges are dependent on the thermal conductivities of the gases that surround them. Because different gases have different thermal conductivity, different gases will indicate different values for the same pressure. It is possible to adjust a gauge reading to a 'calibrated accurate' pressure if you know the constant with which to alter the gauge reading. The real vacuum can be determined with Eq. (7.13) and the appropriate sensitivity constant. For a list of *Sensitivity Constants*, see Table 7-13. It is impossible to achieve either gauge's potential accuracy when measuring a vacuum system filled with unknown gases.

$$\text{Pressure} = \frac{\text{Gauge Reading}}{\text{Sensitivity Constant}} \qquad \text{Eq. (7.13)}$$

ELECTRICAL WARNING: These types of gauges generally do not use high voltage. They do however, use 110-120 V current, which means that common sense should be observed. For instance, do not pull on a cord when unplugging, pull on the cord outlet. Avoid spilling conducting liquids around the gauges. All pieces should be grounded. If the unit has a three-pronged outlet, do not cut the ground off or by-pass the ground by using a two pronged extension cord. Unplug the unit if repair needs to be done. Replace worn and/or frayed cords immediately. Cover all bare (exposed) electrical leads with tape, tubing, shrink tubing, or plastic screw caps. In addition, keep flammable liquids or gases away from electrical devices in case of any sparks and/or electrical arcs.

Table 7-13‡

Properties of Various Gases

Gas	Sensitivity Constant[a]	Viscosity[b]	Thermal Conductivity[c]
Air	1.00	180	0.057
Butane	2.50		
Carbon Monoxide	1.00		
Carbon Dioxide	1.10	145	0.034
Helium	1.00	194	0.344
Hydrogen	1.30	87	0.416
Krypton	0.45		
Mercury (vapor)	0.34		
Neon	0.90	310	0.110
Nitrogen	1.00	173	0.057
Xenon	0.35		

‡ Adapted from Spinks, *Vacuum Technology*, Franklin Publishing Co., p. 22, and from Guthrie, *Vacuum Technology*, John Wiley and Sons., Inc., (1963), p 504.

a *Sensitivity Constant*: These constants are accurate ± 10% over any given range of pressures.

b *Viscosity*: At 15°C., given in micropoises.

c Units are 10^3 K, where K = *Thermal Conductivity* at 0°C, cal/cm/sec/°C.

7.5.15 The Pirani Gauge

The *Pirani gauge* uses the principle that (usually) the hotter a wire gets the greater its electrical resistance. Therefore, if the resistance of a wire is going up, it must be getting hotter. This relationship implies that less air/gas is available to conduct heat away from the wire, and therefore a higher vacuum is being achieved.

The accuracy of a Pirani gauge is typically ±20%, although an individual (clean) gauge properly used over a two-year period may show a sensitivity drift of only 2%.[43]

One immediate complication of the Pirani gauge is ambient temperature: as the room temperature gets hotter, the filament gets hotter, making the gauge read a false 'better vacuum.' To solve this problem, a dummy filament is included in the Pirani gauge. The dummy filament is evacuated and sealed off (at a lower vacuum than what likely to be used with the Pirani gauge) and is used as a standard to help calibrate a zero point.

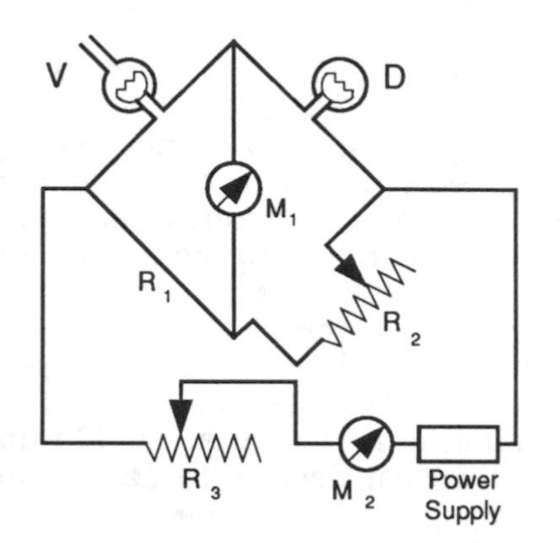

Fig. 7-45 Pirani gauge schematic.‡

‡ From *Vacuum Technology*, p. 163, (© 1963 by John Wiley & Sons, Inc., New York). With permission.

An electrical diagram for a Pirani gauge is shown in Fig. 7-45 where V and D comprise the Pirani tube. D is the dummy filament tube that is sealed off, and V is the tube that is exposed to the vacuum system. The filaments in the V tube are connected to a bridge circuit called a Wheatstone bridge with two resistance units called R_1 and R_2. Power, from the power supply, passes across the Wheatstone bridge and is adjusted to the proper setting by R_3 whose value is read on the milliammeter M_2. The current read on M_1 is proportional to the vacuum. An ammeter will read the current, which is proportional to resistance in ohms.

To set a Pirani gauge, the vacuum on V is set to a pressure lower than what the gauge can normally read. Next M_1 is set on its zero point by adjusting the resistance of R_2. Thereafter M_1 will give proper readings as the pressure is raised to the range of the gauge.

The advantages of a Pirani gauge are:

1) It has a rapid response to changes in pressure.
2) The electrical circuitry in the gauge leads to easy adaptation to recording, automatic devices, and computer sensing.
3) Electrically it is very simple.
4) It measures the pressure of both permanent gases as well as vapors.

The disadvantages of a Pirani gauge are:

1) Because not all gases have the same thermal conductivity, different gases will provide different pressure readings for the same pressure.
2) It is limited to the pressure range of about 10^{-1} to about 10^{-4} torr.
3) If there is any change in the filament wire's surface condition within the Pirani gauge, there will be a change in the heat loss. This change will result in a change of the gauge calibration as well as a change in the zero.

7.5.16 Cleaning Pirani Gauges

If a Pirani gauge becomes contaminated with backstreamed oil, rinsing the gauge with a suitable solvent should be sufficient. Be sure to rinse with distilled water followed by a methanol rinse. Be gentle with the gauge so as not to break the internal wire, which is fragile. Cleaning is likely to change the calibration, so be prepared to re-calibrate the gauge after cleaning.

7.5.17 The Thermocouple Gauge

The thermocouple gauge is more straightforward than the Pirani gauge and less complicated electronically. The thermocouple gauge has a thermocouple attached to a filament under constant electrical load and it measures the temperature at all times. If the filament becomes hotter, it means that there is less air/gas available to conduct heat away from the wire, and therefore there is greater vacuum within the system.

There are two different types of thermocouple gauges, one has three wires and the other has four. Both have a dc meter (or voltmeter) that reads the voltage from the thermocouple. The three-wire unit uses ac to heat the filament wire whereas the four-wire unit may use ac or dc. Although there are essentially no differences in performance between the two, they will likely require different controllers (or different settings) for use.

The advantages of the thermocouple gauge are fairly consistent with the four stated for the Pirani gauge with a few exceptions:

5) Thermocouple gauges can be made smaller and are more rugged than Pirani gauges.
6) Although the thermocouple gauge is subject to the same variations in apparent readings from real pressure (because of variations in the thermal conductivity of different gases), the differences are less apparent than the Pirani gauge.

The disadvantages of the thermocouple gauge are somewhat different from the Pirani gauge:

1) Because not all gases have the same thermal conductivity, you will get different pressure readings for the same pressure with different gases, although the differences for the thermocouple gauge are not as great as for the Pirani gauge.
2) It is limited to the pressure range of about atmospheric to about 10^{-3} torr.
3) The thermocouple gauge scale is non-linear, but the readings can be accurately interpreted by the controller.

Physical abuse, improper cleaning, and age can all cause a thermocouple to break. The symptoms may either be no response to the controller or a jerky twitching of the controller's needle. In either case, the thermocouple is not likely repairable and a new one will be necessary.

7.5.18 Cleaning Thermocouple Gauges

The thermocouple gauge is more durable than the Pirani gauge, which means that after you pour in the appropriate solvents for cleaning, you can shake the gauge for cleaning agitation. This cleaning procedure should be followed by rinses of water, distilled water, and finally, methanol.

7.5.19 The Ionization Gauge Family

All previously mentioned gauges require a certain level of particle density for operation. Once the level of particle density has dropped below a certain level (approximately 10^{18} particles/m^3), it is not possible to detect transfer of momentum forces either between gas to solid wall or gas to gas. On the other hand, it is possible to ionize gas particles, and then "count" the ionized molecules.

A molecule in a 'normal' state has a neutral charge; there is an equal number of electrons and protons. If you subject the molecule to a high amount of energy and knock out one of the electrons, the molecule is now ionized with a positive charge. This charge allows you to force the molecule to travel, bend, be focused (if necessary), and be counted. The number of (positive) ions created is always directly proportional to the molecular number density. It is only proportional to the pressure if the temperature is known and kept constant, and the type of gas being analyzed has known calibration constant for the type of gauge you are using.

There are essentially two types of ionization gauges used in the laboratory, the *hot-* and the *cold-ion gauges*. A third type, the *radioactive ionization gauge* is so limited in both scope and use that it will not be discussed in this book.

The concept of the ionization (ion) gauge is quite simple. Under a given electrical load, the available gas within the vicinity of the vacuum gauge is ionized, either by heat or a high-field (electrical) emission. Then, the ionized gas is collected and counted. From this count you can interpret what you have read as a unit(s) of vacuum and thereby infer the vacuum within the system.

One ironic peculiarity of ion gauges is that the ions collected by the gauges for counting are not re-released to the vacuum system and therefore are bound up as in getter pumps. Therefore an ion gauge also acts as a pump. This feature itself sounds great: what vacuum system couldn't use a little extra pumping? However, this feature adds an accuracy problem as there is no way of knowing whether the vacuum within the confines of the gauge (where active pumping is going on), and the rest of the system is the same. Thus, to maintain accuracy between the pressure within the gauge and within the system, the gauge should not be left on for extended periods of time and the gauge should be connected to the system with large diameter tubing. This setup decreases the opportunities for a pressure gradient to be established and facilitates equalization between the gauge and the system if a gradient condition occurs.

Hot-cathode gauges are fast pumps, but cold-cathode gauges pump ten times faster. At 10^{-10} torr there are only 10^6 molecules per cc. If a hot-cathode gauge's volume is 100 cc, 1% of the gas within the gauge is removed every second![44] It has been shown that if the conductivity of the connecting tube is >10 $^l/_s$, these affects are negligible.

If you are concerned specifically with accuracy, hot-cathode gauges show greater accuracy and reproducibility. However, it must be operated by an experienced technician in controlled conditions with repeated backing and degassing. Otherwise, the cold-cathode gauge will provide greater accuracy without constant attention. This is mostly attributed to the the hot-cathodes heat during operation provides some incidental outgassing as opposed to the cold-cathode gauge, which operates at room temperatures, and therefore has no incidental outgassing.[45]

ELECTRICAL WARNING: Ion gauges require high voltage, so common sense must be observed. For instance, do not pull on the cord when unplugging a gauge from a wall outlet, pull on the electrical plug. Avoid spilling conducting liquids around the gauges. All pieces should be grounded and, if the unit has a three-pronged outlet, do not cut off the ground or by-pass it. Be sure to unplug the unit if it is being repaired. Replace worn and/or frayed cords immediately. Cover all bare (exposed) electrical leads with tape, tubing, shrink tubing, or plastic screw caps. Keep flammable liquids or gases away from electrical devices in case of sparks and/or electrical arcs. Because of the high voltages possible with ion gauges, dirt, and even fingerprints, can cause unexpected arcs. Therefore, keep the gauge surface clean. Turn off the gauge and its controller when helium leak testing because a discharge can be created around high voltage feedthroughs which in turn can destroy the gauges controller. Likewise, turn off the gauge and controller when checking for leaks on a glass system using a Tesla coil. The discharge can destroy the electrical circuits within the controller. Finally, if your vacuum system is mounted on a metal rack, ground the rack with *grounding strap* (braided copper-plated wire) to plumbing or some other substantial ground. For proper contact, file off any paint or corrosion on the rack and the ground before attaching the grounding strap with a threaded hose clamp.

7.5.20 The Hot-cathode Ion Gauge

The most common hot-cathode ion gauge (and the most common high-vacuum gauge used) is the Bayard-Alpert gauge. It can read vacuums between 10^{-3} and 10^{-10} torr. With special gauges, readings as low as 10^{-14} torr can be obtained. A diagram of its general structure is shown in Fig. 7-46. The hot-cathode gauge operates by heating a *filament*, which causes an emission of electrons. These electrons are attracted to a *grid* which is held to a high (+) potential that attracts the (–) electrons. As electrons stream toward the grid, they collide with the gas molecules en-route, ripping off an electron and creating positive ions. The positive ions are attracted to the *ion collector*, collected and counted. The positive ion current is measured in amperes from the gauge tube, but the hot-cathode gauge controller then interprets this count as vacuum reading. Hot-ion gauges are designed to be gas-specific.* Currently, the gauges are accurate to within ± 20% for their specific gas type.

Early versions of the hot-cathode ion gauge used a large cylinder for the ion collector. This design was limited to a vacuum measurement of only approximately 10^{-4} torr. Nottingham[46] proposed that when electrons struck the grid, soft X-rays were created. The X-rays then struck the large ion collector cylinder which in turn caused photo-electrons to flow back to the grid. This action created a current in the external circuit of the ion gauge that was indistinguishable from the ion flow. Thus, the gauge was actually reading lower vacuums, but the excess electronic noise was masking the reading and producing artificially higher readings. By decreasing the size of the ion collector, Bayard and Alpert significantly decreased the electronic noise making the hot-cathode ion gauge a significant high- and ultra-high vacuum gauge.

* Most gauges are designed to read specific pressures in nitrogen atmospheres.

The standard material used for filaments within a hot-ion gauge is Tungsten.* Unfortunately, a Tungsten filament can easily burn out if a gauge is turned on when the pressure is too high within a system. Because of this idiosyncrasy, a thermocouple or Pirani gauge may be connected to a relay that shuts off power to the ion gauge if a loss of vacuum is detected. This addition is strongly recommended. (Some controllers provide automatic switching between a thermocouple and an ion gauge to allow for continuous readings between atmospheric and 10^{-10} torr.)

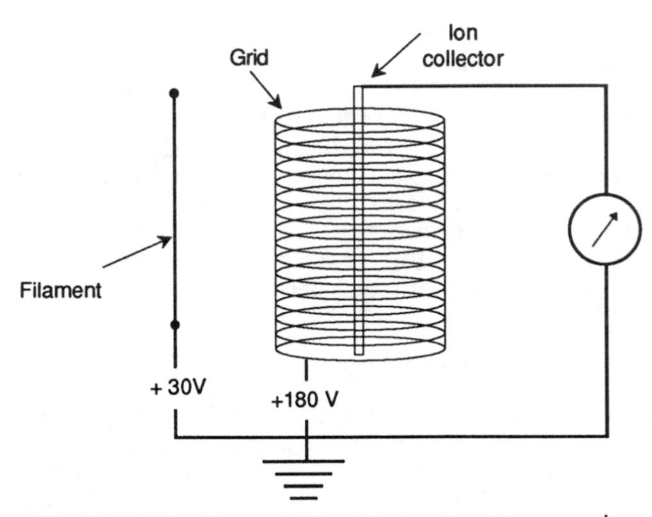

Fig 7-46 The Bayard-Alpert hot-cathode gauge.‡

‡ From *Fundamentals of Vacuum Science and Technology*, p. 92, by G. Lewin, (McGraw-Hill, New York, 1965). With permission.

Another common filament material used in hot-cathode ion gauges is thoriated iridium (ThO_2 on iridium). These filaments are used if there is any likelihood that the system will be exposed to accidental bursts of atmosphere. However, if exposed to hydrocarbons or halocarbons, their emissions can radically change. Re-calibration of these filaments is constantly required.

Studies by McCulloh and Tilford[47] found that dual Tungsten filaments exhibited better linearity as well as sensitivities which were in closer agreement to those stated by the manufacturer. Filaments made out of thoriated iridium demonstrated linearity differences as great as 30%. These differences were evident in standard as well as nude gauges.**

* Tungsten filaments can create large quantities of CO and CO_2 during operation which may, or may not, affect your work. It is possible to obtain Bayard-Alpert gauges that can be preheated, or baked out, prior to use to limit this problem.

** A "nude" gauge has no protective cover and, rather than being attached *to* a vacuum system, it is mounted *in* a vacuum system. This difference prevents any lag in response time from pressure variations through a tube and eliminates the effects of ion-gauge pumping. The main disadvantage is there is no way to shut off the gauge from the line itself, so there is no way of protecting the gauge from any potential contaminating materials. Nude gauges are not available for glass systems.

Sensitivity variations can be due to (at least in part) the proximity of the top of the filament to the cylindrical envelope of the grid spiral (Hirata, et. al.).[48] This sensitivity actor is why (in part) gauges should not be roughly handled or dropped. If the filament changes its location within the gauge, the readings will vary from what they were before the mishandling. Any data that were obtained before such an incident may then need to be redone or re-calibrated to agree with later data.

Magnetic fields, depending on their strength and orientation, can also influence ion gauge readings. Studies by Hseuh[49] showed that a magnetic field can change an electron's path even though changes in the collection of ions were negligible. A magnetic field has the least effect when it is parallel to the gauge. It has the greatest effect when the gauge is perpendicular to the magnetic field. However, the alignment of the electric field relative to a magnetic field is complicated. Optimally, it is best to keep magnetic fields away from Bayard-Alpert gauges.

There are several general rules to follow that will make the operation of a hot-cathode ion gauge as trouble-free as possible. Implementation of these rules cannot guarantee success, but ignoring them will ensure problems:

1) Always connect the gauge as close as possible to the area where measurements need to be made. The farther away the gauge is from the point of measurement, the longer the lag time before the system, and the gauge, come to equilibrium. The greater the vacuum ($> 10^{-5}$ torr), the more pronounced this effect.
2) On glass systems, the ion gauge is attached to the line by a connecting piece of glass, a glass to metal seal, or some type of Swagelok®. Regardless of how the gauge is attached, try to keep this connection as short as possible with as large a diameter of tubing as possible (>1 inch). Do not use any connection with a smaller diameter than that supplied on the gauge. A cold trap placed between the gauge and line will help protect the gauge from condensable vapors. However, it will indicate pressures lower than really exist within your system due to the cryogenic pumping capabilities of the cold trap. In addition, it will also slow the speed required for the system and the gauge to come to equilibrium.
3) On metal systems, you have two choices: connect the ion gauge by connecting tubing (same tubing size rules apply from Point 2), or use a nude gauge.
4) The ion gauge is gas- (see Table 7-14) and temperature-dependent. Therefore, if your lab has temperature swings and/or you vary the gases within your system, constant re-calibration may be required. Table 7-14 (and others like it) can only provide a benchmark for making corrections because your system is not likely to have pure gas samples. In addition, variations between gauges of the same type (but from different manufacturers) can be quite large while the differences between gauges of different designs can be phenomenal. For the most accurate interpretation of your gauges readings, obtain calibration tables from the gauges manufacturer.

Table 7-14‡

Thermionic Ionization Gauge Sensitivity,[a] Relative to that for Nitrogen	
Hydrogen	0.47
	0.53
Helium	0.16
Neon	0.24
Nitrogen	1.00
Argon	1.19
Carbon monoxide	1.07
Carbon dioxide	1.37
Water vapor	0.89
Oxygen	0.85
Krypton	1.9
Xenon	2.7
Mercury vapor	3.4
Narcoil-40 vapor	13

[a] True gauge sensitivity is type and model specific. Accurate sensitivity tables for your gauge should be supplied by the manufacturer. This table is supplied for reference only.

‡ From *Vacuum*, Vol. 13, H.A. Tasman, A.J.H. Boerboom, and J. Kistemaker, "Vacuum Techniques in Conjunction with Mass Spectrometry," 1963, p. 43. Pergamon Press, plc. With permission.

5) Never turn a gauge on until the pressure is below 1 µm or less. You must have a second gauge for higher pressures (for example a Pirani or thermocouple gauge).

6) The hot-cathode gauge must be outgassed every time the gauge is exposed to the atmosphere for an extended period of time, or at pressures near the base pressure of operation (10^{-5} torr). The outgassing cannot be done at any higher pressure than 10^{-5} torr, and can be performed either by providing a current to the grid within the gauge (which can often be supplied by the gauge controller), or by electron bombardment. Note, some gauges can only be resistively heated or electron bombardment degassed. Gauges that require outgassing will read a pressure higher than really exists. The amount of error is dependent on the degree of outgassing required.

   Regardless of the outgassing approach used, filament temperatures of about 800°C are required. Initially, a new gauge may need to be outgassed for some 15-20 minutes. If a gauge is showing evidence (i.e., dirty electrodes) that outgassing may be required again, some 15 seconds should suffice. During degassing, the ion gauge envelope becomes very hot. Be sure that the gauge is mounted in such a fashion that accidental contact with technicians or flammable materials is not possible.

   If a gauge is outgassed at too high a pressure, a layer of metal (from the electrodes) may be deposited outside the gauge envelope [some controllers prevent degassing at too high a pressure]. This condition can cause the insulation to become 'leaky.' A temporary solution can be achieved by grounding the electrical leads and running a Tesla coil on the gauge. However, it is best to simply replace the gauge.[50]

7.5.21 Cleaning Hot-cathode Ion Gauges[51]

The cleaning technique used for hot-cathode ion gauges depends on the shape, structure, and composition of the gauge. Hot-cathode gauges are so sensitive that any cleaning will alter or damage the sensitivity of the gauge. You can minimize gauge damage if the cleaning is performed by electrical heating while the gauge is in a high-vacuum condition. This heat cleaning can be done by grounding the ion collector and applying a Tesla coil to each of the electrodes within the gauge. **Important**: be sure the controller is off when doing this operation. Otherwise the charge from the Tesla coil may destroy the controller.

Chemical cleaning requires the removal of the gauge from the system and a general (*gentle*) rinsing with distilled water. If the gauge has a deposit that is blue or silver gray, the contamination is probably Tungsten oxide or molybdenum oxide. This deposit can be cleaned as follows:

1) Soak the gauge in a 10-20% sodium carbonate solution. Gently heating the solution to 30°C will accelerate the cleaning process, which should take about 15 minutes.
2) Rinse with copious amounts of water, then let the gauge soak for about a half-hour filled with distilled water. Repeat the distilled water soak two more times.

If the deposit is brown, the offending contamination is probably cracked hydrocarbon vapor. This deposit can be cleaned as follows:

1) Soak the gauge in a 10% solution of potassium hydroxide and let it stand for about a half-hour. This soaking may have to be repeated with a fresh potassium hydroxide solution.
2) If there is a discoloration of the electrodes, gently shake some hydrogen peroxide in the gauge until the discoloration disappears.
3) Water spots can be removed with an oxidizer such as any substitute for chromic acid (see Sec. 4.1.8). Do not use chromic acid because it is a toxic waste.
4) Rinse with copious amounts of water, then let the gauge soak for about a half-hour filled with distilled water. Repeat the distilled water soak two more times.

7.5.22 The Cold-cathode Ion Gauge

The cold-cathode (a.k.a. Penning gauge, Philips gauge, or PIG (Philips Ionization Gauge)) requires a high-potential field for removing electrons from the (cold) cathode. This same principle is used in the magnetron and the inverted magnetron. In the cold-cathode gauge, the center loop is an anode (see Fig. 7-47) and it is maintained at a very high electrical potential (2-10 kV). The two plates are cathodes and are grounded. Electrons travel from the cathodes to the anode, ionizing molecules on their way. A U- or circular-shaped magnet is placed over the cathodes. The magnetic field formed by the magnets forces the electrons to take a longer route than they otherwise would.* Thus, they come into contact with a greater number of gas molecules and create a greater number of ions.

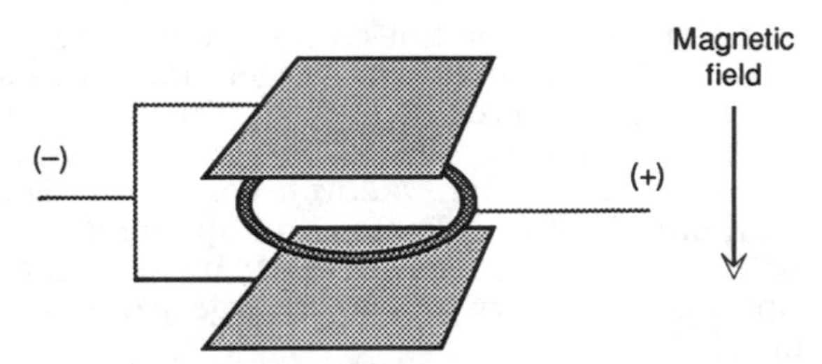

Fig 7-47 Basic electrode layout of the Penning gauge.‡

‡ From *Fundamentals of Vacuum Science and Technology*, by G. Lewin (McGraw-Hill, New York, 1965, (Fig. 5-4). With permission.

The range of the cold-cathode gauge is about 10^{-2} to 10^{-10} torr. Initiating the discharge within the cold-cathode gauge at lower vacuum ranges (>10^{-6} torr) is typically fast, within seconds. However, at lower pressures (10^{-8} to 10^{-10} torr) it can take hours for the discharge to begin. You may wish to start your col-cathode gauge at higher pressures and leave it on as the pressure drops to maintain the discharge.

The cold-cathode gauge can act like a sputtering ion pump just as the hot-cathode gauge. However, it can pump nitrogen 10-100 times faster than a hot-cathode gauge. A cold-cathode gauge can pump nitrogen at rates of 0.1–0.5 $^{l}/_{s}$. Thus, if the connection tube to the gauge is too small in diameter, the gauge will remove the nitrogen from the area of the gauge faster than the system can equilibrate. The gauge will then read a greater vacuum than actually exists within the system.

Like a hot-cathode gauge, the cold-cathode gauge can be maintained longer between cleanings if a cold trap is placed between the gauge and the system. However, the gauge will indicate pressures lower than really exist within your system due to the cryogenic pumping capabilities of the cold trap. In addition, it will also slow the speed required for the system and the gauge to come to equilibrium. When installing the gauge, do not position the opening of the gauge to face the cold trap or momentary warming of the trap will cause evaporating frozen material to contaminate the gauge. When mounting the gauge on a vacuum system, be sure to angle the gauge so that the opening faces down. This angling will prevent particulate matter from falling into the gauge.

Do not leave the gauge on for extended periods of time, especially when the pressure is only about 10^{-2} torr. Otherwise the gauge can contaminate quickly. Likewise, do not let the gauge run continuously while the system is

* At 10^{-10} torr, it can require 20 minutes for an electron to travel from the cathode to the anode.[52]

roughing down. Brief use, when in use, will limit the effects of the pumping action of the gauge.

7.5.23 Cleaning Cold-cathode Ion Gauges[53]

Some cold-cathode gauges can be taken apart. If you have one of these gauges, it is possible to carefully sand any surface contamination off the grid with a fine-grade glass-sanding paper (do not use a metal file that could leave metal filings within the gauge. These filings can cause major problems during use). The gauge should then be rinsed with appropriate solvents followed by rinses of distilled water and then methanol. After the gauge is reassembled, be prepared to re-align the magnet (follow the manufacturer's guidelines).

If your gauge is glass-bodied, it cannot be taken apart and you are limited to the chemical cleaning processes mentioned for the hot-cathode ion gauge.

7.5.24 The Momentum Transfer Gauge (MTG)

The Momentum Transfer Gauge (MTG) was first developed as a lab curiosity in 1962. It uses the principle of gas viscosity to slow a spinning ball bearing that is levitated by magnetism. The levitated ball is rotated at speeds of up to 100,000 rps and then allowed to coast. The only mechanism used to slow down the ball is the friction of air on its surface. The less air in the system, the less friction on the ball.

Currently, MTGs read the vacuum ranges 10^{-2} to 10^{-8} torr, and with modern electronics, have demonstrated remarkable accuracy and sensitivity (±1% accuracy in the 10^{-2} to 10^{-5} torr range).[54]

Because the MTG neither heats nor ionizes the gases within a vacuum system, it is unique in the inferred gauge category. By not altering or changing the composition of the gases within a closed system, it offers special opportunities. However, MTGs require extensive and expensive electronic controls, measuring equipment, and special vibrationless platforms. Thus, very few research labs have the needs that can justify the expense and demands of these gauges.

7.6 Leak Detection and Location

7.6.1 Is Poor Vacuum a Leak or a Poor Vacuum?

Looking for a leak is only fruitful when you know there *is* a leak. Just because you have a poor vacuum does not mean that you have a leak. In the art and science of leak detection, you must first verify whether a problem really exists before you try to do something about the (perceived) problem. Only after you have established that there is a problem can you begin to locate the source of the problem.

It is impossible to obtain a perfect vacuum; there is no way to remove all molecules from a given area. With that idea in mind, the best vacuum of any vacuum system is limited by the quality and design of your vacuum pumps, the composition of material in your vacuum system, and the design of your vacuum system. Ultimately, a leak-tight vacuum system* will be limited by leaks from the outgassing of materials from your system, diffusion of gases from the walls, permeation through the walls, evaporation and desorption from wall surfaces, and backflow from the pumping systems.

Because no vacuum system can be truly leak-free, it is important to determine whether or not you have a leak of consequence. In other words, does any system leak you have affect your work? For example, a common rubber balloon holds water better than it holds air, and it holds air better than it holds helium. If your needs are to contain water, a standard rubber balloon is sufficient. Similarly, if you want to contain helium for a limited time, again a rubber balloon is sufficient. However, if you want a helium balloon to stay up for several days, then a rubber balloon is insufficient and you must spend the money for a Mylar balloon, which can contain helium much better than a rubber balloon.

In addition to knowing the intended use for a vacuum line, you also need to know its history. From the history you can rule out the possible reasons for vacuum failure and proceed to locate the cause. For example, if yesterday a vacuum line was achieving a 10^{-6} torr vacuum, but today it can only achieve a 10^{-2} torr vacuum, something dramatic has obviously occurred, which may or may not be a leak. On the other hand, if the same drop in vacuum occurred gradually over a period of several weeks, it is very unlikely the cause is a leak.

7.6.2 False Leaks

You should not assume that you have a leak just because the vacuum in your system is poor. Reasons for inadequate vacuum can be carelessness (stopcock left open), anxiousness (insufficient pump-down time), or neglect (the diffusion pump was never turned on). You can save a lot of time by eliminating the various reasons why your vacuum is not performing up to expectations before you look for leaks. Factors that can prevent a system from reaching a desired low pressure include:

1) Pumps that are incapable of pulling a greater vacuum because they are too small (or slow) for the given system or need maintenance
2) System components that cannot be baked (to facilitate outgassing)
3) Inaccurate gauges
4) Outgassing of high vapor pressure materials from the system.

Poor Pump Performance. This problem is a common one in labs. It is either caused by purchasing too small a pump for the desired task or, more likely, insufficient maintenance, which causes a perfectly good pump to run poorly. Insufficient trapping can cause short- or long-term damage to pump oils. Running a mechanical pump too long against a 'no-load' condition will

* If such a thing existed, this vacuum system would have no imperfections (holes) in the walls to the atmosphere.

cause pump oils to froth and take more time for outgassing. Probably the greatest cause for poor pump performance is not changing pump oils on a regular or timely basis.

Baking Out a System. Generally, the need to *bake out a system* implies that very high or ultra-high vacuums are required. A system must be baked to about 150°C to remove surface water vapor and greater temperatures (250° to 450°C) to remove enough water required to obtain very low pressures ($<10^{-8}$ torr)[55] for ultra-high vacuums. The materials of the vacuum system's construction must be considered when the tasks of the vacuum system are established. If a vacuum system is constructed with the wrong types of components, proper baking may be impossible, eliminating the possibility of achieving very high or ultra-high vacuums. Because it is not possible to heat glass stopcocks this high (and still expect them to function), a glass system is not recommended for vacuum systems requiring bake-outs.

Inaccurate Vacuum Gauges. Discrepancies can be caused by simple reasons such as an inaccurate gauge, used beyond its range, or a gauge that needs to be calibrated. More complex reasons for vacuum gauge errors could be that the gauge is tuned to a different gas species than what is in the vacuum system, it is poorly placed within the vacuum system, or the gauge is being affected by external interference such as a magnet. The resolutions to these problems can include calibrating the gauge for the specific gas being read, obtaining a gauge which is accurate for the pressure being read, and/or properly locating the gauge within the system (and in the proper alignment). You may need to obtain calibration equipment for your gauge (some controllers have calibration units built into the electronics).

Outgassing. This problem is typically one of user impatience, poor cleaning, or poor choice of materials within the system. Many a beginning vacuum user expects a brand new vacuum system to get to 10^{-7} torr within fifteen minutes of the system's first use. A new system with the usual amounts of atmospheric water vapor within can typically require overnight pumping to reach its lower limits. The same will be required of any system left open to the atmosphere for too long. The longer a system is left open to the atmosphere, the longer it will take for it to outgas with its maximum saturation reached in two to three days (dependent on atmospheric humidity and temperature). Aside from water vapor that has condensed (adsorbed) on the surface of the vacuum system, there is also water that has absorbed into the material of the vacuum system itself. With a glass vacuum system, the depth of absorbed water can be as deep as fifty molecules. Because it is almost impossible to remove absorbed water without baking the system, this water (in a glass system) can be considered non-removable.

O-rings and other flexible seals also require an outgassing period. Neoprene or perbunan rubber O-rings can be used to pressures of 5×10^{-7} torr, but cannot be heated beyond 100°C. These materials give off small quantities of water vapor and CO. Viton O-rings, which show very little propensity for outgassing, can be heated to 250°C and can be used in pressures as low as 10^{-9} torr.[56]

If your system contains materials with high vapor pressures that cannot be baked out, you cannot achieve a vacuum greater than those vapor pressures. For example, a piece of paper placed in a vacuum system will limit the system so that it can never achieve better than a 10^{-5} torr vacuum.[57]

A dirty system takes longer to outgas (and may never outgas*) than a clean system and a rough surface takes longer to outgas than a smooth surface. The former because of extra materials, and the latter because of increased surface area. Rough surfaces tend to hold water tighter than smooth surfaces.

The first and foremost step to solving any vacuum problem is to KNOW YOUR SYSTEM. The more familiar you are with your system, the less time you will waste on wild goose 'leak hunts.' One technique that will help you know your system is to maintain a log book. Kept within this log could be information such as the date the system was used, how long it was used, what vacuum was achieved, what type of work was done, and when pump oils were changed. This data can be invaluable aids to determining if any changes are gradual or sudden, if there is any relation to using new pump oils, and any other indications of change. A log book should be mandatory any time more than one person uses a vacuum system.

7.6.3 Real Leaks

There are two ways to demonstrate that a leak in a vacuum system really exists: Either you are unable to obtain a dynamic** vacuum that previously could be obtained with no problem, or the system is not maintaining a static*** vacuum that it previously could maintain.

Verifying that a real leak exists only tells you part of the story because there are four types of leaks. You must eliminate them by analysis of the symptoms, experience, the history of the vacuum system, and/or trial and error. The following are the four types of real leaks:

1) Virtual leaks
2) Leaks at demountable seals****
3) Real leaks through the walls of the vacuum system
4) Backstreaming from pumps.

Virtual Leaks. A *virtual leak* is an honest-to-goodness leak, but because it is inside the vacuum system, it cannot be detected from outside. A formal definition of a virtual leak is 'a self-contained gas supply within a vacuum system.' There are two types of virtual leaks:

1) Those leaks in which a gas is physically trapped (such as a gas which is poorly, or improperly, frozen or has had insufficient outgassing).
2) Those leaks where the gas is mechanically trapped as shown in Figs. 7-48 and 7-49.

* After reassembling a very powerful vacuum system that had been left open for a weekend, a friend of mine could not obtain a vacuum greater than 10^{-3} torr. Finally, in desperation he dismantled the system to find the remains of a small mouse whose outgassing rate set a limit for a system that normally was capable of significantly greater vacuum.[58]

** A dynamic vacuum is when a pump is actively evacuating a system.

*** A static vacuum is when the passageway from the pump to a system has been closed, and the quality of the vacuum within the system must be self-maintained.

**** Demountable seals are sections of a system that can be removed and replaced without damage to the system, yet maintain integrity of the system.

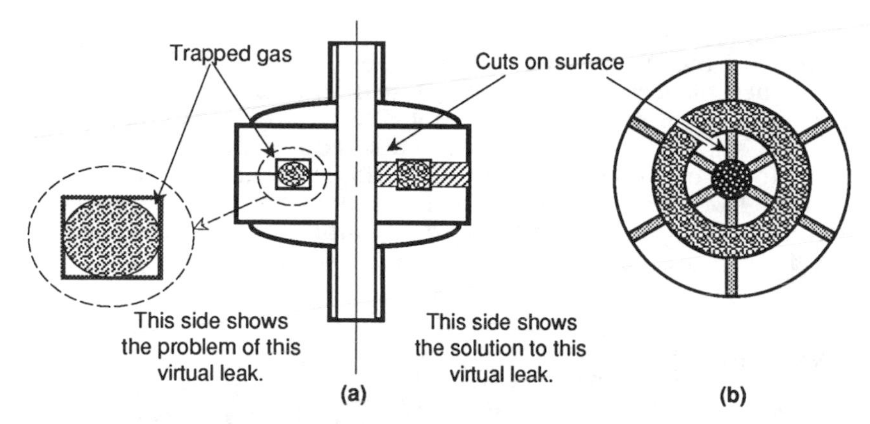

Fig. 7-48 An O-ring caused virtual leak.

If frozen gases within a cold trap are too close to the top, they are likely to evaporate as the liquid nitrogen level in the trap drops. This situation will cause the pressure in the system to rise and/or fluctuate (see Sec. 7.4.3). Different problems are caused when the wrong coolant is used for trapping such as when dry ice, rather than liquid nitrogen, is used as the coolant for a cold trap. If water vapor is in such a trap, the lowest possible achievable pressure would be 10^{-3} torr. This level is equal to the vapor pressure of water at the freezing temperature of carbon dioxide. If liquid nitrogen is used as the coolant, the system could potentially achieve 10^{-15} torr,[59] the vapor pressure of water at liquid nitrogen temperatures.

Although outgassing is one type of virtual leak, not all virtual leaks are results of outgassing. One example of a mechanically-caused virtual leak is the space trapped in the channels of an O-ring joint by a compressed O-ring. As you can see on the left half of Fig 7-48(a), an O-ring is compressed in the O-ring groove. In the blow-up seen at the left of Fig. 7-48(a), you can clearly see the areas of trapped gas in the corners of the groove. Because these areas are at a higher pressure than the system, they will 'leak' into the vacuum. However, because they are contained within the system, it is impossible to find the leak from outside the system. To prevent this type of virtual leak, the groove that supports an O-ring should be made without sharp corners. In addition, radial 'pie cuts' can be made which allow gas passage. These cuts can be seen on the right half of Fig. 7-48(a) and on the top view of an O-ring joint seen in Fig. 7-48(b).

A second example of a virtual leak occurs in metal vacuum systems by a double weld being made instead of a single weld (see Fig. 7-49). Because only one weld should be required, a second weld can only lead to possible problems. The statement, "If one is good, two is better" does not apply to this case. Incidentally, all welds should be made on the vacuum side of any given chamber. When a weld is on the outside, channels are created on the vacuum side. These channels can collect contamination that may be difficult or impossible to remove.

Demountable seals are sections of a vacuum system that are not permanently mounted. These seals include sections or pieces that can be moved or rotated such as traps, stopcocks, or valves. They differ from permanent seals, which are pieces of a vacuum system that are welded, or fused, together and cannot be separated or rotated without damaging the pieces or the system as a whole. Demountable seals require some material between the sections to be compressed, such as O-rings, grease, or gaskets that can prevent gases from passing. If there is inadequate or improperly aligned compression or cutting, or inadequate or improperly placed (or distorted) material to be compressed, or cut into, there will be a leak.

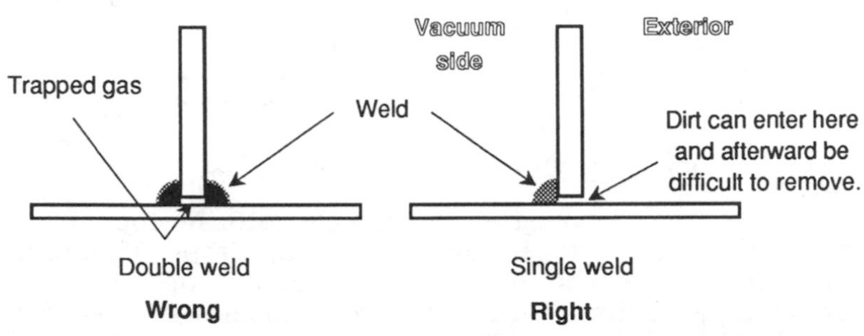

Fig.7-49 Improper and proper welds for a metal vacuum system.

Leaks at Demountable Seals. These leaks are caused by wear, old stopcock grease, poorly-applied grease, poor alignment, twisted or worn O-rings, inadequate stopcocks or joints, mismatched plugs and barrels, or dirt between pieces. This list is by no means complete, but it gives you an idea of the range of problems.

Among the items that may be initially checked on a glass vacuum system are to ensure that all plug numbers of glass vacuum stopcock plugs match their respective barrel numbers. As mentioned in Sec. 3.3.3, these numbers **must** match.* Another factor to consider with vacuum stopcocks is whether they are old, worn, or just defective. Additionally, you should examine that the vacuum stopcocks have been properly and recently greased. I have seen systems where the stopcocks looked perfectly good, but they had not been used in over a year. The user could not attain a vacuum better than 10^{-2} torr. Simply by re-greasing all the stopcocks, the system went down to 10^{-5} torr within half an hour of restarting.**

One specific example of how leaks can originate in a demountable seal is the example of a worn standard taper joint being mated with a new joint. This situation can occur when an old cold trap bottom has been broken. The

* I have seen a system with two plugs and barrels mis-matched that could not achieve a one torr vacuum. Once the plugs and barrels were properly matched, the system went to 10^{-3} torr in five minutes.

** In this situation, the poorly-greased stopcocks could not provide adequate seals for a vacuum, but were able to prevent atmospheric moisture from contaminating the walls of the system. Thus, the system could achieve a reasonable vacuum in a short time.

*** The Kimwipe is not used to prevent your fingers from getting dirty, although that benefit will also be realized,. Rather, it is to prevent your skin oils from contacting the vacuum equipment. Skin oils can create virtual leaks from their outgassing.

new replacement does not fit into the worn areas of the old, original member. If the worn areas (caused by the original members rubbing against each other for years) form a channel, a leak is created. In addition, standard taper joints and glass stopcocks can also be damaged by being left in HF, a base solution, or a base bath for an extended period of time. Unfortunately, the application of excess stopcock grease to cover up these imperfections is likely to create virtual leaks.

The advantage of O-ring joints is their easy demountability. However, if they are mounted unevenly, or torqued unevenly, a leak can be easily created. If a solvent is left in contact with an O-ring for an extended period of time, anything from swelling to destruction of the O-ring can result. This situation, of course, is dependent on the type of solvent and type of O-ring. One way to find out if there will be a problem with an O-ring and particular solvent is to pour some of the solvent in question into a small jar and drop in the type of O-ring you plan to use. Let it sit for several hours. If there are any effects, they should be obvious. A trick that can be used to provide limited protection from solvents is to apply a small dab of stopcock grease (select one not affected by that solvent) on an O-ring before assembly. Place a small amount of the protective grease on a Kimwipe and, while holding the O-ring in another Kimwipe, smear a thin film of grease on the O-ring.*** The grease will provide a protective barrier for the O-ring.

Ball-and-socket joints should not be used for high-vacuum systems as they are not intended for vacuum work. It is possible to obtain ball joints with O-rings that are acceptable for some vacuum work, and some manufacturers can supply sockets that have not been ground when used in tandem with ball joints with O-rings can achieve satisfactory vacuum performance.

Metal flanges require a gasket between their two sections. The bolts of metal flat flanges must be equally torqued by first tightening all bolts finger-tight. Then, using a wrench, you must tighten the bolts with less than a quarter turn each following the pattern shown in Fig. 7-50. When you arrive at the sixth bolt, repeat the same star pattern, but go to bolts 2, 5, 1, 4, 6, and 3 in turn.

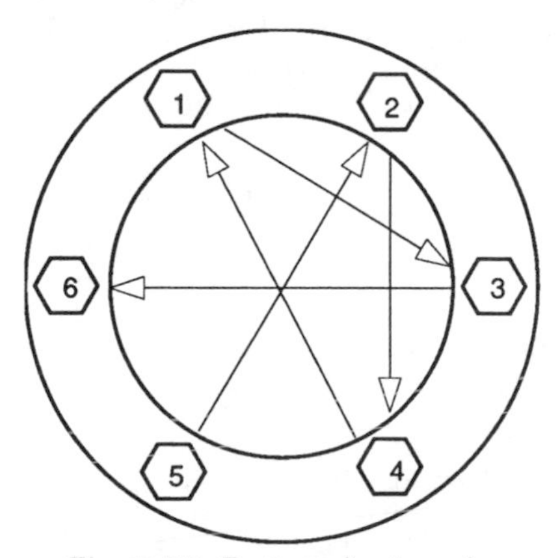

Fig. 7-50 Pattern for torquing bolts on metal flanges.

Some flanges will have eight or more bolts. Use the same alternating pattern for making a proper seal. Continue with this rotational pattern until the bolts are tight. If a metal flange has a knife-edge that cuts into a gasket, the gasket cannot be reused. No matter how well you try to line up the cuts with the knife-edge, it will not work, and it will leak. If you are desperate, you might try to have a machine shop remove the top and bottom surface of the gasket. However, experience has shown that this attempt is less than 50% successful for being leak-tight.[60]

Avoid using pipe threads whenever possible because they create long leak paths. Long thin leak paths require long time periods for leak verifications. If pipe threads must be used, do not use Teflon tape because it voraciously absorbs helium and will later thwart the use of a helium leak detector. Rather, use a small amount of thread sealant paste that contains Teflon (see Fig. 7-51). Glyptol, a common temporary leak sealant, is sometimes used as a pipe sealant. This sealant is not recommended because it ages and, after some six months or so, can develop a leak.

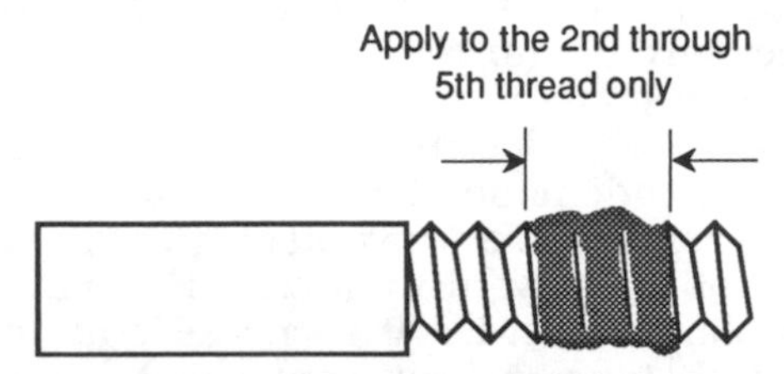

Fig. 7-51 Applying sealant to a pipe thread.‡

‡ From *Introduction to Helium Mass Spectrometer Leak Detection* by Varian Associates, Inc. © 1980 by Varian Associates, Inc., Palo Alto. With permission.

Thermocouples often come with pipe threads for attaching onto vacuum systems. Avoid using these pipe joints entirely by attaching a thermocouple to a glass system by silver soldering* the thermocouple to a pre-manufactured glass-to-metal seal.

Leaks in System Walls. *Real leaks* in the walls of your system can occur along three different passageways:

1) A hole through the wall. A hole can be through a thick or thin wall and follow a straight or circuitous path. It can be represented as a *door* (large diameter compared to length of hole) or a *tunnel* (small diameter compared to length of hole).
2) A crack in the wall. This type of hole differs by having a much greater surface area of the hole channel. It can also show up as crazing of the surface (crazing is a collection of many small cracks).
3) The molecular structure of the wall allows the permeation of gas(es) through the wall itself.

The significance of these three pathways is how they affect leak detection. The longer and thinner a hole is, the more likely it will be plugged up by a probe liquid, or even fingerprint oils. In addition, long, thin holes require lengthy leak testing. Cracks, with their greater surface area, will retain a probe gas or liquid for a longer time. If a gas is affecting your work by permeating the walls of your vacuum system, you need to be more selective of the construction materials of your system.

* Use a cadmium-free silver solder and a suitable silver-solder flux. You will need a gas–oxygen torch to make this seal (see Chapter 8 for information on gas–oxygen torches). Do not use a 'hissy' torch flame as that indicates a flame too rich in oxygen. A soft blue flame is reducing and will prevent oxidation of the metal pieces, which prevents a good seal.

It will be easier to assemble these pieces if you can set up the Thermocouple and the glass-to-metal seal so that they rotate horizontally during the silver-soldering operation.

Permeation of the walls itself is generally not a concern of a vacuum system within a research lab because a lab system can seldom achieve vacuums that are affected by gas permeation. Regardless, it can be resolved by selecting wall material better suited to the desired work.

The size of a leak is measured in the amount of gas (mass) that can be leaked within (per) a given unit of time (time). In the U. S., the standard measurement is std cc/sec, or standard cubic centimeters per second. The term atm (atmospheric) std cc/sec provides the ambient pressure. This level can be important if a specific leak rate is relevant to a particular gas(es), at a particular pressure.

Most leaks have a viscous flow of gas. That is, the length of the gas's mean free path is such that a gas molecule is more likely to hit a wall before hitting another molecule. On a bell-shaped distribution curve of average leaks found, this size leak is between 10^{-5} to 10^{-3} std cc/sec. The next most likely type leak is one in the transitional flow range of gas movement. Here, the molecules of gas are about equally likely to hit a wall as another molecule. This size leak is between 10^{-6} to 10^{-5} std cc/sec. The least likely type of leak found is when the size of leak is so small that a sufficient vacuum is achieved to cause molecular flow in the region of the leak. This size leak is below 10^{-7} std cc/sec.[61]

Leaks up to the size of 10^{-5} std cc/sec are not difficult to find in a laboratory vacuum system achieving a 10^{-6} torr vacuum. Below that range, baking the system may be required to fully open the leak from water vapor, fingerprints, or a host of other possible contaminants. Leaks smaller than 10^{-5} std cc/sec may not affect your system, but the amount of time spent in looking for these small leaks needs to be justified.

In vacuum systems, because measurements are usually made in liters/sec and pressure is stated in torr or Pascal, leak rates are known as torr-liters/sec, or Pa liters/sec. These expressions are not equivalent leak rates, but they are proportional in the following ratios:

1 std cc/sec = 0.76 torr-liters/sec = 101.13 Pa liters/sec
1 torr-liters/sec = 1.3 std cc/sec = 133.3 Pa liters/sec
1 Pa liters/sec = 0.0075 torr-liters/sec = 0.0098 std cc/sec

Backstreaming. *Backstreaming from pumps* is caused by poor baffles, traps, or poor vacuum system design. Once in molecular flow, gas molecules are as likely to drift 'upstream' as 'downstream.' Although a system typically will not provide leak symptoms until it is in the ultra-vacuum range ($>10^{-8}$ torr), at lower vacuum ranges ($<10^{-6}$ torr), other complications from backstreaming can occur such as:

1) Silicon diffusion pump oil that drifts onto the main filament of a mass spectrometer. This oil can coat and damage the filament.
2) Mechanical pump oils (with their high vapor pressure) that drift into the diffusion pump and decrease its pumping ability.

The simple resolution for all backstreaming complications is proper traps and maintaining the traps properly.

7.6.4 Isolation to Find Leaks

When a vacuum system is suspected of having a leak, one of the first tests is to determine whether the leak may be caused by outgassing. An easy way to determine this is to chart the rate of pressure loss versus time. To chart this rate, obtain the lowest vacuum you can in a reasonable amount of time, then close the section in question* from the pumping section by a stopcock or valve. Next, periodically over a few minutes, or an hour or two (or three), note the pressure and elapsed time. As seen in Fig. 7-52, a real leak will indicate a constant rate of pressure rise over time while an outgassing problem will indicate a decreasing rate of pressure rise over time.

Once you have determined that you have a leak (as opposed to outgassing), you must then decide if the leak is virtual or real. If you know your system (**and its history**) you should be able to review your own operations, procedures, and activities to make this determination. On the other hand, if this system is new (to you), or there are a variety of people who work on the same system, then you may have to assume that there is a real leak, and prove that it does or does not exist. Once you have proved that there is no real leak, and all other indications lead you to believe that a leak exists, you can assume that you have a virtual leak.

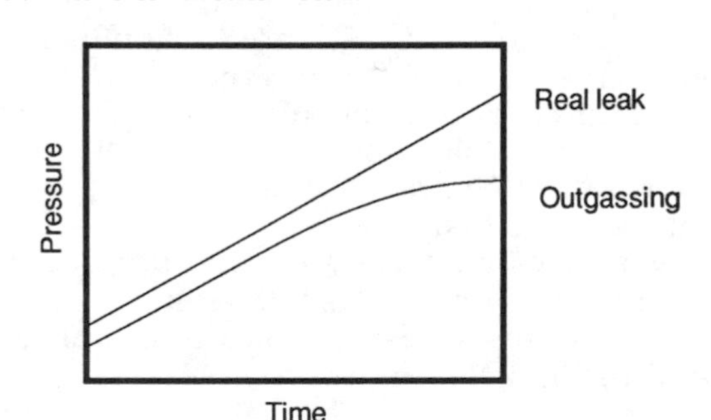

Fig. 7-52 A chart of a real leak versus outgassing as a matter of time.

Regardless of whether a leak is real or virtual, the first step after a leak is verified, is to isolate the area or section of the vacuum system that is leaking. This isolation decreases the time involved in locating the specific leak location. To find a leak's location, first examine major sections followed by smaller and smaller sections until you have the section in question. A stylized version of a vacuum system is shown in Fig. 7-53.

One of the best vacuum gauges to use when looking for a leak is a thermocouple or Pirani gauge. McLeod gauges, although very accurate, take too long to cycle through a reading. The ion gauges can only be used at *good* to *high* vacuums, which make their use irrelevant at the low vacuum levels that exist with large leaks.

In the stylized vacuum system shown in Fig. 7-51, you would go through the following process to locate a leak:

1) Close Stopcocks 5 and 2, and open Stopcocks 1, 3, and 4

* A thermocouple gauge, Pirani gauge, or manometer must be attached to the section being tested.

2) Once a vacuum is showing on the vacuum gauge, close Stopcock 4, and verify that the gauge is not leaking (this procedure may seem silly, but it caught me off-guard once). If the gauge leaks, you will be unable to check the rest of the system with any validity (that is why this step is important). If there is no leak, open Stopcock 4 and leave it open.

3) Close Stopcock 1. If the vacuum gauge shows a drop in the vacuum, the leak is somewhere within this area. If there is no leak, continue.

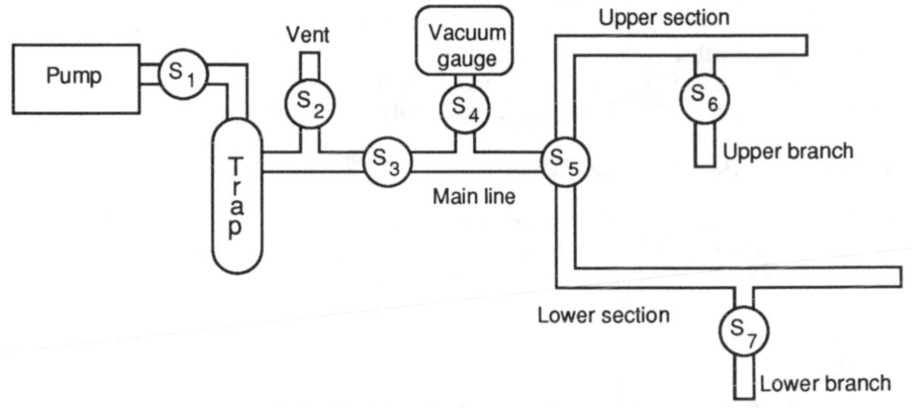

Fig 7-53 A stylized vacuum system.

4) Open Stopcock 1 and then rotate Stopcock 5 so that the upper section is connected to the vacuum. Stopcock 6 should be closed. Wait until a vacuum is obtained and close Stopcock 1. If there is no leak, continue.

5) Open Stopcock 1 and then rotate Stopcock 6 so that the upper branch is connected to the vacuum. Wait until a vacuum is obtained and close Stopcock 1. If there is no leak, continue.

6) Open Stopcock 1 and then rotate Stopcock 5 so that the lower section is connected to the vacuum. Stopcock 7 should be closed. Wait until a vacuum is obtained and close Stopcock 1. If there is a leak, you must locate and repair the leak before you can continue. It is important to continue, because there may still be a leak in the lower branch.

7) Open Stopcock 1 and rotate Stopcock 7 so that the lower branch is connected to the vacuum. Wait until a vacuum is obtained and close Stopcock 1. If there is no leak, go back to work, the system is checked and repaired.

As you can see, this process is aided or thwarted by vacuum system design. Temporary isolation of the various parts of any vacuum system may be impossible by bad design. Therefore, be forewarned: if you are designing a vacuum system, provide a stopcock at every branch* of the system to aid in

* On the other hand, prevent unnecessary constrictions because they can limit gas flow and thereby slow system pumping speed. Therefore, do not use small stopcocks (such as 2 or 4 mm) along main lines, and do not place extra stopcocks along a tube for no purpose.

leak detection (section isolation not only helps in leak detection, but creates a more robust vacuum system allowing greater control and protection).

Once you have isolated the part of the system in question, examine, regrease or tighten (if necessary) the stopcocks, joints, and other connections. In a glass system, look for cracks that may have developed on the tubing. You should look especially where the tubing has been worked, such as places where the glass is joined to another piece, or has been bent.

In searching for leaks, you must always look first for larger leaks followed by progressively smaller leaks. It makes sense: If you have some very large leaks (sometimes called "hissers") there is no way a small leak will show up. However, if you have found and repaired some large leaks, do not assume that you are finished. You still may have leaks that can affect your work. Therefore, after repairing a leak, recheck your system for leaks. Just because you found one leak does not mean you've found them all.

7.6.5 Probe Gases and Liquids

Probe gases and liquids are often used in a variety of leak detection techniques. Probe gases and liquids are materials not normally found within the vacuum system, or at least not in the quantity created when they enter through a leak. Turnbull[62] defines four characteristics of the probe gas, vacuum system, and leak detector that affect the speed and effectiveness of leak detection:

1) The viscosity of the search gas, which governs the rate at which gas enters the leak
2) The speed at which the search gas is removed from the system by the pumping system
3) The sensitivity of the leak-detecting element to the particular search gas used
4) The volume of the system.

Although each of these factors can be considered individually, their effects on the speed of effective leak detection are cumulative. Materials that are less viscous will enter a given leak faster than those with greater viscosity. Materials that can be removed from the system faster will allow for faster verification. Materials that are easy for the detector to notice require less hesitation during detection. Finally, the smaller a system is, the less time that is needed for the probe gas to fill all areas.

When selecting probe gases (or liquids) and techniques to locate leaks, consider how they may affect your leak detection. Probe liquids are easy to see, handle, and can be used with greater control, thereby providing an accuracy that is typically unobtainable with gases. Fiszdon[63] analyzed twelve different liquids that are commonly used in leak detection and developed the following list (in Table 7-15) of their merits for leak detection analysis. Note that the vapor pressures of the various liquids are irrelevant. Rather, low molecular weight and low viscosity are more important.

Probe liquids have their own peculiar problems. For example, although water will pass right through a large hole, it may effectively plug up a small, thin tunnel, giving the illusion that the leak is gone. In addition, a liquid may fill the entire surface of a crack causing a very slow removal, or 'clean-up,' of the indicator. Baking a system can reopen blocked holes and facilitate clean-up, but not all systems can be baked.

Table 7-15[64]

Liquids Used in Leak Detection Rated Best (top) to Worst (bottom)
acetone
ether (diethyl)
methanol
pentane
benzene
toluene
ethyl alcohol
carbon tetrachloride
xylene
isopropyl alcohol
butyl alcohol

On the other hand, gases require special, cumbersome handling and often require you to enclose a given section of a vacuum system in some sort of bag. Bagging can facilitate localizing the *area* of a leak, but cannot help in locating the *exact location* of the leak.

Do not use liquids for leak detection if you are considering using in a mass spectrometer further on in your experimentation. Liquids tend to have slow clean-up times and can severely slow down, or confound, future experimentation. Thus some rules for the use of probe gases and liquids are:

1) When possible, use a gas over a liquid.
2) Do not spray or squirt a liquid on. Use a cotton swab (or Kimwipe®) to wipe it across parts of the system. In addition to safety, this method provides more control in finding leaks.
3) Solvents can damage parts of your system such as O-rings, causing a greater leak than the one you were originally trying to find. In addition, if the solvent drips on a water hose and creates a hole, you then have a water leak and a mess. If it drips on electric wires, you may have a short.
4) All of the chemicals listed in Table 7-15 are dangerous to breathe or have in contact with skin from short to extended periods of time. Some of the chemicals listed have severe OSHA restrictions for use without a fume hood, and some are toxic chemicals and should not be used at all (such as carbon tetrachloride and xylene).
5) Do not have any open flames or sparks while working on leak detection. Unplug ion gauges and other electrical equipment.
6) Be sure to have plenty of ventilation and limit yourself to the first three liquids in Table 7-15. These liquids are (relatively) the safest, and they are the best from the list. Still, you should use a buddy system, and check with OSHA and/or local/state regulations to verify if there are any legal restrictions on the uses of any of these chemicals in your area.

7) When possible, select a low vapor pressure liquid over a high vapor pressure liquid to provide faster clean-up time. In addition, non-polar solvents are more easily removed from glassware than polar solvents (for example, methanol vs. acetone).
8) When spraying probe gases on a vacuum system, be sure to start at the top with gases less dense than air and start at the bottom with gases denser than air.

Other aspects of Turnbull's four factors will be considered further in Sec. 7.6.8.

7.6.6 The Tesla Coil

If you have a glass system, and can achieve a vacuum between approximately 10 to approximately 10^{-3} torr, then you can use a Tesla coil (sometimes called a 'sparker') to look for moderate-size leaks. Because this range is the vacuum range of a mechanical vacuum pump, the Tesla coil provides an excellent tool for examining such systems.

The Tesla coil will ionize the gas molecules remaining in a vacuum system and cause them to glow. Above pressures of about 10 torr the gas molecules quench a discharge. That is, they are so close together they lose their extra energy by bumping into other molecules rather than giving off light. Below 10^{-3} torr, the molecules are too far apart and the mean free paths are too long to maintain a discharge.

As the tip of a Tesla coil is slowly waved within 1-3 cm of an evacuated glass vacuum system, the gases remaining will discharge with a glow characteristic of the gases within the system. If you bring the Tesla coil near a leak, a large white spark will jump from the tip of the coil to the specific leak spot. This event is very dramatic and demonstrates the location of a leak in a very effective manner. The spark is actually seeking ground, and the leak provides a path to the discharge inside the vacuum system. In turn, this discharge provides a path to the mechanical pump for completion of the ground.[65]

Metal components confound Tesla coil use: their ground is easier to obtain than the ground found by passing through a glass leak and the poorer conducting discharge within.

There are a few limitations to the use of a Tesla coil:

1) It cannot be used near metal clamps or glass-to-metal seals on a glass system. The metal provides a ground for the electric discharge, by-passing the ionization of the gas inside the system.
2) A large quantity of very small holes may prevent you from obtaining a decent vacuum. However, none of the holes may be large enough for the Tesla coil to indicate a leak.
3) The self-contained Tesla coils often found in many labs are recommended by the manufacturers not to be used for longer than ten minutes of continuous operation. If your needs require long continuous use on a consistent basis, heavy-duty Tesla coils are available that can be used continuously. These coils are easily identified because they have small boxes

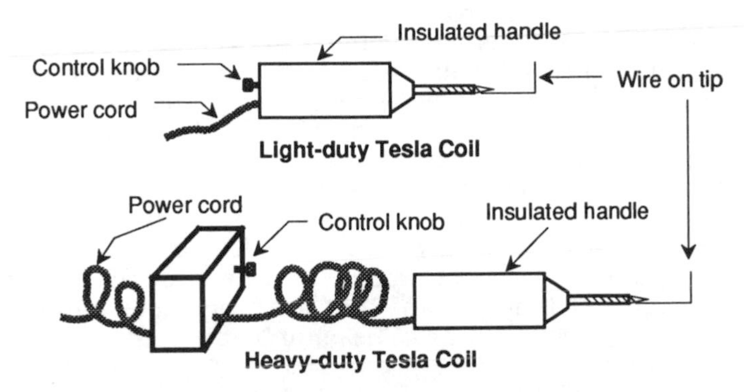

Fig.7-54 The light- and heavy-duty Tesla coil.

(approximately 5 in x 5 in x 8 in) connected to their hand-held sections. These hand-held sections are the same as those found on standard light-duty Tesla coils (see Fig. 7-54).

4) The Tesla coil will not find a leak within a demountable (seals such as stopcocks or joints) that is caused by poor application of grease or old stopcock grease that has sheared.
5) The Tesla coil should not be used near O-ring joints because the coil can destroy the O-ring by burning a strip across its side.
6) The Tesla coil should not be used near intentionally thin sections of glass (such as a break-off) because it can punch a hole through such a section.

It is unlikely that a leak could develop on a glass tube that has not received stress, or has not been worked on by a glassblower. Therefore to save time, simply pass the Tesla coil around areas where glass sections have been joined (see Fig 7-55).* If there is a two- (or three-) fingered clamp in an area that needs to be tested, the clamp must be removed to properly check the seal.

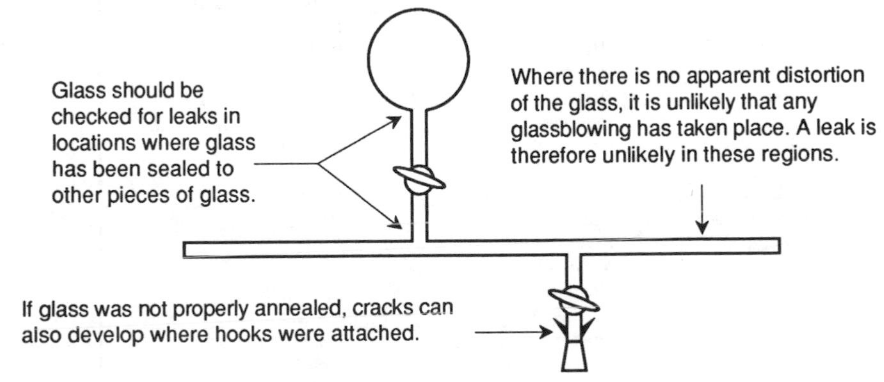

Fig. 7-55 Suggested areas of testing with a Tesla coil.

* An exception to this rule would be if a two- (or three-) fingered support clamp was overly tightened on a glass tube, causing the tube's wall to crack.

The spark from a Tesla coil is very powerful and can punch its way through thin sections of glass (see Point 6 above). Keep the coil away from known thin sections in which you wish to maintain integrity, such as break-offs. Most texts on vacuum technique recommend that the Tesla coil not be allowed to sit over weak areas for fear of "punching" a hole through the glass. I disagree from this reticence because if there is not a hole in a potentially weak area now, there may very likely be one in the area some time in the future. Therefore, go ahead and provide some stress while you are in a position to do something about any holes that develop. Otherwise a hole may develop in the middle of an experiment, or when repair personnel are not available. In addition, It is important that all gauge controllers be turned off before initiating Tesla coil testing. The charge from the Tesla coil can destroy the controllers electronics.

Some extra Tesla coil tips:

1) Turn off some (or all) of the lights in a room so that any discharge from the coil can more easily be seen (it may also be necessary to close doors and/or cover windows). In addition, try not to face an open window.
2) A metal wire (such as copper or nichrome) wrapped around the tip of the Tesla coil (see Fig. 7-52), and bent in a right-angle, can extend your reach and/or reach behind glassware.

The Tesla coil can also be used to make the vacuum line a discharge tube to light up probe gases or liquids. Gases or vapors in a vacuum will light up in specific colors from a discharge caused by a Tesla coil. A list of the colors that can be achieved with *pure* gases in a discharge tube can be seen in Table 7-16.

Table 7-16

Appearance of Discharges in a Gas Discharge Tube at Low Pressures [66]

Gas	Negative Glow	Positive Column
Air	Blue	(Reddish)
Nitrogen	Blue	Yellow (red gold)
Oxygen	Yellowish white	Lemon
Hydrogen	Bluish pink (bright blue)	Pink (rose)
Helium	Pale green	Violet-red
Argon	Bluish	Deep red (violet)
Neon	Red-orange	Red-orange (Blood red)
Krypton	Green	—
Xenon	Bluish white	—
Carbon monoxide	Greenish white	(White)
Carbon dioxide	Blue	(White)
Methane	Reddish violet	—
Ammonia	Yellow-green	—
Chlorine	Greenish	Light green
Bromine	Yellowish green	Reddish
Iodine	Orange-yellow	Peach blossom
Sodium	Yellowish green (whitish)	Yellow
Potassium	Green	Green
Mercury	Green (goldish white)	Greenish blue (greenish)

— indicates no distinctive color.
( ) indicates a different observer's opinion of the color.

Table 7-16 (found in several books,[67, 68, 69] or found as a similar table in other books[70, 71] on vacuum technology) provides the colors of pure gases from discharge tubes—which have very little to do with the discharge from the Tesla coil because of the following conditions:

1) The applied voltage and pressure can vary greatly.
2) The Tesla coil discharge, in leak testing, is a through–glass discharge, whereas a discharge tube is a metal–electrode discharge.
3) The distance between electrodes in a discharge tube is fixed, as opposed to the distance between the Tesla coil and ground (this distance can vary constantly from as little as several centimeters to many meters during leak detection).
4) The colors indicated in the tables are for pure gases. In a working vacuum system there will always be some air and moisture mixed with the probe gases. There will also be a trace amount of hydrocarbon vapors (from the mechanical pump) as well as other gases and vapors that incidentally may be in the system.

Work by Coyne and Cobb[72] provided a more effective list of discharge colors because the colors were observed in the act of actual leak detection. They are presented in Table 7-17. Their work demonstrated that spraying gases had limited success in leak indication, but moderate success when the item being tested was enclosed within the gas (by way of a bag). Vapors from applied liquids provided better indicators and more easily demonstrated the specific location of a leak.

Table 7-17‡

Appearance of Discharges in Gases and Vapors at Low Pressures When Excited by a Tesla Coil

Gas	Color Observed in Discharge	Gas Mixed w/ Air
Air (room)	soft violet	(same)
Argon	pale-magenta	pale-magenta
CO_2	greenish white	washed out magenta
Helium	soft-pink	purplish-magenta
Nitrogen	(too pale for magenta, not quite purple)	purple
Oxygen	white	pale-magenta
Volatile Liquid	**Color Observed in Discharge**	**Gas Mixed w/ Air**
Acetone	turquoise (then quickly to purple)	void* (then quickly to purple)
Dichloro-methane	bluish-white	void* (then quickly to bluish white)
Methanol	void* (then slowly to soft violet)	void* (then slowly to soft violet)

* Void: a lack of discharge due to a loss of vacuum.

‡ From *J. of Chem. Ed.*, 68 (1991), pp. 526-28. With permission.

If you have items with glass-to-metal seals or metal supports you wish to leak-check with a Tesla coil, you can encase them in a bag. Initiate a discharge with the Tesla coil while filling the bag with a test gas, such as oxygen or helium. This procedure can help to verify whether or not there is a leak. However, it will not help locate the leak. To specifically locate the leak, you may wipe acetone (on a cotton swab) over the suspected areas while a discharge is maintained in the system. If the discharge changes color, you have found the leak. If leaks are located far apart on a vacuum system, it is possible to find a leak while looking for others. However, leaks that are close together may appear, or act like one. Therefore, after finding and repairing one leak(s), you should re-examine the area for other leaks.

When using this technique, turn off all lights and shut all windows (and doors) that allow light into the room. If your workplace is dark, more subtle variations in the discharge will be apparent. As a safety precaution, place a small table lamp near where you are working. Then, you will be less likely to bump into equipment as you make your way from the room's light switch to the testing area after turning the room lights off.

If you spray a probe gas lighter than air (such as helium) on a vacuum system, start at the top of the system. Start at the bottom with gases heavier than air. Otherwise, the drifting of the respective gases may provide false or inconclusive leak identification. It may be necessary to close windows, doors, and even to set up baffles to minimize drifting gases to prevent false or inconclusive leak identification.

Once a leak is verified and localized, pinpointing the leak can be done by taking an acetone-soaked cotton swab and rubbing it all around the section in question. Once the acetone is wiped over the leak, there will be an immediate color change of the discharge to white or turquoise.

The shape and angle of a leak can have a significant effect on the time before a vapor is removed from a system and the colors of a discharge turn back to normal. If the glass is very thin and the leak is perpendicular to the glass, the time for recovery can be very short (<10 seconds). On the other hand, if the glass is thick and/or the leak is on the diagonal through the glass wall, a leak can hold the acetone for a period of time, even many minutes. Thus, the length of time necessary for the colors to go back to normal provides a clue as to the nature of the leak.

7.6.7 Soap Bubbles

As stated before, the Tesla coil cannot be used on metal systems. If a leak on a metal system is large enough to prevent you from using a mass spectrometer (or you do not own a mass spectrometer), you may be able to use positive pressure to locate leaks in a vacuum system.* Place the vacuum system under pressure with dry air, nitrogen, or helium up to about 60 psig. Then squirt a soapy solution on areas in question while looking for the formation of bubbles. This technique is the same that is used on all pressure systems and is even used by plumbers when installing gas pipe.

* Positive pressure cannot be used on glass systems because the pressure can cause glass stopcock plugs to blow out and be damaged.

Some important rules are:

1) The gas used to pressurize a system should be clean and dry. Nitrogen and argon are very good because they are non-reactive gases. Do **not** use pure oxygen, especially if there are any greasy areas in the system as an explosion may result.
2) Any components that may be damaged by high pressure (i.e., ion gauges) should not be exposed to any pressure, or limit the pressure to 1 to 2 psig.
3) Watch out for fittings not intended to be used in pressure conditions.
4) Use a very bubbly solution (dish soap in water is good). Professionally-made solutions such as Snoop™ or kid's bubble-blowing solutions work as well.* Window-cleaning solutions are not recommended because they typically have non-foaming additives.
5) Apply the solution slowly and carefully so as not to create bubbles.
6) Do not immediately assume that there is no leak if no bubbles readily form; they may take a few seconds to an hour to form.
7) Use a waterproof marking pen or crayon to mark each leak as you proceed.

7.6.8 Pirani or Thermocouple Gauges

As mentioned in Sec. 7.5.14, many gauges read an *inferred* pressure, not *real* pressure. Some vacuum gauges use the thermal conductivity of gases present in the system to infer the pressure of the system. These gauges are based on the concept that 'less' gas will conduct 'less' heat. Because different gases have different thermal conductivities, the user needs to make allowances if the gas in a system has a different thermal conductivity than the gas a gauge has been calibrated to use.

The standard reading for a thermionic gauge is based on air. If a probe gas is used whose thermal conductivity is higher (or lower) than air, its presence would be detected by the Pirani or thermocouple gauge as a change of indicated pressure. As would be expected, light gases with low viscosity can find smaller leaks. Hydrogen and helium are such gases, and they show a leak by an apparent increase in pressure because they have higher thermal conductivities than air. Other gases, such as argon, show an apparent decrease in pressure because of their lower thermal conductivities.[73] The liquids in Table 7-15 can also be used with Pirani and/or thermocouple gauges because they can provide similar apparent changes in pressure.

Regardless of which direction the gauge varies, the important thing is that the gauge varies because the gauge is detecting a change from what was in the system before the probe gas (or liquid) was introduced. Sometimes there may be a slight drop in pressure caused by the liquid filling up a long thin 'tunnel' type of hole. This drop is then followed by a rise (or depression)** in pressure as the vapors of the liquid affect the thermocouple.[74] Do

* Soap solutions must be rinsed off of stainless steel because soap can corrode the metal.

** If the thermal conductivity of the probe gas (or liquid) is higher than nitrogen (the base standard) the gauge will indicate a drop in pressure. If the thermal conductivity is less than nitrogen, the gauge will indicate an increase in pressure.

not expect equal swings of low followed by high pressure, because the causes of pressure swings are not related.

Another leak detection system that uses thermal conductivity is a *gas leak detector* that samples the air around the system being checked by a simple pump (often a diaphragm pump). The system under examination is filled with any type of gas (except combustible gases). Then, changes in the gas composition surrounding the piece being tested are 'sniffed' by a probe. Any changes detected by a thermocouple within the detector are an indication of a leak and the user is notified by a swing of a dial's needle (one way or another) and/or an audio alarm. The sensitivity of these devices varies depending on the probe gas used. Helium provides the best sensitivity. They are capable of detecting leaks as low as 10^{-5} atm/sec, or, in other words, they can be used with simple vacuum systems.

One particular advantage with gas leak detectors is the constant distance from the leak to the thermal conductivity sensing device. This constant distance makes the time lag consistent wherever you are 'sniffing' with the probe. When relying on installed thermocouples (on the vacuum system), a leak located right near the thermocouple will provide an immediate and full-strength response, whereas a leak some distance from the thermocouple will have a more subtle effect and will require some time before this effect is observed.

As is standard when using probe gases on the inside for detection outside, the greater the pressure exerted within the system being tested, the greater the sensitivity that can be expected from the detection device. With glass systems, it is not recommended to use a pressure greater than can be developed by attaching a balloon filled with the probe gas to the system. It is possible to flow helium through the system before beginning the testing, but it then becomes necessary to wait until all the excess ambient helium in the room drifts off. Otherwise, there is too much helium 'noise' in the room to allow for any sensitive readings. The wait could be minutes to hours depending on the size of the room, ventilation, and sloppiness of administering the helium to the system.

Do not breathe too close to the sniffing probe or its reference chamber as the CO_2 from your breath could cause a deflection of the needle. If you are going to use a helium test gas after doing a bubble test, you must wait until the system is completely dry. The water will cause a negative deflection and the helium will cause a positive deflection, providing a combined weaker deflection. If, coincidentally, the water and helium are perfectly balanced, there will be no deflection.

Once a leak is indicated, move the detector away from the system being tested and allow the needle to come to a normal position. Then you can go back in and zero in on the specific location.

7.6.9 Helium Leak Detection

Imagine a room full of people. Now imagine you want to find the one(s) named Bill. If there was no way to easily identify the Bills, the process would become significantly laborious. However, if you called out and asked all those named Bill to please raise their hands, the task of identification becomes easy. The helium leak detector has the ability to isolate, and count, the helium atoms from a vacuum system full of many other atoms and molecules. In reality, the helium leak detector is nothing more than a highly specialized mass spectrometer. The mass spectrometer was developed in the 1920's, but became a leak detection tool in the 1940's when the need for ultra-vacuum systems became critical on the Manhattan project.*

The operation of a mass spectrometer is fairly straightforward. It places an electrical charge on all the gases passing through its system. This electrical charge allows a flowing path of charged gas molecules to be bent by a magnetic field. The angle of the bend depends on the mass of the molecule** and can be calculated with high precision. Because the particles have an electrical charge, they can be collected on a grid and counted. If a collection grid is placed at a specific point of the possible radius path, it is possible to collect only one type of molecule. A helium leak detector is a mass spectrometer that is permanently tuned to collect and record only the helium ion.

Without the magnet, the grid would collect all charged particles and the total gas pressure could be recorded. With the magnet separating the charged particles by their molecular weights, only helium is detected. This quality allows the helium leak detector a sensitivity of leak detection that is independent of total gas pressure. The leak detector will indicate that it is sensing helium by swinging a needle on a meter, sounding an electronic signal, and/or illuminating some type of display. For quantitative work, a digital or analog readout is required.

When helium leak detectors first became available, they were large, cumbersome, and required highly-trained technicians. They were also very expensive. Now, like most electronic equipment, they are smaller, (relatively) easier to use (for some operations, they can be automated), less expensive, more reliable, more sensitive, and faster. As is typical of most things, helium leak detectors cannot have all the aforementioned attributes at once. However, it is possible to pick and choose which options are more important to you and select a detector that has features that are best suited to your needs. The usual trade-off is sensitivity for speed.

Although there are other leak detectors and leak detection techniques that rely on detection of a change in gas, none have enjoyed the success of the mass spectrometer tuned to helium. There are other types of tuned mass spectrometers that specifically look for oxygen or halogen, but they are not as common. Their operation is similar to the helium leak detector, so the information here will also apply to them.

The use of helium as a probe (or tracer) gas in leak detection is not new. Before the mass spectrometer, it was used with thermocouple and/or Pirani gauges because of the greater thermal conductivity of helium than air. If you list all the attributes of a perfect probe gas, helium obviously does the job:

1) It should be non-toxic
2) It should diffuse readily through minute leaks

* An interesting article on the development of the helium leak detector can be found in the article "History of Helium Leak Detection" by Albert Nerken in the Journal of Vacuum Science, and Technology A, Volume 9, #3, May/June 1991, pg. 2036-38.

** The larger the mass, the larger the radius; the smaller the mass, the more the magnetic field can bend the path and the smaller the radius.

3) It should be inert
4) It should be present in not more than trace quantities in the atmosphere
5) It should be relatively inexpensive.‡

‡ From *Introduction to Helium Mass Spectrometer Leak Detection* by Varian Associates, Inc. ©1980, p. 21. With permission.

Helium is *non-toxic* and will not poison or affect living tissue. That is why it is safe to inhale helium and pretend you're a comedian.* The only danger that helium can present is that it can fill up a room, displacing oxygen so that there is an insufficient amount for human survival. However, because helium floats up and out of any room, this situation is almost impossible.

The evidence of *helium diffusion* through small leaks can be demonstrated by its ability to pass through glass. Everyone has seen how helium leaks out of a rubber balloon. Depending on the quality of the rubber, it can take anywhere from a few, to several, hours of time to render a balloon into pathetic lethargy. The transfer of space-age plastics (Mylar) into balloon construction has added days to a toy balloon's life because Mylar's molecules are more dense and it is more difficult for the helium atoms to diffuse past them. Borosilicate glass is more dense than rubber or Mylar and therefore does not leak as fast as either of them, but it leaks nonetheless. This fact has been successfully used for construction of 'standard leaks' which are used to calibrate helium leak detectors. There is a family of glass called aluminosilicates, one of which (Corning's 1720) is helium leak-tight. Only hydrogen (mass 1) has a smaller mass than helium (mass 4). Hydrogen can diffuse through materials very well, but it is dangerously flammable. Neon (mass 20) could be used, but it is much more expensive to use than helium. Argon (mass 40) is larger in size than the two main gases in the atmosphere (nitrogen: mass 28 and oxygen: mass 32) and therefore would not be a good indicator of leakage potential of smaller materials.

Being one of the noble gases, *helium does not react* with other materials. Its presence will not affect any of the materials of the vacuum system nor will its presence affect any future work within the system. In addition, because it is a non-condensable gas, it will not clog up any cold traps within the system or the detector during leak analysis. Helium, like other noble gases, is completely nonflammable.

Helium is found in extremely small quantities (approximately five parts per million) *in the atmosphere.* Any discovery of concentrations greater than normal are easily discernible. Imagine the challenge of someone looking for water leaking out of a vessel held underwater.

*Helium is relatively inexpensive.*** Neon (mass 20), which could pass many of the qualifying tests for leak detection, fails here because of its greater costs.

* It is important for the comedian to occasionally take a breath of air to replenish some oxygen in his lungs or the joke will be on the comedian as he or she passes out.

** This level of leak detection is not inexpensive: A new mass spectrometer helium leak detector with the accompanying equipment can easily cost between $20,000 to $30,000. Purchasing used equipment can significantly reduce the initial costs, but one should not enter the level of helium leak detection because it *seems* like a good idea. On the other hand, if you *need* a helium leak detector, you cannot afford not to have one.

Finally, it should be pointed out that *helium leaks better than air* because it is a much smaller molecule than air. The kinetic theory of gases predicts that helium will flow 2.7 times faster than air through an ideal leak at molecular flow conditions.[75] Because of this property, helium can indicate a leakage rate that is greater than would be encountered in practice. For instance, because helium can permeate through O-rings, a helium leak detector could indicate a leak (albeit very small) at an O-ring joint despite the fact that oxygen or nitrogen are not likely to leak past the same O-ring.

7.6.10 Helium Leak Detection Techniques

There are four different approaches to helium leak detection. They are displayed in Table 7-18. A graphical representation of these techniques is shown in Figs. 7-56, 7-57, and 7-58(a) and 7-58(b).

Table 7-18

Approaches to Helium Leak Detection

Leak Detection Technique	What it can do	What it cannot do
Detector-probe technique	It can find the location of a leak when the tracer gas is on the inside of the piece in question. This technique is time consuming. Because the probe is collecting both air and helium, it is about 10% as sensitive as the tracer-probe technique.	It cannot tell you the size of the leak.
Tracer-probe technique	It can find the location of a leak when the tracer gas is sprayed on the exterior of the piece in question. This technique is time-consuming.	It cannot tell you the size of the leak.
Inside-out technique	It can verify if there is a leak, and quantify the size of a leak. The tracer gas must be on the inside of the piece in question. It provides a fast go/no-go operation.	It cannot tell you where the leak is.
Outside-in technique	It can verify if there is a leak and quantify the size of a leak. The tracer gas must be on the outside of the piece in question. It can provide a fast go/no-go operation.	It cannot tell you where the leak is.

Some of the other leak detection methods explained earlier in this section are based on these techniques. Methods such as bagging parts of a vacuum system and filling the bagged parts with a probe gas are similar to the outside-in technique. However, it is not possible to quantify a leak with this approach. Alternatively, filling a container with gas and either submerging it in a liquid or covering it with a bubbling solution is similar to the detector–probe technique.

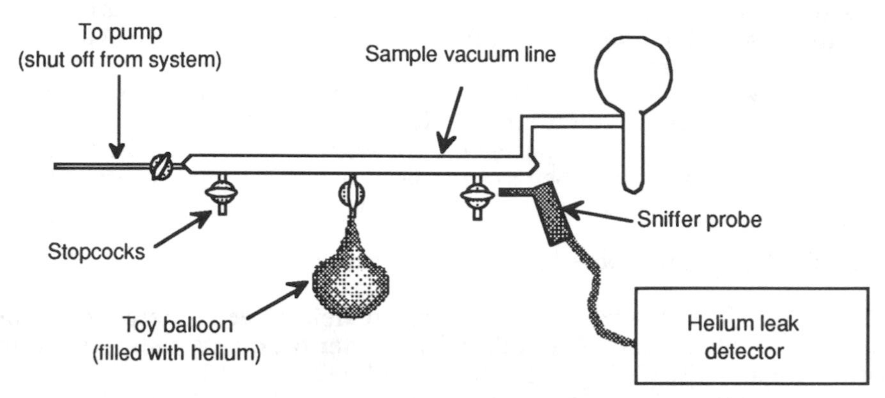

Fig. 7-56 In the *detector-probe technique*, the tested piece is filled with helium and the sniffer probe 'sniffs' the areas in question to detect leaks. Always 'sniff' from bottom to top.

What makes the helium leak detector unique is that it can be calibrated against a *standard leak*. Then, any measurements made (using the inside-out or outside-in techniques) can provide a quantifiable number. This calibration feature is usually more important to the manufacturing industry than to the needs and demands of the laboratory. For example, when the top of a soda pop can is scored to provide its 'pop-top' feature, the depth of the scoring is very important: if the scoring is not deep enough the top will not work as advertised; if it is too deep, the gases within the can can diffuse out, leaving a flat, non-bubbly product. So, by periodically checking to see that the amount of helium diffusion past the scored top is within tolerable limits, the canning industry can ensure that a good product will be delivered to cus-

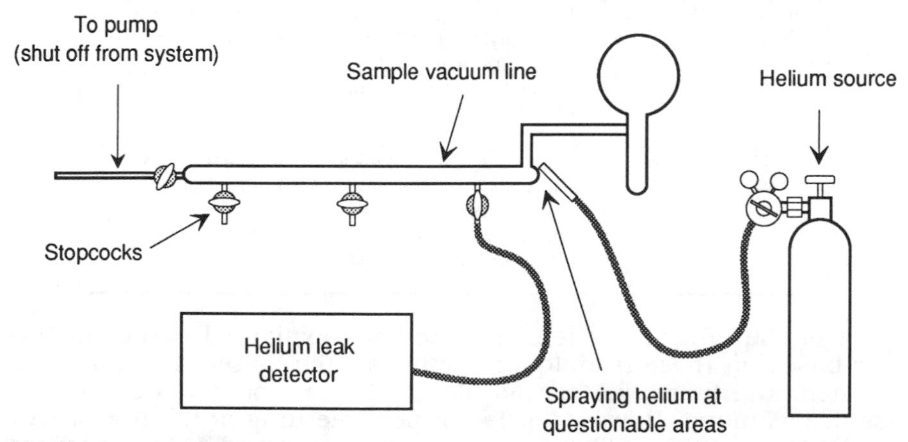

Fig. 7-57 In the *tracer-probe technique*, the helium leak detector is connected to the vacuum system at a central location. Then helium is lightly sprayed on the areas in question. Always spray the helium from top to bottom.

tomers, and the customers will be able to obtain access to the beverage. The canning industry does not care *where* the leak is—it already knows. What it does not know is if the leak is within tolerable limits or not.

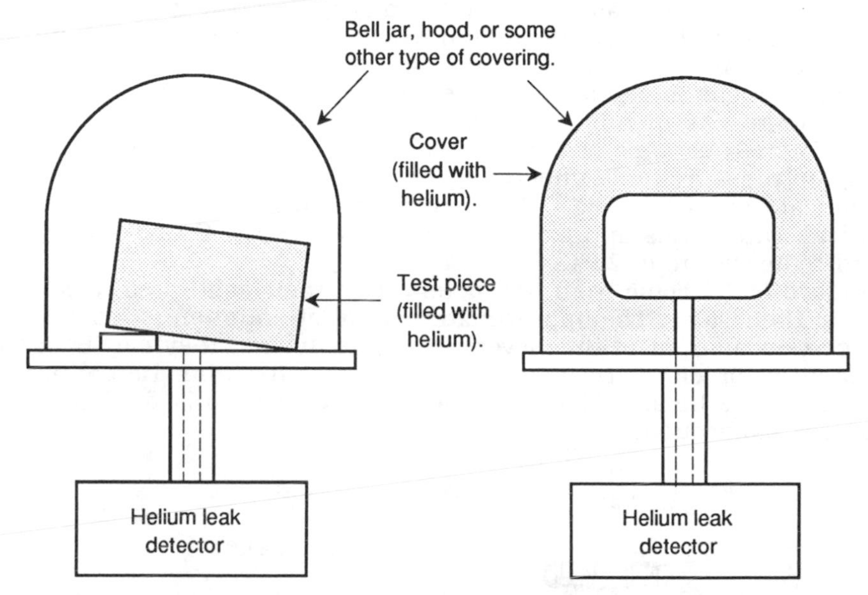

Fig. 7-58(a) Using the *inside-out technique*, the piece in question is filled with helium and then placed in a covering of some type that can be evacuated. Any helium that is released within this covering will be detected by the helium leak detector. Note that the tested piece can have either end lifted up so that all surfaces are exposed to the vacuum.‡

Fig. 7-58(b) With the *outside-in technique*, the piece in question is directly connected to the helium leak detector. A covering, or hood, is placed all around the tested piece and is filled with helium. Any helium escaping into the tested piece will be detected by the helium leak detector.‡

‡ From *Introduction to Helium Mass Spectrometer Leak Detection*, Figs. 3.3 and 3.4, by Varian Associates, Inc. (© 1980 by Varian Associates, Inc.). With permission.

When examining a vacuum system for leaks in the laboratory, calculating the size of a leak is often unnecessary for two reasons:

1) The techniques used to find the size of a leak entail the use of the *inside-out* or *outside-in* techniques, both of which do not allow indication of where the leak is. With a vacuum system, we are trying to find the location of a leak so that the leak can be eliminated.
2) Although on occasion it may be useful to know the relative size of one leak to another, the specific size of a leak is irrelevant. If the leak is large in relation to the *standard leak* it should be

repaired. On the other hand, if the leak is close to the size of the *standard leak* (and your system does not get below 10^{-7} torr) there is little reason to repair the leak.

Even if there is no reason to quantify any given leak, it is still important to establish a leak limit for the helium leak detector before leak hunting begins. This limit is set with a standard leak.

A standard leak is a small container that leaks helium at a very slow and specific amount. As mentioned before, helium diffuses through borosilicate glass (or through ceramic or capillary tube). This diffusion is very consistent at any given pressure and temperature. Within the standard leak is a thimble of borosilicate glass (see Fig. 7-59) around which is the stored helium. The helium leak rate varies approximately ± 3% per degree C above or below its standard temperature of 22°C.

Standard leaks come from the factory with their calibrated leak rate imprinted on their sides. The approximate annual attenuation is also indicated. After a certain amount of time (every five to ten years) they can be returned to the factory for either recharging and/or re-calibration. Re-calibration is not critical if the helium leak detector is being used only for leak location and not leak calibration.

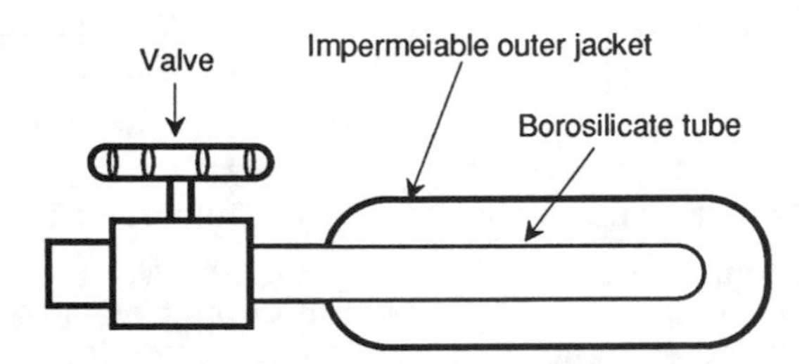

Fig. 7-59 An example of one standard leak design.

Standard leaks should be handled carefully and not dropped because sharp hits can break the glass inserts. They should not be stored with their valves closed. Otherwise, an artificially high accumulation of helium would be created in the area between the valve and glass tube. The valve provides temporary stoppage of helium calibrations between its use during leak readings.

Because of the many different brands and types of helium leak detectors, it is impossible to explain how to operate your particular model use your owner's manual. If you cannot find the owner's manual for your machine, contact the manufacturer and obtain a replacement, possibly for a nominal charge. If you have been using your leak detector on the fly with apparent success, it would still be worthwhile to obtain the manual.

7.6.11 General Tips and Tricks of Helium Leak Detection

Probably the biggest problem with helium leak detector use comes from the construction material of the vacuum system or apparatus being tested. If the materials within the tested piece readily absorb helium, a false and/or confusing reading will exist for a considerable time afterward and will remain until the helium has left the piece.

A second complication can come from materials that helium can *easily* permeate such as the elastomers used in O-ring connections. Helium readings from these materials can distract you from other 'true' leak readings. Once an elastomer has become saturated with helium, it slowly desorbs the gas which creates a large background noise. This background noise interferes with a helium leak detector's sensitivity and performance. All elastomers readily absorb, and slowly desorb, helium. Thus the fewer elastomers used within a system, the less of a problem there will be. Silicone tends to absorb helium the most, followed by natural rubber, Buna-N, and then neoprene. The permeability of several materials is included in Table 7-19.

One additional complication with elastomers has to do with the connection between a vacuum system and a helium leak detector. If it is possible, connect the two with a non-elastomeric connection such as a stainless steel bellows. If this connection is not possible, the best elastomeric connection that you can use is a high-vacuum, thick-walled (½ in I.D., 3⁄16 in wall) vinyl tube of the shortest length possible. Do not use rubber or neoprene tubing. If you have several systems that may need leak detection, it is unlikely that one length of tube will accommodate all systems. It is best to have several connecting tubes, one for each system (or several systems) being tested to allow you to select the shortest one that can be successfully used.

Table 7-19‡

Permeability of Polymeric Materials to Various Gases

std cc cm/cm^2 sec bar x 10^8 (Note bar = 10^4 dynes/cm^2)

Material	Helium	Hydrogen	Nitrogen	Oxygen
Buna S	17.3	30.1	4.7	12.8
Perbunan 18	12.7	18.9	1.89	6.12
Neoprene G	3.38	10.2	0.88	3.0
Rubber, natural	25.0	38.3	7.4	17.5
Rubber, methyl	10.9	12.8	0.36	1.6
Rubber, dimethyl silicone	263	495	210	450
Teflon, FEP	30.1	9.89	1.44	3.37
Teflon, TFE	523	17.8	2.4	7.5
Fused silica	0.75	1.1×10^{-4}	*	*
Vycor	1.13	3.8×10^{-4}	*	*
Pyrex	0.09	*	*	*
Soda-lime	5.6×10^{-4}	*	*	*
X-ray shield	3.1×10^{-7}	*	*	*
Vitreosil	0.480	*	*	*

* data not provided

‡ From *Introduction to Helium Mass Spectrometer Leak Detection*, Tab. 6-1, published by Varian Associates, Inc. ©1980, p. 49. With permission.

The actual connection between the vacuum system being tested and the helium leak detector can be made at a variety of points. One approach is to attach the connecting tube to the foreline of the diffusion pump where the fore pump would normally be connected. This attachment has the advantage of being physically at the end of the system, and therefore it is less likely to encounter confusing signals as you examine the system from one end to the other. The disadvantage is that you need to remove, and then later replace the fore pump hose. In addition, this attachment can only be made if the system being tested is small (i.e., less than five to ten liters). If the system is any larger, the pumping system of the vacuum system being tested will be needed to assist the helium leak detector's pumping system.

Alternatively, you may attach the helium leak detector to any of the demountable connections that you already have on your system, such as standard taper joints (see Fig. 7-60). Always use a hose clamp on both ends of the tube connecting the helium leak detector and the vacuum system being tested to ensure a leak-tight fit.

The simplest way to limit the effects of permeability during leak detection is to work quickly. The longer a material is exposed to helium, the greater the saturation will be, and the greater the amount of time necessary for the helium to outgas. Helium can leak through an O-ring within five minutes, and even faster through a rubber diaphragm. If an O-ring becomes saturated with helium and its saturation is affecting your leak check, try removing the O-ring and replacing it with one that was stored away from any helium. The offending O-ring will be fine to use again within several days. Another approach is to use the permeability of the helium in the elastomer as a base level, sort of a constant level of noise. Then all helium found beyond that level would indicate a leak. Although this process eliminates the problem of changing elastomers, it significantly decreases the sensitivity of the leak detector.

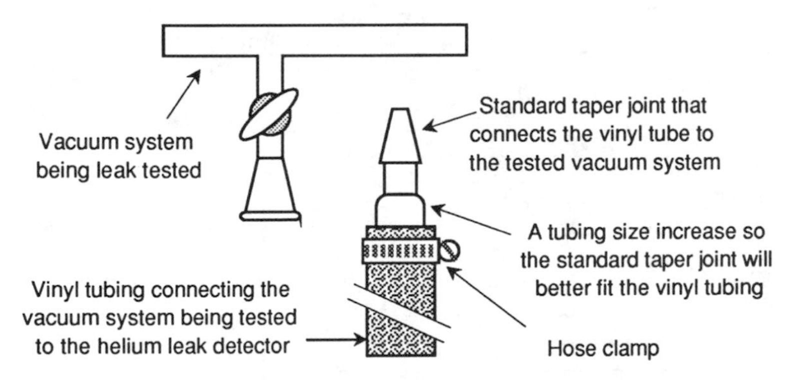

Fig. 7-60 Using Standard taper joints to connect a helium leak detector to the vacuum system being tested.

Aside from permeability and absorbancy complications, other universal concerns of helium leak detection are factors such as *source operating pressure*, *spraying patterns* (for tracer-probe technique), *response time*, *cleanup time*, and *cold trap usage*. Pump use and general helium leak detector maintenance operations are also fairly universal.

The *source operating pressure* is the vacuum necessary to operate the leak detection device. This pressure is not specific, rather it is a pressure range within which the leak detector will work. Optimistically, we want the helium leak detector, and the system to which it is connected, to have the greatest possible vacuum when using the tracer-probe technique to provide the maximum sensitivity with the quickest response time. As an added benefit, when one is operating at a very high vacuum, the amount of maintenance required for the leak detector is at a minimum. (There is, of course, an amusing contradiction here: if you require a leak detector, you are not operating at the highest possible vacuum.)

Most helium leak detectors will not operate with pressures above 10^{-4} torr to 10^{-5} torr. At these greater pressures, the main element to the mass spectrometer will burn out. Fortunately most, if not all, helium leak detectors have various safety check mechanisms that automatically shut off the current to the main filament if the pressure goes above a set limit. So, you must depend on alternate leak detection methods, or use the detector-probe technique to discover large leaks. Once large leaks have been discovered and closed, you can concentrate on the smaller leaks that can be found with the *tracer-probe technique.*

The detector-probe technique is, at best, about 10% as sensitive as the tracer-probe technique.[76] It can be made more sensitive by placing a greater pressure of the probe gas within the system. This greater pressure cannot be used on glass systems for fear of blowing out stopcock plugs or standard taper joint pieces. One technique for introducing helium to your vacuum system (with a moderate pressure) is simply to fill a balloon with helium and attach it to the vacuum system. The pressure within a helium balloon is not great enough to do any damage to a glass vacuum system. However, be careful how you 'sniff' around the balloon to avoid misreading helium permeating from the balloon walls. Using a Mylar balloon will provide less permeation through the walls, but it will be more difficult to attach to the vacuum system.

By placing a small vinyl tube over the end of the 'sniffer probe' (see Fig. 7-61), it is possible to press the tip of the probe over the area of the possi-

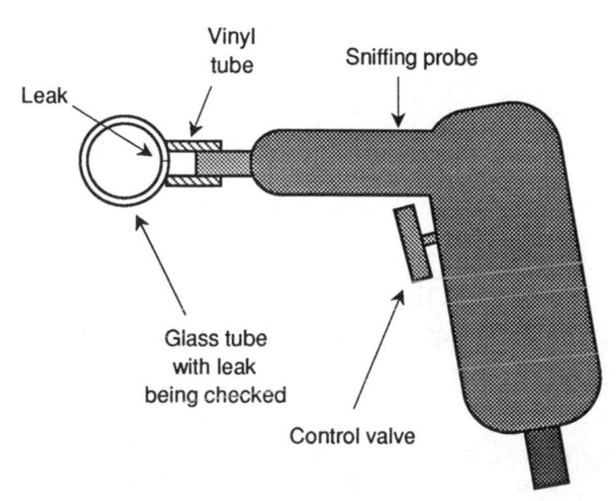

Fig. 7-61 Using a plastic tube to increase the sensitivity of a sniffing probe.

ble leak. This method closes the leak off from ambient air, which both increases the sensitivity at this location and helps to precisely pinpoint the leak site. If you hold the vinyl tube too long near the helium, it will become saturated, and your 'sniffer' will be reading the 'leak' from the plastic on the nose of the probe. If this saturation occurs, have another plastic tube (that has been kept away from any helium) ready to slip on. A saturated plastic tube will lose its helium in several days.

If you ever wish to change from the detector-probe technique to the tracer-probe technique, you must remove all the helium already in the vacuum system by running the leak detector's vacuum system and/or the pumps of the vacuum system being tested for a period of time. Otherwise false and/or artificially high readings may affect any serious or useful leak testing. A dry nitrogen purge is recommended to help sweep the helium out of the system.

The spraying pattern used with the tracer-probe technique is very important. Improper spraying can significantly affect the amount of time and quality of the leak detection. If there is too much ambient helium in the air (caused by sloppy helium spraying), false readings from unknown origins can result. You can blow helium from a pipette tip, or a blunt hypodermic needle inserted in the end of a tube connected directly to a regulator on a helium compressed gas tank. To control the flow rate, you will need to adjust the needle valve on the regulator. To see the amount of helium spraying out of the tip, place the pipette tip into a beaker of water. The flow rate can then be set to allow for a slow, but steady stream of bubbles.

The location of where to spray is a combination of 1) spraying the system parts close to the helium leak detector and then moving further out, **and** 2) spraying the highest parts of the system and then moving on down toward the floor. The reason for the former is to keep clean-up time and response time to a minimum. Otherwise, you must wait for all the helium to be removed before you can check for leaks closer to the detector. The reason for the latter is because helium rises in air. If you spray down low first, the he-

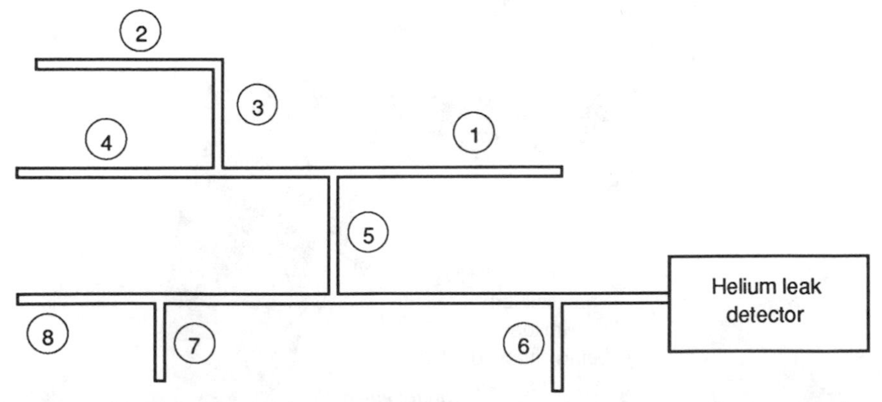

Fig. 7-62 A suggested pattern for spraying helium on a vacuum system when using the tracer-probe technique. The pattern is a compromise between proximity to the leak detector and spraying high before low areas.

lium that drifts up may get sucked into an unknown leak in a higher section of the vacuum system, thus creating confusion as to whether the leak is in the area you are studying, *or anyplace above* the area you are studying. See Fig. 7-62 for a suggested spraying pattern.

Response time can be easily explained with a hot water analogy: When you turn on the hot water in your shower, water immediately comes out of the shower fixture, but there is a time lag before *hot* water arrives. Time is a function of water pressure, the amount you have the valve open, the pipe diameter, and the distance between the hot water tank and the shower. It is important to keep in mind this delay factor, or response time, when operating a helium leak detector. For example, say your setup takes one minute from the time you introduce helium to a leak until it is acknowledged by the helium leak detector. In addition, say your spray probe is being moved at four inches per minute (which is pretty slow if you think about it). You will therefore be four inches away from a leak when it is first identified. Generally, moving the spray probe at a rate of one foot per minute provides an adequate time response which of course can be made faster or slower as conditions warrant. However, if you know the response time, you will know about how much to back up to re-test the area in question to allow pinpointing the specific leak spot.

The response time for a helium leak detector is "the time required for a known leak rate to be indicated on the leak detector from zero to 63% of its maximum equilibrium level."[77] Of course the helium leak detector must be calibrated with a standard leak so that the maximum reading can be properly set on the leak detector. The response time is dependent on the quality of vacuum, the size of the leak, physical barriers (such as constrictions, bends, or traps) between the leak and the leak detector, and the proximity of the leak to the leak detector. Because of so many factors, there can be no set specific response time.

The *clean-up time* is the length of time for the detection signal to stop indicating that it is sensing a leak. More specifically, it is "the time required after removal of a tracer gas from a leak for the initial leak indication to decay 63% of the indication at the time the leak was removed."[78]

Clean-up time is almost always longer than response time because of the difficulty in desorption of helium from a vacuum system, compounded by helium permeation into porous materials. So, when using the tracer-probe technique do not overspray helium onto your system. The more helium that enters your system, the more that can permeate into porous materials, and the longer the clean-up time.

Liquid-nitrogen cold traps stop condensable vapors* from traveling between mechanical pumps, diffusion pumps, and the rest of the system. They also protect the helium leak detector from possibly contaminating materials (such as silicon-based diffusion pump oil).

Any coating that develops on the ion source of the helium leak detector will hinder operation. This coating is not easy to prevent because during op-

* Helium is a non-condensable gas. If you are using an oxygen or halogen probe gas, you have a new complication because these gases are condensable. Thus, using a cold trap will likely prevent the leak detector from detecting the tracer gas. Because a liquid nitrogen cold trap would not be practical in this circumstance, consult the manufacturer for assistance on how to protect the electronics of your leak detector.

eration, the ion source behaves like a getter-ion pump, drawing materials to its surface. Therefore anything that can be done to protect the ion source without hindering normal operation is desirable.

During operation, maintain the liquid nitrogen at a constant level to prevent the evaporation of condensable vapors that could in turn collect on the ion source. Use the trap-filling method suggested in Sec. 7.4.3.

During operation, if a heavy build-up of vapors collects on the cold trap, an incidental backlog of probe gas can develop, thus causing a rise of background probe gas noise, which in turn slows down the speed of detection ability and decreases the sensitivity of the leak detector. If this situation develops, close the valves to the vacuum system being checked, turn off all gauges within the leak detector,* and let the cold trap heat up. Meanwhile, the pumps should remain on to draw the condensable vapors from the leak detector. Do not run the mechanical pump with the diffusion pump off under a no-gas load (nothing coming in) condition because mechanical pump oils will backstream into the diffusion pump.

The following are several extra general maintenance tips:

1) If you need (or desire) to run the mechanical pump of a helium leak detector with the diffusion pump off, the best means to protect the diffusion pump oils is with a dry nitrogen gas purge. This purge will prevent mechanical pump oils from backstreaming into the diffusion pump.** The leaking dry nitrogen maintains a low level laminar flow toward the mechanical pump. The nitrogen (leak) pressure should be maintained between approximately 1 to 10^{-2} torr.
2) Never use anything but hydrocarbon oils in the diffusion pumps of helium leak detectors. Silicon oils can deposit insulating layers on the ion sources of the mass spectrometers and ion gauges. However, if hydrocarbon oils are allowed to migrate to the guts of the leak detector's mass spectrometer section, they can be cracked by ions and electrons. The cracked hydrocarbon remains can deposit on the filaments and other system parts causing a loss of sensitivity and resolution. Admittedly the effects of hydrocarbons are not good, but they are better than the effects of silicon oils.
3) Within a helium leak detector, there are valves that isolate the mass spectrometer and gauge sections, the diffusion pump, and the trap from the rest of the system. These valves are used to prevent contamination or damage to these sections when cleaning, adjusting, or venting is required during use. Find and use them.
4) When shutting down a system, it is best to remove a cold trap *while it is still full of liquid nitrogen.* This removal procedure will help decrease the amount of material that could pass into

*This procedure should always be done when running any vacuum system without the use of traps because of the damage some condensable vapors can do to vacuum gauges.

** On a regular vacuum system, this can be done by either separating the diffusion pump from the rest of the system by stopcocks or valves, or by the use of a nitrogen cold-trap between the mechanical and diffusion pump.

the leak detector, or the leak detector's pumping system, after the system is turned off. The process for removing a trap is as follows:

a) Be sure that all protective valves around the cold trap are closed before venting the area around the trap to atmospheric pressure.
b) Wear protective gloves.
c) Pour the liquid nitrogen into another dewar, or some safe receptacle. Do not pour the liquid nitrogen down a sink, on equipment, or on another co-worker.
d) Allow the trap to warm up (you may pour room temperature distilled water on it to facilitate the warming process).
e) Gently wipe clean with a rag or Kimwipe. Use tolulene for a final wipe (these traps are made of very thin metal to facilitate heat transfer. They are very vulnerable to dings, bangs, and bumps, so be very careful).
f) When replacing a trap, be sure that any gasket (if present) is clean and in place. If there are any bolts to tighten, use a rotational tightening pattern as shown in Fig. 7-50.

Alternatively, you can run a dry nitrogen gas purge at a pressure of approximately 200-300 μm. This purge should be left on after the diffusion pump is turned off, and remain on while the mechanical pump is left to pump overnight. As the trap warms up, all the contaminants will be discharged out of the system. The next day, the leak detector can be restarted by adding liquid nitrogen to the trap, starting the diffusion pump, and turning off the nitrogen purge.[79]

Calibration can be done every time you start a system up, but for general leak detection, calibration is less important. However, whenever any general repair or replacement maintenance is performed, such as replacing the filament on a mass spectrometer, calibration becomes mandatory. See your owner's manual for specific instructions.

7.6.12 Repairing Leaks

If a leak has been located with a Tesla coil, you may find a (very) small white dot where the spark went through the glass. This dot may be hard to re-locate, so it is best to clearly mark the exact location of the hole with a grease pencil or any other marker that can write on glass. Other leak detection techniques do not leave any mark and some marking is essential.

The best possible repair* for a leak within a glass system is for the leak to be removed (see Fig. 7-63), the whole area to be heated and blown smooth, and finally, the worked area should be flame-annealed to prevent later cracking.

A temporary solution to vacuum system leaks that has been used for quite some time is the application of Glyptol. Although this approach may work in

* The repair of a leak on a glass system should be done by someone with glassblowing experience.

some isolated circumstances, covering a hole with Glyptol, or any other temporary patch should not be relied on because:

1) There is a chance that you may create a virtual leak.
2) Once a system is (temporarily) repaired, there is a good chance that the need for a permanent repair will drop ("out of sight, out of mind"). Temporary repairs are, at best, temporary and should not be relied on for future needs.
3) Glyptol, or other temporary leak repair material, can be very difficult to properly remove from a glass surface. However, it must be totally removed before any glassworking can be done.

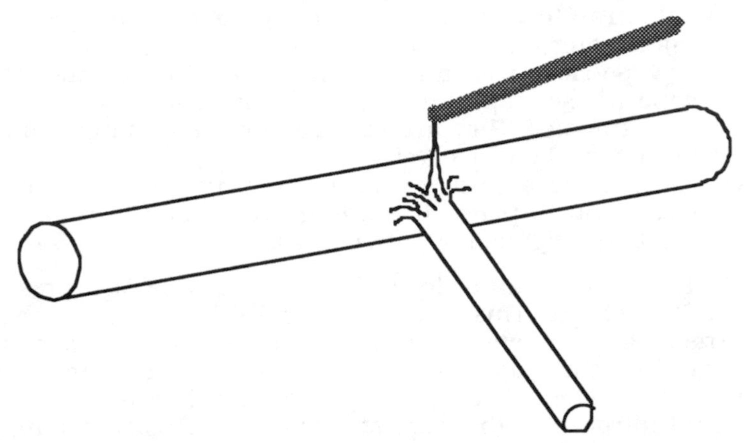

Fig. 7-63 Removing a hole from a glass seal by removing the glass around the hole with a glass rod and a gas–oxygen torch (not shown).

Obviously there will be times when the temporary repair of a leak is essential, but otherwise it is best to make all repairs permanent.

After a leak has been discovered, removed, and the glass repaired, flame annealed, and cooled, the system must be leak checked again not only to verify that the repair was successful, but to see if there are any other leaks that need repair. It is not uncommon for another leak to be immediately adjacent to the first leak. An adjacent, smaller may be unable to attract the spark from a Tesla coil away from a larger leak. Or, a larger leak can cause too much helium noise for a smaller leak to be pinpointed. However, once larger leaks are repaired, smaller leaks can be more readily identified.

7.7 More Vacuum System Information

7.7.1 The Designs of Things

As you work with vacuum systems, you may be called upon to design new parts and/or sections for the system. Your success will depend on your experience with the materials you select, your choice of materials, and your access to experienced technicians.

If you require metal components on your vacuum system, select metals with low gas permeation, such as 300 series stainless steel. It is non-magnetic, and like glass, is a poor conductor of heat and electricity. Stainless steel, also like glass, it is relatively non-reactive, and therefore is less likely to rust or be affected by chemicals. If welding the stainless steel is required, select 304L stainless steel, which is low in carbon. Otherwise, at welding temperatures, the carbon will combine with the chromium (within the stainless steel) to form chromium carbide and the corrosion protection of the chromium will be lost. Type 303 stainless steel should not be used for vacuum work because it contains selenium, which has a high vapor pressure.

The metals zinc and cadmium should be avoided because of their high vapor pressures. These metals includes alloys that contain zinc and cadmium such as brass (copper and zinc) and some silver solders (cadmium). It is possible to obtain cadmium-free silver solder and brazing materials that use tin, lead, and indium for vacuum use. Some steel screws are cadmium coated and also must be avoided.

Copper is often used as gasket material, such as in Conflat (Varian) flanges, however, standard copper is often permeated with oxygen. If copper is heated in a hydrogen environment, the oxygen combines with the hydrogen to form water, which causes the copper to become brittle. Thus, when selecting copper, obtain OFHC (Oxygen-Free High-Conductivity) copper.

Besides copper, indium is often used as a metal gasket material. Despite its lack of elasticity and its limited operating temperature (100°C), indium wire (0.8 - 1.6 mm) can easily be bent into a circle of any desired size and, if the ends are overlapped, no soldering is required. Aluminum can be made into a gasket as well by using a wire (0.5 - 1 mm), bending it into a circle of the desired size, and soldering it together using a flux. The flux must be washed off before use. Aluminum can be used to temperatures as high as 400°C. Gold is probably the best material for gaskets because it requires no flux to join the ends of a circled wire and can be used to temperatures as high as 450-500°C.

Parts are fairly easy to attach and change with metal systems; it is mostly a matter of unbolting one unit and replacing it with another. Be sure to follow the bolting techniques described in Sec. 7.6 to insure even cutting on gaskets and equal torque on flange surfaces.

Glass systems are a little different; if you have not worked with glass before, **do not** try to learn glassblowing **on** a vacuum system. If you want to learn some glassblowing techniques, it is imperative that you start with small items at a bench before risking the destruction of a very valuable piece of equipment. Watching an experienced glassblower is deceiving because it looks easy. A concert pianist makes playing piano look just as easy, but playing the piano is difficult as well.

This book is not an instruction book on glassblowing. Chapter 8 has some tips on gas–oxygen torch use and Appendix D contains some recommended books for your consideration. The following discussion will also provide some of the dynamics of glass vacuum systems so that common errors and problems can be avoided. Consider the following questions before doing any glasswork and/or additions to a vacuum line:

Are the Glasses You Have to Work With Compatible (see Sec. 1.1.3)? Trying to repair a small leak in a vacuum line made of borosilicate glass by filling the hole with soft glass will result in an entire day of repair by a qualified person instead of the half-hour originally required.

Do Any of the Items Being Added onto the Vacuum System Have any Concave Surfaces? Convex items can support themselves against atmospheric forces with a vacuum on the other side. Concave-shaped items, on the other hand, cannot deal with the stress unless they have been specially-built. For example, a standard Erlenmeyer flask cannot withstand atmospheric forces against a vacuum because the base is somewhat concave. However, a filter flask can withstand these pressures because it is constructed out of heavy wall glass. There should be the same concern of implosion, whether a vacuum is created by a small single-stage vacuum pump or with a vacuum system capable of ultra-high vacuum. The greatest percentage of change in force against an walls of a container occur when the item is brought from atmospheric to about 1 torr because the primary force at work is atmospheric pressure, not the vacuum. Remember, a vacuum of about 1 torr can pull water about 33 ft into the air, but no matter how much more vacuum is applied, the water cannot be brought vertically any higher.

Are Your Vacuum Line Supports on Vertical or Horizontal Rods of the Vacuum Rack? Vacuum lines are supported by two- and/or three-fingered clamps, which in turn are held onto the vacuum rack by clamp supports that grip the rack by screws. Regardless of how tightly a clamp support is screwed down, it's not difficult to swing a support one way or another. This movement can be an advantage if you wish to swing away a support for a Dewar, but could be a disaster for a vertical axis support of a vacuum line (see Fig 7-64). A disaster could also happen when an entire system is being supported on horizontal bars or even just a single item.

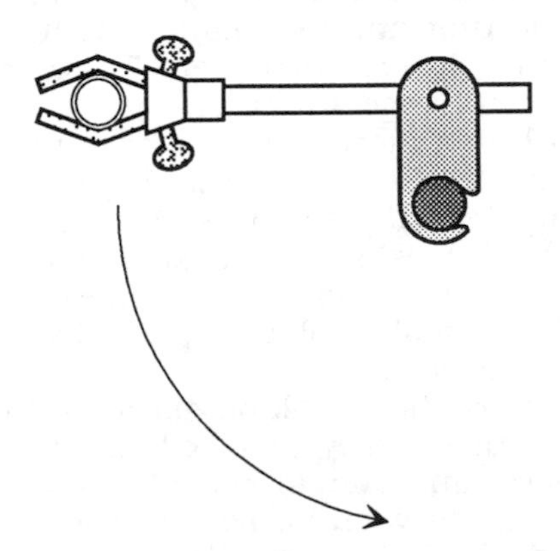

Fig. 7-64 Support on vacuum lines must be connected to the vertical bars of a support rack, not the horizontal bars.

Are Sections Between Vacuum Parts Breaking as You Tighten Two- and/or Three-fingered clamps, or Clamp Supports? Glass is perfectly elastic until the point of fracture and it is easy to place too much torque against glass with the tightening screws of clamp fingers and/or clamp supports. During and/or after placing parts of a vacuum system in these devices, relieve the tension by heating the glass sections between the clamps (see Fig 7-65). The heating should be done with a gentle, bushy flame. Be careful not to get the flame near larger diameter tubing or near glass seals that could cause strain within the glass. These situations could cause later cracks or breaks.

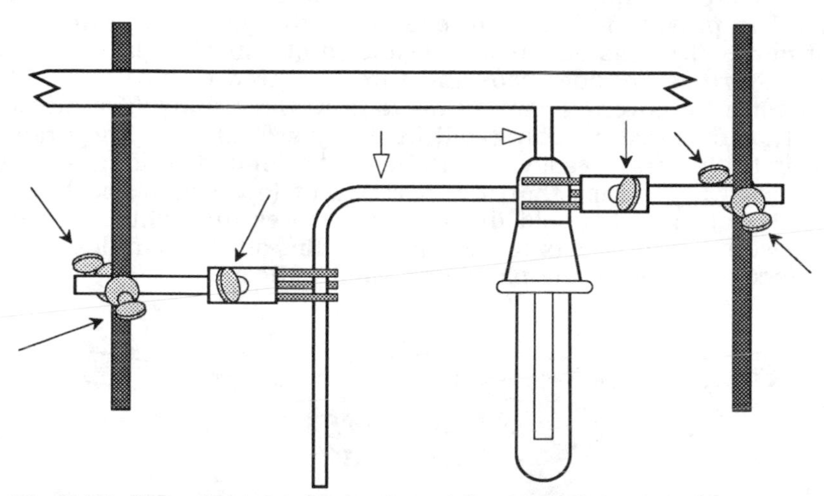

Fig. 7-65 If you tighten two- (or three-) fingered clamps or clamp supports (dark arrows) it is important to relieve the stress that is created on the glass by heating the connecting glass (hollow arrows).

Are There Rubber Cushions Supporting Round Bottom Flasks on Support Rings? By simply taking pieces of rubber tubing and slitting them open along

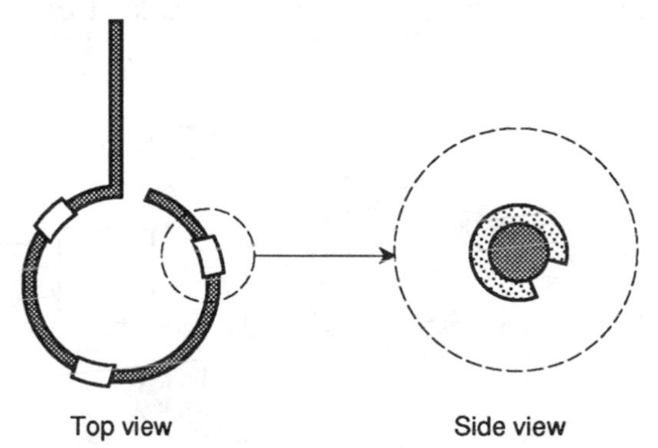

Fig. 7-66 Placing cut rubber tubing on a support ring makes a good support for a round bottom flask.

one side, protective surfaces are made on which to lay round bottom flasks (see Fig. 7-66). This method of protection should **not** be used if a round bottom flask is warmed greater than 100°C, or if you plan to heat the round bottom flask in its support (see Sec. 1.1.7 for information on how to heat round bottom flasks). Such heating can burn or melt the rubber or vinyl tubing, releasing dangerous fumes.

Is There Fiberglass, Kevlar, or Ceramic Tape Protecting the Glass from Two- or Three-fingered Clamp Arms? When purchasing new two- or three-fingered clamps, they will be supplied with either a plastic coating or fiberglass covers on the fingers. Asbestos covers are no longer commercially available. The plastic provides protection for the glass, but cannot survive any heating or direct flames. It is possible to obtain fiberglass covers to replace the coverings on older two- and three-fingered clamps (see Fig. 7-67). Another choice for covering clamp fingers is Kevlar tubing. The advantage of Kevlar is that one size of tubing will fit many size fingers. The drawback is that Kevlar tubing frays easily, so it should be limited to long-term holding and not used with clamps that are in constant (open and close) use. There are also several products sold in hardware stores into which you dip your tool handles. These products leave a plastic film on the handles and can be used to place, or replace, the plastic on clamp fingers.

Fig. 7-67 Use Kevlar or glass wool tape on two- and three-fingered clamps, *especially* if they will be subjected to any heat.

Are you Unable to Squeeze the Fingers of a Two-fingered Clamp Down Sufficiently to Obtain a Good Grip on Small Diameter Tubing? There are two choices to resolve this problem depending on the conditions surrounding the section of tubing in question. The easiest solution is to take a short piece of flexible tubing, slit open the side, and wrap this tubing piece around the glass tubing where the clamp will support the item. This slitting of the flex-

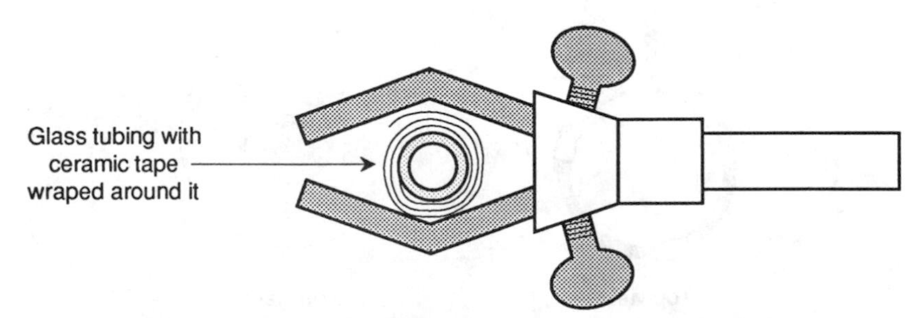

Fig. 7-68 Wrapping ceramic tape around glass tubing makes it easier to be held by two- or three-fingered clamps.

ible tube is exactly the same process described in Fig. 7-66, except the slitted tube goes onto straight tubing instead of the support ring. If, on the other hand, the glass tubing will be heated, the piece of flexible tubing could be damaged. In these environments, wrap ceramic tape* (do not use asbestos tape) around the tubing that would otherwise be too narrow for a fingered clamp to obtain a firm grip (see Fig. 7-68).

Are You Snapping Stopcocks Off at the Base While Rotating the Plug? Whenever possible, stopcocks should be supported from both ends either by fusion to other glass and/or a two- or three-fingered clamp, although this support cannot always be supplied. One of the purposes of an extra clamp is to provide support and bracing against the torque from accidental knocks. The main purpose of an extra stopcock is to protect it from heavy-handed rotation. As stopcock grease gets old, it becomes thicker and loses its slippery nature. When rotating a stopcock plug with old grease, tremendous torque can be created on the stopcock and it can snap off at the base (see Fig. 7-69).

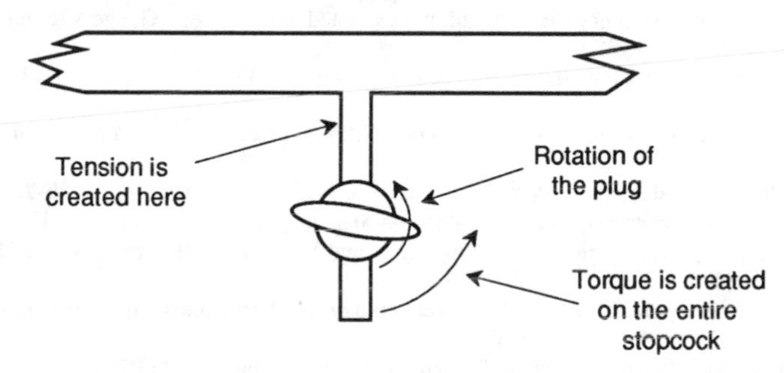

Fig. 7-69 By rotating the stopcock plug without proper support, torque is created on the stopcock. If the tension is great enough, the stopcock may break.

This breakage can be prevented by maintaining fresh stopcock grease in the stopcock, **and** by holding the stopcock with one hand while rotating the plug with the other. In fact, it is always best to hold the body of a stopcock with one hand while rotating the plug with the other.

References

[1] H.G. Tompkins, *An Introduction to the Fundamentals of Vacuum Technology*, (New York: American Vacuum Society, American Institute of Physics, Inc., 1984), p. 2.

[2] R.J. Naumann, "Prospects for a Contamination-free Ultravacuum Facility in Low-Earth Orbit," *J. of Vacuum Science and Technology* , 7 (1989), pp. 90-9.

[3] J.F. O'Hanlon, *A User's Guide to Vacuum Technology* (Wiley-Interscience, N.Y.: John Wiley & Sons, 1980), p. 11.

[4] A. Guthrie, Vacuum Technology, (Wiley-Interscience, New York: John Wiley & Sons, 1963), p. 6.

[5] T.E. Madey, "Early Applications of Vacuum, From Aristotle To Langmuir," *J. Vac. Sci. Tech. A*, 2 (1984), pp. 110-7.

* Ceramic tape can be obtained from the various glassblowing supply houses mentioned in Appendix C, for Chapter 8.

[6] M.H. Hablanian, and B.B. Dayton, "Comments on the History of Vacuum Pumps," *J. Vac. Sci. Tech. A*, 2 (1984), pp. 118-25.

[7] J.H. Singleton, "The Development of Valves, Connectors and Traps for Vacuum Systems During The 20th Century," *J. Vac. Sci. Tech. A*, 2 (1984), pp. 126-31.

[8] P.A. Redhead, "The Measurement of Vacuum Pressures," *J. Vac. Sci. Tech. A*, 2 (1984), pp. 132-8.

[9] J.F. Peterson, "Vacuum Pump Technology; A Short Course on Theory and Operations, Part I," *Solid State Technology*, 24 (1981), pp. 83-6.

[10] R. Barbour, *Glassblowing for Laboratory Technicians* (Elmsford, N. Y.: Pergamon Press Ltd., Pergamon Press, 1978), pp. 184-6.

[11] A.M. Russel, "Use of Water Aspirator in Conjunction with Sorption Pumping on an Ultrahigh Vacuum System," *Review of Sci. Instruments*, 36 (1965), p. 854.

[12] P. Sadler, "Comparative Performance Characteristics Between a Vane and a Rotary Piston Type of Mechanical Vacuum Pump," *Vacuum*, 19 (1969), pp. 17-22.

[13] M.H. Hablanian, "Performance of Mechanical Vacuum Pumps in the Molecular Flow Range," *J. Vac. Sci. Technol.*, 19 (1981), pp. 250-52.

[14] Welch Vacuum Technology, Inc., 1988, p. 125.

[15] Z.C. Dobrowolski, "Mechanical Pumps" in *Methods of Experimental Physics*, Vol. 14, ed. G.L. Weissler and R.W. Carlson (New York: Academic Press, Inc., 1979), pp. 468-71.

[16] Carstens, Hord, and Martin, "Mercury Contamination Associated with McLeod Gauge Vacuum Pumps," *Rev. Sci. Instrum.*, 36 (1972), pp. 1385-6.

[17] M.H. Hablanian, "Comments on the History of Vacuum Pumps," *J. Vac. Sci. Technol. A*, 2 (1984), pp. 174-81.

[18] J.F. O'Hanlon, *A User's Guide to Vacuum Technology* (Wiley-Interscience, New York: John Wiley & Sons, 1980), p. 165.

[19] M.H. Hablanian, "Mechanical Vacuum Pumps," *J. Vac. Sci. Technol.*, 19 (1981), pp. 250-2.

[20] R. Rondeau, "Design and Construction of Glass Vacuum Systems," *J. of Chem. Ed.*, 42 (1965), pp. A445-59.

[21] N.S. Harris, "Practical Aspects of Constructing, Operating, and Maintaining Rotary Vane and Diffusion-pumped Systems," *Vacuum*, 31 (1981), pp. 173-82.

[22] J.F. Peterson and H.A. Steinherz, "Vacuum Pump Technology; A Short Course on Theory and Operations, Part I," *Solid State Technology*, 24 (1981), pp. 83-6.

[23] G.F. Weston, "Pumps for Ultra-high Vacuum," *Vacuum*, 28 (1978), pp. 209-233.

[24] *Ibid.*

[25] L. Laurenson, "Vacuum Fluids," *Vacuum*, 30 (1980), pp. 275-281.

[26] J.F. O'Hanlon, *A User's Guide to Vacuum Technology* (Wiley-Interscience, New York: John Wiley & Sons, 1980), p. 198.

[27] *Ibid.*

[28] D.J. Santeler, "Use of Diffusion Pumps for Obtaining Ultraclean Vacuum Environments," *Journal of Vacuum Science and Technology*, 8 (1971), pp. 299-307

[29] W.W. Roepke and K.G. Pung, "Inexpensive Oil Vapor Trap for use with Rotary Vacuum Pumps," *Vacuum*, 18 (1968), pp. 457-58.

[30] Brown, A.B., Proceedings of the Thirtieth Symposium and Exhibition on the Art of Glassblowing, © 1985 by the American Scientific Glassblowers Society, pp. 80-2

[31] W. Jitschin, "Accuracy of Vacuum Gauges," *J. Vac. Sci. Technol. A* , 8 (1990), pp. 948-56.

[32] *International Vocabulary of Basic and General Terms in Metrology*, (ISO, Geneva, 1984).

[33] P. Nash, "The Use of Hot Filament Ionization Gauges," *Vacuum*, 37 (1987), pp. 643-9.

[34] W. Jitschin, "Accuracy of Vacuum Gauges," *J. Vac. Sci. Technol. A* , 8 (1990), pp. 948-56.

[35] N.S. Harris, "Practical Aspects of Constructing, Operating, and Maintaining rotary Vane and Diffusion-Pumped Systems," *Vacuum*, 31 (1981), pp. 173-82.

[36] S.T. Zenchelsky, "Pressure Measurement, Part One," *J. of Chem. Ed.*, 40 (1963), pp. A611-32.

[37] H.F. Carroll, "Avoid Parallax Error When Reading a Mercury Manometer," *J. of Chem. Ed.*, 44 (1967), p. 763.

[38] W.G. Brombacher, D.J. Johnson, and J.L. Cross, "Errors in Mercury Barometers and Manometers," *Instruments & Control Systems*, 35 (1962), pp. 121-2.

[39] S.O. Colgate and P.A. Genre, "On Elimination of the Mercury Pumping Error Effect in McLeod Gauges," *Vacuum*, 18 (1968), pp. 553-8.

[40] J.K.N. Sharma, et al., "A Simple Graphical Method of Pressure Determination in a McLeod Gauge," *Vacuum*, 31 (1981), pp. 195-7.

[41] K.B. Wear, "Condensable Gases in a McLeod Gauge," *Review of Scientific Instruments*, 39 (1968), pp. 245-50.

[42] C.J. Carstens, C.A. Hord, and D.H. Martin, "Mercury Contamination Associated with McLeod Gauge Vacuum Pumps," *Review of Sci. Instruments*, 43 (1972), pp. 1385-6.

[43] W. Jitschin, "Accuracy of Vacuum Gauges," *J. Vac. Sci. Technol. A* , 8 (1990), pp. 948-56.

[44] J.C. Snaith, "Vacuum Measurement," *J. of the British Society of Scientific Glassblowers*, 7 (1969), pp. 3-7.

[45] R.N. Peacock, N.T. Peacock, and D.S. Hauschulz, "Comparison of Hot Cathode and Cold Cathode Ionization Gauges," *J. of Vac. Sci. and Technology*," 9 (1991), pp. 1977-85.

[46] W.B. Nottingham, *7th Ann. Conf. on Phys. Electron.*, M.I.T., (1947).

[47] K.E. McCulloh and R. Tilford, "Nitrogen Sensitivities of a Sample of Commercial Hot Cathode Ionization Gas Tubes," *J. Vac. Sci. Technol.*, 18 (1981), pp. 994-6.

[48] M. Hirata, M. Ono, H. Hojo, and K. Nakayama, "Calibration of Secondary Standard Ionization Gauges," *J. Vac. Sci. Technol.*, 20 (1982), pp. 1159-62.

[49] H.C. Hseuh, "The Effect of Magnetic Fields on the Performance of Bayard-Alpert Gauges," *J. Vac. Sci. Technol.*, 20 (1982), pp. 237-40.

[50] N.S. Harris, "Practical Aspects of Constructing, Operating, and Maintaining Rotary Vane and Diffusion-Pumped Systems," *Vacuum*, 31 (1981), pp. 173-82.

[51] *Ibid.*

[52] R.N. Peacock, N.T. Peacock, and D.S. Hauschulz, "Comparison of Hot Cathode and Cold Cathode Ionization Gauges," *J. of Vac. Sci. and Technology*," 9 (1991), pp. 1977-85.

[53] N.S. Harris, "Practical Aspects of Constructing, Operating, and Maintaining Rotary Vane and Diffusion-Pumped Systems," *Vacuum*, 31 (1981), pp. 173-82.

[54] J.J. Sullivan, "Research Efforts Boost Range and Accuracy of Vacuum Gages," *Industrial Res. & Dev.*, 25 (1983), pp. 161-9.

[55] H.A. Tasman, et al., "Vacuum Techniques in Conjunction With Mass Spectrometry," *Vacuum*, 15 (1963), pp. 33.

[56] *Ibid.*

[57] D.F. Klemperer, "Recent Developments in Ultrahigh Vacuum," *J. of the B.S.S.G.*, 2 (1965), pp. 28-38.

[58] Tom Orr, personal conversation.

[59] A.H. Turnbull, "Leak Detection and Detectors," *Vacuum*, 15 (1965), pp. 3-11.

[60] Dr. Cathy Cobb, personal conversation.

[61] D. Santeler, "Leak Detection - Common Problems and Their Solutions," *J. Vac. Sci. Technology A*, 2 (1984), pp. 1149-56.

[62] A.H. Turnbull, "Leak Detection and Detectors," *Vacuum*, 15 (1965), pp. 3-11.

63 W. Fiszdon, "'Ad Hoc' Liquid Spray Vacuum Leak Detection Method," *Phys. Fluids*, 22 (1979), pp. 1829-31.

[64] *Ibid.*

[65] G. Coyne and C. Cobb, "Efficient, Inexpensive, and Useful Techniques for Low Vacuum Leak Detection with a Tesla Coil, *J. of Chem. Ed.*, 68 (1991), pp. 526-8.

[66] A. Guthrie, *Vacuum Technology*, (New York: John Wiley and Sons, Inc., 1965), p. 514.

67 L.H. Martin and R.D. Hill, *Manual of Vacuum Practice*, (Melbourne,: Melbourne University Press, 1946), p. 112.

[68] A. Guthrie, *Vacuum Technology*, (New York: John Wiley and Sons, Inc., 1965), p. 514.

69 E.L. *Wheeler, Scientific Glassblowing*, (New York: Interscience Publishers, Inc., 1958), p. 348

70 W. Espe, *Materials for High Vacuum Technology, Vol. 3*, (New York: Pergamon Press, 1968), p. 393.

71 J.F. O'Hanlon, *A User's Guide to Vacuum Technology*, (New York: John Wiley & Sons, 1980), p. 365.

[72] D. Santeler, "Leak Detection - Common Problems and Their Solutions," *J. Vac. Sci. Technology A*, 2 (1984), pp. 1149-56.

[73] A. Guthrie, *Vacuum Technology*, (New York: John Wiley and Sons, Inc., 1965), p. 468.

[74] C.C. Minter, "Vacuum Leak Testing with Liquids," *Rev. Sci. Inst.*, 31 (1960), pp. 458-9.

[75] Varian Associates, Inc. *Introduction to Helium Mass Spectrometer Leak Detection*, Published by Varian Associates, Inc. ©1980.

[76] Wilson, N.G. and Beavis, L.C., *Handbook of Vacuum Leak Detection*, Published by the American Institute of Physics, Inc., © 1976, 1979, p. 26.

[77] *Ibid*, p. 37.
[78] *Ibid*, pg. 39.
[79] D. Santeler, "Leak Detection," *J. Vac. Sci. Technology A*, 2 (1984), pp. 1149-56.

Chapter 8

The Gas–Oxygen Torch

8.1 The Dynamics of the Gas–Oxygen Torch

8.1.1 Types of Gas–oxygen Torches

There are operations in the laboratory that require more flame heat than a Bunsen burner can provide. For example, if you wish to fire-polish the end of a chipped glass tube or "seal off" (also called "tip off) a sample for NMR study, you will need to use a gas–oxygen torch.

A Bunsen burner burns methane (lab gas) and the oxygen available in air (≈21%), and cannot provide sufficient energy to effectively work borosilicate glass, which makes up most of the glassware used in a laboratory.* To obtain more energy you must use a flammable gas (methane or propane), pure oxygen, and a torch that is specifically designed for this type of use. There are two torch designs for gas-oxygen combustion, one is known as a *premix* torch and the other is known as a *surface-mix* torch.

The designations premix and surface-mix refer to the region of the torch where the gas and the oxygen are mixed (in relation to the tip) before combustion. A premix torch design combines the gas and oxygen in the neck of the torch before they reach the tip. A surface-mix torch keeps the gas and oxygen separated until they are mixed at the moment of combustion. Both types of torches are represented in Fig. 8-1. In addition to these designs, there are larger table models (of both designs) typically used by glassblowers. These table models are impractical for laboratory use because they are not portable and cannot be used for operations such as sealing off tubes.

Superficially, the torches look similar. However, there are distinct advantages and disadvantages to both. The premix design is less expensive, and because it has one flame emitted from the tip (see Fig. 8-2), it is possible to obtain precise heating in one specific location. On the other hand,

* Typically, the first opportunity one has to do any glass work is making an eye dropper or bending a tube with a Bunsen burner. This is possible because the glass used was a soda-lime glass. Most of the glassware used in the laboratory is borosilicate glass, which has a higher working temperature than soda-lime glass. For more information on the different types of glass used in the laboratory, see Sec. 1.1.

although it is possible to control the size and character of the flame to some degree by controlling the gas and oxygen flow, radical flame size changes require changing the removable tip with one of five different available sizes. In addition, the premix torch's flame is louder than the surface-mix torch.

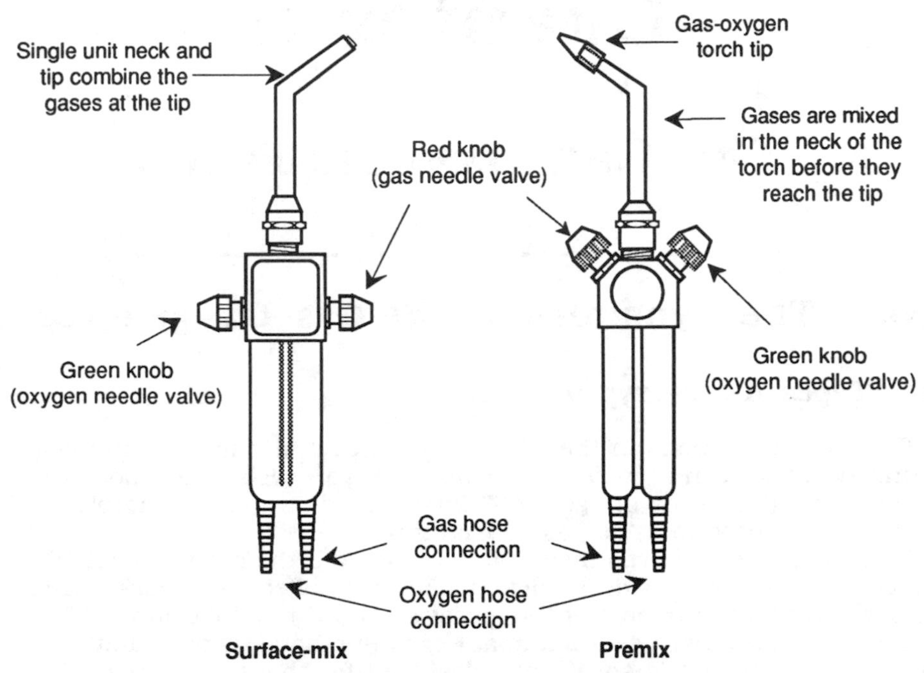

Fig.8-1 Surface-mix and premix torch designs.‡

‡ The illustration of the surface-mix torch is based on the Sharp Flame Hand Torch made by the Bethlehem Apparatus Company, Inc. (Hellertown, PA 18055). With permission. The illustration of the premix torch is based on the National Torch made by the Veriflow Corp. (Richmond, CA 94804). With permission.

Probably the biggest disadvantage of a premix torch is the characteristic of the torch to POP loudly when it changes flame size or when you turn the flame off. This popping sound will occur *if* the rate of gas combustion is faster than the gas flow rate from the tip. If the gas speed is too slow, the flame can enter the neck where premixed gas and oxygen are ready to explode with a POP. This noise invariably does more damage to your nerves than anything else. If your torch has popped, turn off both oxygen and gas needle valves, then open the oxygen valve until you hear for a loud hissing noise for several seconds. Finally, close the oxygen valve and relight the torch. To prevent the POP, maintain steady and full gas and oxygen flow rates. For more POPing prevention, see Sec. 8.1.3.

When using the premix torch to work on borosilicate glass, use a *gas-oxygen* tip design, not a *gas-air* tip* (both tip designs are represented in Fig.

* To use a premix torch with a gas-air tip, connect the hose connection for the green knob to a house air supply, or a tank of compressed air, rather than to the regulator of an oxygen tank.

8-2). Using a gas-air tip with oxygen will cause a great deal of popping and may cause damage to the tip or torch. The gas-oxygen tip comes in a variety of types and sizes: "1" is the smallest size, and "5" is the largest. A size "3" tip is a good, general size tip for most needs around the lab.

The surface-mix torch (See Fig. 8-2) has a non-removeable, one-piece tip with multiple holes, or channels, that run the length of the neck. The oxygen passes to the tip using the area surrounding the gas tubes. This design prevents any mixing of the oxygen and gas until they meet at the tip where combustion takes place. These torches are impossible to pop, and glass tubing from sizes 4 to 35 mm can all be heated without changing the tip (there are no tips to change). The surface-mix torch cannot be used as a gas–air torch.

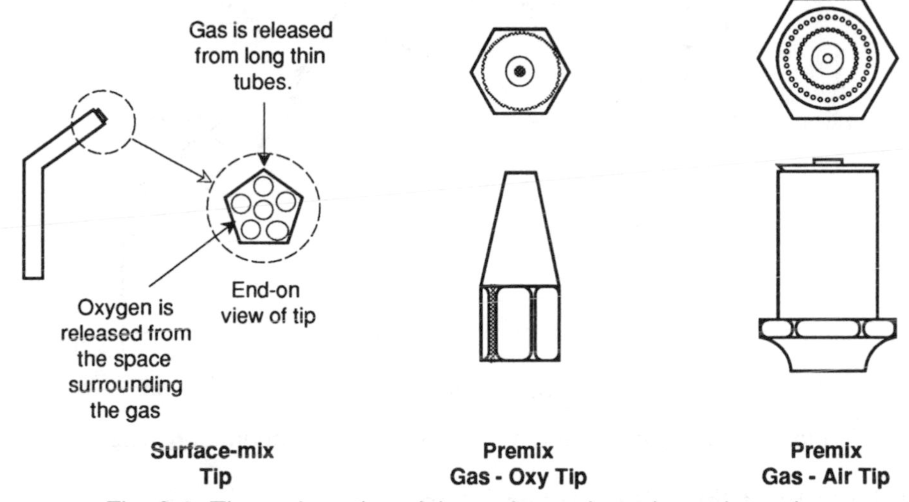

Fig. 8-2 The various tips of the surface-mix and premix torches.

In both torches, different flame sizes are obtained by controlling the amount of gas and oxygen. However, in the surface-mix design, you do not need to change tips to achieve the same range of flame sizes that are capable with the premix torch. On the other hand, the premix torch can achieve significantly smaller, and/or more concentrated, narrower flames than the surface-mix torch. The premix torch's flame has a cooler hole in the center of the bushy flame while the surface-mix's bushy flame is more uniform throughout (see Fig. 8-3). Some people like the subtle heat variations that are possible in the premix design, while others prefer the uniformity of the surface-mix. All in all, the decision of which type of torch to use (surface- or premix) is often based on the personal likes and dislikes of the user. For most laboratory uses, either is sufficient.

The connection between the *gas hose* connection of the torch and the gas source can be made using Tygon, an amber latex, or any similar tubing used to connect gas lines in the lab (or a propane tank).* The *oxygen hose*

* Acetylene should never be used on glass because it is a dirty gas and can leave undesirable deposits.

connection can also be connected using Tygon, a heavy-wall amber latex ($1/4$ in x $1/8$ in), any or similar tubing used to connect a regulator attached to an oxygen tank.* Open any needle valve on the regulator fully to control the gas flow from the needle valve on the torch, not from the regulator. Paired tubing is available where one tube is red or orange and the other tube is green. This tubing can be handy to know which tube is carrying gas (the orange or red), and which is carrying oxygen (green).

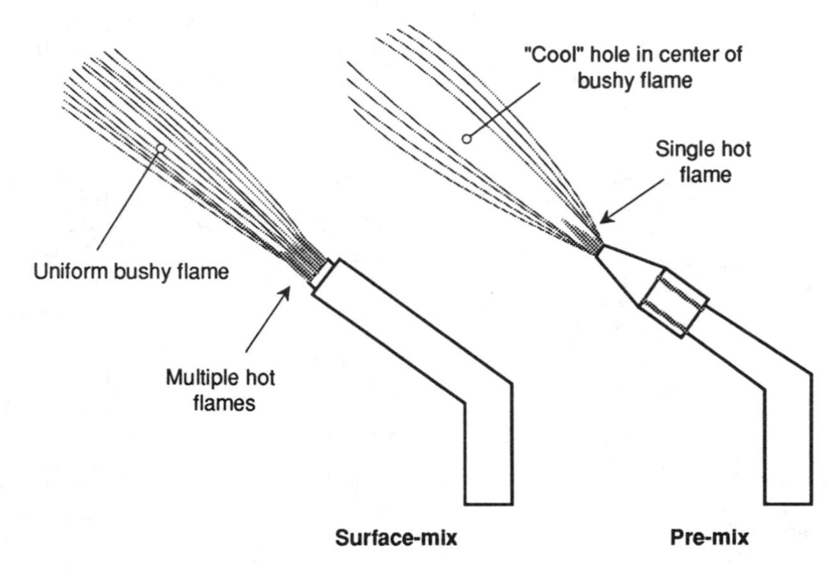

Fig. 8-3 Surface-mix and premix design flame characteristics.

8.1.2 How to Light a Gas-Oxygen Torch

Although it may seem straightforward, lighting a gas–oxygen torch seems to provide a challenge for a surprisingly large number of people. Probably the biggest mistakes in torch-lighting are either not allowing enough time for the gas to reach the torch tip, or flooding the torch region with too much gas. The former is similar to turning on the shower water but not waiting for the hot water to reach the tub. The latter is simply not allowing enough air in the region to support combustion. A simple step-by-step process is as follows:

1) Be sure that any valves from the gas and oxygen supplies to the torch are open and that both gases are available.
2) Open the torch's gas and then oxygen valves for a second or two to purge any air that might be in the lines, then close them. There is a definable change of noise when air has passed

* Before opening the main tank valve, set the outgoing pressure of the oxygen regulator to zero, then bring the pressure up to 5 to 10 lbs. Too much pressure on the hoses can blow them off the hose connections or cause the tubing to expand like a balloon.

through the torch and gas has arrived. In addition, you can smell the gas. Neither of these aids are available with oxygen.

3a) *If you are using a match or lighter*, first start the flame on the match (or the lighter), hold it near the tip of the torch, and then slowly open the gas (red) needle valve until you have a three- to four-inch flame.

3b) *If you are using a striker*, slowly open the gas (red) needle valve about 1/4 to 1/3 of a revolution, then strike the striker several inches from the tip of the torch. Once the flame has ignited, set the size of the flame to three to four inches.

4) *Slowly* open the oxygen (green) needle valve until the flame is lightly hissing. If the flame is flickering and jumping around, increase the oxygen. If the flame is *very* thin and hissing loudly, decrease the amount of oxygen.

Note: You control the *size* of the flame with the gas and the *character* of the flame with the oxygen. The character of a flame refers to its shape of either a cool-bushy flame (not enough oxygen) or hot-hissy flame (too much oxygen). If you make a major change of flame size, you will probably need to re-adjust the oxygen for the new flame size. Minor changes of size will maintain the same character.

8.1.3 How to Prevent a Pre-mix Torch from Popping

Popping is caused when the flow rate of the gas leaving the tip is slower than the flow rate of the flame burning its way back into the tip. Once the flame enters the tip, it can ignite the gas and oxygen within in an uncontrolled manner, resulting in an explosion and a loud "POP."

The chance of popping a torch can be decreased by making sure that you always have a "hissing" flame especially when decreasing the rate of oxygen or gas. Although annoying, the popping does not do any damage to the torch, although it is possible for the explosion to blow the flexible tubing off the torch's hose connections.

If a torch pops, a flame may continue to burn inside the torch, although this situation is not likely (a light whistling noise indicates that this condition exists). To extinguish an internal flame, first turn off the gas and oxygen needle valves on the torch. Then turn the oxygen valve forcefully open for a few seconds and close it. Finally, relight the torch and continue. Although they are common occurrences, experience will help you avoid "pops."

Because of the propensity for the premix torch to "pop," there is a recommended procedure for shutting the torch off. Even if you are capable of controlling the gas flow so proficiently that you never pop your torch, this procedure is still recommended.

1) Open the oxygen needle valve to increase the flow rate of the oxygen. Keep increasing the flow rate until the flame is blown out. A good half to three-quarter turn is usually sufficient.
2) Turn off the oxygen.
3) Turn off the gas.
4) Turn off the source of the gas
5) Turn off the oxygen tank.

Note: It is better to blow the flame out, and then turn off the gases than to turn off the gases before the flame is out to reduce the chance of the torch "popping."

The surface mix torch does not require special shut-off procedures because the surface-mix torch cannot "pop." Therefore, it does not make a difference which gas you turn off first. There is also no reason to blow out the flame of a surface-mix torch—besides, you cannot.

8.2 Using the Gas-Oxygen Torch

8.2.1 Uses for the Gas–Oxygen Torch in the Lab

The most common use for the gas–oxygen torch in the laboratory is to work on glass, and as anyone who has ever played with glass has found out, glass is not very forgiving. There are however, a few operations that can be mastered by a non-glassblower, and "tipping off" (sealing off) a sample in an NMR tube is one of them. It is a simple operation that can be mastered in a short time by most students and technicians.

In addition, it is possible for a beginner to fire-polish laboratory glassware that has chipped or cracked edges. However, be forewarned: any actions on broken or damaged glassware should be done with an "it is lost anyway" attitude. When inexperienced technicians try their hand at glassblowing, they learn fast that it's not as easy as it looks. But that does not mean you should not try. By accepting that any broken glassware is lost anyway, anything that survives the repair is bonus glassware. With time, as more experience is gained, more glassware will be saved. One of the most important pieces of knowledge that one needs to obtain with glass repair is the awareness of what you cannot repair.

8.2.2 How to Tip Off a Sample

It is a common practice to seal a sample in a closed glass tube for storage or for further study by means of EPR (Electron Paramagnetic Resonance) or NMR (Nuclear Magnetic Resonance). Although it is not difficult to tip off a sample, if done poorly, the NMR tube will spin very badly. If done badly, the sample may be destroyed along with hours, weeks, or months of work.

When you heat glass to its softening temperature, two separate forces begin to control its behavior: surface tension (the material clings to itself, which is seen as glass beading up), and gravity (soft glass will drip down). A bead of glass will drip when gravity on the glass is stronger than the surface tension forces that hold it together.

There are two additional forces involved when tipping off a sample: gas pressure buildup and vacuum from a vacuum system. Positive pressure is caused by inadvertently heating a sample (causing a pressure buildup), which makes it very difficult to seal off the sample. The possibility of a 'glass aneurysm' or a rupture of the glass wall is likely from a positive pressure buildup (see Fig. 8-4, Samples A and B). On the other hand, if the sample is prepared on a vacuum line, and too much heating is done on one spot for too

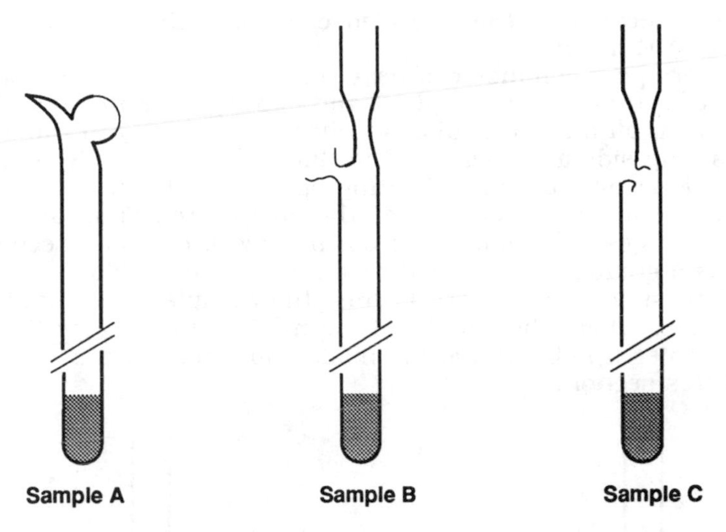
Sample A
Sample B
Sample C
A bubble and a blown-out hole caused by the pressure of the sample being heated by the tipping off process.
A sucked in hole caused by the vacuum in the system line when tipping off.

Fig. 8-4 Problems in tipping off samples.

long, the tube will burst in rather than out (see Fig. 8-4, Sample C). Not only may this bursting destroy the sample, but allowing air into a vacuum line can damage or destroy parts or sections of the vacuum system.

Whether a sample tube is attached to a vacuum line or not, the easiest way to prevent a positive buildup of pressure is to freeze the sample with a slush bath or liquid nitrogen (depending on the solution's freezing temperature or what coolants you have available). By freezing the sample solution you accomplish three objectives: one, you protect the sample by protecting it from the heat of the flame; two, the sample is not likely to evaporate and thereby cause a buildup of pressure (and cause a pressure burst as shown in Fig. 8-4, Samples A and B); and three, the freezing will make a cryogenic pump that facilitates the tipping off process by creating a vacuum. The sample must remain frozen during the entire tipping off procedure, so do not remove it from the coolant until the tipping off process is complete.

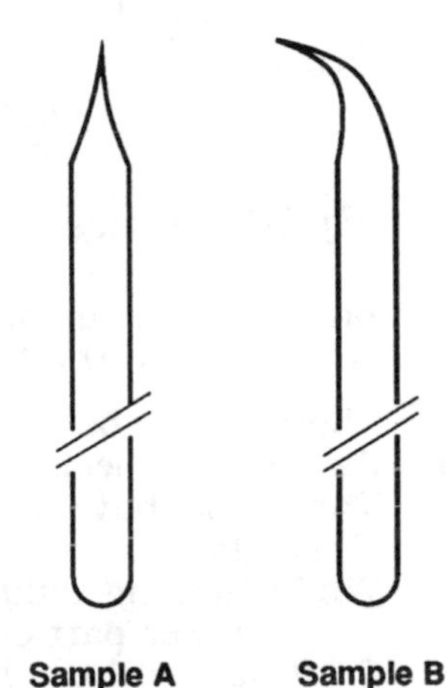
Sample A
Sample B
A good tip off
A poor tip off

Fig. 8-5 Alignment is important when "tipping off."

Figure 8-5 shows two examples of a tipped off tube: Sample A is what you want, with the tipped off section concentrically and linearly straight. Sample B is not acceptable because the tipped off area is bent off from center. Because of this lopsided condition, the tube will not spin properly in an NMR machine.

Glass is a very poor conductor of heat, and therefore, it is not possible to heat one side of a tube to be tipped off and expect to be finished. The only way to properly tip off a sample tube is to heat it uniformly in the area to be tipped, that is, one side and then 180° to that area (see Fig. 8-6). The gradual heating is done by alternately heating opposite sides in a frequent and uniform manner. You must try to heat the entire area gradually and uniformly. Glass becomes soft before it becomes juicy. If one side becomes juicy too fast there is a greater chance for the wall to suck in or blow out.

Whenever possible, before work begins, the sample tube should have a constriction made, where the tip-off will be made. The smaller the internal diameter, the easier tip-off will be. Pre-made constrictions are shown on all the figures in this section.

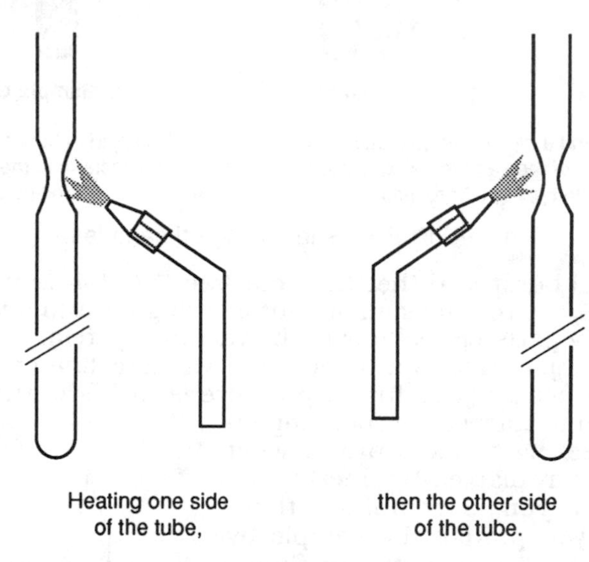

Heating one side
of the tube,
then the other side
of the tube.

Fig. 8-6 You must heat both sides of a tube to obtain an even tip-off.

The proper sequence of steps for making a tip-off are shown in series by Fig. 8-7 and listed in the following sequence:

1) The sample, before beginning.
2) As the tube begins to collapse, the walls begin to thicken
3) The walls have collapsed to the point of closing off the lower section.
4) While heating *only the middle of the closed section*, start to pull the lower part of the tube down.
5) Continue pulling the lower section down while heating the middle of the section.
6) The middle of the section will become thin enough to be cut by the force of the flame.

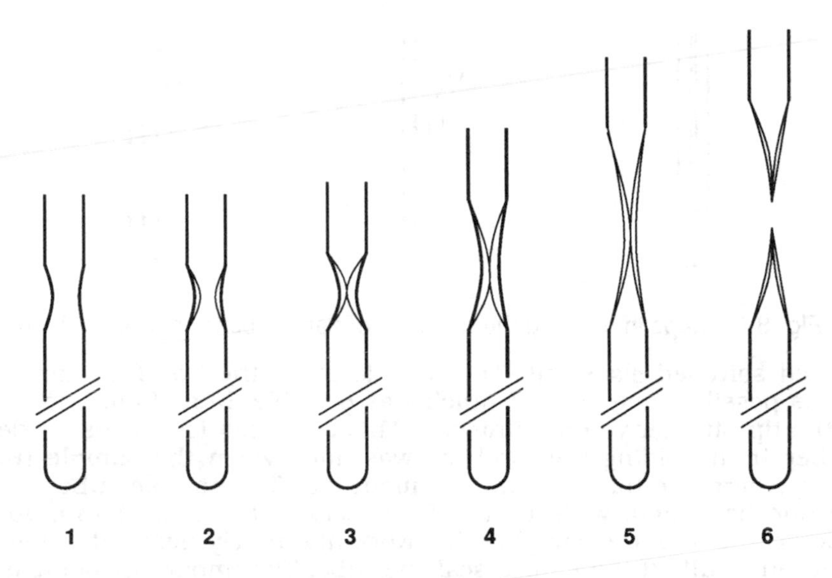

Fig. 8-7 Steps to a good tip off.

It is possible to shorten the length of the point on the sample tube as long as you heat well above the closed section of the tipped off area as seen in Fig. 8-8.

It is seldom possible to hold the lower portion of a sample tube with your hand as you tip it off. There is too much danger of burning your hand in any coolant or when heating the sample. However, by bending a pair of tweezers as shown in Fig. 8-9, it is easy to support any tube. Good quality tweezers are made with hardened metal and attempting to bend tweezer tips while they are cold may cause breakage. It may be necessary to heat the part of the tweezers that you wish to bend in a gas–oxygen torch flame to prevent breakage. After all bending is complete, reheat the entire section in the flame until it glows red and then immediately quench in water to re-temper the metal for hardness.

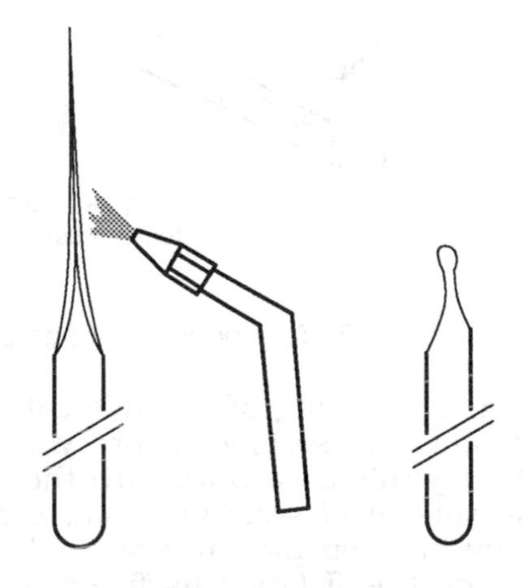

Fig. 8-8 Keep heat away from the sample.

There are times when you might do a tip-off of a tube that is not connected to a vacuum system and it may not be practical to freeze the sample within the tube. In these situations you will not have a vacuum drawing the walls of the tubing in. Instead there will be only the pressure expanding

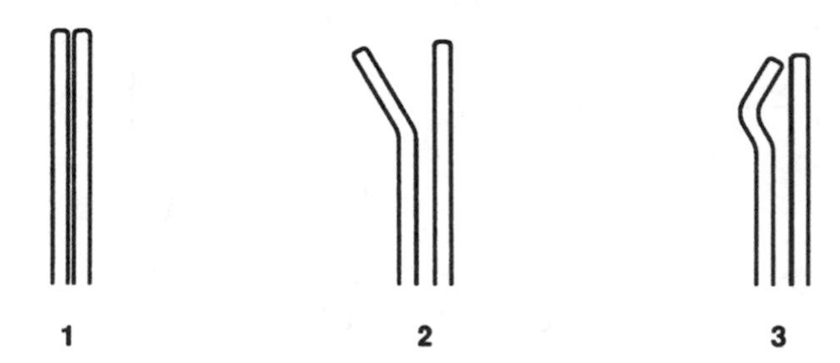

Fig. 8-9 Steps in bending the tips of tweezers to better support tubing.

the heated softened glass out. The trick is to do the tip-off as far from the sample as possible, and do it as quickly as possible. In addition, do not try to repair the tip-off after you are finished. This operation is a one-shot deal.

Rather than holding the torch as was done when the sample tube was stationary, here the torch should be mounted. The sample tube should be held in one hand and, with a pair of tweezers in the other hand, you must grab the open end of the sample tube. Now, alternately heat both sides of the tube end and pull off the end to seal the tube. The important point is not to dwell the flame on the tube and heat the solution, thereby raising the pressure inside the tube (see Fig. 8-10).

This "open tube technique" should never be done on flammable solutions unless the sample is frozen!

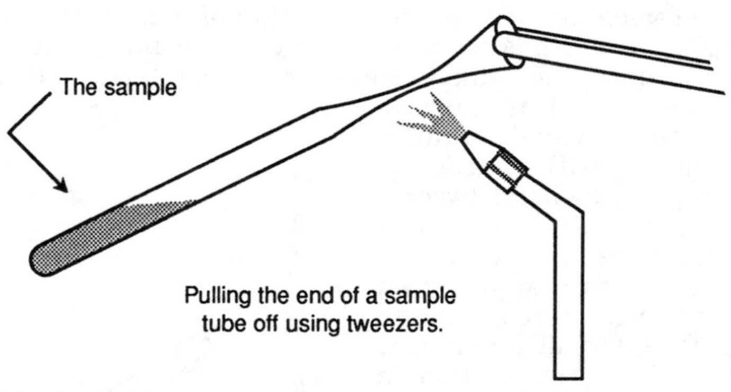

Fig. 8-10 Tipping off a sample not connected to a vacuum line.

Practice tipping off sample tubes that mimic the condition and environment of a real sample before you try with an important sample. In other words, try this procedure with the solvent that you plan to use (i.e., if water is the solvent of your material, practice with water), attached to the same vacuum line you plan to use, and in the state it will be in when you will be doing the tip-off (such as frozen in liquid nitrogen). Remember to keep the flame of the torch away from the walls of the Dewar containing the liquid nitrogen.

8.2.3 How to Fire-polish the End of a Glass Tube

Chipped and broken ends of tubes are common sights in the average laboratory. Because it is easy to repair this dangerous and unnecessary situation, there is no reason not to take a torch in hand and repair the problem. However, there are a few limitations involved:

1) If the glass is thicker than 2 to 3 mm (3⁄32 to 1⁄8 in) do not fire-polish the glass because you are likely to cause more damage than already exists. The heat-induced strain caused by the gas–oxygen torch is likely to crack the glass after it cools. Some examples of glass that should not be fire-polished are bases of graduated cylinders or rims of funnels (see Fig. 8-11).

2) If there are crack lines radiating from a broken section (see Fig. 8-11), fire-polishing by an inexperienced person is not advised, or should not be done with great expectations. Under the heat of a flame, cracks tend to spread far beyond your wildest imagination or your worst fears.

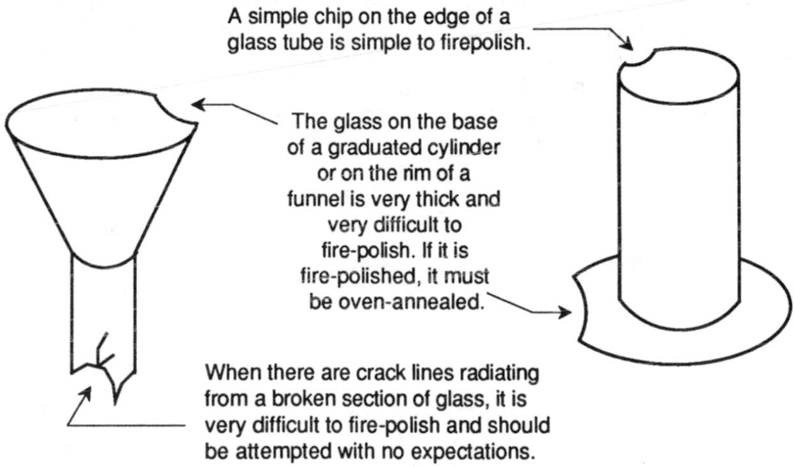

A simple chip on the edge of a glass tube is simple to firepolish.
The glass on the base of a graduated cylinder or on the rim of a funnel is very thick and very difficult to fire-polish. If it is fire-polished, it must be oven-annealed.
When there are crack lines radiating from a broken section of glass, it is very difficult to fire-polish and should be attempted with no expectations.

Fig. 8-11 Problem repairs of broken glassware.

(**Safety Note**: This manual is not meant to teach glassblowing as there are already several excellent books for that purpose (see Appendix D). I should mention one safety aspect: it is easy to blow the softened end of a glass tube into a bubble. However, a little more pressure can easily cause the bubble to burst leaving little "glass angels" floating around the room. Aside from being difficult to clean up, these "angels" can be very dangerous if inhaled. Do not try to make them.)

If you have an oven that can obtain temperatures of 565°C (1049°F) or greater and is big enough to contain the entire piece that you have fire-polished, you should formally anneal any piece you repair. After placing the entire piece in the oven, bring the oven's temperature up to 565°C, hold that temperature for about fifteen minutes, and turn the oven off. It is best to leave the piece in the oven until it reaches room temperature, at least three to four hours.

8.2.4 Brazing and Silver Soldering

The high heat that the gas–oxygen torch can deliver allows it to be used for both brazing and silver soldering. However, because you can vary the gas and oxygen content of the flame, you can provide either an oxidizing or reducing environment to your work. Because brazing and silver soldering cannot be done with oxide on the material's surface, using an oxidizing flame will prevent a good seal.

Normally a flux is used to remove oxides from surfaces when brazing or soldering, and a gas-oxide torch also requires the use of a flux. However, if the flame is oxygen-rich, as the flux is boiled or burnt off, the excess oxygen from the torch will oxidize the metal's surface before the solder can wet the surface. In a reducing flame, as the flux is boiled or burnt off, the reducing flame will maintain the pure metal until the solder can flow over and wet the surface.

To make a reducing flame on a Gas–oxygen torch, simply cut back the oxygen flow until there is a bit of yellow in the flame. There may be a desire to increase the oxygen flow to obtain a hot 'hissy' flame to get the metal hotter so the solder will flow. Unfortunately, this flame is the oxidizing flame that will prevent the solder from wetting the metal and completing the brazing or soldering operation.

Appendix

Appendix A Preparing Drawings for the Glass Shop

A.1 The Problems with Requests to Glass Shops

There are many brilliant men and women who are excellent teachers, can analyze the most complex data, and can draw the most complex molecules or mathematical equations; yet, many of these persons cannot communicate the shapes of even simple items on a piece of paper. In addition to this complication, when an object is made of glass, new complications are created. These new problems typically derive from the fact that working with glass is like working with no other material on earth. Thus, the average person will try to rationalize construction with materials they are more familiar with, such as wood or metal. In addition, it is unlikely that this average person will have had the opportunity to learn the parameters and limitations of glassware construction.

I am in no way trying to belittle the problems that machine, wood, or electrical shops have with poorly-made or inadequate drawings. However, sawing, turning, gluing, soldering, drilling, and cutting are common terms that are understood by most people. The terminology of glass is different: fire-polishing, fusing, blowing open a hole, and spinning a base are as foreign to the average person as the tools used. Because of misunderstandings of the tools and operations used with glass, what a person asks for is often not what they want. Over the years, I've had requests for 'drilling a hole in a tube and sticking a tube over the hole in the tube so that material could go several ways,' or 'welding a flat glass plate on the end of a tube so that it could stand up.' In these two examples, what the requestors wanted were a "T" seal and a base to a column, respectively (see Fig. A-1). If I had done exactly what was requested, the costs could have been phenomenal.

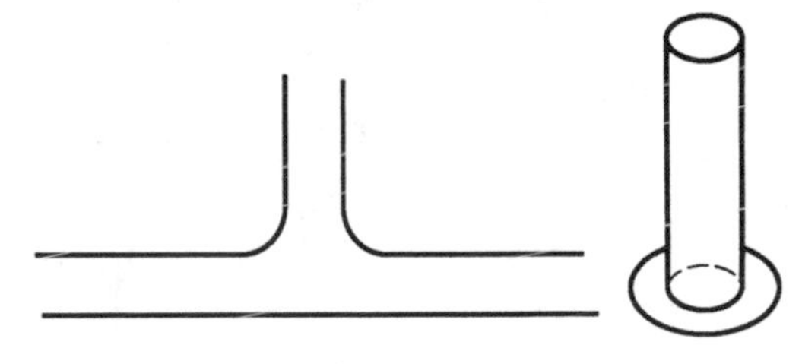

Fig. A-1 Examples of requests to a glass shop.

A.2 Suggestions for Glassware Requests

Whether you have a glass shop in your facility, or send your glassware requests to an outside glass shop, by improving your apparatus requests you will save time and money. By simply stating what is wanted, providing necessary dimensions and related construction materials, and avoiding discussions on how to make the item, much time and confusion can be avoided.

The Sketch. Prepare a reasonably accurate sketch. You do not have to be a draftsman or an artist. However, you are not a two-year old with a crayon. It is likely that the more time you spend on trying to figure out what your drawing is trying to represent, the more you will be charged.

The Measurements. Measurements should preferably be given in centimeters (or millimeters). English measurements are not recommended, and tenths of English measurements are not acceptable.

Tubing Identification. Commercially available tubing is only available in metric sizes. These tubing sizes are shown in Chapter 3. American-made medium and heavy wall tubing is made to English measurement dimensions, but are now listed in metric equivalent sizes. For example, 1-in tubing is listed as 25.4 mm. Commercially-available sizes (both metric and English) of medium and heavy wall tubing are shown in Chapter 3.

Demountable Joint Identification. Joints should be drawn *and* identified as being either an inner (male) or outer (female) section. Size should be identified as well. Ball-and-socket joints should be equally identified (see Fig. A-2). Because O-ring joints do not have inner and outer sections, only their sizes are is required.

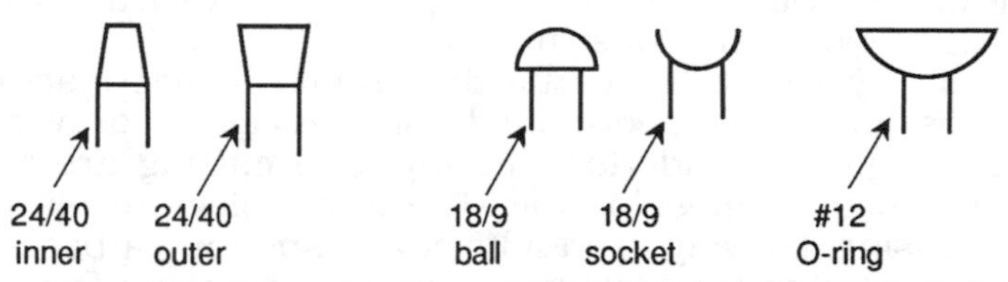

Fig. A-2 Examples of complete joint identification.

In addition, hooks always point away from joints and cannot be attached to joints (see Fig. A-3). The former is obvious because hooks are used to attach springs used to keep pieces together. If hooks pointed the wrong way, they cannot function. The location of hooks is more subtle because problems can occur if a hook is placed on the outside of an outer joint or too close to ground glass. By misplacing a hook, there is a risk of warping the shape of the ground section, which may cause a poor fit resulting in a leaky joint.

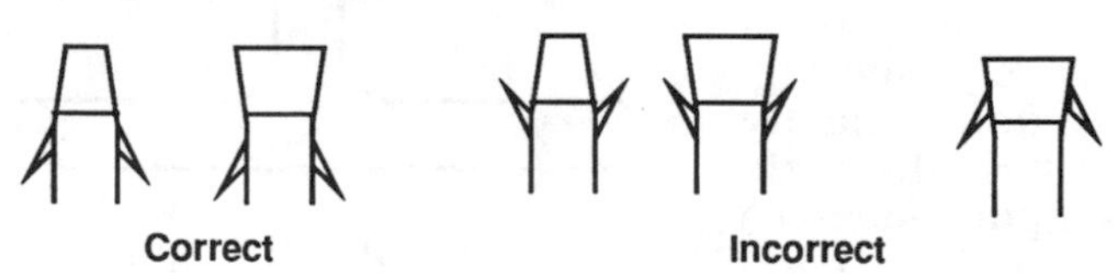

Fig. A-3 Proper orientation and placement of hooks.

"But you never told me." Leave nothing understood or assumed unless you are willing to accept whatever is handed back to you. Therefore, leave your name and how (and where) the glassblower may get in touch with you if additional questions should develop. If you are working with (or for) someone else, leave a name and how to get in touch with that person, as well as your own information.

Provide the Right Measurements. Give measurements from specified areas that cannot vary. For example, see Fig. A-4(a), where the length of the cold trap's internal tube is ambiguous by the "A" measurement because "C" can vary extensively. On the other hand, the measurement from the ground area of the joint to the end of the inside tube is specific. The "B" (or "B' ") measurement is preferred.

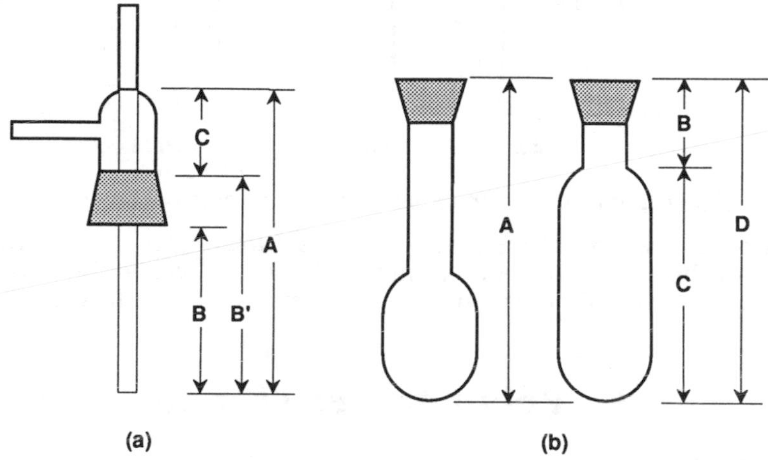

Fig. A-4 Where measurements are made is important.

Provide measurements when a change occurs, such as the seal of one size tubing to another (see Fig. A-4(b)). The length of the internal tube of the cold trap is ambiguous with the "A" measurement because the location of "B" and "C" can vary extensively. Two preferred measurements would be either "B" and "C," or "B" and "D."

Do Not Make Your Primary Drawing in 3-D. Most of these types of drawings are incomprehensible or confusing, and are difficult to place measurements. An *extra* sketch in 3-D can sometimes be helpful.

Accuracy. Most glasswork is done within an accuracy of ±1.5 mm ($^{1}/_{16}$th of an inch). If you require greater accuracy—say so. Similarly, if the glassware fits into some other device, or some other device fits into the glassware—say so. For an example, see Fig. A-5. If item "A" is ordered, and it is required to fit into item "B," it is important to provide the following specifications of item "A" to the glassblower:

1) The maximum O.D. of the (lower) tube.
2) The minimum length of the (lower) tube.
3) The maximum O.D. of the ball.

Whenever possible, provide item "B" to the glassblower so that he or she can test the item before it is delivered.

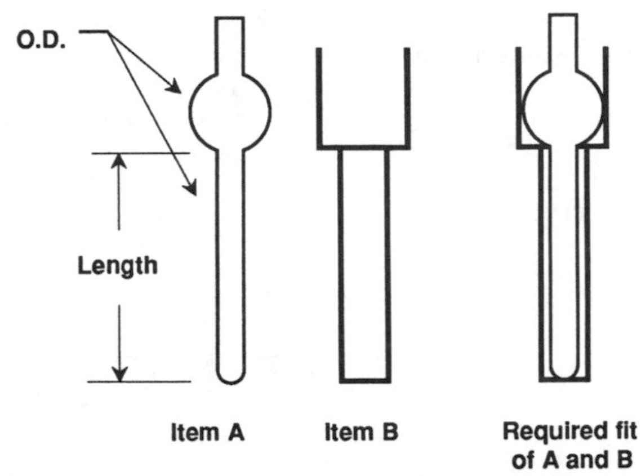

Fig. A-5 If your piece needs to fit within something, say so.

Listen to Any Suggestions from the Glassblower. He or she knows what glass can (and how to do it), and cannot do (and how to avoid it). In addition, he or she makes, and often repairs items (perhaps similar to what you want) every day.

B Polymer Resistance

B.1 Introduction

The ability of polymers (plastics) to resist chemical attack is generally good. Obviously, some plastics can better resist some chemical types better than others. Unfortunately, the results obtained from testing one chemical with one polymer is not going to provide conclusive information on the survivability of an experimental setup because other factors such as heat, pressure, and internal (or external) stresses may play predominant parts. A polymer that normally indicates good compatibility (with a given chemical) may fail outright. Thus, caution is advised, and pretesting, where possible, is recommended. These problems notwithstanding, knowing the compatibility of a given polymer with a given chemical can significantly help a researcher or technician avoid early pitfalls.

The Nalge Company, manufacturer of Nalgene Labware, has performed extensive testing with the majority of industrial and laboratory plastics. With permission of the Nalge Company, I have included the *Chemical Resistance Chart* from their catalog (see Table B 1). In addition, I have preceded this table with a summary of their analysis on the various plastics tested. For more information please contact:

Technical Service
Nalge Company
Box 20365
Rochester, New York 14602-0365
(716) 586-8800
Fax (716) 586-8431

The following are various plastic families and their respective specific plastics that are included in the *Chemical Resistance Chart*. The emboldened uppercase letters correspond to the plastics in the chart. These acronyms are generally accepted, but are not universal.

B.2 Polyolefins

The polyolefins are high molecular weight hydrocarbons. They include low-density, linear low-density and high-density polyethylene, polyallomer, polypropylene, and polymethylpentene. All are break-resistant, nontoxic, and non-contaminating. These plastics are the only one lighter than water. They easily withstand exposure to nearly all chemicals at room temperature for up to 24 hours. Strong oxidizing agents eventually cause embrittlement. All polyolefins can be damaged by long exposure to uv light. The Chemical Resistance Chart details the following polyolefin types:

Polyethylene: The polymerization of ethylene results in the essentially branched-chain, high molecular weight hydrocarbon polyethylene. The polyethylenes are classified according to the relative degree of branching (side chain formation) in their molecular structures, which can be controlled with selective catalysts.

Like other polyolefins, the polyethylenes are chemically inert. Strong oxidizing agents will eventually cause oxidation and embrittlement. They have no known solvent at room temperature. Aggressive solvents will cause softening or swelling, but these effects are normally reversible.

The following are a list of commonly used polyethylenes used in the lab:

LDPE (low-density polyethylene).* These have more extensive branching, resulting in a less compact molecular structure.

HDPE (*high-density polyethylene*).** These plastics have minimal branching, which makes them more rigid and less permeable than LDPEs.

PP/PA (*polypropylene/polyallomer*). These plastics are similar to polyethylene, but each unit of their chains has a methyl group attached. They are translucent, autoclavable, and have no known solvents at room temperature. They are slightly more susceptible than polyethylene to strong oxidizing agents. They offer the best stress-cracking resistance of the polyolefins. Products made of polypropylene are brittle at ambient temperature and may crack or break if dropped from benchtop height.

* Formerly referred to as conventional polyethylene (CPE).
** Formerly referred to as linear polyethylene.

PMP (*polymethylpentene, "TPX™"*).* These plastics are similar to polypropylene, but have isobutyl groups instead of methyl groups attached to each monomer group of their chains. Their chemical resistances are close to those of PP. They are more easily softened by some hydrocarbons and chlorinated solvents. PMP is slightly more susceptible than PP to attack by oxidizing agents. PMP's excellent transparency, rigidity, and resistance to chemicals and high temperatures make it a superior material for labware. PMP withstands repeated autoclaving, even at 150°C. It can withstand intermittent exposure to temperatures as high as 175°C. Products made of polymethylpentene are brittle at ambient temperature and may crack or break if dropped from benchtop height.

PVC (*polyvinyl chloride*). These plastics are similar to polyethylene, but each of their units contain a chlorine atom. The chlorine atom renders PVC vulnerable to some solvents, but also makes it more resistant in many applications. PVC has extremely good resistance to oils (except essential oils) and very low permeability to most gases. Polyvinyl chloride is transparent and has a slight bluish tint. Narrow-mouth bottles made of this material are relatively thin-walled and can be flexed slightly. When blended with phthalate ester plasticizers, PVC becomes soft and pliable, providing the useful tubing found in every well-equipped laboratory.

PS (*polystyrene*). These plastics are rigid and non-toxic, with excellent dimensional stability and good chemical resistance to aqueous solutions. This glass-clear material is commonly used for disposable laboratory products. Products made of polystyrene are brittle at ambient temperature and may crack or break if dropped from benchtop height.

B.3 Engineering Resins

Engineering resins offer exceptional strength and durability in demanding lab applications. For specific uses, they are superior to the polyolefins. Typical products are centrifuge ware, filterware, and safety shields.

The following are a list of commonly used engineering resins used in the lab:

NYL (*nylon*). Polyamide is a group of linear polymers with repeating amide linkages along their backbones. These polymers are produced by an amidation of diamines dibasic acids, or polymerization of amino acids. Nylon is strong and tough. It resists chemicals with negligible permeation rates when used with organic solvents. However, it has poor resistance to strong mineral acids, oxidizing agents, and certain salts.

* Registered trademark of Mitsui & Co., Ltd.

PC (*polycarbonate*). These plastics are window-clear and amazingly strong and rigid. They are autoclavable, non-toxic, and the toughest of all thermo-plastics. PC is a special type of polyester in which dihydric phenols are joined through carbonate linkages. These linkages are subject to chemical reaction with bases and concentrated acids, hydrolytic attack at elevated temperatures (e.g., during autoclaving), and make PC soluble in various organic solvents. For many applications, the transparency and unusual strength of PC offset these limitations. Its strength and dimensional stability make it ideal for high-speed centrifuge ware.

PSF (*polysulfone*). Like polycarbonate, PSF is clear, strong, non-toxic, and extremely tough. PSF is less subject than PC to hyrolytic attack during autoclaving and has a natural straw-colored cast. PSF is resistant to acids, bases, aqueous solutions, aliphatic hydrocarbons, and alcohols. PSF is composed of phenylene units linked by three different chemical groups — isopropylidene, ether, and sulfone. Each of the three linkages imparts specific properties to the polymer, such as chemical resistance, temperature resistance, and impact strength.

B.4 Fluorocarbons

Teflon® FEP (*fluorinated ethylene propylene*).* This material is translucent, flexible, and feels heavy because of its high density. It resists all known chemicals except molten alkali metals, elemental fluorine, and fluorine precursors at elevated temperatures. It should not be used with concentrated perchloric acid. FEP withstands temperatures from -270°C to + 205°C, and may be sterilized repeatedly by all known chemical and thermal methods. It can even be boiled in nitric acid.

Teflon® TFE (*tetrafluoroethylene*).* This material is opaque, white, and has the lowest coefficient of friction of any solid. It makes superior stopcock plugs and separatory funnel plugs.

Teflon® PFA (*perfluoroalkoxy*).* This plastic is translucent and slightly flexible. It has the widest temperature range of the fluoropolymers — from -270°C to +250°C — with superior chemical resistance across the entire range. Compared to TFE at + 277°C, it has better strength, stiffness, and creep resistance. PFA also has a low coefficient of friction, outstanding antistick properties, and is flame-resistant.

Halar® ECTFE (*ethylenechlorotrifluoroethylene*).** This material is an alternating copolymer of ethylene and chlorotrifluroethylene. This fluoropolymer withstands continuous exposure to extreme temperatures and maintains excellent mechanical

* Registered trademark of E.I. du Pont de Nemours & Co. (Inc.).
** Registered trademark of Allied Corporation.

properties across this entire range (from cryogenic temperatures to 180°C). It has excellent electrical properties and chemical resistance, having no known solvent at 121°C. It is also non-burning and radiation-resistant. Its ease of processing affords a wide range of products.

Tefzel® ETFE (*ethylenetetrafluoroethylene*).* This plastic is white, translucent, and slightly flexible. It is a close analog of the Teflon fluorocarbons, an ethylene tetrafluoroethylene copolymer. ETFE shares the remarkable chemical and temperature resistance of Teflon TFE and FEP, and has even greater mechanical strength and impact resistance.

PVDF (*Kynar® polyvinylidene fluoride*).** This plastic is a fluoropolymer with alternating CH_2 and CF_2 groups. PVDF is an opaque white resin. Extremely pure, it is superior for non-contaminating applications. Mechanical strength and abrasion resistance are high, similar to ECTFE. It resists UV radiation. The maximum service temperature for rotationally-molded PVDF tanks is 100°C. Up to this temperature, PVDF has excellent chemical resistance to weak bases and salts, strong acids, liquid halogens, strong oxidizing agents, and aromatic, halogenated, and aliphatic solvents. However, organic bases and short chain ketones, esters, and oxygenated solvents will severely attack PVDF at room temperature. Fuming nitric acid and concentrated sulfuric acid will cause softening. At temperatures approaching the service limit, strong caustic solutions will cause partial dissolution. Autoclavable if tanks are empty and externally supported.

B. 5 Chemical Resistance Chart

The following table lists a wide variety of chemicals and their effects on a variety of polymers. In addition, several concentrations of a chemical are included if each affect the same polymer differently. Chemicals affect polymers by one (or more) of three different mechanisms:

1) *Chemical attack on the polymer chain*, with resultant reduction in physical properties, including oxidation, reaction of functional groups in or on the chain, and depolymerization.
2) *Physical change in the polymer*, including absorption of solvents (resulting in softening and swelling of the plastic), permeation of solvent through the plastic, and dissolution in the solvent.
3) *Stress-cracking from the interaction of a "stress-cracking agent,"**** with molded-in or external stresses.

* Registered trademarks of E.I. du Pont de Nemours & Co. (Inc.).

** Registered trademarks of Pennwalt Corp.

*** Environmental stress-cracking is the failure of plastic materials in the presence of certain chemicals types. This failure is not a result of chemical attack, but rather the simultaneous presence of three factors: tensile stress, a stress-cracking agent, and inherent susceptibility of the plastic to stress-cracking.

Because heat typically enhances chemical reactions, each polymer indicates the compatibility of two temperatures: the first one shows the compatibility at room temperature (20°C) and the second shows the compatibility at an elevated temperature (50°C). Thus, "GN" indicates that at 20°C the polymer is "good" to use, but at 50°C, the polymer is "not recommended." The following codes are used in Table B 1:

E = excellent G = good F = fair N = not recommended -- = (not tested)

Table B 1

Chemical Resistance Chart

Chemical	LDPE	HDPE	PP/PA	PMP	NYL	FEP/TFE/PFA	ECTFE/ETFE	PC	Rigid PVC	PSF	PS	PVDF
Acetaldehyde	GN	GF	GN	GN	EG	EE	GF	FN	GN	NN	NN	EE
Acetamide, Sat.	EE	EE	EE	EE	EE	EE	EE	NN	NN	NN	EE	–
Acetic Acid, 5%	EE	EE	EE	EE	FN	EE	EE	EG	EE	EE	EG	EE
Acetic Acid, 50%	EE	EE	EE	EE	NN	EE	EG	EG	EG	GG	GG	EE
Acetic Anhydride	NN	FF	GF	EG	NN	EE	EE	NN	NN	NN	NN	NN
Acetone	EE	EE	EE	EE	EE	EE	GF	NN	NN	NN	NN	NN
Acetonitrile	EE	EE	FN	FN	EE	EE	EE	NN	NN	NN	NN	EE
Acrylonitrile	EE	EE	FN	FN	EG	EE	EG	NN	NN	NN	NN	GF
Adipic Acid	EG	EE	EE	EE	EF	EE	EE	EE	EG	GG	EE	–
Alanine	EE	EE	EE	EE	EG	EE	EE	NN	NN	NN	EE	–
Allyl Alcohol	EE	EE	EE	EG	NN	EE	EE	GF	GF	GF	GF	–
Aluminum Hydroxide	EG	EE	EG	EG	EE	EE	EE	FN	EG	GG	GG	EE
Aluminum Salts	EE	EE	EE	EE	NN	EE	EE	EG	EE	EE	GG	EE
Amino Acids	EE	EE	EE	EE	EG	EE	EE	EE	EE	EE	EE	EE
Ammonia	EE	EE	EE	EE	FF	EE	EE	NN	EG	GF	GF	EE
Ammonium Acetate, Sat.	EE	EE	EE	EE	EG	EE	EE	EE	EE	EE	EE	EE
Ammonium Glycolate	EG	EE	EG	EG	GG	EE	EE	GF	EE	GG	EE	EE
Ammonium Hydroxide, 5%	EE	EE	EE	EE	GF	EE	EE	FN	EE	GG	EF	EE
Ammonium Hydroxide, 30%	EG	EE	EG	EG	FN	EE	EE	NN	EG	GG	GF	EE
Ammonium Oxalate	EG	EE	EG	EG	GF	EE	EE	EE	EE	EE	EE	EE
Ammonium Salts	EE	EE	EE	EE	NN	EE	EE	EG	EG	EE	GG	EE
n-Amyl Acetate	GF	EG	GF	GF	EE	EE	EE	NN	NN	NN	NN	EE
Amyl Chloride	NN	FN	NN	NN	EG	EE	EE	NN	NN	NN	NN	EE
Aniline	EG	EG	GF	GF	GF	EE	GN	FN	NN	NN	NN	EF
Aqua Regia	NN	NN	NN	NN	NN	EE	EG	NN	NN	NN	NN	FF
Benzaldehyde	EG	EE	EG	EG	EG	EE	EF	FN	NN	FF	NN	EE
Benzene	FN	GG	GF	GF	EE	EE	EG	NN	NN	NN	NN	EE
Benzoic Acid, Sat.	EE	EE	EG	EG	NN	EE	EE	EG	EG	FF	GG	EE
Benzyl Acetate	EG	EE	EG	EG	EG	EE	EG	FN	NN	NN	NN	–
Benzyl Alcohol	NN	FN	NN	NN	NN	EE	EE	GF	GF	NN	NN	EE
Bromine	NN	FN	NN	NN	NN	EE	EG	FN	GN	NN	NN	EE
Bromobenzene	NN	FN	NN	NN	EG	EE	GN	NN	NN	NN	NN	EE

Table B 1

Chemical Resistance Chart (cont.)

Chemical	LDPE	HDPE	PP/PA	PMP	NYL	FEP/TFE/PFA	ECTFE/ETFE	PC	Rigid PVC	PSF	PS	PVDF
Bromoform	NN	NN	NN	NN	FF	EE	GF	NN	NN	NN	NN	EE
Butadiene	NN	FN	NN	NN	FF	EE	EE	NN	FN	NN	NN	EE
Butyl Chloride	NN	NN	NN	FN	EG	EE	EE	NN	FN	NN	NN	EE
n-Butyl Acetate	GF	EG	GF	GF	EE	EE	EG	NN	NN	NN	NN	EE
n-Butyl Alcohol	EE	EE	EE	EG	NN	EE	EE	GF	GF	GF	EG	EE
sec-Butyl Alcohol	EG	EE	EG	EG	NN	EE	EE	GF	GG	GF	GG	EE
tert-Butyl Alcohol	EG	EE	EG	EG	NN	EE	EE	GF	EG	GF	EE	EE
Butyric Acid	NN	FN	NN	NN	FN	EE	EE	FN	GN	GG	NN	EE
Calcium Hydroxide, Conc.	EE	EE	EE	EE	NN	EE	EE	NN	EE	GG	GG	EE
Calcium Hypochlorite, Sat.	EE	EE	EE	EG	NN	EE	EE	FN	GF	EE	GF	EE
Carbazole	EE	EE	EE	EE	EE	EE	EE	NN	NN	NN	EE	–
Carbon Disulfide	NN	NN	NN	NN	EG	EE	EF	NN	NN	NN	NN	EE
Carbon Tetrachloride	FN	GF	GF	NN	EE	EE	EE	NN	GF	NN	NN	EE
Cedarwood Oil	NN	FN	NN	NN	EG	EE	EG	GF	FN	FF	NN	EE
Cellosolve Acetate	EG	EE	EG	EG	EE	EE	EG	FN	FN	NN	NN	EG
Chlorobenzene	NN	FN	NN	FF	EG	EE	EG	NN	NN	NN	NN	EE
Chlorine, 10% in Air	GN	EF	GN	GN	NN	EE	EE	EG	EE	NN	FN	EE
Chlorine 10% (Moist)	GN	GF	FN	GN	NN	EE	EE	GF	EG	NN	NN	EE
Chloroacetic Acid	EE	EE	EG	EG	NN	EE	EE	FN	FN	NN	GN	E-
p-Chloroacetophenone	EE	EE	EE	EE	EG	EE	EE	NN	NN	NN	NN	–
Chloroform	FN	GF	GF	NN	FF	EE	GF	NN	NN	NN	NN	EE
Chromic Acid, 10%	EE	EE	EE	EE	NN	EE	EE	GF	EG	NN	EE	EE
Chromic Acid, 50%	EE	EE	GF	GF	NN	EE	EE	FN	EF	NN	FF	EG
Cinnamon Oil	NN	FN	NN	NN	GF	EE	EG	GF	NN	FF	NN	–
Citric Acid, 10%	EE	EE	EE	EE	NN	EE	EE	EG	GG	EE	EG	EE
Cresol	NN	FN	GF	NN	NN	EE	EG	NN	NN	NN	NN	EE
Cyclohexane	FN	FN	FN	NN	EE	EE	EG	EG	GF	NN	NN	EE
Cyclohexanone	NN	FN	FN	GF	EG	EE	EE	NN	NN	NN	NN	FN
Cyclopentane	NN	FN	FN	FN	EE	EE	EE	NN	FN	NN	NN	EE
DeCalin	GF	EG	GF	FN	EE	EE	EE	NN	EG	NN	NN	–
n-decane	FN	FN	FN	FN	EE	EE	EE	FN	EG	GF	FN	EE
Diacetone Alcohol	FN	EE	EF	EE	NN	EE	EG	NN	NN	NN	GN	NN
o-Dichlorobenzene	FN	FF	FN	FN	EG	EE	EF	NN	NN	NN	NN	EE
p-Dichlorobenzene	FN	GF	GF	GF	EG	EE	EF	NN	NN	NN	NN	EE
1,2-Dichloroethane	NN	NN	NN	NN	EE	EE	EE	NN	FN	NN	NN	EE
2,4-Dichlorophenol	NN	NN	NN	FN	NN	EE	EE	NN	NN	NN	NN	EE
Diethyl Benzene	NN	FN	NN	NN	EE	EE	EG	FN	NN	NN	NN	–
Diethyl Ether	NN	FN	NN	NN	EE	EE	EG	NN	FN	NN	NN	EG
Diethyl Ketone	GF	GG	GG	GF	EE	EE	GF	NN	NN	NN	NN	NN
Diethyl Malonate	EE	EE	EE	EG	EE	EE	EE	FN	GN	FF	NN	EG
Diethylamine	NN	FN	GN	FF	EG	EE	EG	NN	NN	GF	GG	NN
Diethylene Glycol	EE	EE	EE	EE	EE	EE	EE	GF	FN	GG	GG	EE
Diethylene Glycol Ethyl Ether	EE	EE	EE	EE	EE	EE	EE	FN	FN	FF	NN	–
Dimethyl Acetamide	FN	EE	EE	FG	NN	EE	EG	NN	NN	NN	NN	NN

Table B 1

Chemical Resistance Chart (cont.)

Chemical	LDPE	HDPE	PP/PA	PMP	NYL	FEP/TFE/PFA	ECTFE/ETFE	PC	Rigid PVC	PSF	PS	PVDF
Dimethyl Formamide	EE	EE	EE	EE	GF	EE	GG	NN	FN	NN	NN	NN
Dimethylsulfoxide	EE	EE	EE	EE	EE	EE	EG	NN	NN	NN	EG	–
1,4-Dioxane	GF	GG	GF	GF	EF	EE	EF	GF	FN	GF	NN	NN
Dipropylene Glycol	EE	EE	EE	EE	EE	EE	EE	GF	GF	GG	EE	–
Ether	NN	FN	NN	NN	EE	EE	EG	NN	FN	NN	NN	EG
Ethyl Acetate	EE	EE	EE	FN	EE	EE	EE	NN	NN	NN	NN	NN
Ethyl Alcohol (Absolute)	EG	EE	EG	EG	NN	EE	EE	EG	EG	EG	FN	EE
Ethyl Alcohol, 40%	EG	EE	EG	EG	NN	EE	EE	EG	EE	EG	GF	EE
Ethyl Benzene	FN	GF	FN	FN	EE	EE	GF	NN	NN	NN	NN	–
Ethyl Benzoate	FF	GG	GF	GF	EE	EE	EG	NN	NN	NN	NN	NN
Ethyl Butyrate	GN	GF	GN	FN	EE	EE	EG	NN	NN	NN	NN	NN
Ethyl Chloride Liquid	FN	FF	FN	FN	GF	EE	EE	NN	NN	NN	NN	EE
Ethyl Cyanoacetate	EE	EE	EE	EE	GF	EE	EE	FN	FN	FF	GN	NN
Ethyl Lactate	EE	EE	EE	EE	EG	EE	EE	FN	FN	FF	FN	NN
Ethylene Chloride	GN	GF	FN	NN	EG	EE	EE	NN	NN	NN	NN	EE
Ethylene Glycol	EE	EE	EE	EE	EE	EE	EE	GF	EE	EE	EE	EE
Ethylene Glycol Methyl Ether	EE	EE	EE	EE	EE	EE	EE	FN	FN	FF	NN	–
Ethylene Oxide	FF	GF	FF	FN	EE	EE	EE	FN	FN	EE	NN	EE
Fatty Acids	EG	EE	EG	EG	EE	EE	EG	GF	EE	GG	EF	EE
Fluorides	EE	EE	EE	EE	EE	EE	EE	EE	EE	EE	GG	EE
Fluorine	FN	GN	FN	FN	NN	EG	EF	GF	EG	NN	NN	–
Formaldehyde, 10%	EE	EE	EE	EG	GF	EE	EE	EG	GF	GF	FN	EE
Formaldehyde, 40%	EG	EE	EG	EG	GF	EE	EE	EG	GF	GF	NN	EE
Formic Acid, 3%	EG	EE	EG	EG	NN	EE	EE	EG	GF	GG	EG	EE
Formic Acid, 50%	EG	EE	EG	EG	NN	EE	EE	EG	GF	GG	FF	EE
Formic Acid, 98-100%	EG	EE	EG	EF	NN	EE	EE	EF	FN	FF	FF	EE
Freon TF	EG	EG	EG	FN	–	EE	EG	GF	GF	EG	FN	EE
Fuel Oil	FN	GF	EG	GF	EE	EE	EE	EG	EE	EG	NN	EE
Gasoline	FN	GG	GF	GF	EE	EE	EE	FF	GN	FF	NN	EE
Glacial Acetic Acid	EG	EE	EG	EG	NN	EE	EE	NN	EG	FN	NN	EG
Glutaraldehyde (Disinfectant)	EG	EE	EE	FF	EG	EE	EG	EF	EE	GG	EF	EE
Glycerine	EE	EE	EE	EE	EE	EE	EE	EE	EE	EE	EE	EE
n-Heptane	FN	GF	FF	FF	EE	EE	EE	EG	GF	EG	NN	EE
Hexane	NN	GF	GF	FN	EE	EE	EE	FN	GN	EG	NN	EE
Hydrazine	NN	NN	NN	NN	NN	EE	GF	NN	NN	NN	NN	NN
Hydrochloric Acid, 1-5%	EE	EE	EE	EG	NN	EE	EE	EE	EE	EE	EE	EE
Hydrochloric Acid, 20%	EE	EE	EE	EG	NN	EE	EE	GF	EG	EE	EE	EE
Hydrochloric Acid, 35%	EE	EE	EG	EG	NN	EE	EE	NN	GF	EE	FF	EE
Hydrofluoric Acid, 4%	EG	EE	EG	EG	NN	EE	EE	GF	GF	GF	GF	EE
Hydrofluoric Acid, 48%	EE	EE	EE	EE	NN	EE	EE	NN	GF	FN	NN	EE
Hydrogen Peroxide, 3%	EE	EE	EE	EE	NN	EE	EE	EE	EE	EE	EG	EE
Hydrogen Peroxide, 30%	EG	EE	EG	EG	NN	EE	EE	EE	EE	EE	EG	EE
Hydrogen peroxide, 90%	EG	EE	EG	EG	NN	EE	EE	EE	EG	EE	EG	E-
Iodine Crystals	NN	NN	FN	GN	NN	EE	EG	NN	NN	NN	NN	EE

Table B 1

Chemical Resistance Chart (cont.)												
Chemical	**LDPE**	**HDPE**	**PP/PA**	**PMP**	**NYL**	**FEP/TFE/PFA**	**ECTFE/ETFE**	**PC**	**Rigid PVC**	**PSF**	**PS**	**PVDF**
Isobutyl Alcohol	EE	EE	EE	EG	NN	EE	EE	EG	EG	EG	GG	EE
Isopropyl Acetate	GF	EG	GF	GF	EE	EE	EG	NN	NN	NN	NN	–
Isopropyl Alcohol	EE	EE	EE	EE	NN	EE	EE	EE	EG	EE	EG	EE
Isopropyl Benzene	FN	GF	FN	NN	EG	EE	EG	NN	NN	NN	NN	–
Isopropyl Ether	NN	NN	NN	EE	EG	EE	EG	NN	NN	NN	NN	EE
Jet Fuel	FN	FN	FN	FN	EE	EE	EF	NN	EG	FN	GF	EE
Kerosene	FN	GG	GF	GF	EE	EE	GF	EE	EE	GF	NN	EE
Lacquer Thinner	NN	FN	FN	FF	EE	EE	EE	NN	NN	NN	NN	EE
Lactic Acid, 3%	EG	EE	EG	EG	NN	EE	EE	EG	GF	EE	GG	EG
Lactic Acid, 85%	EE	EE	EG	EG	NN	EE	EG	EG	GF	EE	GG	GF
Mercury	EE	EE	EE	EE	EE	EE	EG	NN	EE	EE	NN	EE
2-Methoxyethanol	EG	EE	EE	EE	EE	EE	EG	NN	GN	NN	NN	EE
Methoxyethyl Oleate	EG	EE	EG	EG	EG	EE	EE	FN	NN	NN	NN	–
Methyl Acetate	FN	FF	GF	EE	EE	EE	EG	NN	NN	NN	NN	NN
Methyl Alcohol	EE	EE	EE	EE	NN	EE	EE	GF	EF	GF	FN	EE
Methyl Ethyl Ketone	EG	EE	EG	NN	EE	EE	GF	NN	NN	NN	NN	NN
Methyl Isobutyl Ketone	GF	EG	GF	FF	EE	EE	GF	NN	NN	NN	NN	GN
Methyl Propyl Ketone	GF	EG	GF	FF	EE	EE	EG	NN	NN	NN	NN	NN
Methyl-t-butyl Ether	NN	FN	FN	EE	EG	EE	EG	NN	NN	NN	NN	EE
Methylene Chloride	FN	GF	FN	FN	GF	EE	GG	NN	NN	NN	NN	NN
Mineral Oil	GN	EE	EE	EG	EE	EE	EE	EG	EG	EE	EE	EE
Mineral Spirits	FN	FN	FN	EE	EE	EE	EG	NN	GN	FN	FF	EE
Nitric Acid, 1-10%	EE	EE	EE	EE	NN	EE	EE	EG	EG	EF	GN	EE
Nitric Acid, 50%	GG	GN	FN	GN	NN	EE	EE	GF	GF	GF	NN	EG
Nitric Acid, 70%	FN	GN	NN	GF	NN	EE	EE	NN	FN	NN	NN	GF
Nitrobenzene	NN	FN	NN	NN	FF	EE	EG	NN	NN	NN	NN	EN
Nitromethane	NN	FN	FN	EF	EE	EE	EF	NN	NN	NN	NN	GF
n-Octane	EE	EE	EE	EE	EE	EE	EE	GF	FN	GF	NN	EE
Orange Oil	FN	GF	GF	FF	GF	EE	EE	FF	FN	FF	NN	EE
Ozone	EG	EE	EG	EE	EG	EE	EE	EG	EG	EE	FF	EE
Perchloric Acid	GN	GN	GN	GN	NN	GF	EG	NN	GN	NN	GF	EE
Perchloroethylene	NN	NN	NN	NN	EE	EE	EE	NN	NN	NN	NN	EE
Phenol, Crystals	GN	GF	GN	FG	NN	EE	EE	NN	FN	FF	NN	EE
Phenol, Liquid	NN	NN	NN	NN	NN	EE	EF	NN	FN	NN	FN	EE
Phosphoric Acid, 1-5%	EE	EE	EE	EE	NN	EE	EE	EE	EE	EE	GG	EE
Phosphoric Acid, 85%	EE	EE	EG	EG	NN	EE	EE	EG	EG	EE	EG	EE
Picric Acid	NN	NN	NN	EE	NN	EE	GF	NN	NN	NN	GF	EE
Pine Oil	GN	EG	EG	GF	GF	EE	EG	GF	FN	FF	NN	EE
Potassium Hydroxide, 1%	EE	EE	EE	EE	FF	EE	EE	FN	EE	EE	GG	EE
Potassium Hydroxide, Conc.	EE	EE	EE	EE	FF	EE	EE	NN	EG	EE	GG	EG
Propane Gas	NN	FN	NN	NN	FF	EE	EE	FN	EG	FF	NN	EE
Propionic Acid	FN	EF	EG	EF	NN	EE	EF	NN	GN	GG	GN	EE
Propylene Glycol	EE	EE	EE	EE	EE	EE	EE	GF	FN	GG	EE	–
Propylene Oxide	EG	EE	EG	EG	EE	EE	FN	GF	FN	GG	NN	FN

Table B 1

Chemical Resistance Chart (cont.)

Chemical	LDPE	HDPE	PP/PA	PMP	NYL	FEP/TFE/PFA	ECTFE/ETFE	PC	Rigid PVC	PSF	PS	PVDF
Resorcinol, Sat.	EE	EE	EE	EE	NN	EE	EE	GF	FN	NN	GF	–
Resorcinol, 5%	EE	EE	EE	EE	NN	EE	EF	GF	GN	NN	GF	–
Salicylaldehyde	EG	EE	EG	EG	EG	EE	EN	GF	FN	FF	NN	EG
Salicyclic Acid, Powder	EE	EE	EE	EG	EG	EE	EE	EG	GF	EE	EE	EE
Salicyclic Acid, Sat.	EE	EE	EE	EE	NN	EE	EE	EG	GF	EE	EG	EE
Salt Solutions, Metallic	EE	EE	EE	EE	FF	EE	EE	EE	EE	EE	GG	EE
Silicone Oil	EG	EE	EE	EE	EE	EE	EE	EE	EE	EE	EG	EE
Silver Acetate	EE	EE	EE	EE	EF	EE	EE	EG	GG	EE	GG	EE
Silver Nitrate	EG	EE	EG	EE	NN	EE	EE	EE	EG	EE	GF	EE
Skydrol LD4	GF	EG	EG	EG	EG	EE	EE	NN	NN	NN	NN	EF
Sodium Acetate, Sat.	EE	EE	EE	EE	FF	EE	EE	EG	GF	EE	GG	EE
Sodium Hydroxide, 1%	EE	EE	EE	EE	EE	EE	EE	FN	EE	EE	GG	EE
Sodium Hydroxide, 50% to Sat.	GG	EE	EE	EE	GF	EE	EE	NN	NN	EG	EE	EG
Sodium Hypochlorite, 15%	EE	EE	GF	EE	NN	EE	EE	GF	EE	EE	EE	EE
Stearic Acid, Crystals	EE	EE	EE	EE	EF	EE	EE	EG	EG	GG	EG	EE
Sulfuric Acid, 1-6%	EE	EE	EE	EE	NN	EE	EE	EE	EG	EE	EG	EE
Sulfuric Acid, 20%	EE	EE	EG	EG	NN	EE	EE	EG	EG	EE	EG	EE
Sulfuric Acid, 60%	EG	EE	EG	EG	NN	EE	EE	GF	EG	EE	GN	EE
Sulfuric Acid, 98%	GG	GG	FN	GG	NN	EE	EE	NN	GN	NN	NN	EG
Sulfur Dioxide, Liq., 46 psig	NN	FN	NN	NN	NN	EE	EG	GN	FN	GG	NN	EE
Sulfur Dioxide, Wet or Dry	EE	EE	EE	EE	NN	EE	EE	EG	EG	GG	FN	GE
Sulfur Salts	FN	GF	FN	FN	NN	EE	EG	FN	NN	GG	NN	GF
Tartaric Acid	EE	EE	EE	EE	EF	EE	EE	EG	EG	EE	GG	EE
Tetrahydrofuran	FN	GF	GF	FF	EE	EE	GF	NN	NN	NN	NN	FN
Thionyl Chloride	NN	NN	NN	NN	NN	EE	EE	NN	NN	NN	NN	–
Toluene	FN	GG	GF	FF	EE	EE	EE	FN	NN	NN	NN	EE
Tributyl Citrate	GF	EG	GF	GF	EG	EE	EG	NN	FN	FF	NN	EF
Trichloroacetic Acid	FN	FF	FN	EE	NN	EE	EF	FN	FN	GG	FN	EG
1,2,4-Trichlorobenzene	NN	NN	NN	GF	GG	EE	EG	NN	NN	NN	NN	EE
Trichloroethane	NN	FN	NN	NN	EE	EG	NN	NN	NN	NN	NN	
Trichloroethylene	NN	FN	NN	NN	EE	EE	EG	NN	NN	NN	NN	EE
Triethylene Glycol	EE	EE	EE	EE	EE	EE	EE	EG	GF	EE	EG	–
2,2,4-Trimethylpentane	FN	FN	FN	FN	EE	EE	EG	NN	NN	GF	NN	EE
Tripropylene Glycol	EE	EE	EE	EE	EE	EE	EE	EG	GF	EE	EE	–
Tris Buffer, Solution	EG	EG	EG	EG	EE	EE	EE	GF	GF	GF	GN	EG
Turpentine	FN	GG	GF	FF	EE	EE	EE	FN	GF	NN	NN	EE
Undecyl Alcohol	EF	EG	EG	EG	EE	EE	EG	GF	EF	FF	GG	EE
Urea	EE	EE	EE	EG	EE	EE	EE	NN	GN	FF	EG	EE
Vinylidene Chloride	NN	FN	NN	NN	NN	EE	GF	NN	NN	NN	NN	EE
Xylene	GN	GF	FN	FN	EE	EE	EG	NN	NN	NN	NN	EE
Zinc Stearate	EE	EE	EE	EE	EE	EE	EE	EE	EE	EE	EE	EE

Appendix C Manufacturers

This appendix lists of some of the major U. S. manufacturers of the materials, components, equipment, and/or support described in this book. The list is organized by chapter.

Most laboratories order materials through laboratory supply-houses. This standard approach has developed because it is easy and efficient. It is also much easier to look through one or two catalogs than a room that is full of catalogs. However, there may be occasions when you need specific information that a supply-house cannot provide, or materials that a supply-house does not carry.

Any product or manufacturer mentioned here is not necessarily an endorsement of that product or manufacturer. Similarly, the lack of any mention of a product or manufacturer is not meant to withhold an endorsement. Often, the specific materials mentioned in this book were selected by my familiarity with them in my academic environment. Thus, materials peculiar to medical, industrial, and manufacturing environments may have been overlooked. Fortunately, because many specific laboratory equipment pieces have a variety of applications, a piece of equipment used in a chemistry laboratory may also be used in a geology laboratory.

If you prefer to purchase from a laboratory supply-house, the following are the most common supply-houses in the U. S. The listing shown below gives information for each supplier's main office. Because all of these suppliers have offices all across the U.S. and Canada, simply call the main office for the local office near you.

Baxter Diagnostics, Inc.
Scientific Products Division
1430 Waukegan Rd.
McGaw Park, IL 60085-6787
(708) 689-8410

CMS
General Offices
P.O. Box 1546
Houston, Texas 77251-1546
(713) 820-9898

Fisher Scientific Headquarters
711 Forbes Ave.
Pittsburgh, PA 15219
(412) 562-8300

Thomas Scientific
99 High Hill Rd
P.O. Box 99
Swedesboro, NJ 08085-0099
1-800-345-2100

VWR
P.O. Box 7900
San Francisco, CA 94120
(415) 468-7150
Telex: 3729921 VWR UC
Cable: VANROG
Fax: (415) 330-4185

Chapter 1

This list contains companies that specifically manufacture glass tubing and rod. Some of these companies specialize in speciality glasses (such as quartz glass), others manufacture a variety of glass types. There are a number of companies that sell glass tubing as well, but they were not listed here because the tubing they sell was probably made by one of the following companies.

Corning Glass Works
Science Products
Corning, N.Y. 14831
607) 974-4001 (N.Y. and Canada)
1-800-222-7740
(For all glass types)

Kimble Industrial Sales Office
537 Crystal Avenue
Vineland, NJ 08360-3257
(609) 692-0824 (in NJ)
FAX: 609-692-1964 (in NJ)
1-800-331-2706 (outside NJ)
FAX: 1-800-331-2714 (outside NJ)
(For all glass types)

GE Quartz Products
21800 Tungsten Rd.
Cleveland, OH 44117
(216) 266-3702
FAX: (216) 266-3702
(For quartz glass and a few electrical specialty glasses)

Heraeus Amersil
650 Jernees Mill Rd.
Sayreville, NJ 08872
(201) 254-2500
(For quartz glass)

Schott America
3 Odell Plaza
Yonkers, NY 10701
(914) 968-8900
FAX (914) 968-4422
(For borosilicate, quartz glass, and a few specialty glasses)

The companies listed below specialize in the products described in Sections 1.2 through 1.4 of Chapter 1 (flexible tubing, corks, rubber stoppers and enclosures, and O-rings).

The Gates Rubber Co.
999 South Broadway
P.O. Box 5887
Denver, CO 80217
(303) 744-5151
(for flexible tubing)

Hygenic Corporation
1245 Home Avenue
Akron, Oh. 44310
(216) 633-8460 (in OH)
1-800-321-2135 (outside OH)
(for flexible tubing)

Kent Latex Products, Inc.
1500 St. Clair Ave.
P.O. Box 668
Kent, Ohio 44240-0668
Telephone: (216) 673-1011
(for flexible tubing)

Nalge Company
A Subsidiary of Sybron Corp.
P.O. Box 20365
Rochester, NY 14602-0365
(716) 586-8800
FAX: (716) 586-8431
Telex: 97-8242
(for flexible tubing)

Norton Industrial Plastics
P.O. Box 350
Akron, OH 44309
(216) 798-9240
Telex: 433-8012
TWX: 810-431-2015
(for flexible tubing)

Precise Rubber Manufacturing Co.
19 Spielman Rd.
Fairfield, NJ 07006
(201) 227-4747
1-800-237-5678
FAX: 201-227-3938
(for flexible tubing)

Primeline Industries, Inc.
2508 East Bailey Road
Cuyahoga Falls, OH 44221
Telephone: (216) 929-2857
(for flexible tubing)

Stevens Inc.
395 Pleasant
P.O. Box 658
North Hampton, MA
(201) 782-8400
(for rubber stoppers)

Rhoades Rubber Corp.
150 Pleasant
P.O. Box 110
East Hampton, MA 01027
(413) 527-2100)
(for rubber stoppers)

Parco
2150 Parco Ave.
Ontario, CA 91761
(714) 947-2200
TWX: 910-581-1206
Cable: Parco Ontario
(for O-rings)

Chapter 2*

The companies and organizations listed below can help you with your lab's measurement needs.

NIST
National Institute of Standards and Technology
U.S. Department of Commerce
Publications and Programs Inquiries
Room E128, Administration Building
Gaithersburg, MD 20899
(301) 975-3058

ASTM
American Society for Testing and Materials
1916 Race Street
Philadelphia, Pa. 19103-1187
(215) 299-5585
FAX: 215-977-9679

McMaster-Carr
P.O. Box 54960
Los Angeles, CA 90054
(213) 692-5911

Capitol Tool & Supply
14530 Carmenita Road
Norwalk, CA 90650
(714) 523-5582 (in CA)
1-800-421-2541 (outside California)

Mettler Instrument Corporation
P.O. Box 71
Hightstown, NJ 08520-9944
1(800) METTLER

Sartorius Instruments
1430 Waukegan Road
McGaw Park, IL 60085
1-800-645-3108

Ohaus Corporation
29 Hanover Road
Florham Park, NJ 07932
1-800-526-0659

Cahn Instruments
16207 S. Carmentia
Cerritos, CA 90701
1-800-423-6641

Omega International Corporation
Post Office Box 2721
Stamford, Ct. 06906
1-800-622-2378
FAX: (203) 359-7807
Telex: 996404

Extech Instruments
150 Bear Rd.
Waltham, MA 02154
(617) 890-7440

Chapter 3

The following companies manufacture stopcocks, joints and custom glass items. Corning tubing and rod can be obtained from Corning Glass Works,

*Most of the glassware companies listed in Chapter 3 sell, or can provide. standard or custom volumetric ware as described in Chapter 2.

Kimble tubing and rod can be obtained from Kimble Industrial Sales Office, and Schott tubing and rod can be obtained both from the Glass Warehouse and Witeg Scientific.

Ace Glass Inc.
P. O. Box 688
1430 Northwest Boulevard
Vineland, NJ 08369
609) 692-3333
FAX: 1-800-543-6752

Andrews Glass Co.
Division of Fischer Porter Co.
410 S. Fourth St.
Vineland, NJ 08360
(609) 692-4435

Chemglass, Inc.
3861 North Mill Road
Vineland, NJ 08360
(609) 696-0014 (in NJ)
1-800-843-1794 (outside NJ)
FAX: 1-609-696-9102

Corning Glass Works
Science Products
Corning, NY 14831
(607) 974-4001 (N.Y. and Canada)
1-800-222-7740

Glass Warehouse
800 Orange Street
P. O. Box 1039
Millville, NJ 08332
(609) 327-5228
FAX: 1-609-825-9014

GM Associates
9803 Kitty Lane
Oakland, CA94603
(415) 430-0806
FAX: (415) 562-9809

Kimble Industrial Sales Office
537 Crystal Avenue
Vineland, NJ 08360-3257
(609) 692-0824 (in NJ)
1-800-331-2706 (outside NJ)
FAX: 609-692-1964 (in NJ)
FAX: 1-800-331-2714 (outside NJ)

Kontes
Spruce Street
P. O. Box 729
Vineland, NJ 08360
609) 692-8500 (in NJ)
1-800-223-7150 (outside NJ)
TWX: 510-687-8967
FAX: 609-692-3242

Lab Glass
P.O. Box 610
1172 Northwest Boulevard
Vineland, N.J 08360
(609) 691-3200
FAX: 1-800-522-1329

Wheaton
1501 N. Tenth Street
Millville, NJ 08332
1-800-225-1437
TLX: 55-1295 (WHEATON US)
FAX: 1-609-825-1368

Witeg Scientific
14235 Commerce Drive
Garden Grove, CA 92643
(714) 265-1855
FAX: (714) 265-1860

Chapter 4

Biddle Instruments
(for Apiezon greases and oils)
510 Township Lane Road
Blue Bell, PA 19422
(215) 646-9200

DuPont
DuPont Chemicals
Wilmington, Deleware 19898
(302) 774-2099 (in Wilmington DE)
1-800-441-9442

Chapter 5

Linde Division of Union Carbide
National Specialty Gases Office
100 Davidson Avenue
Somerset, NJ 08873
(201) 271-2600
TLX: 833-199
TWX: 710-997-9550

Matheson Gas Products
932 Paterson Plank Road
P.O. Box 85
East Rutherford, NJ 07073
(201) 933-2400
FAX: 201 933-1928
Telex: 424546 Matson

Chapter 6

Compressed Gas Association, Inc.
1235 Jefferson Davis Highway
Arlington, VA 22202
(703) 979-4341

Cryogenic Services, Inc.
P.O. Box 1312
Canton, GA 30114
1-800-241-7452

Linde Division of Union Carbide
National Specialty Gases Office
100 Davidson Avenue
Somerset, NJ 08873
(201) 271-2600
TLX: 833-199
TWX: 710-997-9550

Chapter 7

There are *many* companies that make equipment for vacuum systems, and most of them can provide a full range of materials from pumps to oils, and from gauges to leak detection equipment. Unfortunately, most of these vacuum systems are metal. For glass components, such as vacuum stopcocks, joints, and even glass diffusion pumps, see the suppliers listed for Chapter 3.

A&N Corporation
P.O. Box 878
Route 40 West
Inglis, FL 32649
(904) 447-2411
FAX: (904) 447-2322

CTI-Cryogenics
Helix Technology Corporation
266 Second Ave.
Waltham, MA 02254-9171
1-800-622-5400

Dalton Electric Heating Co.
28 Hayward St.,
Ipswich Business Park
Ipswich, MA 01938
(617) 356-9844

Edwards High Vacuum, Inc.
3279 Grand Island Blvd.
Grand Island, NY 14072
1-800- 462-2550 (in NY)
1-800- 828-6691 (outside NY)
Telex: 6854319 EDHIVAC
FAX: (716) 773-3864

Granville-Phillips
5675 E. Arapahoe
Boulder, CO 80303-1398
(303) 443-7660
1-800-222-5577
Telex: 45791 GPVAC
FAX: (303) 443-2546

High Vacuum Apparatus Mfg. Inc.
1763 Sabre St.
Hayward, CA 94545
(415) 785-2744
FAX: 1(415) 732-9853
TWX: 910-383-2045

Kurt J. Lesker Company
1515 Worthington Avenue
Clairton, PA 15025
1-800- 245-1656
FAX: (412) 233-4275

Larson Electronic Glass
2840 Bay Road
P.O. Box 371
Redwood City, CA 94064
(415) 369-6734
FAX: (415) 369-0728

Leybold Vacuum Products Inc.
5700 Mellon Road
Export, PA 15632
(412) 327-5700
FAX: 1(412) 733-5960

McMaster-Carr Supply Company
P.O. Box 54960
Los Angeles, CA 90054
213) 692-5911
TWX: 910-586-1882
Telex: 69-8573

MDC Vacuum Products Corporation
23842 Cabot Boulevard
Hayward, Ca. 94545-1651
(415) 887-6100 (CA)
1-800-443-8817 (outside CA)
FAX: 415-887-0626
TWX: 910-383-2023

Perkin Elmer
6509 Flying Cloud Dr.
Eden Prairie, MN 55344
(612) 828-6157 (in MN)
1-800- 237-3603 (outside MN)
FAX: (612) 828-6322
TLX: 29 0407 PHY ELECT ENPE

Televac
P.O. Box 67
Huntingdon Valley, PA 19006
(215) 947-2500
FAX: 215-947-7464
Telex: 5101003532

Thermionics Laboratory, Inc.
22815 Sutro St.
Hayward, CA 94541
(415) 538-3304
FAX: 1(415) 538-2889

Varian
Vacuum Products Division
121 Hartwell Ave.
Lexington, MA 02173
1-800- 882-7426

Willson Scientific Glass, Inc.
528 E. Fig Street
Monrovia, CA 91016
(818) 303-1656
FAX: 1(818) 303-0599

Chapter 8

Bethlehem Apparatus Company, Inc.
Front & Depot Streets
Hellertown, PA 18055
(215) 838-7034

Heathway
903 Sheehy Drive
Horsham, PA 19044
(215) 443-5128
FAX: 895-71068
Telex: 902775

Laboratory Supply & Equipment Co.
P.O. Box 668812
Charlotte, NC 28266
Telephone: (704) 394-9616

Wale Apparatus Company
400 Front Street
P.O. Box D
Hellertown, PA18055
Telephone: (215) 838-7047
FAX: 215-838-7440

Wilt Glass Shop Equipment
Route 8
Lake Pleasant, NY 12108
(518) 548-4961
1-800-232-WILT (9458)
FAX: 518-548-5504

Appendix D Recommended Reading

Because this book tries to fill so many shoes, it cannot do justice to all subjects. Because of this limitation, some readers may wish to examine other books to provide more depth in a given subject. A number of the books listed in this appendix can be found in the citations of their respective chapters. However, not all of the texts listed in this appendix gave me citable material. This appendix therefore, provides me with the opportunity of directing your attention to other fine sources of information.

Laboratory Safety

Although I have tried to provide proper safety information throughout this book, safety is one area that cannot be covered thoroughly enough. The American Chemical Society recognizes this problem and has assembled a wonderful booklet on safety in the laboratory titled:

SAFETY in Academic Chemistry Laboratories, ©1990 by the American Chemical Society, Washington, D.C.

This excellent booklet on safety is available from the American Chemical Society. To obtain a copy, call 1-800- 227-5558 and ask Sales and Distribution for further information.

Chemical Safety

Another aspect of safety is the handling and disposal of toxic, poisonous, hazardous, and dangerous materials. However, the number of chemicals used in the laboratory are so vast that any attempt of my selecting chemicals or abbreviating information could create a potential hazard. Three excellent books on this subject are:

Dangerous Properties of Industrial Materials, 6th Edition, by N. Irving Sax Van Nostrand Reinhold Co., Inc., New York, N.Y., © 1984 .

Prudent Practices for Handling Hazardous Chemicals in the Laboratory by the Committee on Hazardous Substances in the Laboratory National Academy Press, Washington D.C., © 1981.

Safe Storage of Laboratory Chemicals Ed. by David A. Pipitone John Wiley & Sons, New York, N.Y. © 1984.

Chapter 1

Glass Engineering Handbook, 2nd Ed. by E.B. Shand, McGraw-Hill Book Co., Inc. New York, N.Y., ©1958.

For many years this has been a definitive source for glass information.

Glass Engineering Handbook, 3rd Ed. , by G.W. McLellan and E.B. Shand, McGraw-Hill Book Co., Inc. New York, N.Y., ©1984.

The Physical Properties of Glass
D.C. Holloway, Wykeham Publications, London G.B., LTD, ©1973.
Although good on many levels, this book has an excellent treatment of glass fracture.

The Physical Properties of Glass Surfaces by L. Holland, John Wiley & Sons, Inc., New York, N.Y. © 1964.
A very intensive examination of all aspects of glass surfaces, from damage to cleaning to chemistry.

Chapter 2

Volumetric Glassware, by V. Stott, Northumberland Press Ltd., Newcastle-on Tyne, G.B., © 1928, H. F. & G. Witherby, London.
Although somewhat dated, this text contains very interesting information on volumetric ware.

Mettler, Dictionary of Weighing Terms, by Dr. L. Biétry, Printed in Switzerland © 1983, Mettler Instruments AG .
Excellent general reference of weighing terms. This book is available free from Mettler Instruments Corp. (P.O. Box 71, Hightstown, NJ 08520-9944)

Scales and Weights, a Historical Outline, by Bruno Kisch, New Haven and London, Yale University Press, ©1965.
Good historical background on the development of weights and weighing.

A History of the Thermometer and Its Use in Meteorology, by W.E. Knowles Middleton, The John Hopkins Press, Baltimore, Maryland, ©1966.
Good historical background on the development of temperature reading.

Liquid-in-Glass Thermometry, by J. A. Wise, U.S. Government Printing Office, Washington, D.C., ©1976.
An excellent NBS (now NIST) publication on the use of liquid-in-glass thermometers.

Chapter 3

For information on glass, see the suggested books on glass in Chapter 1.
For information on stopcocks, joints, and glass tubing, see the suggested books in Chapter 8.

Chapter 4

There are no books on cleaning glass. There is an excellent chapter on glass cleaning in *The Physical Properties of Glass Surfaces* by L. Holland, (see further information in in the Chap. 1 listing). Regardless, the needs for safety from the various chemicals often used in cleaning is paramount. Therefore, see the recommended books in the section titled "Chemical Safety."

For additional information, you may wish to see "Standard Practice for Designing a Process for Cleaning Technical Glasses," an ASTM Publication, Designation C912-79 (Reapproved 1984).

For some basic ideas on waste management, see *Journal of Chemical Education*, 65 (1988), pp. A64-68. Although not comprehensive, this journal provides a good general description on the topic. However, because many localities have specific requirements on toxic substance control, always check with local authorities.

Chapter 5

Handbook of Compressed Gases, 2nd Ed., by The Compressed Gas Association, Inc., Van Nostrand Reinhold Publishing Co., New York, N. Y., © 1981.

This book extensively documents the various containment and handling techniques involved with compressed gases, their containers, and connections.

In addition to the above book, both the Linde (*Specialty Gases & Equipment*) and Matheson (*Gas Products*) Companies have excellent information about compressed gases, compressed gas containers, and regulators in their catalogs.

Union Carbide Industrial Gases Inc.
Linde Division
200 Cottentail Lane
Somersset, NJ 08875-6744
(201) 271-2600

Matheson Gas Products
932 Paterson Plank Rd.
P.O. Box 85
East Rutherford, NJ 07073
(201) 933-2400

Chapter 7

The Design of High Vacuum Systems and the Application of Kinney High Vacuum Pumps, C.M. Van Atta Kinney Manufacturing Division, New York, N.Y., © 1955.

A good practical reference booklet on pumps.

Vacuum Science and Engineering, C.M. Van Atta, McGraw-Hill Book Company © 1965 by McGraw-Hill, Inc.

Like his other book *The Design of High Vacuum Systems and the Application of Kinney High Vacuum Pumps*, this book is full of clear information.

Vacuum Technology, A. Guthrie, John Wiley & Sons, Wiley-Interscience, New York, N.Y., © 1963.

An older book, but excellent for the fundamentals of vacuum technology.

A User's Guide to Vacuum Technology J.F. O'Hanlon, John Wiley & Sons, Inc., Wiley-Interscience, New York, © 1980.

One of the top books on vacuum technology. Unfortunately it is limited to metal systems.

Introduction to Helium Mass Spectrometer Leak Detection, Varian Associates, Inc., published by Varian Associates, Inc., Palo Alto, CA © 1980.

Although understandingly biased toward Varian products, there is excellent information on the art and science of leak detection in this text.

Materials for High Vacuum Technology, Vol. 3, W. Espe, Pergamon Press, New York, N. Y., © 1968.

Somewhat dated, but has excellent information on vacuum science.

The Manipulation of Air-Sensitive Compounds, 2nd Ed., D.F. Shriver and M.A. Drezdzon, John Wiley & Sons, Inc., Wiley-Interscience, New York, © 1986.

Not purely a vacuum book, but working in an oxygen-free environment is one of the objectives for using a vacuum. This book is the leading book in its field.

Total Pressure Measurements in Vacuum Technology, A. Berman, Academic Press, Inc. (Harcourt Brace Javanovich, Publishers), Orlando FL, © 1985.
Extremely up-to-date and intensive study on all aspects of pressure measurement.

Methods of Experimental Physics: Volume 14: Vacuum Physics and Technology, Ed. by G. L. Weissler and R.W. Carlson, Academic Press, Inc. (Harcourt Brace Javanovich, Publishers) Orlando FL, © 1979.
This particular volume of this excellent series deals with vacuum technology. Each chapter is written by an expert resulting in an excellent, total book.

The following monographs were published by, and are available from the American Vacuum Society. They were all written by experts in their respective fields and are excellent resource materials. For further information, and prices, please contact the American Vacuum Society, 335 E. 45th St., New York, NY 10017.

Dictionary of Terms for the Areas of Science and Technology Served by the American Vacuum Society, 2nd Ed., Ed. H.G. Tompkins, © 1980 by the American Vacuum Society.

Vacuum Hazards Manual, 2nd, Ed., L.C. Beavis, V.J. Harwood, and M.T. Thomas, Coordinated by M.T. Thomas, © 1979 by the American Vacuum Society.

An Introduction to the Fundamentals of Vacuum Technology, H.G. Tompkins, © 1984 by the American Vacuum Society.

An Elementary Introduction to Vacuum Technique, G. Lewin, © 1984 by the American Vacuum Society.

Diffusion Pumps, Performance and Operation, M.H. Hablanian, © 1984 the American Vacuum Society.

Handbook of Vacuum Leak Detection N.G. Wilson and L.C. Beavis, American Institute of Physics, Inc., New York, N.Y., © 1976, 1979.

Partial Pressure Analyzers and Analysis, M.J. Drinkwine and D. Lichtman, © 1984 by the American Vacuum Society.

History of Vacuum Science and Technology, a Special Volume Commemorating the 30th Anniversary of the American Vacuum Society, 1953-1983, Ed. T.E. Madey and W.C. Brown. © 1983.

Chapter 8

Scientific Glassblowing, E. Wheeler, Interscience Publishers, Inc. , New York, N.Y., ©1958.
This, and the following book are considered the two best books on scientific glassblowing.

Scientific and Industrial Glassblowing and Laboratory Design, Barr and Anthorn, Instruments Publ. Co., Pittsburgh, PA., © 1959.
This, and the previous book are considered the two best books on scientific glassblowing.

Glassblowing, an Introduction to Artistic and Scientific Flameworking, 2nd Ed., Edward Carberry, MGLS Publishing, Marshall, MN., © 1989.
This book has excellent basic information on glassblowing, and setting up glassblowing facilities. The majority of the book is on artistic glassware with some information on Scientific glassblowing.

Glassblowing, and Introduction to Solid and Blown Glass Sculpturing,
Homer L. Hoyt, Crafts and Arts Publishing Co., Inc., Golden, CO., © 1989.

This book has excellent basic information on glassblowing ,and setting up glassblowing facilities. Although this book has no information on scientific glassblowing, it has some of the best illustrations on what to look for when heating glass.

Index

A

H

I

J

N

Q

R

W

X

Z

Notes

Notes

Notes

Notes

Notes

Notes

Notes

Notes

Notes